Eco-friendly Functional Polymers

AF179977

Advances in Green and Sustainable Chemistry

Series Editors

Béla Török

Timothy Dransfield

Eco-friendly Functional Polymers

An Approach from Application-Targeted Green Chemistry

Manuel Palencia
GI-CAT, Department of Chemistry, Universidad del Valle, Cali, Colombia

Tulio A. Lerma
GI-CAT, Department of Chemistry, Universidad del Valle, Cali, Colombia
Polymer and Advanced Materials Section, Mindtech-RG, Mindtech s.a.s, Cali, Colombia

Viviana Garcés
GI-CAT, Department of Chemistry, Universidad del Valle, Cali, Colombia
Polymer and Advanced Materials Section, Mindtech-RG, Mindtech s.a.s, Cali, Colombia

Mayra A. Mora
GI-CAT, Department of Chemistry, Universidad del Valle, Cali, Colombia
Analytical Chemistry and Chemometrics Section, Mindtech-RG, Mindtech s.a.s, Cali, Colombia

Jina M. Martínez
GI-CAT, Department of Chemistry, Universidad del Valle, Cali, Colombia
Analytical Chemistry and Chemometrics Section, Mindtech-RG, Mindtech s.a.s, Cali, Colombia

Sixta L. Palencia
Applied Microbiology Section, Mindtech-RG, Mindtech s.a.s, Concepción, Chile

ELSEVIER

Elsevier
Radarweg 29, PO Box 211, 1000 AE Amsterdam, Netherlands
The Boulevard, Langford Lane, Kidlington, Oxford OX5 1GB, United Kingdom
50 Hampshire Street, 5th Floor, Cambridge, MA 02139, United States

Copyright © 2021 Elsevier Inc. All rights reserved.

No part of this publication may be reproduced or transmitted in any form or by any means, electronic or mechanical, including photocopying, recording, or any information storage and retrieval system, without permission in writing from the publisher. Details on how to seek permission, further information about the Publisher's permissions policies and our arrangements with organizations such as the Copyright Clearance Center and the Copyright Licensing Agency, can be found at our website: www.elsevier.com/permissions.

This book and the individual contributions contained in it are protected under copyright by the Publisher (other than as may be noted herein).

Notices
Knowledge and best practice in this field are constantly changing. As new research and experience broaden our understanding, changes in research methods, professional practices, or medical treatment may become necessary.

Practitioners and researchers must always rely on their own experience and knowledge in evaluating and using any information, methods, compounds, or experiments described herein. In using such information or methods they should be mindful of their own safety and the safety of others, including parties for whom they have a professional responsibility.

To the fullest extent of the law, neither the Publisher nor the authors, contributors, or editors, assume any liability for any injury and/or damage to persons or property as a matter of products liability, negligence or otherwise, or from any use or operation of any methods, products, instructions, or ideas contained in the material herein.

Library of Congress Cataloging-in-Publication Data
A catalog record for this book is available from the Library of Congress

British Library Cataloguing-in-Publication Data
A catalogue record for this book is available from the British Library

ISBN: 978-0-12-821842-6

For information on all Elsevier publications visit our website at https://www.elsevier.com/books-and-journals

Publisher: Candice Janco
Acquisitions Editor: Kathryn Eryilmaz
Editorial Project Manager: Ruby Smith
Production Project Manager: Kumar Anbazhagan
Cover Designer: Alan Studholme

Typeset by TNQ Technologies

ELSEVIER Book Aid International Working together to grow libraries in developing countries
www.elsevier.com • www.bookaid.org

Contents

Introduction: The current problems associated with functional polymers and plastics xv

Part I
Polymer fundamentals in the current environmental context

1. Polymer green chemistry: principles of polymer synthetic green chemistry

1.1 Green chemistry 3
1.2 Principles of polymer synthetic green chemistry 5
1.2.1 Prevention: design of processes minimizing the waste generated 6
1.2.2 Atom economy 8
1.2.3 Less hazardous chemical synthesis 9
1.2.4 Designing safer chemicals 10
1.2.5 Use safer solvents and auxiliaries 11
1.2.6 Design for energy efficiency 12
1.2.7 Use of renewable feedstocks 12
1.2.8 Reduce derivatives 13
1.2.9 Catalysis 14
1.2.10 Design for degradation 15
1.2.11 Real-time analysis for pollution prevention 16
1.3 Conclusions and remarks 17
References 17

2. Biopolymers and bioplastics: from green chemistry to eco-friendly polymers

2.1 Biopolymers and bioplastics 23
2.2 Biopolymer classification and their eco-friendly approach 25
2.2.1 Biopolymers extracted from biomass 26
2.2.2 Biopolymers synthesized from biomass monomers 28
2.2.3 Biopolymers produced directly by living organisms 29
2.3 Conclusions and remarks 30
References 31

3. Recyclable polymer chemistry: structural, physicochemical aspects, uses, and transformation technologies

3.1 Basic principles of polymer recycling 35
3.2 Plastic classification according to chemical recycling 38
3.3 Conclusions and remarks 46
References 46

4. Biodegradable polymer chemistry: structural, physicochemical aspects, uses, and transformation technologies

4.1 Environmental overview of biodegradable functional polymers 51
4.2 Structural and morphological characteristics 54
4.3 Physicochemical properties 58
4.4 Transformation technologies 60
4.5 Uses of biodegradable polymers and their potential applications 62
4.6 Conclusion 66
References 66

5. Market, economic impact, and perspectives of eco-friendly polymers: biological, biodegradable, recyclable, and reusable polymers

5.1 An overview of the polymer market 73
5.2 Ecofriendly polymer market 75
5.2.1 Biologic polymer market 75
5.2.2 Biodegradable polymer markets 77
5.2.3 Recyclable polymer markets 78
5.3 Economic impact 78
5.4 Perspectives 80
5.5 Conclusions 81
References 82

Part II
Theoretical and experimental aspects in the design of eco-friendly polymers

6. Polymer biosynthesis and biotransformations

6.1 Biosynthesis and biotransformations for obtaining bio-based monomers and polymers 89
6.2 Biosynthesis of biopolymers 92
6.2.1 Cellulose biosynthesis 93
6.2.2 Biosynthesis of starches 94
6.2.3 Chitin biosynthesis 96
6.2.4 Biosynthesis of polyhydroxyalkanoates (PHAs) 97

6.3 Biotransformations of biopolymers 97
6.4 Conclusions and remarks 98
References 99

7. Eco-friendly composites and nanocomposites

7.1 Composites and nanocomposites 105
7.2 Classification of composites 107
7.2.1 Classification based on the type of matrix 107
7.2.2 Classification according to the shape and size of the reinforcement agent 109
7.3 Eco-friendly composites and nanocomposites 111
7.4 Natural polymer matrices and reinforcements used in eco-friendly composites 112
7.5 Chemical and physical treatments for the development of eco-friendly composites 113
7.6 Conclusions and remarks 116
References 117

8. Eco-friendly supramolecular systems: polymer geo- and biomimicry

8.1 Supramolecular systems: a fundamental approach and critical analysis 121
8.1.1 Description of matter organization levels 121
8.2 Eco-friendly supramolecular polymeric systems 123
8.3 Biomimicry: biomimetic supramolecular systems 125
8.3.1 Approaches to biomimetic design 126
8.3.2 Polymer-based biomimetic supramolecular systems 127
8.4 Geomimicry: a bioinspired approach to the development of artificial soils 129
8.4.1 Soil degradation as a current global problem 129
8.4.2 A brief description of soil composition and interactions 130
8.4.3 Clay–polymer supramolecular systems 131
8.4.4 Geomimicry: mimetic systems of the structure and/or function of soils 132
8.5 Conclusions and remarks 134
References 135

9. Eco-friendly hydrogels

9.1 Hydrogels: concepts and characteristics 141
9.2 Classification of hydrogels 143
9.2.1 Hydrogels according to the origin 143
9.2.2 Hydrogels according to the polymer composition 144
9.2.3 Hydrogels according to the nature of cross-linking 145
9.2.4 Hydrogels according to the nature of the monomer 145
9.2.5 Hydrogels according to the function and nature of the external stimulus 146

9.3 Synthesis of hydrogels 146
9.4 Applications of hydrogels 149
9.5 Conclusions 149
References 149

10. Eco-friendly chemical transformations: strategies and technologies based on green chemistry principles, click chemistry, and eco-sustainable development

10.1 Chemical transformations in the current environment 155
10.2 Definitions of green chemistry, click chemistry, and eco-sustainable development 156
10.3 Strategies and technologies based on the principles of green chemistry 158
10.4 Strategies and technologies based on the principles of click chemistry 159
10.5 Conclusions 162
References 163

11. Eco-friendly particles, fibers, and surfaces 167

11.1 Particles, fibers, and surfaces: an overview 167
11.2 Polymer-based particles, fibers, and surfaces 169
11.3 Conclusions and remarks 173
References 173

Part III
Applications of eco-friendly functional polymers

12. Functional and eco-friendly polymers in agriculture

12.1 An overview of current problems in agriculture 181
12.2 Functional polymers in agriculture 182
12.2.1 Availability of suitable soils for agriculture 182
12.2.2 Availability and efficient use of water resources 183
12.2.3 Adequate control of plagues 184
12.2.4 Efficient fertilization 185
12.2.5 Adequate seed and crop conservation 187
12.3 Conclusions and remarks 187
References 188

13. Functional and eco-friendly polymers for environmental applications

13.1 An overview of current problems in the environmental context 193
13.2 Functional polymers for environmental monitoring 194

13.2.1 Monitoring of chemical species in soils 194
13.2.2 Monitoring of air 196
13.2.3 Monitoring of water 198
13.3 Functional polymers for environmental remediation 200
13.3.1 Remediation of terrestrial ecosystems 201
13.3.2 Remediation of aquatic ecosystems 202
13.4 Conclusions and remarks 204
References 204

14. Functional and eco-friendly polymers in food: edible polymers

14.1 Introduction: an overview of polymers in the food industry 209
14.2 Types of edible polymers according to composition 210
14.2.1 Polysaccharides 211
14.2.2 Proteins 212
14.2.3 Synthetics and composites 213
14.3 Applications 215
14.3.1 Functional edible films and coatings on food packaging 215
14.3.2 Functional food 222
14.3.3 Edible polymer-like emulsifying food additives 225
14.3.4 Synthetic meat 226
References 227

15. Functional and eco-friendly polymers in structural applications

15.1 Functional polymers versus structural polymers: a conceptual approach 235
15.2 An overview of multifunctional structural polymers for technological applications 238
15.3 Conclusions and remarks 242
References 242

16. Functional and eco-friendly polymers in coatings and surfaces

16.1 General aspects about surface, coatings, and adhesion 247
16.2 Eco-friendly polymer surfaces and coatings 250
16.2.1 Eco-friendly polymer surfaces 250
16.2.2 Eco-friendly polymer coatings 252
16.3 Conclusions and remarks 253
References 254

17. Functional and eco-friendly polymers in medical and biomedical applications

17.1	Polymer materials in medicine and biomedicine: a general overview	257
17.2	Biopolymer materials in biomedical applications	258
	17.2.1 Biopolymers in tissue engineering	260
	17.2.2 Wound dressing	264
	17.2.3 Diagnosis and detection	265
17.3	Antimicrobial surfaces and platforms	266
17.4	Conclusions and remarks	267
	References	267

18. Functional and eco-friendly polymers in pharmaceutical applications

18.1	Pharmaceutical industry: an environmental look	271
18.2	Biopolymers in the pharmaceutical industry	272
	18.2.1 Biopolymer-based drug-delivery systems and delivery pathways	273
	18.2.2 Stimuli-responsive biopolymers	277
18.3	Conclusions and remarks	279
	References	279

19. Functional and eco-friendly polymers in textile applications

19.1	Background to the textile industry and advanced fiber applications	285
19.2	Eco-friendly polymer-based fibers	288
19.3	Conclusions	291
	References	291

Part IV
Selected applications of eco-friendly functional polymers

20. Passive sampler of organochloride compounds in water and air

20.1	Sampling of pollution in water and air: an overview	297
20.2	General context of organochloride contaminants	299
	20.2.1 Organochlorine pesticides (OCPs)	300
	20.2.2 Polychlorinated biphenyls	301
	20.2.3 Dioxins and furans	301

20.3 Brief description of accumulation of chemicals by passive samplers 302
20.4 Functional polymers using passive samplers 303
20.4.1 Monophasic samplers 304
20.4.2 Biphasic and triphasic samplers 306
20.5 Factors involved in the capture process during passive sampling 309
20.6 Chemical analysis of passive samplers 314
20.7 Conclusions and remarks 316
References 317

21. Removal of emergent pollutants of waters

21.1 Emerging pollutants: an overview 327
21.2 Remediation of emerging contaminants in water 329
21.3 Removal of pharmaceuticals in water 331
21.4 Other emerging pollutants: nanomaterials, and metallic and metalloid ions 332
21.5 Conclusions and remarks 334
References 335

22. Polymer-based restoration of functional properties of degraded soils

22.1 Soil as a geomimetic model of new multifunctional materials 341
22.2 Hydrogels as artificial systems of microbial growth 343
22.2.1 Methodological strategy 344
22.2.2 Relevant results and conclusions 344
22.3 Nanostructured hydrogels based on hybrid polymer–clay composites 346
22.3.1 Synthesis strategy 346
22.3.2 Relevant results and conclusions 347
22.4 Multi- and bio-functional hybrid polymer hydrogel geomimetics of soil particles 348
22.4.1 Synthesis strategy 349
22.4.2 Relevant results and conclusions 350
22.5 Conclusions 352
References 352

23. Immunopharmacological control of blood and hemoderivatives

23.1 Bacteremia during blood storage: an overview 355
23.2 Starch-based hydrogels for the controlled release of nanostructured hybrid antibiotics 358

23.2.1 Synthesis strategy of hydrogels based on starch–poly(acrylic acid) 359
23.2.2 Relevant results and conclusions 359
23.3 Functional materials with antibacterial properties for surface-contact pathogen inactivation 360
23.3.1 Methodological approach for making of the PVC-based antibacterial surface 361
23.3.2 Relevant results and conclusions 362
23.4 Antibacterial and nonhemolytic materials based on eco-friendly cationic polyurethanes 364
23.4.1 Methodological approach 364
23.4.2 Relevant results and conclusions 365
References 366

24. Surface for spectral sensors in microbiological analysis

24.1 Introduction: polymer-based spectral sensors 371
24.2 Study of the mid-infrared vibrational spectrum of pathogenic bacterial biofilms: context and problematics 372
24.3 Methodological strategy 373
24.4 Results and discussion: analysis of bacterial biofilms by mid-IR spectroscopy 374
24.5 Conclusions and remarks 378
References 378

25. Separation of secondary metabolites and bioactive substances from agricultural residues

25.1 Introduction: from agricultural wastes to bioactive raw materials 381
25.1.1 Agricultural waste problem 381
25.1.2 Polymeric membrane technology 382
25.1.3 Secondary metabolites and bioactive substances 384
25.2 Applications 384
25.2.1 Use of artichoke waste 384
25.2.2 Methodological strategy for the use of artichoke residues 385
25.2.3 Results of the separation of phenolic compounds present in artichoke residues 387
25.3 Use of waste from wine industries 388
25.3.1 Methodological strategy for the use of wine waste 388
25.3.2 Results of the separation of phenolic compounds present in red wine lees 389
25.4 Conclusions 391
References 391

26. Hydrogels for separation and delivery of antibacterial inorganic nanoparticles

26.1 Inorganic nanoparticles 395
26.1.1 Antibacterial inorganic nanoparticles: silver nanoparticles (AgNPs) 398
26.2 Hydrogels for delivery of antibacterial inorganic nanoparticles 400
26.2.1 Methodological strategy and results 401
26.3 Development of hydrogels for massive synthesis of AgNPs 402
26.4 Conclusions and remarks 403
References 404

27. Polymeric supports for grown of beneficial microorganisms in agriculture

27.1 Sustainable agriculture: food security and conservation of the environment 407
27.2 Plant growth-promoting microorganisms (PGPMs) 409
27.3 Polymer-based bacterial inocula 410
27.3.1 Development of bacterial inocula from bio-based hydrogels: a methodological approach 412
27.4 Conclusions and remarks 413
References 414

28. Polymeric supports for growth of probiotic microorganisms

28.1 Microbiota and probiotics 419
28.2 Probiotics used for health and disease treatment 420
28.3 Biopolymers for the encapsulation of probiotic strains 421
28.4 Immobilization and encapsulation of probiotic strains 423
28.5 Conclusions and remarks 428
References 428

29. Polymeric drug carriers against *Helicobacter pylori*

29.1 Introduction: microbiological characteristics of *H. pylori* and its importance in health 433
29.1.1 Drugs against *H. pylori*: an overview of pharmacological treatments 434
29.2 Polymers as carriers of anti-*H. pylori* drugs 436
29.3 Conclusions and remarks 436
References 439

Index 441

Introduction: The current problems associated with functional polymers and plastics

The term plastic is colloquially used to refer to all materials of a polymeric nature which are commonly used in numerous applications, for example, packaging material (Zhong et al., 2020), coating of biomedical devices (Palencia et al., 2020), electronic components (Cao and Uhrich, 2019), separation systems (Palencia et al., 2014), and in a wide variety of other applications; therefore, all plastics are polymers. It is estimated that, in 2017, the world production of plastics reached 348 Mt (Shen et al., 2020). Standing out among the polymers most used were poly(vinyl chloride), poly(ethylene), and poly(propylene), which recorded consumption rates in 2018 of 44.3, 99.6, and 56 Mt, respectively (Alsabri and Al-Ghamdi, 2020). Undoubtedly, their excellent durability, lightness, stability, and low cost have made them ubiquitous materials. Unfortunately, many of the commercially used polymers are obtained from nonrenewable sources, and as a consequence, they are ecologically harmful, due to their low biodegradability and high accumulation rate, making them problematic on a global scale (Shen et al., 2020).

As a result of the foregoing, different government strategies based on comprehensive waste management have been implemented. However, in many cases the implementation of these programs is inefficient, due to the lack of coverage, infrastructure, resources, or community interest in the proper management of waste (Azeez, 2020). At the same time, physical and chemical processes have been developed for the transformation and use of polymer waste. Thus, in thermoplastic polymers mechanical recycling is commonly used to produce the raw materials to manufacture new products, through the selection, washing, compaction, and pelletization of plastic waste, however, after several cycles of recycling, the mechanical properties of the polymers can be altered to the point of losing their practical applicability, and also, depending on the initial type of application, the reuse of material is limited, for example, recycled food packaging for health reasons should not be used in other applications; the same occurs with biomedical packages for collection of biological samples, etc. (Okan et al., 2019). Meanwhile, in thermoset polymers, chemical transformation is common with the purpose of generating the structural degradation of the material into simpler molecules, for example, depolymerization by glycolysis, aminolysis, hydrolysis, or promoted by ionic liquids etc.; in certain cases, these processes are expensive, have low yields, or generate a greater environmental impact due to the use of polluting chemicals (Rahimi and Garcia, 2017).

According with the above, it is clear that there is an imperative need to create and promote actions and legislations aimed at guaranteeing the maximum environmental preservation, however, these initiatives must be accompanied by a impactful social engineering, focused exclusively on the environmental awareness of people and all entities within society. However, at the same time, new advances in science and technology must be promoted to increase the efficiency of processes, reduce the residuality of products as much as possible, and find intelligent solutions in the broad context of the problem and not only in particular cases. In line with the above, in the early 1990s, the concept of "green chemistry" emerged as a novel proposal for a set of principles driven by the design of sustainable chemical processes with low environmental impact (Sheldon, 2018). In general, the concept of green chemistry bases its principles on the search for the sustainability of our civilization, which will depend on our success in meeting and satisfying the requirements for food, energy, and raw material demands, without compromising the environment, and for a constantly growing population. The problem becomes particularly difficult when we reflect on the finiteness of our planet, in terms of space, capabilities, and resources, but also on the fragility of ecosystem dynamics and balances. In this way, 12 principles of green chemistry have been postulated. These are: (1) prevention, (2) use of renewable raw materials, (3) atomic economy, (4) reduction of synthetic derivatives, (5) synthesis of less dangerous chemical substances, (6) development and use of catalysts, (7) design of safer chemicals, (8) design of degradable products, (9) use of safer solvents and auxiliaries, (10) real-time analysis of processes for the prevention of contamination, (11) design of processes with high energy efficiency, and (12) intrinsically safer chemistry for accident prevention. Therefore, green chemistry seeks the design of "benign" chemical products and processes to reduce or eliminate the use or generation of hazardous substances; and is applied throughout the life cycle of a substance or material, including its design, manufacture, use, and final disposal (de Marco et al., 2019). As a consequence, green chemistry is one of the most attractive philosophies in the field of chemistry to achieve the sustainability of its processes with a low environmental impact.

In this way, *Eco-friendly Functional Polymers: An Approach from Application-Targeted Green* provides a holistic overview of the chemistry, development, and applications of polymers in the context of sustainability driven by these principles. Several concepts are analyzed from their applied context, and not only under a line of thought focused on basic science. As a result, this book contains an interdisciplinary look at diverse topics.

Researchers are encouraged to consider, in their investigations, green chemistry principles, direct and indirect environmental impacts on the short, medium, and long term, the real end-user needs, as comprehensive aspects in the development of science and technology of new materials, and, in particular, functional and structural polymers. This book develops its content in four sections, in the first, basic aspects in the environmental context, and of polymers, are developed. This first section seeks to give the reader the argumentative basis for the eco-friendly development of polymer science and technology, touching on key aspects such as recycling, biodegradability, economic importance, and the bases of green chemistry that underlie the context of polymers. The second section presents a practical view of polymers, from a fundamental approach. Finally, Sections 3 and 4 delve into the applied components, with a general approach, based on trends in research in a variety of fields, including agriculture, environment, food, coatings, textiles, biomedicine, among others.

References

Alsabri, A., Al-Ghamdi, S.G., 2020. Carbon footprint and embodied energy of PVC, PE, and PP piping: perspective on environmental performance. Energy Rep. 6, 364–370. https://doi.org/10.1016/j.egyr.2020.11.173.

Azeez, T.O., 2020. Thermoplastic recycling: properties, modifications, and applications. In: Evingur, G., Pekcan, O., Achilias, D. (Eds.), Thermosoftening Plastics. IntechOpen. https://doi.org/10.5772/intechopen.81614.

Cao, Y., Uhrich, K.E., 2019. Biodegradable and biocompatible polymers for electronic applications: a review. J. Bioact. Compat. Polym. 34 (1), 3–15. https://doi.org/10.1177/0883911518818075.

de Marco, B.A., Rechelo, B.S., Tótoli, E.G., Kogawa, A.C., Salgado, H.R.N., 2019. Evolution of green chemistry and its multidimensional impacts: a review. Saudi Pharm. J. 27 (1), 1–8. https://doi.org/10.1016/j.jsps.2018.07.011.

Okan, M., Aydin, H.M., Barsbay, M., 2019. Current approaches to waste polymer utilization and minimization: a review. J. Chem. Technol. Biotechnol. 94 (1), 8–21. https://doi.org/10.1002/jctb.5778.

Palencia, M., Lerma, T.A., Arrieta, Á.A., 2020. Antibacterial and non-hemolytic cationic polyurethanes with N-carboxymethyl-N, N, N-triethylammonium groups for bacteremia-control in biomedical-using materials. Mater. Today Commun. 22, 100708. https://doi.org/10.1016/j.mtcomm.2019.100708.

Palencia, M., Vera, M., Rivas, B.L., 2014. Modification of ultrafiltration membranes via interpenetrating polymer networks for removal of boron from aqueous solution. J. Membr. Sci. 466, 192–199. https://doi.org/10.1016/j.memsci.2014.05.003.

Rahimi, A., García, J.M., 2017. Chemical recycling of waste plastics for new materials production. Nat. Rev. Chem. 1 (6), 1–11. https://doi.org/10.1038/s41570-017-0046.

Shen, M., Song, B., Zeng, G., Zhang, Y., Huang, W., Wen, X., Tang, W., 2020. Are biodegradable plastics a promising solution to solve the global plastic pollution? Environ. Pollut. 263, 114469. https://doi.org/10.1016/j.envpol.2020.114469.

Sheldon, R.A., 2018. Metrics of green chemistry and sustainability: past, present, and future. ACS Sustain. Chem. Eng. 6 (1), 32–48. https://doi.org/10.1021/acssuschemeng.7b03505.

Zhong, Y., Godwin, P., Jin, Y., Xiao, H., 2020. Biodegradable polymers and green-based antimicrobial packaging materials: a mini-review. Adv. Ind. Eng. Polym. Res. 3 (1), 27–35. https://doi.org/10.1016/j.aiepr.2019.11.002.

Part I

Polymer fundamentals in the current environmental context

Chapter 1

Polymer green chemistry: principles of polymer synthetic green chemistry

1.1 Green chemistry

Currently, the progressive advance of environmental problems, including climate change, pollution of water, soil, and air, deforestation, reduction of biodiversity, and global warming, among others, has been caused largely by anthropogenic activities. Chemical and physical processes accompanying the human lifestyle, characterized by a consumer economy as an integral part of its well-being, the indiscriminate use of natural resources and inappropriate waste management have been identified as determining factors of environmental problems. The above has generated the need to create and promote actions and laws directed at preserving the environment in order to try to warrant long-term survival of humans in a finite resource environment. In this way, due to the intimate relationship between today's global society and chemical products and processes in all aspects of modern life, a set of adverse interactions for the environment have been described since the beginning of the last century. Typical examples are the heavy metals, plastics, and gases used in the making of high-technology products, cosmetics, food, and everyday products among others. The negative impact of the large-scale and uncontrolled chemical industry is well known. This impact can be direct as a result of the emissions generated within a process, but in also can be indirect as a consequence of degradation processes, poor disposal of the waste generated after its use, or as a result of accidents. It is therefore the awareness of the need for chemical transformations to sustain our civilization, as well as the recognition of the risks associated with these transformations throughout the life cycle of a product, that lead to the construction of a concept that involves a philosophy of work and a vision of our life as a species in a unique ecosystem called planet Earth: "green chemistry." Therefore, it has been declared that green chemistry emerged as a novel proposal grouping a set of criteria or guidelines for the design of sustainable industrial chemical processes and reducing the environmental deterioration produced not only by productive processes but also by the

Eco-friendly Functional Polymers
https://doi.org/10.1016/B978-0-12-821842-6.00012-9
Copyright © 2021 Elsevier Inc. All rights reserved.

products (Linthorst, 2009; Anastas and Eghbali, 2010; Gałuszka et al., 2013; Mülhaupt, 2013; Sheldon, 2014; De Marco et al., 2019).

Green chemistry bases its principles on the search for social and environmental sustainability of our civilization, which will depend on the success in being able to meet the food, energy, and chemical demands of a constantly growing population without compromising the environment. Thus, according to the United States Environmental Protection Agency (EPA), "green chemistry" can be defined as the design of chemical products and processes reducing or eliminating the use or generation of hazardous substances, the efficient use of energy within chemical processes, and that minimize the general impact of chemical transformations on the environment. The above is applied throughout the life cycle of a chemical, including its design, manufacture, use, and final disposal, turning this into one of the most attractive concepts in chemistry to achieve the sustainability of its processes. The concept of green chemistry has had a great impact on society in general, because its application has gone beyond the walls of research laboratories and been adopted by industry to validate their products or in building a favorable imagen in a society that is each time more preoccupied by environmental questions. As a consequence, having as an antecedent the positive environmental vision of green chemistry, in 1998, Paul Anastas and Jhon Warner proposed for the first time a series of 12 principles or guidelines that provide a template for the design of sustainable processes (Fig. 1.1). These principles are visualized as an integrated network of concepts for the design and improvement of products and chemical processes, which contemplate their entire life cycle, from the synthesis design, appropriate selection of precursors, energy efficiency of the process, and minimization of waste generation, among others (Linthorst, 2009; Anastas and Eghbali, 2010). The importance of these postulates has led to the widely accepted view that by adopting these principles the development of truly sustainable processes is possible.

At present, the green chemistry approach has been attractive in the specialized chemistry community, not only because products and processes that are cost-effective and environmentally friendly can be designed, but also because it has been evidenced that, in reality, the chemical sciences can be a tool for industrial transformation supported in the design substances by sustainable processes. However, it is clear that this required a high degree of awareness of humanity about its role in the environment and a clear-cut view about the vulnerability of everything that supports us as a species in, as previously mentioned, a planet with finite resources. The concept "green chemistry" has been successfully applied in different industries, for example, automobile, cosmetic, energetic, pharmaceutical, among many others (Gałuszka et al., 2013; Mülhaupt, 2013; Sheldon, 2014). However, it is evident that its success in the foray into the industrial field is not due to the elevation of humans as an environmentally friendly species, but rather that the superposition of factors such as the recognition of the possibility of extinction,

FIGURE 1.1 The twelve principles of green chemistry.

the connection of the adjective "environmentally friendly" as a positive characteristic of products, and a population increasingly identifying with the importance of preserving the environment; in addition, increasingly strict regulations influence the productive sectors. Thus, from market and economic perspectives, "environmentally friendly" is undoubtedly a commercial label for the positive promotion of products. However, without a doubt, faced with the threat to the survival as a species, the concept of green chemistry has served as a tool to stimulate governmental and private participation through the creation of pacts and alliances, educational initiatives, research centers, etc. (Anastas and Kirchhoff, 2002).

1.2 Principles of polymer synthetic green chemistry

Polymer chemistry undoubtedly plays an important role in our society, currently, polymers have become "ubiquitous materials" immersed in a large number of processes and applications to such an extent that our society is inconceivable without synthetic, natural, or modified polymers. However, their extensive and frequent use, inadequate final disposal of polymeric residues, and the low biodegradability of many of these polymeric materials, mainly those derived from the petrochemical industry, have led to their inevitable accumulation in very diverse ecosystems (De la Guardia and Armenta, 2011; Wang et al., 2012; Hartmann et al., 2017; Song et al., 2014; Suaria et al., 2016). However, although millions of tons of polymeric waste are dumped into the environment throughout the planet daily, in the same way, millions of tons of these materials are required daily to support our civilization, and as a consequence, it is clear that the polymer synthetic chemistry must be thought about from a green chemistry approach in order to contribute to the sustainable development of hundreds of industries and productive sectors. Next, the 12 principles of green chemistry are described and analyzed for their application in the polymer chemistry.

1.2.1 Prevention: design of processes minimizing the waste generated

The starting premise is a well-known argument, which is rationalized by the following sentence: it is better to take the necessary measures to prevent waste than to invest time, money, effort, and human resources in treating or cleaning it after it has formed, or to correct the effects and consequences resulting from not having foreseen it. In this context, the "production of wastes" is understood as the generation of any type of substance without added value, or the loss of energy within a process. Accordingly, wastes should not be produced because these are potential sources of environmental pollution; however, the added value of a substance is a subjective feature depending on context, and as a consequence, in order to adequately characterize any no-added-value residue, the overall context should be taken into consideration. An interesting approximation to solve this problem is the introduction of the degree of pollution, which is directly related with its chemical nature, toxicity, quantity, and routes by which it is released into the environment (Kaur et al., 2018).

In the 1990s, the concept of the environmental impact factor, also called the E-factor, was introduced (Sheldon, 2017) in order to estimate the amount of waste produced in a chemical process from the ratio between the total waste generated, w (in mass), and the amount of product produced, m (in mass):

$$\text{E-factor} = \frac{\sum w}{\sum m}\left[\frac{\text{kg}}{\text{kg}}\right] \tag{1.1}$$

The E-factor has been widely accepted as a very simple but reliable metric for estimating the efficiency of a chemical process in relation to the generation of products and wastes. However, among the limitations of the E-factor are that it can be considered an oversimplification of the system to be described,

TABLE 1.1 Relationship between E-factor and the chemical industry.

Industry	Production in tons	E-factor
Oil refining	10^6–10^8	<0.1
Bulk chemicals	10^4–10^6	1–5
Special chemicals	10^2–10^4	5–50
Pharmaceutical products	10–10^3	25–100

so it could be adequate for some systems and completely inappropriate for others. Furthermore, it does not consider aspects such as size, stoichiometry and the nature of the residue, the use of water and energy in the process, and it does not provide an indication of the dangerousness of the waste generated (Tobiszewski et al., 2015; Sheldon, 2017). In general, the E-factor is related to the type of industry, as indicated in Table 1.1.

In the production of polymers, the wastes generated are directly associated with the synthesis processes due to the presence of residual monomers and reaction auxiliaries, as well as to secondary products generated from inadequate control of the synthesis parameters. In this way, the control over residual monomers presents a particular interest due to its usually reactive chemical nature representing a high toxicity risk. Currently, there are two approaches to classifying the techniques of elimination of residual monomers and reaction auxiliaries, known as the chemical and physical approaches (Dubé and Salehpour, 2014). Thus, the choice will be directly related to the characteristics of the polymer synthesis process, for example, type of polymerization carried out, strength of polymer–impurity interaction, nature of the impurity, etc.

In general, chemical methods are based on: (1) the use of residual monomer for additional polymer production by controlling process variables or through the use of catalysts and (2) the generation of inactive or nontoxic compounds that are easily removable, through side reactions that do not compromise the properties of the final product. On the other hand, the physical methods are based on the use of various separation techniques. Thus, the evaporation of the residual monomer is applied in those cases where the residues have high vapor pressures. Solvent extraction techniques are also usually used. However, several of the green chemistry pillars would be violated when these methods are used. In the same sense, extraction techniques through the use of supercritical fluids have been used (Araújo et al., 2002; Aerts et al., 2011). Based on the above, strategies for polymerization waste disposal have been varied and their choice will depend on each particular process, even in the combination of various purification strategies. These additional stages of purification in the polymer synthesis do not only have a negative impact on the process economic viability, but also at the environmental level, increasing the use of chemical inputs and energy demand. The foregoing demonstrates the need for an adequate polymer synthesis process design, in which the reaction yields are optimized and the generation of secondary products is minimized.

1.2.2 Atom economy

Atom economy refers to the efficient design of synthesis processes characterized by the nonproduction of secondary products; under this approach, synthetic procedures should be designed in order to maximize the incorporation of all the chemical components used in the final product. In this context, the term atomic economy is directly related to optimization in the use of raw materials with the aim of minimizing the generation of waste. The term was introduced by Barry Trost (1991) and is calculated through the ratio between the molecular weight of the product, MW-product, and the sum of all the molecular or atomic weights of the reagents, MW-reagent, used for its production. From Eq. (1.2) it can be seen that this is a simple concept which can be easily calculated. In green chemistry, processes that are 100% atomically efficient are strongly desired because the final result permits obtaining of the expected product with the waste production during the making process.

$$\text{Atom Economy} = \frac{\text{MW} - \text{product}}{\sum \text{MW} - \text{reagent}} \times 100 \left[\frac{\text{g/mol}}{\text{g/mol}}\right] \quad (1.2)$$

However, to achieve a high atom economy, that is, ~100%, in many cases the possibilities are limited.

In the case of polymers, atomic economy will depend on the process nature and the impossibility of alternative methods of synthesis. In some cases, the loss of atoms and therefore the generation of waste is inevitable because it is directly related with the stoichiometry of the chemical reaction. Polymer synthesis mechanisms are classified from a wide context in step or chain growth polymerizations. Generally, step-growth polymerizations are associated with processes that usually produce by-products, low-molecular-weight molecules, which leads to a process with low atomic efficiency, for example, esterification and amidation reactions (Billiet et al., 2009; Tabone et al., 2010). In these cases, the studies focus on the search for alternative synthesis routes promoting the generation of harmless by-products. An example is the synthesis of poly(ethylene terephthalate), which can be produced from the reaction of n-dimethyl terephthalate or n-terephthalic acid with ethylene glycol, producing methanol and water as reaction waste, respectively (Pang et al., 2006). When these reactions are analyzed it is found that, with the use of n-dimethyl terephthalate, 87.6% of atomic economy is achieved, while the poly(ethylene terephthalate) obtained from n-terephthalic acid reaches 92.1%. This demonstrates the improvement in atomic efficiency of the process by using n-terephthalic acid as the synthesis precursor. However, in many processes of step-growth polymerization an atomic economy close to 100% is limited to the availability of synthesis monomers, process costs, and energy demand. Lastly, and importantly, not all reactions mediated by a step-growth mechanism produce waste and are 100% efficient as is the case with the production of polyurethanes (Zhang et al., 2016; Lerma et al., 2016; Palencia et al., 2016) and synthesis of some polyester hydrophilic resins obtained in the absence of solvent (Garces et al., 2017).

In contrast to the step-growth reactions, the chain growth reactions are, from a conceptual point of view, generally atomically efficient and all atoms present in the monomers are incorporated into the product throughout the polymerization process. Traditionally, the increase in atomic efficiency in this type of process is achieved through the use of catalysts increasing the efficiency of conversion of monomers to polymers. However, it should be noted that there are polymerization processes involving addition polymers that lose atoms through secondary reactions. Some examples include the production of poly(vinyl alcohol), which is produced from vinyl acetate through chain growth polymerization and subsequent hydrolysis of acetate groups, generating residual acetic acid (Aslam et al., 2018), and the functionalization reaction of side chains of PVC modified with diethylene amine to achieve the formation of quaternary ammonium side groups (Palencia et al., 2019). In conclusion, the atomic economy in polymer synthesis processes is closely related to the reaction mechanism and chemical transformation routes. Generally, the processes of obtaining polymers tend to be atomically profitable, in those cases where this does not happen, the objectives will be directed toward the search for alternative processes ideally characterized to be sustainable with the minimal generation of wastes.

1.2.3 Less hazardous chemical synthesis

The premise on which this principle is based consists of the use and generation of harmless substances. Although this is not always possible, it is clear that every effort should be made to minimize the risk of toxicity to human health and the environment. The ability of a process to use and generate products of low or no toxicity will be strongly related to the ability of synthetic chemists to improve the yields of current processes and generate alternative routes of synthesis that are environmentally friendly and economically profitable. In the field of polymers, an important aspect to highlight is the fact that the generation of polymeric materials is not connected exclusively with the conversion of one or more monomers into a final product. These processes involve the use of various additives such as solvents, plasticizers, flame retardants, pigments, catalysts, lubricants, stabilizers, fillers, and reinforcement materials, which can be released into the environment and cause pollution problems (Ambrogi et al., 2017).

On the other hand, monomers used in polymer synthesis can represent great challenges in the search for safe synthesis processes due to their high reactivity, making them potential sources of contamination. This has led to the use of reaction inhibitors such as 4-tert-butylcatechol, in order to reduce the reactivity of the monomers and allow their handling in a safe way until they are converted into polymers, a state in which the degree of toxicity has decreased or disappeared (Nesvadba, 2012). In this sense, at present, efforts are focused on the search for natural sources of monomers or polymers, in order to take advantage of their inherent safety, and thus improve current processes. Thus, materials based on cellulose and its derivatives, in lignocellulosic compounds, lactic acid, and starch, among many others, are just a few examples of raw materials from natural sources with which we are currently seeking to replace the use of polymers derived from the petrochemical industry (Vroman and Tighzert, 2009; Afanasjeva et al., 2017).

During polymer production it is common to incorporate one or more additives for the modification of their properties in order to obtain the final products with desired characteristics. Generally, the incorporation of additives in polymer synthesis processes does not imply the generation of chemical reactions, that is, a covalent additive–polymer bond. In contrast, these interactions usually are physical and can be easily overcome, consequently, specific conditions can promote their release into the environment. For example, a particular case is flame retardants that are widely used in the plastics industry. These additives are based on the use of organohalogen compounds, which can present bioaccumulation problems and pose risks to health and the environment (Bar et al., 2015; Laoutid et al., 2009).

1.2.4 Designing safer chemicals

A major aspect of green chemistry is directed toward the design of functional and safe products. Under this guideline, chemical products must be designed to maintain efficacy in fulfilling their function with the least possible toxicity. The design of safe chemicals goes beyond the search for new efficient and economically viable synthetic routes. This concept is based on the concept that the design of a process to obtain a product involves aspects of synthesis, use, final disposition, and possible chemical and physical transformations throughout its useful life cycle. In this way, through adequate design of chemical production processes materials of high functional efficiency are sought that are nonpolluting and that once they have completed their life cycle, are degraded into products that are harmless to the environment. In this context, the development of computational chemistry has contributed to the in silico study of various chemical processes and the possibility of creating correlations, equations, and models that involve the variables of the processes, allowing a priori to predict their impact on the environment (de Ruyck et al., 2016; Martin et al., 2014; Raises and Bajic, 2016).

Polymeric materials generally tend to be stable and low-toxicity products compared with their synthesis precursors. However, their degradation, incineration, and inadequate final disposition can generate negative environmental impacts (Avio et al., 2017; Webb et al., 2013). Added to this is the use of solvents generating a significant fraction of waste; solvents are usual inputs in many synthesis processes and in the final application of polymeric products such as paints or adhesives (Welton, 2015). As mentioned earlier, the addition of secondary compounds in order to improve the properties of polymeric materials also represents a critical point in the sustainability of the processes due to their relatively easy release from the polymer matrix and the toxicity of these compounds (Hahladakis et al., 2018). In this sense, the optimization of synthesis processes, the use of alternative solvents, and the search for new eco-friendly additives are areas of ongoing research.

1.2.5 Use safer solvents and auxiliaries

In the previous section, the impact of the use of solvents and other auxiliary substances or additives in the design of synthetic processes was briefly described; however, it is important to emphasize that the unnecessary use of this type of auxiliary substance should be limited, and if unavoidable, the strategy should be directed to the use of harmless chemicals.

One of the main problems associated with the chemical industry is possibly the use of large amounts of solvents, which contribute in large proportion to the waste generated in industrial processes. In addition, the high toxicity, flammability, corrosivity, volatility, solubility, among other properties of many common solvents, not only contribute to the environmental contamination of soil, air, and water, but also represent a health risk to workers (Lipshutz et al., 2016). The problem of the environmental impact of solvents is generally addressed through the implementation of recovery techniques that allow their reuse. However, in situations where solvent recovery is possible, energy-intensive distillation techniques are commonly used, and so the profitability of chemical processes is adversely affected. Therefore, the search for processes that do not require solvents or use less polluting inputs is one of the main areas of research in the field of green chemistry.

The laboratory and industrial production of polymers is not exempt from the use of solvents. In fact, solution polymerization methods are still widely used today in the polymer synthesis industry. The advantages of solution processes are mainly based on the control of process variables, such as viscosity, reagent concentration, and reaction rate, as well as greater efficiency in heat transfer processes. However, the possible generation of chain transfer processes and their elimination from the final polymeric product lead to a decrease in the efficiency of the process due to the need for additional processing steps, and consequently, a negative environmental impact is generated. In response to this, the development of solvent-free processes and the use of alternative polymerization solvents with less impact on processes and the environment are sought. In this regard, the use of supercritical fluids, ionic liquids, monoterpenes, fatty acids, among others, have been studied as alternative methods of polymerization in solution (De Simone, 2002; Dubé and Salehpour, 2014). The synthesis of hydrogels based on sorbitol and citric acid are a particular example of solvent removal as an auxiliary reagent in polymer synthesis. In this case, citric acid is used as a solvent (Palencia et al., 2017a,b).

1.2.6 Design for energy efficiency

As has already been indicated and surely concluded, green chemistry is not only aimed at risk mitigation and waste reduction, but also contemplates the efficient use of resources from an energy point of view. In accordance with the above, it is clear that the design of processes is required under environmental conditions, that is, pressure and temperature. The design of a highly energy-efficient process not only contemplates the use of alternative energy sources to fossil fuels to meet its demand, for example, photovoltaic, thermal, hydroelectric, and wind energy, but also the optimization of its synthesis processes in order to achieve more efficient processes with a low carbon footprint. From a chemical point of view, in the design of highly energy-efficient processes, two of the main objectives of research in this field are focused toward the development of chemical reactions at room temperature and low pressures, and the use of the energy resulting from exothermic processes.

The energy demand in the manufacture of polymers depends on the nature of the reaction. Often, the chain-growth polymerization reactions are exothermic, allowing the energy produced by the reaction to be used for self-heating of the system and minimizing of the energy required during the process. The opposite happens with step-growth polymerization in which the use of high temperatures is generally required. Due to the above, alternative polymerization methods have been developed that do not involve the use of thermal energy, for example, photopolymerization and microwave-assisted polymerization, which use electromagnetic radiation as an energy source for the polymerization process (Yagci et al., 2010; Kempe et al., 2011; Obi, 2018). However, there are limitations associated with these methods, for instance, by photopolymerization, polymeric materials obtained can undergo volume contraction, altering their mechanical properties, presence of residual monomer in the polymer matrix, among others (Tasdelen et al., 2012).

1.2.7 Use of renewable feedstocks

The inclusion of renewable raw materials leads to processes being somewhat more sustainable, consequently, if the technical and economic feasibility for obtaining raw materials from renewable sources is satisfied, these must be implemented in the relevant processes. However, the simultaneous compatibility of technical and economic aspects is not always directly or easily feasible. At present, many of the polymer products used around the world are derived from fossil sources such as oil and natural gas. However, although they come from natural sources, their recovery speed is very low compared to their consumption, which leads to their inevitable depletion and the need to search for alternative raw material sources that have a rapid recovery rate and ensure the sustainability of the processes. In this sense, in recent years, interest in developing polymeric materials based on natural and renewable sources has been enhanced; having as the main objective the transformation and use of the waste generated from current production processes (Iwata, 2015; Wong et al., 2015).

Nature is a source of renewable raw materials such as polymers, monomers, or monomer precursors that can be used or modified to obtain value-added products. This is the reason why biomass is not only attractive as an energy source, but also as a source of metabolites, polymer precursors, or auxiliary reagents (Tapasco et al., 2016; Afanasjeva et al., 2018). In general, the processes for obtaining renewable monomers and polymers can be classified as "direct" when they are obtained from separation processes that do not involve a chemical transformation of the products, for example, chitin, cellulose, starch, humic acids, and collagen, among others (García-Quintero et al., 2018; Anaya-Tatis et al., 2019), or "indirect" when they are the result of chemical and/or biological processes, for example, chitosan, coke, pyrolytic bio-oil modified cellulose fibers, hydrogels from citric acid, and sorbitol (Afanasjeva et al., 2017; Kristufek et al., 2017; Muxica et al., 2017; Palencia et al., 2017a, 2017b, 2020). The structural complexity of natural systems is often challenging to achieve high purity of monomers or polymers, which affects the material properties, efficiency, and process costs, and includes steps that can trigger waste production. Consequently, research into obtaining and using sources of renewable raw materials is constantly expanding, given the growing need to obtain new polymeric products from renewable sources that not only satisfy current needs, but also replace those materials derived from fossil sources.

1.2.8 Reduce derivatives

The generation of derivatives in the synthesis and characterization processes is often unavoidable, and these can result directly as a consequence of the chemical transformation or be the result of necessary steps in the production process; in any case, their production should be avoided as far as possible. However, in many synthesis processes the use of blocking groups, protection/deprotection, and temporary modification of physical-chemical processes is necessary. As the complexity of the product or the specificity of its function increases, the complexity of the stages increases and the need for auxiliary stages is often prevalent.

Derivatization involves the transformation of a compound through simple and generally reversible reactions in a product with similar structural characteristics, but with different chemical properties compared to its precursor. In general, the derivatization technique is widely used both in analytical processes and in organic synthesis including the polymer synthesis. In particular, for polymer manufacture it is necessary, either temporarily or permanently, to modify monomers to protect functional groups, make phases compatible, or impart specific properties to the materials (Pagliano et al., 2018; Lavilla et al., 2014; Zang et al., 2019). Examples of processes involving protective groups include the production of poly(vinyl alcohol) (Aslam et al., 2018), the processes of controlled radical synthesis, for example, atom transfer radical polymerization (ATRP) and more commonly in the production of specialized polymeric materials for biomedical applications (Matyjaszewski, 2012). However, given the need for derivatization processes to incorporate additional reagents, such as blocking or protective groups, it is inevitable that there is an increase in the steps in the synthesis processes and the generation of waste that directly affects the yields of the processes. Thus, given that these methods are harmful to the environment, the search for alternative polymer production systems or derivatization methodologies that generate harmless by-products, arises as a problem that must be addressed when designing chemical processes.

1.2.9 Catalysis

The need to increase the efficiency of the processes has been shown in several of the principles previously described. One way to achieve this goal is through catalytic reactions. This alternative is based on the use of catalysts that should be as selective as possible. At present, the need to increase the efficiency of chemical processes from energy, economic, and environmental points of view has led to the use of catalysts, which not only increase the reaction rates but also intervene in the physical-chemical properties of the final product (Kakaei et al., 2019). It should be noted that for a reagent to be considered catalytic it must be used in small quantities, in addition, once the reaction is completed, the catalyst must be regenerated for later recovery or recycling in the process. A desired characteristic of these substances is that they are not toxic, recalcitrant, or difficult to eliminate.

In the polymer industry, the use of catalysts during synthesis processes is necessary and widely accepted (Walther, 2014). The catalysts used in polymerization can be classified as homogeneous or heterogeneous. In those processes that use homogeneous phase catalysts, the separation of the catalyst from the product is usually difficult, so in most cases deactivation is chosen given its low concentration (Cornils and Herrmann, 2003). However, these often remain in the product, and the impact on subsequent stages of use must be taken into consideration. On the other hand, heterogeneous phase catalysts have had great appeal in polymer synthesis, thanks to their physicochemical properties that give them the ability to be used in a wide range of conditions, and the ease with which they can be used and removed from the product (Shen et al., 2004; Sun et al., 2015). Thanks to the advancement of science in the field of polymeric catalysts, high performance, high activity, selectivity, and specificity have been achieved during processes with their use in low concentrations. Among the main examples of the use of catalysts in the polymer industry are Ziegler—Natta type catalysts (Abedi and Abdouss, 2014), metallocene-based catalysts (Kaminsky, 2016), and catalysts used in controlled radical synthesis processes (Skubi et al., 2016). Recently, the use of enzymatic catalysis has aroused the interest of researchers as an ecological alternative due to its interesting properties of inducing stereo-, regio-, and chemo-selective reactions with high reaction rates and under relatively mild conditions (Douka et al., 2018). Generally, the transformations are carried out at ambient pressures and temperatures, which eliminates the use of high temperatures in the polymerization processes typical of polycondensation processes.

1.2.10 Design for degradation

One of the main aspects of green chemistry is to remember that products have a function and a life cycle, therefore it is important to keep in mind the stages after their use. In this area, the design of easily degradable products acquires great relevance. Chemicals must therefore be designed in such a way that, at the end of their function, they do not persist in the environment and break down into harmless degradation products.

The degradability of a material refers to its ability to decompose into energy and more elemental substances such as water, carbon dioxide, methane, etc. This degradation can be mediated by the individual or joint action of several agents such as: (1) chemical agents, for example, acids or bases, oxides, oxidizing agents, etc.; (2) natural agents, for example, solar radiation, temperature, etc.; or (3) biological agents, for example, bacteria, fungi, enzymes, etc. (Ammala et al., 2011). However, recently the concept of biodegradability has gained greater importance and much research in the area of polymeric materials has been directed toward the development of biodegradable products with low environmental persistence. Polymeric materials are used in countless applications to replace traditional materials such as metals and ceramics. However, as a result of the high durability that polymeric materials can present, their persistence times in nature can be very long and extend beyond the useful life of the material, so it is widely recognized that these are a focus of generation of persistent waste. An example is single-use polymeric materials, for example, plastic bags, cutlery, and disposable cups made of polyethylene, polypropylene, or polystyrene (Hakkinen and Vares, 2010). Furthermore, in certain cases, the inadequate recycling and final disposal of these wastes translates into unavoidable environmental contamination (Erni-Cassola et al., 2019). Therefore, the need to develop biodegradable polymeric materials to replace traditionally used polymers prevails. It is important to highlight that the concept of biodegradability does not imply that the materials are based on biopolymers; therefore, biodegradable polymers can also be made from fossil fuels or renewable raw materials. Thus, butanediol, dicarboxylic acids, and succinic acid, among others, are some of the monomers from fossil sources that are used in the development of biodegradable polymers (Papageorgiou, 2018). Among the polymers of biological origin are poly(lactic acid), polyhydroxyalkanoates, starches, polyesters, poly(aspartate), etc. (Zhu et al., 2016). Currently, in the area of the science of biodegradable polymers, their study is of particular interest as a consequence of the environmental problems reported by the indiscriminate use and inappropriate disposal of man-made waste.

1.2.11 Real-time analysis for pollution prevention

A fundamental aspect of prevention is the monitoring and control of processes at all stages of production. Adequate monitoring allows the identification of failures, critical points, and aspects to improve, and guides decision-making. Green chemistry actively promotes the implementation of monitoring and control of synthesis processes. Therefore, analytical methodologies must be improved to allow real-time monitoring and control of synthesis processes in order to prevent or eliminate the formation of hazardous substances.

At present, the constant advances in sensor technology, the development of portable, compact, multiparametric analytical instruments, and the optimization of analytical methodologies have undoubtedly allowed monitoring of the various variables that can alter the yields of chemical transformations (Saleh and Fadillah, 2019; Yin et al., 2018). In the polymer industry it is essential to apply analysis techniques for monitoring of process variables and product properties, mainly during the polymerization stage. Usually, the main variables analyzed during polymerization include the flow, reactor level, pressure and temperature, monomer concentration, and polymerization grade, among others (Fonseca et al., 2009).

1.2.11.1 Inherently safer chemistry for accident prevention

Accidents are random and uncontrolled events that must be avoided where possible, from economic, environmental, and human health perspectives. However, the possibility of their occurrence always remains, consequently, necessary measures must be taken to minimize their occurrence, and at the same time, due to its fortuitous nature, it must be ensured that in the event of an accident its impact is minimized. In this context, safe processes must be used and designed to minimize risks.

In the polymer industry, the main causes of accidents occur in the stages of transportation and handling of chemical products, for example, solvents, monomers, initiators, catalysts, etc. Added to this, the high reactivity that monomers generally show, the exothermic nature of many of the polymerization reactions, and the high production volume increase the risk of accidents due to the strict control required in process variables such as temperature.

Within the framework of risk prevention, research should not only focus on the development of advanced chemical processes that implement analysis techniques in real time and strict control of the process variables in its different stages, but also toward the search for process biotechnology for the production of polymeric materials that can be carried out under "soft" conditions, that is, at ambient temperature and pressure (Gustin and Laganier, 2005; Sales et al., 2007; Kao and Hu, 2002; Casson et al., 2014).

1.3 Conclusions and remarks

Green chemistry defines 12 principles that can be used for the development of sustainable chemical processes. These principles are directed toward the minimization of waste and by-products, the atomic economy, the reduction of dangers from synthetic chemical products, either through the mitigation of their effects or through safer design and accident prevention, a reduction in the use of solvents and auxiliary substances, an increase in energy efficiency and the yields of the process, the use of renewable, degradable raw materials, and rigorous control of the process from product design to beyond their useful life cycle.

These principles make it possible to create an approximation to the definition of eco-friendly functional polymers, as an eco-friendly polymeric material should comply, as far as possible, with the principles previously enunciated. However, the satisfactory fulfillment of all of these statements is a difficult goal, and, from chemical and practical aspects, it could be considered that these guidelines allow the building of an ideal instead of technical specifications for a product or process. Nevertheless, it should be recognized that moving toward that ideal allows the gradual transformation of technology for sustainable production systems and eco-friendly products.

Note that a polymeric material based on the "principles of green chemistry" is itself eco-friendly in terms of its design, origin, life cycle, disposal, reuse ,or disposal. On the other hand, this eco-friendly feature is not restricted to the above aspects, and it could be considered as a broader concept. In particular, the eco-friendly concept transcends the focus of the material and includes its function. In subsequent chapters we will delve into specific aspects of the chemistry of eco-friendly functional polymers both in general and specific aspects.

References

Abedi, S., Abdouss, M., 2014. A review of clay-supported Ziegler–Natta catalysts for production of polyolefin/clay nanocomposites through in situ polymerization. Appl. Catal. Gen. 475, 386–409. https://doi.org/10.1016/j.apcata.2014.01.028.

Aerts, M., Meuldijk, J., Kemmere, M., Keurentjes, J., 2011. Residual monomer reduction in polymer latex products by extraction with supercritical carbon dioxide. Macromol. Symp. 302, 297–304. https://doi.org/10.1002/masy.201000052.

Afanasjeva, N., Castillo, L., Sinisterra, J., 2017. Lignocellulosic biomass. Part I: biomass transformation. J. Sci. Technol. Appl. 3, 27–43. https://doi.org/10.34294/j.jsta.17.3.22.

Afanajeva, N., Castillo, L., Sinisterra, J., 2018. Lignocellulosic biomass. Part II: tendence in the biomass pyrolysis. J. Sci. Technol. Appl. 5, 4–22. https://doi.org/10.34294/j.jsta.18.5.31.

Ambrogi, V., Carfagna, C., Cerruti, P., Marturano, V., 2017. 4 - additives in polymers. In: Jasso-Gastinel, C.F., Kenny, J.M. (Eds.), Modification of Polymer Properties. William Andrew Publishing, pp. 87–108. https://doi.org/10.1016/B978-0-323-44353-1.00004-X.

Ammala, A., Bateman, S., Dean, K., Petinakis, E., Sangwan, P., Wong, S., Yuan, Q., Yu, L., Patrick, C., Leong, K.H., 2011. An overview of degradable and biodegradable polyolefins. Prog. Polym. Sci. 36, 1015–1049. https://doi.org/10.1016/j.progpolymsci.2010.12.002.

Anastas, P., Eghbali, N., 2010. Green chemistry: principles and practice. Chem. Soc. Rev. 39, 301–312. https://doi.org/10.1039/B918763B.

Anastas, P., Kirchhoff, M.M., 2002. Origins, current status, and future challenges of green chemistry. Acc. Chem. Res. 35, 686–694. https://doi.org/10.1021/ar010065m.

Anaya-Tatis, L., Libreros, K., Palencia, V., Atencio, V., Palencia, M., 2019. Mid-infrared spectral characterization of fish scales: "Bocachico" (*Prochilodus magdalenae*) by functionally-enhanced derivative spectroscopy (FEDS) - a methodological approach. J. Sci. Technol. Appl. 6, 28–39. https://doi.org/10.34294/j.jsta.19.6.39.

Araújo, P.H.H., Sayer, C., Giudici, R., Poço, J.G.R., 2002. Techniques for reducing residual monomer content in polymers: a review. Polym. Eng. Sci. 42, 1442–1468. https://doi.org/10.1002/pen.11043.

Aslam, M., Kalyar, M.A., Raza, Z.A., 2018. Polyvinyl alcohol: a review of research status and use of polyvinyl alcohol based nanocomposites. Polym. Eng. Sci. 58, 2119–2132. https://doi.org/10.1002/pen.24855.

Avio, C.G., Gorbi, S., Regoli, F., 2017. Plastics and microplastics in the oceans: from emerging pollutants to emerged threat. Mar. Environ. Res. 128, 2–11. https://doi.org/10.1016/j.marenvres.2016.05.012.

Bar, M., Alagirusamy, R., Das, A., 2015. Fibers Polym. 16, 705–717. https://doi.org/10.1007/s12221-015-0705-6.

Billiet, L., Fournier, D., Du Prez, F., 2009. Step-growth polymerization and 'click' chemistry: the oldest polymers rejuvenated. Polymer 50, 3877–3886. https://doi.org/10.1016/j.polymer.2009.06.034.

Casson, V., Snee, T., Maschio, G., 2014. Investigation of an accident in a resins manufacturing site: the role of accelerator on polymerisation of methyl methacrylate. J. Hazard Mater. 270, 45–52. https://doi.org/10.1016/j.jhazmat.2014.01.038.

Cornils, B., Herrmann, W.A., 2003. Concepts in homogeneous catalysis: the industrial view. J. Catal. 216, 23–31. https://doi.org/10.1016/s0021-9517(02)00128-8.

De la Guardia, M., Armenta, S., 2011. Chapter 10 - practical consequences of green analytical chemistry. In: De la Guardia, M., Armenta, S. (Eds.), Comprehensive Analytical Chemistry, vol. 57. Elsevier, pp. 219–232. https://doi.org/10.1016/B978-0-444-53709-6.00010-0.

De Marco, B.A., Rechelo, B.S., Tótoli, E.G., Kogawa, A.C., Nunes, H.R., 2019. Evolution of green chemistry and its multidimensional impacts: a review. Saudi Pharmaceut. J. 27, 1–8. https://doi.org/10.1016/j.jsps.2018.07.011.

De Ruyck, J., Brysbaert, G., Blossey, R., Lensink, M.F., 2016. Molecular docking as a popular tool in drug design, an in silico travel. Comput. Biol. Chem. Adv. Appl. 9, 1–11. https://doi.org/10.2147/AABC.S105289.

De Simone, J.M., 2002. Practical approaches to green solvents. Science 297, 799–803. https://doi.org/10.1126/science.1069622.

Douka, A., Vouyiouka, S., Papaspyridi, L.M., Papaspyrides, C.D., 2018. A review on enzymatic polymerization to produce polycondensation polymers: the case of aliphatic polyesters, polyamides and polyesteramides. Prog. Polym. Sci. 79, 1–25. https://doi.org/10.1016/j.progpolymsci.2017.10.001.

Dubé, M.A., Salehpour, S., 2014. Applying the principles of green chemistry to polymer production technology. Macromol. React. Eng. 8, 7–28. https://doi.org/10.1002/mren.201300103.

Erni-Cassola, G., Zadjelovic, V., Gibson, M.I., Christie-Oleza, J.A., 2019. Distribution of plastic polymer types in the marine environment; A meta-analysis. J. Hazard Mater. 369, 691–698. https://doi.org/10.1016/j.jhazmat.2019.02.067.

Fonseca, G.E., Dubé, M.A., Penlidis, A., 2009. A critical overview of sensors for monitoring polymerizations. Macromol. React. Eng. 3, 327–373. https://doi.org/10.1002/mren.200900024.

Gałuszka, A., Migaszewski, Z., Namieśnik, J., 2013. The 12 principles of green analytical chemistry and the significance mnemonic of green analytical practices. Trac. Trends Anal. Chem. 50, 78–84. https://doi.org/10.1016/j.trac.2013.04.010.

Garces, V., Palencia, M., Combatt, E., 2017. Development of bacterial inoculums based on biodegradable hydrogels for agricultural applications. J. Sci. Technol. Appl. 2, 13–23. https://doi.org/10.34294/j.jsta.17.2.11.

Garcia-Quintero, A., Combatt, E., Palencia, M., 2018. Structural study of humin and its interaction with humic acids by Fourier-transform mid-infrared spectroscopy. J. Sci. Technol. Appl. 4, 28–39. https://doi.org/10.34294/j.jsta.18.4.28.

Gustin, J.L., Laganier, F., 2005. Understanding vinyl acetate polymerization accidents. Org. Process Res. Dev. 9, 962–975. https://doi.org/10.1021/op050097f.

Hahladakis, J.N., Velis, C.A., Weber, R., Iacovidou, E., Purnell, P., 2018. An overview of chemical additives present in plastics: migration, release, fate and environmental impact during their use, disposal and recycling. J. Hazard Mater. 344, 179–199. https://doi.org/10.1016/j.jhazmat.2017.10.014.

Häkkinen, T., Vares, S., 2010. Environmental impacts of disposable cups with special focus on the effect of material choices and end of life. J. Clean. Prod. 18, 1458–1463. https://doi.org/10.1016/j.jclepro.2010.05.005.

Hartmann, N.B., Rist, S., Bodin, J., Jensen, L., Schmidt, S.N., Mayer, P., Meibom, A., Baun, A., 2017. Microplastics as vectors for environmental contaminants: exploring sorption, desorption, and transfer to biota. Integrated Environ. Assess. Manag. 13, 488–493. https://doi.org/10.1002/ieam.1904.

Iwata, T., 2015. Biodegradable and bio-based polymers: future prospects of eco-friendly plastics. Angew. Chem. Int. Ed. 54, 3210–3215. https://doi.org/10.1002/anie.201410770.

Kakaei, K., Esrafili, M.D., Ehsani, A., 2019. Chapter 1 - introduction to catalysis. In: Kakaei, K., Esrafili, M.D., Ehsani, A. (Eds.), Interface Science and Technology, vol. 27. Elsevier. https://doi.org/10.1016/B978-0-12-814523-4.00001-0, 1-201921.

Kaminsky, W., 2016. The discovery and evolution of metallocene-based olefin polymerization catalysts. Rendiconti Lincei 28, 87–95. https://doi.org/10.1007/s12210-016-0588-5.

Kao, C.S., Hu, K.H., 2002. Acrylic reactor runaway and explosion accident analysis. J. Loss Prev. Process. Ind. 15, 213–222. https://doi.org/10.1016/s0950-4230(01)00070-5.

Kaur, G., Uisan, K., Ong, K.L., Ki Lin, C.S., 2018. Recent trends in green and sustainable chemistry & waste valorisation: rethinking plastics in a circular economy. Curr. Opin. Green Sustain. Chem. 9, 30–39. https://doi.org/10.1016/j.cogsc.2017.11.003.

Kempe, K., Becer, C.R., Schubert, U.S., 2011. Microwave-assisted polymerizations: recent status and future perspectives. Macromolecules 44, 5825–5842. https://doi.org/10.1021/ma2004794.

Kristufek, S.L., Wacker, K.T., Tsao, Y.Y.T., Su, L., Wooley, K.L., 2017. Monomer design strategies to create natural product-based polymer materials. Nat. Prod. Rep. 34, 433–459. https://doi.org/10.1039/c6np00112b.

Laoutid, F., Bonnaud, L., Alexandre, M., Lopez-Cuesta, J.M., Dubois, P., 2009. New prospects in flame retardant polymer materials: from fundamentals to nanocomposites. Mater. Sci. Eng. R Rep. 63, 100–125. https://doi.org/10.1016/j.mser.2008.09.002.

Lavilla, I., Romero, V., Costas, I., Bendicho, C., 2014. Greener derivatization in analytical chemistry. Trac. Trends Anal. Chem. 61, 1–10. https://doi.org/10.1016/j.trac.2014.05.007.

Lerma, T., Collazos, S., Córdoba, A., 2016. Effect of side chain length of carbamates on the surface properties of porous interpenetrating networks. J. Sci. Technol. Appl. 1, 30–38. https://doi.org/10.34294/j.jsta.16.1.3.

Linthorst, J.A., 2009. An overview: origins and development of green chemistry. Found. Chem. 12, 55–68. https://doi.org/10.1007/s10698-009-9079-4.

Lipshutz, B.H., Gallou, F., Handa, S., 2016. Evolution of solvents in organic chemistry. ACS Sustain. Chem. Eng. 4, 5838–5849. https://doi.org/10.1021/acssuschemeng.6b01810.

Martin, R.L., Simon, C.M., Smit, B., Haranczyk, M., 2014. In silico design of porous polymer networks: high-throughput screening for methane storage materials. J. Am. Chem. Soc. 136, 5006–5022. https://doi.org/10.1021/ja4123939.

Matyjaszewski, K., 2012. Atom transfer radical polymerization (ATRP): current status and future perspectives. Macromolecules 45 (2012), 4015–4039. https://doi.org/10.1021/ma3001719.

Mülhaupt, R., 2013. Green polymer chemistry and bio-based plastics: dreams and reality. Macromol. Chem. Phys. 214, 159–174. https://doi.org/10.1002/macp.201200439.

Muxika, A., Etxabide, A., Uranga, J., Guerrero, P., de la Caba, K., 2017. Chitosan as a bioactive polymer: processing, properties and applications. Int. J. Biol. Macromol. 105, 1358–1368. https://doi.org/10.1016/j.ijbiomac.2017.07.087.

Nesvadba, P., 2012. Radical polymerization in industry. In: Chatgilialoglu, C., Studer, A. (Eds.), Encyclopedia of Radicals in Chemistry, Biology and Materials. https://doi.org/10.1002/9781119953678.rad080.

Obi, B.E., 2018. 2 - polymer chemistry and synthesis. In: Obi, B.E. (Ed.), Polymeric Foams Structure-Property-Performance. William Andrew Publishing, pp. 17–40. https://doi.org/10.1016/B978-1-4557-7755-6.00002-1.

Pang, K., Kotek, R., Tonelli, A., 2006. Review of conventional and novel polymerization processes for polyesters. Prog. Polym. Sci. 31, 1009–1037. https://doi.org/10.1016/j.progpolymsci.2006.08.008.

Pagliano, E., Campanella, B., D'Ulivo, A., Mester, Z., 2018. Derivatization chemistries for the determination of inorganic anions and structurally related compounds by gas chromatography - a review. Anal. Chim. Acta 1025, 12–40. https://doi.org/10.1016/j.aca.2018.03.043.

Palencia, M., Mora, M., Palencia, S.L., 2017a. Biodegradable polymer hydrogels based in sorbitol and citric acid for controlled release of bioactive substances from plants (polyphenols). Curr. Chem. Biol. 11, 36–43. https://doi.org/10.2174/2212796810666161028114432.

Palencia, M., Mora, M.A., Lerma, T., Afanasjeva, N., Isaza, J.H., 2020. Reinforcing agents based on cellulose fibers modified by insertion of end-alkyl groups obtained from pyrolytic bio-oil of sugarcane bagasse. Polym. Bull. 77, 5711–5724. https://doi.org/10.1007/s00289-019-03050-0.

Palencia, M., Lerma, T., Combatt, E., 2017b. Hydrogels based in cassava starch with antibacterial activity for controlled release of cysteamine-silver nanostructured agents. Curr. Chem. Biol. 11, 28–35. https://doi.org/10.2174/2212796810666161108152319.

Palencia, M., Lerma, T., Córdoba, A., 2016. Polyurethanes with boron retention properties for the development of agricultural fertilization smart systems. J. Sci. Technol. Appl. 1, 39–52. https://doi.org/10.34294/j.jsta.16.1.4.

Palencia, M., Lerma, T., Afanasjeva, N., 2019. Antibacterial cationic poly(vinyl chloride) as an approach for in situ pathogen-inactivation by surface contact with biomedical materials. Eur. Polym. J. 115, 212–220. https://doi.org/10.1016/j.eurpolymj.2019.03.041.

Papageorgiou, G., 2018. Thinking green: sustainable polymers from renewable resources. Polymers 10, 952. https://doi.org/10.3390/polym10090952.

Raies, A.B., Bajic, V.B., 2016. In silico toxicology: computational methods for the prediction of chemical toxicity. Wiley Interdiscipl. Rev. Comput. Mol. Sci. 6, 147–172. https://doi.org/10.1002/wcms.1240.

Saleh, T.A., Fadillah, G., 2019. Recent trends in the design of chemical sensors based on graphene-metal oxide nanocomposites for the analysis of toxic species and biomolecules. Trac. Trends Anal. Chem. 120, 115660. https://doi.org/10.1016/j.trac.2019.115660.

Sales, J., Mushtaq, F., Christou, M.D., Nomen, R., 2007. Study of major accidents involving chemical reactive substances. Process Saf. Environ. Protect. 85, 117–124. https://doi.org/10.1205/psep06012.

Shen, Y., Tang, H., Ding, S., 2004. Catalyst separation in atom transfer radical polymerization. Prog. Polym. Sci. 29, 1053–1078. https://doi.org/10.1016/j.progpolymsci.2004.08.002.

Sheldon, R.A., 2014. Green and sustainable manufacture of chemicals from biomass: state of the art. Green Chem. 16, 950–963. https://doi.org/10.1039/C3GC41935E.

Sheldon, R.A., 2017. The E factor 25 years on: the rise of green chemistry and sustainability. Green Chem. 19, 18–43. https://doi.org/10.1039/c6gc02157c.

Skubi, K.L., Blum, T.R., Yoon, T.P., 2016. Dual catalysis strategies in photochemical synthesis. Chem. Rev. 116, 10035–10074. https://doi.org/10.1021/acs.chemrev.6b00018.

Song, Y.K., Hong, S.H., Jang, M., Kang, J.H., Kwon, O.Y., Han, G.M., Shim, W.J., 2014. Environ. Sci. Technol. 48, 9014–9021. https://doi.org/10.1021/es501757s.

Suaria, G., Avio, C.G., Mineo, A., Lattin, G.L., Magaldi, M.G., Belmonte, G., Moore, C.J., Regoli, F., Aliani, S., 2016. The mediterranean plastic soup: synthetic polymers in mediterranean surface waters. Sci. Rep. 6, 37551. https://doi.org/10.1038/srep37551.

Sun, Q., Dai, Z., Meng, X., Wang, L., Xiao, F.S., 2015. Task-specific design of porous polymer heterogeneous catalysts beyond homogeneous counterparts. ACS Catal. 5, 4556–4567. https://doi.org/10.1021/acscatal.5b00757.

Tabone, M.D., Cregg, J.J., Beckman, E.J., Landis, A.E., 2010. Sustainability metrics: life cycle assessment and green design in polymers. Environ. Sci. Technol. 44, 8264–8269. https://doi.org/10.1021/es101640n.

Tapasco, J., Isaza, J.H., Torres, H., 2016. Gas chromatography and mass spectrometry of hydrocarbons in hexane extract from miconia acuminifera triana (melastomataceae). J. Sci. Technol. Appl. 1, 87–96. https://doi.org/10.34294/j.jsta.16.1.8.

Tasdelen, M.A., Çiftci, M., Uygun, M., Yagci, Y., 2012. Possibilities for photoinduced controlled radical polymerizations. In: Matyjaszewski, K., Sumerlin, B.S., Tsarevsky, N.V. (Eds.), Progress in Controlled Radical Polymerization: Mechanisms and Techniques. American Chemical society, pp. 59–72. https://doi.org/10.1021/bk-2012-1100.ch005.

Tobiszewski, M., Marc, M., Galuszka, A., Namiesnik, J., 2015. Green chemistry metrics with special reference to green analytical chemistry. Molecules 20, 10928–10946. https://doi.org/10.3390/molecules200610928.

Trost, B.M., 1991. The atom economy-A search for synthetic efficiency. Science 254, 1471–1477. https://doi.org/10.1126/science.1962206.

Vroman, I., Tighzert, L., 2009. Biodegradable polymers. Materials 2, 307–344. https://doi.org/10.3390/ma2020307.

Walther, G., 2014. High-performance polymers from nature: catalytic routes and processes for industry. ChemSusChem 7, 2081–2088. https://doi.org/10.1002/cssc.201402379.

Wang, H., Yan, Z., Li, H., Yang, N., Leung, K.M.Y., Wang, Y., Yu, R., Zhang, L., Wang, W., Jiao, C., Liu, Z., 2012. Progress of environmental management and risk assessment of industrial chemicals in China. Environ. Pollut. 165, 174–181. https://doi.org/10.1016/j.envpol.2011.12.008.

Webb, H.K., Arnott, J., Crawford, R.J., Ivanova, E.P., 2013. Plastic degradation and its environmental implications with special reference to poly(ethylene terephthalate). Polymers 5, 1–18. https://doi.org/10.3390/polym5010001.

Welton, T., 2015. Solvents and sustainable chemistry. Proc. Royal Soc. A 471, 20150502. https://doi.org/10.1098/rspa.2015.0502.

Wong, S.L., Ngadi, N., Abdullah, T.A.T., Inuwa, I.M., 2015. Current state and future prospects of plastic waste as source of fuel: a review. Renew. Sustain. Energy Rev. 50, 1167–1180. https://doi.org/10.1016/j.rser.2015.04.063.

Yagci, Y., Jockusch, S., Turro, N.J., 2010. Photoinitiated polymerization: advances, challenges, and opportunities. Macromolecules 43, 6245–6260. https://doi.org/10.1021/ma1007545.

Yin, M., Gu, B., An, Q.F., Yang, C., Guan, Y.L., Yong, K.T., 2018. Recent development of fiber-optic chemical sensors and biosensors: mechanisms, materials, micro/nano-fabrications and applications. Coord. Chem. Rev. 376, 348–392. https://doi.org/10.1016/j.ccr.2018.08.001.

Zhang, L., Ji, Y., Gu, X., Zhang, W., Zhou, N., Zhang, Z., Zhu, X., 2019. Synthesis of sequence-controlled polymers with pendent "clickable" or hydrophilic groups via latent monomer strategy. React. Funct. Polym. 138, 96–103. https://doi.org/10.1016/j.reactfunctpolym.2019.03.002.

Zhang, Y., Ying, H., Hart, K.R., Wu, Y., Hsu, A.J., Coppola, A.M., Kim, T.A., Yang, K., Sottos, N.R., White, S.R., Cheng, J., 2016. Malleable and recyclable poly(urea-urethane) thermosets bearing hindered urea bonds. Adv. Mater. 28, 7646–7651. https://doi.org/10.1002/adma.201601242.

Zhu, Y., Romain, C., Williams, C., 2016. Sustainable polymers from renewable resources. Nature 540, 354–362. https://doi.org/10.1038/nature21001.

Chapter 2

Biopolymers and bioplastics: from green chemistry to eco-friendly polymers

2.1 Biopolymers and bioplastics

Biopolymers are biologically originated macromolecules containing monomeric units covalently linked among them. The prefix "bio" means that they are molecules produced by living organisms, therefore, the term biopolymer can be used to describe polymers generally derived from biological sources such as microorganisms, animals, or plants (IUPAC, 1997). However, biopolymers can also be produced by chemical synthesis from biological sources such as vegetable oils, sugars, fats, resins, proteins, amino acids, among others. Compared to synthetic polymers with a simple and random structure, biopolymers are a set of macromolecules characterized by specific three-dimensional shapes and structures. This three-dimensional ordering is an important property that makes biopolymers active in vivo molecules (Bilal et al., 2019; Kanmani et al., 2017). Today, the field of applications for biopolymers is truly multidisciplinary, encompassing research areas ranging from biology, chemistry, and medicine to bioengineering and materials science. Its versatility is due to properties such as biodegradability, biocompatibility, and adaptability in different research areas. For example, in areas associated with biomedicine, the use of biopolymers has allowed the development of systems for the administration of drugs, products for the treatment of wounds, promising substitutes for plastic, and scaffolding for tissue engineering (Bano et al., 2017; Jacob et al., 2018; Murphy et al., 2018; Davidenko et al., 2019; Palencia et al., 2019). Although some of these technologies can be commercially expensive, their specialized biomedical application justifies their high value.

On the other hand, the development of synthetic polymers is considered one of the great advances of the 20th century, due to their innumerable possibilities of use, not only in industry but in everyday life. These synthetic polymers are obtained commercially from petroleum and created for specific functions. For example, plastics are inexpensive, lightweight, oxidation-resistant, and highly versatile materials due to mechanic and thermal

Eco-friendly Functional Polymers
https://doi.org/10.1016/B978-0-12-821842-6.00028-2
Copyright © 2021 Elsevier Inc. All rights reserved.

properties, however these same advantages, the success of their implementation in multiple applications, and their high production volume create their own problems (Karan et al., 2019; Shruti and Kutralam-Muniasamy, 2019).

A bioplastic is essentially a biopolymer, and its conceptual differentiation is very subtle. The main difference is that not all biopolymers are considered to be bioplastic. The above may be the result of the fact that although its nature is polymeric and its origin is biological, its properties do not necessarily correspond to the characteristics of a plastic material, without the foregoing meaning that they cannot be considered functional materials. Examples of this are wood, DNA, and humic acids. Wood is an organic material with a vegetable origin made mainly by cellulose, DNA is a biomolecule seen to be a biological information encoding system rather than a material susceptible to large-scale processing, and humic acids are biomacromolecules resulting from biogeochemical processes (Hung et al., 2012; Graca, 2015; Garcia et al., 2018). However, neither wood, DNA, or humic acids are materials that can be subjected to the processing and shaping processes typical of plastics. If we go a little deeper into the previous analysis, for example, for the case of humic acids, these can be easily extracted from soil and characterized by techniques similar to those used for the characterization of plastic materials. Even many of its chemical properties such as the interaction of ions in such a way that they act as ion exchangers, analogously to anionic polyelectrolytes, can be observed (Garcia et al., 2018; Mora et al., 2018; Restrepo et al., 2018). However, the above similarities are not sufficient to transform humic acids in a practical way into plastic bottles, exterior coatings, or fibers. In conclusion, the concept "plastic" is directly related to processability.

In addition to biopolymers, bioplastics and plastics precursors can be obtained from residual biomass (Afanasjeva et al., 2017; Palencia et al., 2020). The possibility of obtaining the precursors from renewable sources allows the synthesis of materials completely analogous to those from oil industry by-products. For example, polymeric hydrogels can be obtained from oil by-products as acrylic acid and vinyl acetate, thus, poly(vinyl alcohol) is obtained through a sequence of reactions involving polymerization of vinyl acetate followed by the corresponding deacetylation. In an analogous way, acrylic acid can be polymerized to obtain poly(acrylic acid). Subsequently, the esterification of the respective polymers allows the obtaining of a three-dimensional network that, depending on the molecular weights, proportion of reagents, and reaction conditions, will behave like a hydrogel. However, from citric acid and low-molecular-weight polyols (e.g., sorbitol, glucamine, etc.), hydrogels with similar properties, and even superior from the point of view of their biodegradability, can be obtained by a one-step reaction (Palencia et al., 2017a). Similar results are obtained using a biopolymer such as starch (Garcés et al., 2017; Palencia et al., 2017b). Another interesting approach to the formation of plastic materials is possible by combining precursors of different origin, that is, precursors whose origin is the petrochemical industry and precursors whose

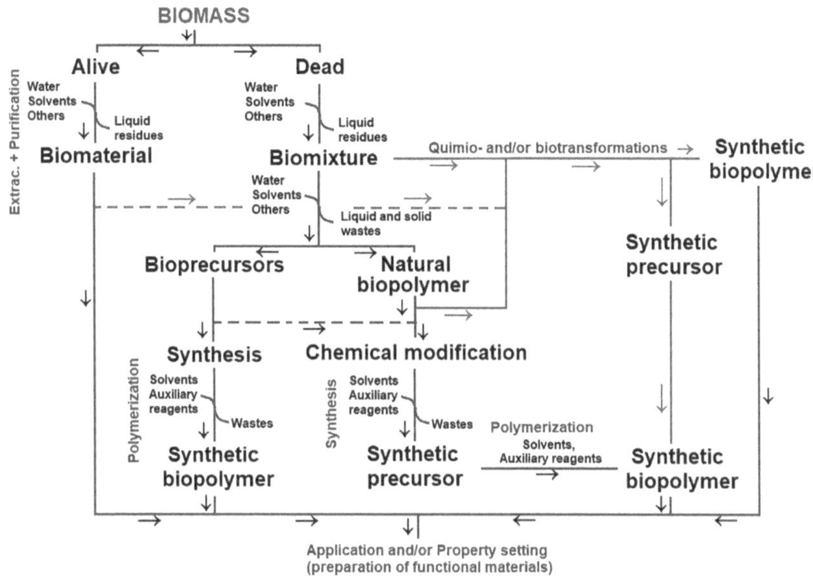

FIGURE 2.1 Possible routes for obtaining of biopolymers from biomass.

origin is biological. An example of this is the synthesis of hydrogels based on polyureas obtained from 1,6-hexamethylenediisocyanate, sorbitol, and cysteine (Otálora et al., 2019), or polyurethanes based on N-methyl-D-glucamine (Palencia et al., 2016). In any case, obtaining plastics and new materials from renewable sources, such as crops and agro-industrial waste, and not from petroleum derivatives, is a strategy that promotes the development and use of more ecological materials, undoubtedly offering a great opportunity to develop green chemistry in industries. Possible routes for obtaining of biopolymers from biomass are shown in Fig. 2.1. The availability of raw material for the development and research of these biomaterials is of such magnitude that, according to the report "Global food losses and food waste—Extent, causes and prevention" the FAO estimates that 30% of the world's food is lost or wasted every year, and predicted that by 2030 this figure could translate to roughly 2.1 billion tons a year and economically to more than $750 billion annually (Gustavsson et al., 2011).

2.2 Biopolymer classification and their eco-friendly approach

To understand a little more in-depth the importance of biopolymers as biomaterials from green chemistry and in the context of eco-friendly polymer materials, it is useful to take your classification as a starting point. Biopolymers are classified according to their source of origin in: (1) biopolymers extracted from biomass; (2) biopolymers synthesized from biomass monomers; and (3) biopolymers produced directly by living organisms (Garcés and Palencia, 2020). Other approaches classify biopolymers as first-class biopolymers (bio-based biopolymers) and second-class biopolymers (bioengineered biopolymers) (Nakajima et al., 2017). However, many considerations can be derived from this classification because a biopolymer can be synthetic, although its precursors come from some type of biomass or bioprocess, but also a bioengineered biopolymer can be understood as those that are the result of a biotechnological process of bioengineering (transgenic plant-derived polyhydroxyalkanoates) (Snell and Peoples, 2002), although for obtaining these they do not undergo a chemical transformation.

2.2.1 Biopolymers extracted from biomass

Currently, materials made from biopolymers extracted from biomass have received great attention because they are a profitable and economical alternative to produce materials with added value and, in principle, because of their origin they are often considered more environment-friendly materials compared to materials obtained from the oil industry. The above statement should be read carefully since, as discussed in Chapter 1, origin is not the only criterion for defining the central objective in this book: "environmentally friendly." Typical biopolymers with industrial uses are starch, collagen, chitin, rubber or *cis*-1,4-polyisoprene, and cellulose (Ross et al., 2017). They are used in conjunction with other substances for the production of new materials with a spectrum wide of applications. An important advantage of this type of biopolymer is biodegradability, which is a property that must be controlled for many applications. The biodegradability of these materials must inevitably be associated with an inherent susceptibility to being colonized by microorganisms. Although in principle this is positive, microbial colonization also can affect the properties of the material, its function, storage, transport, and shelf-life. Therefore, on many occasions, it is necessary to incorporate additives that allow adequate modulation of its microbial colonization and biodegradation. At the same time, these biopolymers tend to be hydrophilic, so their application in outdoor applications involves their chemical modification, for example, by acetylation of carboxylic or hydroxyl acid groups (Laurichesse and Averous, 2014; Chen et al., 2018), or by preparing polymeric composites with other natural or synthetic materials (Gardner et al., 2015; Rogovina et al., 2019).

The environmentally critical stages, in terms of impact, for working with biopolymers directly extracted from biomass are the extraction, purification, and chemical modification in cases where this is necessary. With regard to extraction, in general it is a process that involves the use of solvents and therefore generates a significant waste stream. Even when this procedure is carried out from sources rich in the biopolymer of interest, the impact of secondary currents within the extraction process and its discharges must always be carefully considered. In general, the lower the amount of biopolymer

TABLE 2.1 Examples of widely used biopolymers, biomass, and residues generated from extraction processes.

Biopolymer	Biomass	Residues from extraction	References
Starch	Corn (*Zea mays*), potato (*Solanum tuberosum*), cassava (*Manihot esculenta*), etc.	HCN, solid residues (peels and bagasse containing residual starch), water with high content of dissolved organic substances	López et al. (2015); Abouelezz et al. (2018)
Cellulose	Wood (e.g., *Pinus radiata*)	Chlorophenols, chlorocatechols, heavy metals, etc.; solid residues, water with a high content of dissolved organic substances	Antonkiewicz et al. (2018); Kumar and Chandra (2019)
Collagen	Fish scale	Salts, hydroxyapatite, and organic substances in small amounts	Jafari et al. (2020)
cis-1,4-Polyisoprene	Natural rubber (*Hevea brasiliensis*)	Polycyclic aromatic hydrocarbons, solid residues, and several fungicides, bacteriostatics, or pesticides	Tekasakul and Tekasakul (2006)

in the raw material the more it is associated with higher chemical residuals, however, some wastes in a lower proportion can have high toxicity or a high impact in especially sensitive environments such as water bodies.

Some examples of biopolymers, biomass, and residues from extraction processes are summarized in Table 2.1. Extraction of starch from cassava requires about 15–25 m^3 of water per ton of cassava fresh roots and produces approximately 342–570 g of HCN, 15–25 m^3 of waste water, and 18 g of residual solids (López et al., 2015; Abouelezz et al., 2018). Cellulose industries associated with the obtaining of pulp paper release wastewater containing very complex organic and inorganic pollutants such as chlorophenols, chlorocatechols, and heavy metals, among others (Antonkiewicz et al., 2018; Kumar and Chandra, 2019). Collagen is extracted using different techniques, and depending on the technique the waste generated varies. For example, the solvent extraction process can be carried out using, after washing with water, 0.5 M acetic acid with an efficiency of 16 mg of collagen per gram of fish skin, which means not only that the waste stream is enriched with acetic acid but also a high amount of other wastes. In contrast, when the source is fish scales, thermal dissolution accompanied by purification through the use of inorganic acids can be employed. In this case, most of the residue will be hydroxyapatite and dissolved salts after neutralization of excess acid (Jafari et al., 2020). The natural rubber extraction industry produces aerosol particles (polycyclic aromatic hydrocarbons such as naphthalene, acenaphthene, pyrene, benzopyrene, etc.), wastewater with a high amount of organic substances, tetramethyl triuram disulfide, zinc oxide, diammonium phosphate, and solid residues (Tekasakul and Tekasakul, 2006).

2.2.2 Biopolymers synthesized from biomass monomers

Biomass is undoubtedly a suitable renewable source for obtaining chemical precursors which can be used in the synthesis of biopolymers. In particular, the biomolecules used as precursors can be simple or of high structural complexity. In the case of complex structures, biomass offers an eco-friendlier strategy, not only due to the sustainable nature of the raw material, but also because the biochemical processes that give rise to the precursors of interest are the result of transformations occurring normally in the natural surroundings and by very specific enzymatic routes. The obtaining in the laboratory of many of these precursors, or structurally analogous molecules, would require an enormous synthetic effort, characterized by multiple stages of reaction, purification, and the use of auxiliary reagents and solvents. For example, it has been reported that from leaves of *Miconia acuminifera Triana* (Melastomataceae), using hexane as a solvent, it is possible to obtain at least 25 compounds, such as methylated phenols (e.g., tocopherol), which could be used for the production of phenol-formaldehyde resins, or if they have at least two hydroxyl groups, then they could also be used for the synthesis of polyurethanes or polyesters (Tapasco et al., 2016). In contrast, the synthesis of this type of compound from phenol would involve multiple reactions, including ring substitution and cyclization, and the management of a racemic mixture in one or more synthesis stages (Kuzakov and Shmidt, 2000; Ito et al., 2007; Denmark and Kornfilt, 2017).

Although the structural richness is enormous, two important limitations must be taken into account: (1) the contents of biomolecules are often low, which implies that their use on a large scale is only viable with enormous quantities of available raw material, therefore, the residual biomasses from crops with a large area of extension are ideal for the extraction of precursors, and (2) the extraction and purification processes are not always simple due to the enormous complexity of the biomass. Consequently, an enormous amount of waste can be inevitably produced during the extraction and purification. Due to these limitations, the ideal precursors should be in high content, be easy to extract and purified, and come from biomass generated in large quantities and readily available.

There are many biomolecules obtained from residual biomass that can be used for the formation of polymers; however, strategies used for the obtaining of precursors can be grouped into two routes, the first is based on extraction–purification or extraction–purification–modification, while the second is directed toward the initial transformation of the biomass followed by subsequent obtaining of the precursors. Examples of the first route are: synthesis of polymers directly from biomolecules like sorbitol and citric acid

(Palencia et al., 2017) and the obtaining of precursors such as glucose, lactic acid, and 3-hydroxypropionic acid, which can be used as a synthetic platform for the obtaining of new polymerizable precursors. Thus, glucose can be obtained by enzymatic hydrolysis of starches, cellulose, sucrose, and lactose; lactic acid can be obtained by fermentation of carbohydrates such as glucose, maltose, sucrose, lactose, etc.; 3-hydroxypropionic acid can be obtained from residual biomass and be used as a synthetic platform for several compounds with synthetic interest or directly polymerized to obtain poly(3-hydroxypropionic acid) (Corma et al., 2007; Msuya et al., 2017). In addition, depolymerization of polysaccharides and proteins can be performed by catalytic methods (Jiang et al., 2015). On the other hand, examples of the second route are pyrolytic transformation, thermal conversion, and bio-assisted transformation of biomass (Meng et al., 2014; Chan et al., 2017; Kumar et al., 2016; Palencia et al., 2020).

2.2.3 Biopolymers produced directly by living organisms

This classification includes biopolymers obtained from living organisms, without losing their living condition. In a broad sense, all biomass is the result of biotransformations mediated by living organisms; thus, for instance, plants can use carbon, oxygen, water, and sunlight to generate biopolymers such as starch and cellulose. Other well-known examples, but of animal origin, are the production of keratin, constituting wool, which is obtained from sheep and other animals, and the production of fibroin, constituting silk, which is obtained from the cocoons of the larvae of the mulberry silkworm *Bombyx mori*. However, the most widespread cases are based on the use of microorganisms (mainly bacteria, fungi, and yeasts). Consequently, a subclassification of live-organism biopolymers can be made according to their origin: animal biopolymers (e.g., keratin, fibroin), plant biopolymers (e.g., starch, cellulose, rubber), and microbial biopolymers (e.g., polysaccharides, polyamides, polyesters, peptides, DNA, polyphosphates, etc.) (Nakajima et al., 2017; Garcés and Palencia, 2020; Wankhade, 2020; Fata et al., 2020).

Microorganisms are perhaps the living organisms best adapted to the many conditions on our planet. From an industrial point of view, they are bioreactors capable of carrying out chemical reactions with a high degree of specificity (intracellular biotransformations), but at the same time, they are catalysts of biological origin capable of transforming biomass into a great variety of biopolymers (extracellular biotransformations). Among the bacterial biopolymers, the polysaccharides, polyamides, polyesters, and polyphosphates stand out (Garcés and Palencia, 2020; Wankhade, 2020; Fata et al., 2020). Some of the main biopolymers are summarized in Table 2.2. As in the previous sections, the critical stages in environmental terms for the production of these biopolymers are mainly extraction and purification, mainly for plant biopolymers or microbial biopolymers.

TABLE 2.2 Examples of biopolymers produced directly by living organisms (Snell et al., 2015; Garcés and Palencia, 2020; Wankhade, 2020; Fata et al., 2020).

Origin	Biopolymer type	Synthesis site	Example of biopolymer
Animals	Polyamides or polypeptide	External production	Wool, silk
Plants	Polysaccharides	Roots and seeds, leaves, and branches	Starch, cellulose
	Polyesters	Roots and seeds, leaves, and branches	Polyhydroxyalkanoates
Microorganisms	Polysaccharides	Intracellular	Glycogen
		Extracellular	Alginate, cellulose, hyaluronate, alginate, phosphoethanolamine cellulose, poly(N-acetylneuraminic acid)
	Polyamides or polypeptide	Intracellular	Cyanophycin (copolymer of L-aspartic acid and L-arginine)
		Extracellular	Poly(γ-D-glutamic acid), poly(ε-L-lysine)
	Polyesters	Intracellular	Polyhydroxyalkanoates such as poly((R)-3-hydroxybutyrate)
	Polyphosphates	Intracellular	Polymer of inorganic phosphates

2.3 Conclusions and remarks

Biopolymers are a promising alternative to make the plastics industry a more environmentally friendly sector. However, it should be kept in mind that substituting the use of precursors derived from oil, without achieving a substitution of the materials derived from them, does not achieve a significant effect in terms of sustainability, however, it is a clear step in that direction. On the other hand, biopolymers have gradually managed to open space at a technological level, at an industrial level, and as a specialized area of knowledge, which makes this sector a promising field for advancements in the development of more environmentally friendly production systems.

In general terms, the generation of solid residues and discharge rich in organic matter is common both in the bio-fabrication of polymers and in the obtaining of precursors. In addition, these discharges often can be accompanied by toxic substances that can include pesticides, heavy metals, or bioactive metabolites; therefore, new research should be directed in these areas. Given that the generation of these wastes is inevitable, their proper management through circular economy techniques, waste recovery, or new disposal technologies are alternatives to which attention should be focused. On the other hand, the development of biopolymers must be planned in a manner consistent with the principles of sustainability and green chemistry. The overproduction of biopolymers through unfriendly techniques would carry the risk of negatively impacting the environment, since although the waste, products, and by-products are of biological origin, an excessive and uncontrolled release of these wastes into ecosystems would inevitably create new problems to be solved.

References

Abouelezz, K., Yuan, J., Wang, G., Bian, G., 2018. The nutritive value of cassava starch extraction residue for growing ducks. Trop. Anim. Health Prod. 50, 1231–1238. https://doi.org/10.1007/s11250-018-1549-z.

Afanasjeva, N., Castillo, L., Sinisterra, J., 2017. Lignocellulosic biomass. Part I: biomass transformation. J. Sci. Technol. Appl. 3, 27–43. https://doi.org/10.34294/j.jsta.17.3.22.

Antonkiewicz, J., Pelka, R., BIk-Malodzinska, M., Zukowska, G., Glen-Karolczyk, 2018. The effect of cellulose production waste and municipal sewage sludge on biomass and heavy metal uptake by a plant mixture. Environ. Sci. Pollut. Control Ser. 25, 31101–31112. https://doi.org/10.1007/s11356-018-3109-5.

Bano, I., Arshad, M., Yasin, T., Ghauri, M.A., Younus, M., 2017. Chitosan: a potential biopolymer for wound management. Int. J. Biol. Macromol. 102, 380–383. https://doi.org/10.1016/j.ijbiomac.2017.04.047.

Bilal, M., Iqbal, H., 2019. Naturally derived biopolymers: potential platforms for enzyme immobilization. Int. J. Biol. Macromol. 130, 462–482. https://doi.org/10.1016/j.ijbiomac.2019.02.152.

Chan, S.P., Partho, S.R., Su, H.K., 2017. Current Developments in Thermochemical Conversion of Biomass to Fuels and Chemicals. https://doi.org/10.5772/intechopen.71464.

Chen, Y.F., Kaur, L., Singh, J., 2018. Chemical modification of starch. In: Sjoo, M., Nilsson, L. (Eds.), Starch in Food: Structure, Functions and Applications, pp. 283–321. https://doi.org/10.1016/B978-0-08-100868-3.00007-X.

Corma, A., Iborra, S., Velty, A., 2007. Chemical routes for the transformation of biomass into chemicals. Chem. Rev. 107, 2411–2502. https://doi.org/10.1021/cr050989d.

Davidenko, N., Cameron, R., Best, S., 2019. Natural Biopolymers for Biomedical Applications. Biomedical Sciences, pp. 162–176. https://doi.org/10.1016/B978-0-12-801238-3.11026-8.

Denmark, S.E., Kornfilt, D.J., 2017. Catalytic, enantioselective, intramolecular sulfenofunctionalization of alkenes with phenols. J. Org. Chem. 82, 3192–3222. https://doi.org/10.1021/acs.joc.7b00295.

Fata, M., Rehm, B.H., 2020. Bacterial biopolymers: from pathogenesis to advanced materials. Nat. Rev. Microbiol. 18, 195–210. https://doi.org/10.1038/s41579-019-0313-3.

Garcés, V., Palencia, M., 2020. Biosynthesis and Biotransformations of Biopolymers: Starch and Poly(hydroxyalkanoates). J. Sci. Technol. Appl. 9, 4–9. https://doi.org/10.34294/j.jsta.20.9.60.

Garcés, V., Palencia, M., Combatt, E., 2017. Development of bacterial inoculums based on biodegradable hydrogels for agricultural applications. J. Sci. Technol. Appl. 2, 13–23. https://doi.org/10.34294/j.jsta.17.2.11.

Garcia, A., Combatt, E., Palencia, M., 2018. Structural study of humin and its interaction with humic acids by Fourier-transform mid-infrared spectroscopy. J. Sci. Technol. Appl. 4, 28–39. https://doi.org/10.34294/j.jsta.18.4.28.

Gardner, D.J., Han, Y., Wang, L., 2015. Wood–plastic composite technology. Curr. Forest. Rep. 1, 139–150. https://doi.org/10.1007/s40725-015-0016-6.

Graca, J., 2015. Suberin: the biopolyester at the frontier of plants. Fornt. Chem. 3, 1–11. https://doi.org/10.3389/fchem.2015.00062.

Gustavsson, J., Cederberg, C., Sonesson, U., van Otterdijk, R., Meybeck, A., 2011. Global Food Losses and Food Waste: Extent. Causes and Prevention FAO, Rome, p. 29.

Hung, Y.C., Lin, T.Y., Hsu, W.T., Chiu, Y.W., Wang, Y.S., Frunk, L., 2012. Functional DNA biopolymers and nanocomposite for optoelectronic applications. Opt. Mater. 34, 1208–1213. https://doi.org/10.1016/j.optmat.2012.01.035.

Ito, Y., Kato, R., Hamashima, K., Kataoka, Y., Oe, Y., Ohta, T., Furukawa, I., 2007. Intramolecular cyclization of phenol derivatives with C=C double bond in a side chain. J. Organomet. Chem. 692, 691–697. https://doi.org/10.1016/j.jorganchem.2006.05.060.

IUPAC, 1997 (the "Gold Book"). Compiled by A. D. McNaught and A. Wilkinson. In: Compendium of Chemical Terminology, second ed. Blackwell Scientific Publications, Oxford, ISBN 0-9678550-9-8. https://doi.org/10.1351/goldbook. Online version (2019) created by S. J. Chalk.

Jacob, J., Haponiuk, J., Thomas, S., Gopi, S., 2018. Biopolymer based nanomaterials in drug delivery systems: a review. Mater. Today Chem. 9, 43–55. https://doi.org/10.1016/j.mtchem.2018.05.002.

Jafari, H., Lista, A., Mafosso, M., Ghaffari-Bohlouli, P., Nie, L., Alimoradi, H., Shavandi, A., 2020. Fish collagen: extraction, characterization, and applications for biomaterials engineering. Polymers 12, 2230. https://doi.org/10.3390/polym12102230.

Jiang, Y., Wang, X., Cao, Q., Dong, L., Guan, J., Mu, X., 2015. Chemical conversion of biomass to green chemicals. In: Xian M. Sustainable Production of Bulk Chemicals, pp. 19–49. https://doi.org/10.1007/978-94-017-7475-8_2.

Kanmani, P., Aravind, J., Kamaraj, M., Sureshbabu, P., Karthikeyan, S., 2017. Environmental applications of chitosan and cellulosic biopolymers: a comprehensive outlook. Bioresour. Technol. 242, 295–303. https://doi.org/10.1016/j.biortech.2017.03.119.

Karan, H., Funk, C., Grabert, M., Oey, M., Hankamer, B., 2019. Green bioplastics as part of a circular bioeconomy. Trends Plant Sci. 24, 237–249. https://doi.org/10.1016/j.tplants.2018.11.010.

Kumar, A., Chandra, R., 2019. Pollutants released from the pulp paper industry: aquatic toxicity and their health hazards. Aquat. Toxicol. 211, 202–216. https://doi.org/10.1016/j.aquatox.2019.04.007.

Kumar, A., Gautam, A., Dutt, D., 2016. Biotechnological transformation of lignocellulosic biomass in to industrial products: an overview. Adv. Biosci. Biotechnol. 7 https://doi.org/10.4236/abb.2016.73014.

Kuzakov, E.V., Shmidt, E.N., 2000. Syntheis of trpenophenols via direct alkylation of phenols by terpenes. Chem. Nat. Compd. 36, 245–257.

Laurichesse, S., Averous, L., 2014. Chemical modification of lignins: towards biobased polymers. Prog. Polym. Sci. 39, 1266–1290. https://doi.org/10.1016/j.progpolymsci.2013.11.004.

López, O.V., Versino, F., Villar, M.A., García, M.A., 2015. Agro-industrial residue from starch extraction of Pachyrhizus ahipa as filler of thermoplastic corn starch films. Carbohydr. Polym. 134, 324–332. https://doi.org/10.1016/j.carbpol.2015.07.081.

Meng, D.C., Shen, R., Yao, H., Chen, J.C., Wu, Q., Chen, G.Q., 2014. Engineering the diversity of polyesters. Curr. Opin. Biotechnol. 29, 24–33. https://doi.org/10.1016/j.copbio.2014.02.013.

Mora, M., Palencia, M., Combatt, E., 2018. Effect of dissolved organic matter and humic substances on transport of ions and low molecular weight molecules by liquid membranes. J. Sci. Technol. Appl. 4, 40–48. https://doi.org/10.34294/j.jsta.18.4.29.

Msuya, N., Katima, J., Masanja, E., Temu, A.K., 2017. Poly(lactic-acid) production-from monomer to polymer: a review. SciFed J. Polym. Sci. 1.

Murphy, C.A., Costa, J., Silva-Correira, J., Oliveira, J., Reis, R., Collins, M., 2018. Biopolymers and polymers in the search of alternative treatments for meniscal regeneration: state of the art and future trends. Appl. Mater. Today 12, 51–71. https://doi.org/10.1016/j.apmt.2018.04.002.

Nakajima, H., Dijkstra, P., Loos, K., 2017. The recent developments in biobased polymers toward general and engineering applications: polymers that are upgraded from biodegradable polymers, analogous to petroleum derived polymers, and newly developed. Polymers 9, 523. https://doi.org/10.3390/polym9100523.

Otálora, A., Lerma, T., Palencia, M., 2019. Synthesis and characterization of polyurea-based Hydrogels by Multicomponent Polycondensation of 1,6-Hexamethylenediisocyanate, sorbitol and cysteine. J. Sci. Technol. Appl. 7, 5–16. https://doi.org/10.34294/j.jsta.19.7.47.

Palencia, M., Berrio, M.E., García-Quintero, A., Lerma, T., 2019. Cellulose-Based nanocompsite for pharmaceutical and biomedical applications. In: Kumar, T. (Ed.), Ghosh B. Polysaccharide-Based Nano-Biocarrier in Drug Delivery. Taylor & Francis Group, pp. 39–58.

Palencia, M., Lerma, T., Combatt, E., 2017a. Hydrogels based in cassava starch with antibacterial activity for controlled release of cysteamine-silver nanostructured agents. Curr. Chem. Biol. 11, 28–35. https://doi.org/10.2174/2212796810666161108152319.

Palencia, M., Lerma, T., Córdoba, A., 2016. Polyurethanes with boron retention properties for the development of agricultural fertilization smart systems. J. Sci. Technol. Appl. 1, 39–52. https://doi.org/10.34294/j.jsta.16.1.4.

Palencia, M., Mora, M.A., Lerma, T., Afanasjeva, N., Isaza, J.H., 2020. Reinforcing agents based on cellulose fibers modified by insertion of end-alkyl groups obtained from pyrolytic bio-oil of sugarcane bagasse. Polym. Bull. 77, 5711–5724. https://doi.org/10.1007/s00289-019-03050-0.

Palencia, M., Mora, M., Palencia, S.L., 2017b. Biodegradable polymer hydrogels based in sorbitol and citric acid for controlled release of bioactive substances from plants (polyphenols). Curr. Chem. Biol. 11, 36–43. https://doi.org/10.2174/2212796810666161028114432.

Restrepo, D., Palencia, M., Palencia, V., 2018. Study by attenuated total reflectance spectroscopy of structural changes of humified organic matter by chemical perturbations via alkaline dissolution. J. Sci. Technol. Appl. 4, 49–59. https://doi.org/10.34294/j.jsta.18.4.30.

Rogovina, S.Z., Prut, E.V., Aleksanyan, K.V., Krasheninnikov, V.G., Perepelitsina, E.O., Shashkin, D.P., Berlin, A.A., 2019. Composites based on starch and polylactide. Polym. Sci. 61, 334–340. https://doi.org/10.1134/S1560090419030114.

Ross, G., Sukunya, S., Tighe, B.J., 2017. Bioplastics: New Routes, New Products. Brydson's Plastics Materials. Butterworth-Heinemann., pp. 631–652. https://doi.org/10.1016/B978-0-323-35824-8.00023-2

Shruti, V.C., Kutralam-muniasamy, G., 2019. Bioplastics: missing link in the era of microplastics. Sci. Total Environ. 697, 134139. https://doi.org/10.1016/j.scitotenv.2019.134139.

Snell, K.D., Peoples, O.P., 2002. Polyhydroxyalkanoate polymers and their production in transgenic plants. Metab. Eng. 4, 29–40. https://doi.org/10.1006/mben.2001.0214.

Snell, K.D., Singh, V., Brumbley, S.M., 2015. Production of novel biopolymers in plants: recent technological advances and future prospects. Curr. Opin. Biotechnol. 32, 68–75. https://doi.org/10.1016/j.copbio.2014.11.005.

Tapasco, J., Isaza, J.H., Torres, H., 2016. Gas chromatography and mass spectrometry of hydrocarbons in hexane extract from Miconia acuminifera Triana (melastomataceae). J. Sci. Technol. Appl. 1, 87–96. https://doi.org/10.34294/j.jsta.16.1.8.

Tekasakul, P., Tekasakul, S., 2006. Environmental problems related to natural rubber production in Thailand. Earozoru Kenkyu 21, 122–129. https://doi.org/10.11203/jar.21.122.

Wankhade, V., 2020. Chapter 6 - animal-derived biopolymers in food and biomedical technology. In: Pal, K., Banerjee, I., Sarkar, P., Kim, D., Deng, W., Kuar, N., Majumder, K. (Eds.), Biopolymer-Based Formulations: Biomedical and Food Applications, pp. 139–152. https://doi.org/10.1016/B978-0-12-816897-4.00006-0.

Chapter 3

Recyclable polymer chemistry: structural, physicochemical aspects, uses, and transformation technologies

3.1 Basic principles of polymer recycling

To date, synthetic polymers are predominantly based on petroleum by-products, with an estimated global average production of around 6% of fossil raw materials (Ellen MacArthur Foundation et al., 2016; Hatti-Kaul et al., 2020). By 2016, world production of synthetic polymers reached a volume of 335 million tons and production is expected to triple by 2050. On the other hand, the global generation of post-consumer plastic waste is currently estimated at approximately 150 million tons per year, where only 2% of this waste is recycled due to the fact that, generally, reprocessed or recycled polymers have lower properties than virgin materials due to the physical transformations that must be carried out for their reuse, in addition, the recycling industry incurs new costs for the development of the reprocessing stages, which includes collection, storage, and transportation logistics to access these materials in the optimal quantities and operating conditions (Rahimi and García, 2017; Ragaert et al., 2017). Low recycling rates result in large losses in the value of the material to the economy and the environment. It is estimated that 95% of the value of plastic packaging material (€70 million−€105 billion) is lost to the economy after a very short cycle of first use. Thus, plastics are now at the top of the global political agenda, and various initiatives are being put forward to minimize polymer leakage and move toward a circular plastics economy (Ellen MacArthur Foundation et al., 2016; UNEP, 2018).

Recycling is understood as a process aimed at converting waste into new products or raw materials for later use. However, this does not always mean that recycling is an eco-friendly process since this will depend on the balance between the impact generated by waste and the impact of its retransformation and reuse. Currently, four main techniques are implemented for the recycling

Eco-friendly Functional Polymers
https://doi.org/10.1016/B978-0-12-821842-6.00005-1
Copyright © 2021 Elsevier Inc. All rights reserved.

of plastics: primary (closed loop), secondary (mechanical), tertiary (chemical), and energy recovery (incineration) recycling. In the latter, the energy from plastics is recovered in the form of heat, consequently generating new emissions, in addition, the energy generated by the massive burning of plastics is substantially lower than the energy conserved by recycling (Morris, 1996).

Primary recycling, or closed-loop recycling, refers to the reprocessing of polymers to give a product with the same purpose as the original polymer. This is usually implemented in the same production plants to reduce excess raw material losses as a result of design characteristics. Its implementation is part of productive innovation strategies, and on many occasions it is associated with a reduction in costs that justifies the investment. Unlike secondary recycling, or mechanical recycling, where recycled materials are for uses other than those for which the original material was manufactured, the products acquire a lower commercial value than the initial one due to the loss of intrinsic properties of the material. This stage is called "degradation" or "cycle down" In this case, before its commercialization, in some cases, the material can be used in the same factory in different manufacturing processes or commercialized to other sectors where they can be useful (La Mantia, 2004; Solis and Silveira, 2020). However, the situation for both primary and secondary recycling changes as the material leaves the production factory. For example, a sawmill can produce wood columns (cellulose), sawdust, and other wood residues that can be used as raw material in the manufacture of wood–plastic composites, however, the reuse of the residual material of the composite will depend on the characteristics of the same and the process carried out. Both in the cases of wooden columns and wood–plastic composites, after marketing or leaving the factory, the reuse of these materials in their respective processes is limited by the loss of properties necessary for their use, or due to the inclusion of additives, paints, etc., during the initial processing. Although here it was exemplified with a biopolymer, the situation is completely analogous in the manufacture of products based on plastic materials.

On the other hand, tertiary or chemical recycling, as its name indicates, uses chemical processes that are based on the decomposition of polymers by heat, chemical agents, and catalysts for the reuse or depolymerization of the material. This methodology is not commonly implemented on an industrial scale because current methods require considerable energy input. However, if the pure monomer could be recovered through chemical recycling, polymer prices would be decoupled from oil prices, potentially saving ~3.5 billion barrels of oil, in addition to monetary savings that have been estimated to be of the order of 176 billion dollars. This is why this form of reprocessing is currently on the rise to develop new technologies that allow the classification and recycling of solid plastic waste into raw materials efficiently (Thompson et al., 2009; Hahladakis et al., 2018; North and Halden, 2013; Hong and Chen, 2017). However, it should be taken into account that, in addition to being an oversimplification, the visualization of the process in three stages,

(1) monomer–polymer, (2) polymer–material, and (3) material–monomer, is different from understanding this process as the transformation of the monomer into a polymer, to end its initial use, transforms the polymer into a monomer. The key point is that the process focuses not on obtaining the monomer from the polymer, but on obtaining the monomer from the material, where the latter includes additives, and other auxiliary reagents required for the proper performance of its function. In this sense, for example, it should be taken into account that the post-use processing of the constituent material of a plastic bottle for water consumption is not the same as the post-use processing of plastic materials used in computers. In the latter, in addition to being used for multiple materials and mixtures, its separation process is expensive, so its recovery usually has a very low recovery value, to the point that it is not economically viable.

The previous example allows visualization of an important aspect: the classification of materials for the homogenization of the properties, as much as possible, within the output streams of materials in a recycling process. In the case of plastic bottles for water consumption, as is easily visible, the outflow of material can be obtained largely homogeneously, however, in the case of plastic materials used in electronic devices the situation is completely different. Before chemical processing, it is necessary to classify the waste, taking into account the intrinsic properties of the polymer, whether physical and/or chemical. Technologies that take physical properties into account are known as mechanical technologies and examples include magnetic density separation, triboelectric separation, foam flotation methods, and speed accelerators (Bakker et al., 2009; Tilmatine et al., 2009; Maio et al., 2010). Meanwhile, technologies based on chemical properties are known as optical technologies. Some examples are hyperspectral imaging, laser-induced decay spectroscopy, X-ray fluorescence, and infrared spectroscopy (Gondal and Siddiqui, 2007; Hu et al., 2013).

Chemical recycling can be classified into two methods: reuse recycling, where the structure and composition of the polymer is partially modified in order to provide new properties and applications, and chemical recycling by depolymerization in which the polymer is degraded to the monomer of origin, or others, with commercial and functional value (Garcia and Robertson, 2017; Merrington, 2017). Chemical recycling is mainly directed toward discarded polymers that are available in large quantities or that can be reused and reintegrated particularly efficiently in industrial processes, such as poly(ethylene terephthalate) (PET), poly(propylene) (PP), poly(vinyl chloride) (PVC), poly(styrene) (PS), polycarbonate (PC), poly(urethane) (PU), poly(methyl methacrylate) (PMMA), polyamides, and poly(ethersulfone) (PES) (Rochman et al., 2013; Özkan et al., 2015). For this, the Society of the Plastic Industry (SPI) has assigned seven main categories of plastic that are described with codes from 1 to 7. Next, the structural and physicochemical aspects of each plastic category are described, as well as their uses and technologies of transformation.

3.2 Plastic classification according to chemical recycling

The **SPI code 1** corresponds to PET, a resistant and moldable polymer that is synthesized from the polycondensation of ethylene glycol and terephthalic acid and that is used in the production of beverage bottles, fibers, and filaments. PET ductility during mechanical recycling drops from ~310% to ~218% after one cycle and is 2.9% in the third cycle, which is the reason why only a small portion of PET is recycled for its original application (primary recycling), and the majority (50%–77%) becomes fibers used for the production of mixed materials, such as carpets by secondary recycling (Singh et al., 2017). For this same reason, recycling alternatives are currently being sought, such as chemical recycling by depolymerization of PET, which can proceed without a catalyst or with a catalyst such as 1,5,7-triazabicyclo[4.4.0]dec-5-ano (TBD) carrying out transesterification in the polymer skeleton by oxygen-centered nucleophiles such as ethylene glycol (glycolysis), methanol (methanolysis), water (hydrolysis), or hydroxide (alkaline hydrolysis) to produce monomers or small terephthalic oligoesters such as bis(hydroxyethyl) terephthalate, dimethyl terephthalate, and terephthalic acid (Goto et al., 2002; Fukushima et al., 2011). Transesterification can also be carried out with nitrogen-centered molecules (aminolysis) such as amines and anilines such as guanidine-1,5,7-triazabicyclo[4.4.0]dec-5-ene to depolymerize PET to ethylene glycol, and N,N′-bis(4-hydroxyphenyl)terephthalamide, valuable monomers used, for example, in the case of amide, in nucleophilic aromatic substitution polymerizations with bisphenol A and bis(4-fluorophenyl)sulfone to give hybrid materials of poly(aryl ether sulfone amide), as exemplified in Fig. 3.1A (Fukushima et al., 2013; Gardea et al., 2014). The chemical recycling of PET is challenging, since it is less efficient when the entrance is dyed as an additional step (discoloration or dye removal) is required in the process (Rahimi and García, 2017).

SPI codes 2, **4**, and 5 correspond to high-density polyethylene (HDPE), low-density polyethylene (LDPE) and PP, respectively, which are polyolefins or polyalkenes obtained from the polyaddition of ethylene with catalysts based on transition metals type Ziegler–Natta (Peacock, 2000; Merrington, 2017). HDPE is widely used in electrical insulators, toys, bottles, pipes, and films, and is designated as a high-density polymer because it is synthesized at low pressure by copolymerizing ethylene with 1-olefins (such as 1-butene, 1-hexene, and 1-octene), producing polymers with short branches and densities of approximately 0.94 and 0.97 g/cm^3, densities higher than LDPE which has densities between 0.91 and 0.94 g/cm^3, since it is a more branched polymer due to intra- and intermolecular transfer reactions during ethylene polyaddition at high pressures (Xu et al., 2016). The properties of the different polyethylenes are governed by the ranges of densities and melting temperatures. Polypropylene, on the other hand, has a wide use in industrial environments due to its hardness and high chemical resistance. As a result, it is

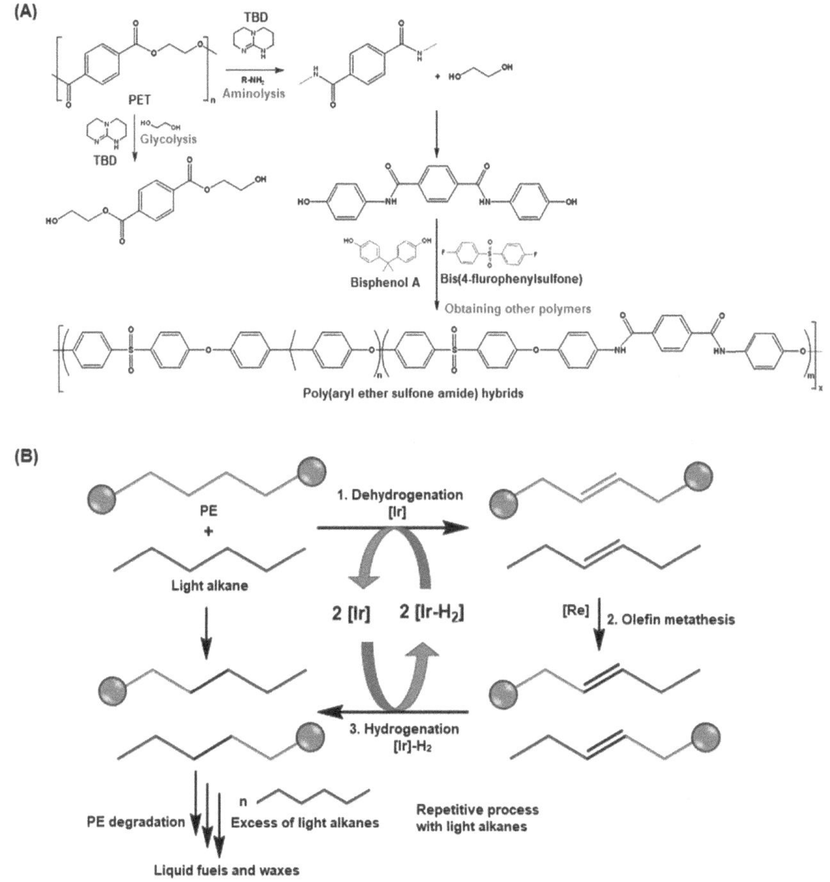

FIGURE 3.1 (A) PET depolymerization processes (glycolysis and aminolysis) and transformation to poly(aryl ether sulfone amide) hybrids. (B) Degradation of PE on a tandem catalytic cross-alkane metathesis (for example, n-hexane).

commonly found in fiber manufacturing, injection molding, wheel covers, instrument panels, and liners of pipes (Jia et al., 2016; Ronca, 2017).

In general, polyolefins have a high thermal stability that allows them to undergo multiple melting and remodeling cycles in mechanical recycling processes. However, from a certain number of cycles (40 extractions, in the case of LDPE), polyalkenes undergo significant changes in the mechanical properties of the polymer due to chain cleavage by random homolysis of the C–C bond, the recombination of the resulting radicals, and cross-linking during heat treatment producing aldehydes, organic acids, and polymers with lower molecular weights (including a solid wax), which leads to decreased values for elongation at breakage and fracture toughness (Achilias, 2012; Ignatyev et al., 2014). Therefore, a new strategy for depolymerization of polyethylene (PE) based on a tandem catalytic process of cross-alkane metastasis in the presence of an excess of a light, low-cost, and readily available alkane such as ether is implemented with oil (see Fig. 3.1B; Jia et al., 2016). The catalysts used are heterogeneous high-surface-area catalysts, which are based on transition metals such as rhodium, rhenium, iridium, and zirconium immobilized in Al_2O_3 particles. The first step in the sequence is dehydrogenation of the polymer and alkane by the catalyst to form internal alkenes, where a regal-selective catalyst ($Rh_2O_7{-}Al_2O_3$) will then act by cleaving the polymer into smaller segments through the olefin metathesis cross, where alkenes in their degradation process go through an intermediate of carbenium ions (Abbas-Abadi et al., 2015). They are then rearranged with the dehydrogenated PE chains to provide a new distribution of shorter chain alkenes that continue to react in an excess of alkenes. They are then hydrogenated to produce suitable short hydrocarbons that can be used in liquid fuels, oils, or waxes (Sharma et al., 2014; Rajendran et al., 2020).

On the other hand, the depolymerization of PP to propylene is best achieved through nonpyrolytic approaches, such as ionization. For example, when polypropylene is subjected to an inductively coupled plasma, it becomes a gas containing 94% propylene. Chemical approaches to depolymerization of PE are based on modification before controlled decomposition, with the main chain of saturated hydrocarbons first functionalized, then depolymerized (Guddeti et al., 2000).

In the event that PE is in a mixture of insoluble polymers such as isotactic polypropylene, block copolymerization processes can be carried out using $B(C_6F_5)_3$ as the activator and organocomplexes of hafnium as the catalyst. Exfoliation studies showed that the incorporation of tetrablock copolymers in contrast to the diblocks, brought a considerably improved adhesive resistance, thus transforming fragile materials into mechanically resistant mixtures, with added value equal to or greater than the initial one (Eagan et al., 2017).

The **SPI code 3** has been designated polyvinyl chloride (PVC), a polymer with the formula $(C_2H_3Cl)_n$, a product of the polymerization by addition of the vinyl chloride monomer, also known as chloroethene, which can be polymerized by four different processes, whether in suspension, emulsion, bulk, or solution. It is presented as a white material with a glass transition temperature of around 80°C and decomposition temperature above 140°C. It has a very good electrical resistance and flame, is economical, durable, high performance, and nonbiodegradable, and so it contributes approximately 12% (55 million tons of PVC) of the total demand for plastics due to its high rigidity, which can be modulated by adding a wide range of plasticizing additives such as di-(2-ethylhexyl)phthalate to be used for frames, windows, and pipes for construction, to blood bags, and plasma and pharmaceutical ampoules (Sadat-Shojai and Bakhshandeh, 2011). However, the accumulation of this highly chlorinated material (56% of the total weight) requires recycling processes, since the incineration (200–360°C) in addition to requiring large amounts of energy, leads to the formation of emissions of toxic chlorine with adverse impacts on public health and the environment. All the degradation processes of PVC include dehydrochlorination for the production of HCl, this being the main problem with recycling it, since the presence of this acid in the course of the reaction will impose severe metallurgical restrictions on the equipment material, so that current research focuses on developing technologies that allow the removal of chlorine from any PVC derivative before it can be used. The little progress in these technologies is reflected in the scarcity of literature on chemical recycling of PVC, also explaining why the recycling of raw materials is mainly limited to small-scale pilot projects or large-scale industrial projects under very strict conditions (Garforth et al., 2004; Keane, 2007; Ragaert et al., 2017).

One of the alternatives for the elimination of HCl is through a mechanochemical approach in which a planetary ball mill grinds the PVC and the remaining chloride in the effluent is neutralized by the addition of $CaCO_3$, CaO, $NaHCO_3$, $Na_2(CO_3)_2$, or NH_3; this is a process that does not require heat and provides a by-product of salt that can be washed (Miskolczi et al., 2004). However, there remains a need for better methods, catalysts, and HCl inhibitors to recycle PVC waste effectively. The inhibitors should hinder the formation of HCl, and the catalysts should have the efficacy to facilitate the decomposition of PVC into monomers in the absence of chlorine (Ragaert et al., 2017). The catalytic processes that have been developed, such as heat treatment using iron-based catalysts, are effective in the short term since the catalyst subsequently loses stability (Sakata et al., 2003).

Hydrogen-catalyzed dechlorination is an emerging methodology for the low-energy treatment of a range of chlorinated waste streams that involve the selective cleavage of one or more C—Cl bonds, reducing toxicity and generating reusable raw material. For this, hydrochlorination processes (HDC) can be carried out where hydrogenation of C—Cl bonds is involved due to the presence of an external source of H_2, or dehydrochlorination processes where internal HCl elimination is described, therefore it is applicable to PVC dechlorination and an external source of H_2 is not necessary, although it may serve to limit catalyst deactivation (Keane, 2007).

In addition, the use of catalysts of different nature such as Pd, Pt, Rh, and Ni in dehydrohalogenation processes has been reported, where the most active metal for dehydrochlorination is Pd (De Jong and Louw, 2004). However, these reactions, like many chemical processes, depend on conditions such as the temperature, concentration, and residence time of the catalyst. For example, Pd catalysts undergo significant deactivation over time in the stream, to which the formation of metal halides on the catalyst surface has been attributed (Yu et al., 2016).

On the other hand, **SPI code 6** was assigned to PS, which is a thermoplastic, economical, durable, and chemically inert polymer that is obtained from the polymerization of styrene monomer, so that it is used in many products, including appliances, auto parts, electronic products, food services, medicines, and packaging (Shah et al., 2015).

Depending on the synthesis process, four main types of PS can be obtained. When polymerizing pure styrene, the general-purpose crystal polystyrene or polystyrene (GPPS) is obtained, general-purpose polystyrene, which is a transparent, rigid, and brittle solid, with a glass transition temperature of 100°C, which is why it is useful in the manufacture of packages through extrusion and thermoforming processes. In order to reduce the fragility of the GPPS, during the polymerization a rubber, usually polybutadiene, can be added to create high-impact polystyrene or HIPS, which has a greater impact resistance, however, it is opaque, which is its main disadvantage. It is used for the production of various objects using injection molding. Expanded polystyrene or EPS can also be synthesized, which consists of 95% polystyrene and 5% of a gas that forms bubbles to reduce the density of the material, making it lighter, so that it is applied as a thermal insulator in construction, fish boxes, and vaccine transport. Finally, by melting the glass polystyrene and injecting it with gas, a rigid foam called extruded polystyrene (XPS) is obtained. Its properties are similar to those of EPS, but it is denser and waterproof, and so is often used in the manufacture of inverted roofs (Marczewski et al., 2013; Aksit et al., 2019).

Due to the diversity of polystyrene manufactured to date, its recycling process has been hindered, since materials with densities from approximately 0.016 to 1.04 g/cm^2 can be obtained. However, current chemical recycling processes are described below (Rahimi and García, 2017). The most commonly used method for chemical recycling of PS is catalyzed thermolysis, either with solid catalysts impregnated with magnesium to obtain a liquid mixture of hydrocarbons such as benzene, toluene, ethylbenzene, styrene, and α-methylstyrene, or, basic catalysts, such as Al_2O_3 catalysts containing NaOH or metal oxides (for example, MgO, BaO, CaO, or K_2O), can be used, which allows for more effective depolymerization and approximately 70% styrene production. However, the selectivity of magnesium-catalyzed thermolysis can be improved by modulating the pH of the solution, in such a way as to obtain monomers, dimers, or oligomers (Ali et al., 2020; Singh et al., 2020). In 2015, a biological alternative was reported as a recycling option, which consists of using mealworms to carry out the digestion of polystyrene to enable its composting (Yang et al., 2015).

To finish, **SPI code 7** is attributed to "other" thermoplastic and thermosetting polymers used in high-performance engineering, including PC, PU, PMMA, nylon-6, PES, and epoxides (Rochman et al., 2013; Özkan et al., 2015; Merrington, 2017).

Some polymers in this category have degradable linkers in their spine, so that they are susceptible to chemical decomposition. This recent technology is known as "dynamic covalent materials" that is, they are polymers joined by covalent bonds that form and break reversibly. Examples of these are disulfides, imines, olefins, Diels–Alder adducts, thiol-eno adducts, alkoxyamines, hydrazines, triazoles, boronic esters and, recently, hemiaminals and thioaminals (Takahashi et al., 2016).

The PC, for example, by having the carbonate group in its structure, can undergo hydrolysis in the presence of bases such as NaOH, K_2CO_3, or Cs_2CO_3, and generate biphenolate due to the nucleophilic attack of the salt to the PC. Bis(aryl fluoride) is subsequently added, a polycondensation reaction being carried out for 18 h at 190°C to finally produce poly(aryl ether sulfone) (PAES), a material commonly used as thermoplastic or water purification membranes and reverse osmosis (Antonakou and Achilias, 2013; Zhu et al., 2017). This product is obtained through a one-step reaction, where the carbonate salt plays two important catalytic roles in this reaction, in the decomposition of the PC by nucleophilic attack, and in the subsequent process of polyether formation, where it promotes the reaction of phenolate dimers formed in situ with the aryl fluorides present. This facilitates the transformation of PC-rich materials that are generally not recycled using methods 1 or 2 since it does not have a unique identification code, so that it is included in the category of "others" and recycled chemically, such as compact discs, glass lenses, cabin awnings, smartphones, LCD screens, and car headlights in optically transparent thermoplastics, that are high value and impact resistant, such as PAES (see Fig. 3.2A) (Jones et al., 2016; Westhues et al., 2018).

On the other hand, PUs, which are polymers obtained from the condensation of polyols with multifunctional isocyanates, generate low-cost materials with high versatility, whose properties can be optimized for applications ranging from soft and flexible foams to hard and compact materials (Akindoyo et al., 2016). However, the nondegradability of post-consumer PU products has caused a massive waste management problem. In addition, the high nitrogen content in its structure (approximately 4% by weight) prevents recycling by oxidative techniques such as incineration and gasification due to the formation of significant amounts of NO in gaseous effluents (Sheel and Pant, 2018). Therefore, current PU recycling processes include hydrolysis of PU to produce polyols, diamines, and carbon dioxide where the first two can be reused as starting materials for the synthesis of new PUs, or polyols can also be used as fuel. However, PU hydrolysis is not economical, mainly due to the high energy consumption since the temperature during the process must be above 280°C. Therefore, it is preferred to carry out glycolysis processes, that is, the reaction with an excess of diols such as diethylene glycol, at a temperature higher than 200°C and at atmospheric pressure (Ugarte et al., 2018; Alberti et al., 2019; Shin et al., 2019). The depolymerization reaction lasts approximately 3–10 h and in this one polyols of low reactivity and high viscosity are obtained in comparison with the original polyols, but can be mixed with new polyols up to 40%, to manufacture rigid and semirigid foams. The reaction time and temperature can be reduced to 2 h and 180–220°C using as tertiary amines, bismuth or tin organic compounds, inorganic salts, carboxylic acids or alkaline

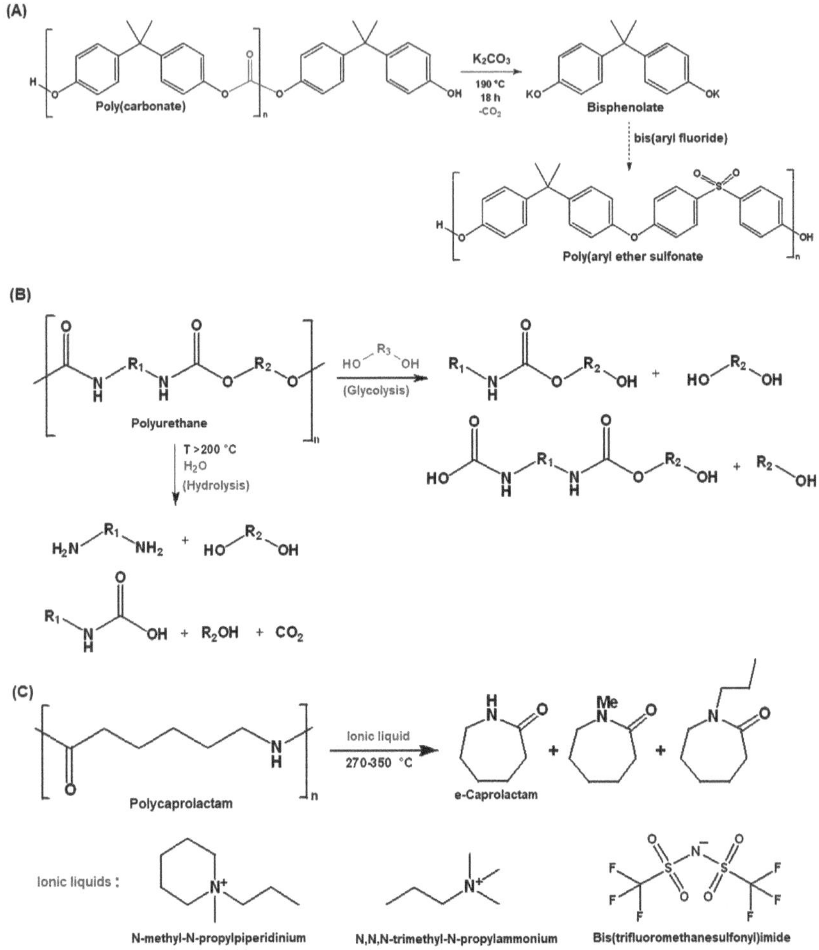

FIGURE 3.2 (A) Representation of the PC transformation in PAES. (B) PU depolymerization by hydrolysis and glycolysis. (C) Depolymerization of Nylon-6 (polycaprolactam) using ionic liquids.

hydroxides, inorganic acids, phosphorous organic compounds, and phosphoric esters, among others. Where alkoxylation processes are carried out, initially, primary amines formed by the reactions of the urea group with diethylene glycol are obtained as products. They are then reacted with an alkylene oxide such as ethylene oxide for another 2 h to finally obtain the recovered polyols and the unreacted alkylene oxide is removed by degassing as vacuum distillation (see Fig. 3.2B) (Sheel and Pant, 2018).

PMMA and polyamides such as Nylon-6 (polycaprolactam) can be efficiently depolymerized to monomer using thermal and photochemical approaches (Nishida, 2011; Suresh et al., 2017). Explicitly, the depolymerization of Nylon-6 is achieved by leaving the material in contact with ionic liquids such as solutions of quaternary ammonium salts, such as N-methyl-N-propylpiperidinium and N,N,N-trimethyl-N-propylammonium with bis(trifluoromethanesulfonyl) imide as a counter-anion, to subsequently heat the mixture at 300°C for approximately 5–6 h and collect the caprolactam monomers by distillation of the reaction mixture (see Fig. 3.2 C). According to the literature, this process achieves recovery yields of 43%–55%. Meanwhile, adding N,N-dimethylaminopyridine as a catalyst improves the efficiency of depolymerization and the monomer yield, which it increases to more than 86% (Zheng et al., 2016; Kamimura et al., 2019).

Another polymer that is included in **SPI 7** is PES, a high-performance thermoplastic that is one of the high T_g polymers, transparent, and amorphous. In addition, it has poor fatigue characteristics and is prone to cracking due to environmental stress, but it has good resistance to long-term thermal aging and reasonable radiation resistance. In general, the PES has high mechanical intensities, thermal stability, and chemical resistance. In addition, it has low creep properties, flame retardants, excellent insulation properties, high dielectric strengths, and lower smoke emission ratings among plastics. It is used in high-temperature electrical applications, bakery oven windows, medical components, structural aircraft and aerospace components, and corrosion-resistant applications in chemical plant packaging and hardware. PES is also widely used in the fields of membrane separation, for the manufacture of ultrafiltration, reverse osmosis, and nanofiltration membranes (Ran, 2015). Due to the excellent mechanical properties of this polymer, it has been unnecessary to carry out PES chemical recycling processes (Rahimi and García, 2017).

Finally, the epoxy is thermostable and typically formed by reacting a bis(epoxide) monomer, such as diglycidyl ether of bisphenol A (DGEBA), with an amine hardener, such as triethylenetetramine. This product has an irreversible nature in thermostable production, and so it cannot be melted and remolded as part of a conventional recycling process, with the extraction of this resin from the filler being necessary. For this, supercritical n-propanol has been used. Even the bonds of the epoxide with acid have been affected as a treatment at the end of the useful life. For this reason, epoxy cleavages are synthesized by radical copolymerization of 2,4-dimethyl-1,3-pentadiene with maleic anhydride and DGEBA (Tsujii et al., 2014; La Rosa et al., 2016). The trisubstituted olefin filtered in the skeleton repeat unit introduced a degradable bond with ozone. This by-product has applications in electronics, automotives, aerospace, adhesives, insulators, and metallic coatings; epoxy also has chemical resistance and desirable mechanical properties that commonly facilitate its use as a matrix resin for composites with fiberglass or carbon fiber charges. The compounds represent a light and attractive engineering alternative to metals, so much so that a recyclable compound was recently developed with properties comparable to nonrecyclable analogs (Jiang et al., 2017).

3.3 Conclusions and remarks

Recycling is undoubtedly a strategy that seeks to mitigate the impact of polymers on the environment. Thus, if a polymer is recyclable then it is usually valued positively in terms of its environmental impact. However, a gray area is drawn throughout this strategy. In general, four aspects should be considered: (1) the loss of properties of the polymers as they are subjected to reprocessing may make it impossible to use them in the same application, therefore, in many cases, new applications must be identified for a new set of properties; (2) the recycling of polymer mixtures implies the separation of the materials until obtaining secondary materials with the greatest homogeneity of properties, consequently, homogeneity usually implies an effort in terms of processes and costs while heterogeneity limits the potential secondary applications of these mixtures as raw material; (3) some alternatives for reuse do not necessarily imply a beneficial impact either because discharge streams can be produced with unwanted residues which should be treated to minimize the impact of its discharge on the environment, but also, irreversible chemical transformation of the polymeric material with the consequent production of emissions and the loss of the material for the next cycle of reuse should be avoided, for example, thermal decomposition of polymers, and (4) the recycling process must be economically viable to become a real alternative.

Finally, it can be concluded that the ideal recycling process should allow the polymer to be reused indefinitely, or at least through a very high number of use cycles, and at the same time be able to transfer added value to the product without generating emissions or by-products.

References

Abbas-Abadi, M.S., Nekoomanesh-Haghighi, M., McDonald, A.G., Yeganeh, H., 2015. Estimation of pyrolysis product of LDPE degradation using different process parameters in a stirred reactor. Polyolefins J. 2, 39–47. https://doi.org/10.22063/poj.2015.1102.

Achilias, D., 2012. Material Recycling: Trends and Perspectives. Intech, p. 406. https://doi.org/10.5772/2003.

Akindoyo, J.O., Beg, M.D.H., Ghazali, S., Islam, M.R., Jeyaratnam, N., Yuvaraj, A.R., 2016. Polyurethane types, synthesis and applications – a review. Royal Soc. Chem. Adv. 6, 114453–114482. https://doi.org/10.1039/C6RA14525F.

Aksit, M., Zhao, C., Klose, B., Kreger, K., Schmidt, H.W., Altstädt, V., 2019. Extruded polystyrene foams with enhanced insulation and mechanical properties by a benzene-trisamide-based additive. Polymers 11, 268. https://doi.org/10.3390/polym11020268.

Alberti, C., Scheliga, F., Enthaler, S., 2019. Recycling of end-of-life poly(bisphenol A carbonate) via alkali metal halide-catalyzed phenolysis. Chem. Open 8, 822–827. https://doi.org/10.1002/open.201900149.

Ali, G., Nisar, J., Iqbal, M., Shah, A., Abbas, M., Shah, M.R., Rashid, U., Bhatti, I.A., Khan, R.A., Shah, F., 2020. Thermo-catalytic decomposition of polystyrene waste: comparative analysis using different kinetic models. Waste Manag. Res. 38, 202–212. https://doi.org/10.1177/0734242X19865339.

Antonakou, E.V., Achilias, D.S., 2013. Recent advances in polycarbonate recycling: a review of degradation methods and their mechanisms. Waste Biomass Valor 4, 9–21. https://doi.org/10.1007/s12649-012-9159-x.

Bakker, E.J., Rem, P.C., Fraunholcz, N., 2009. Upgrading mixed polyolefin waste with magnetic density separation. Waste Manag. 29, 1712–1717. https://doi.org/10.1016/j.wasman.2008.11.006.

De Jong, V., Louw, R., 2004. Performance of supported nickel and other metal catalysts in the hydrodechlorination of chlorobenzene and 1-chlorohexane. Appl. Catal. Gen. 271, 153–163. https://doi.org/10.1016/j.apcata.2004.02.055.

Eagan, J.M., Xu, J., Di Girolamo, R., Thurber, C.M., Macosko, C.W., LaPointe, A.M., Bates, F.S., Coates, G.W., 2017. Combining polyethylene and polypropylene: enhanced performance with PE/iPP multiblock polymers. Science 355, 814–816. https://doi.org/10.1126/science.aah5744.

Ellen MacArthur Foundation, McKinsey and Company, World Economic Forum, 2016. The New Plastics Economy—Rethinking the Future of Plastics. Ellen MacArthur Foundation, pp. 1–36.

Fukushima, K., Coulembier, O., Lecuyer, J.M., Almegren, H.A., Alabdulrahman, A.M., Alsewailem, F.D., Mcneil, M.A., Dubois, P., Waymouth, R.M., Horn, H.W., Rice, J.E., Hedrick, J.L., 2011. Organocatalytic depolymerization of poly(ethylene terephthalate). J. Polym. Sci. Polym. Chem. 49, 1273–1281. https://doi.org/10.1002/pola.24551.

Fukushima, K., Lecuyer, J.M., Wei, D.S., Horn, H.W., Jones, G.O., Al-Megren, H.A., Alabdulrahman, A.M., Alsewailem, F.D., McNeil, M.A., Rice, J.E., Hedrick, J.L., 2013. Advanced chemical recycling of poly(ethylene terephthalate) through organocatalytic aminolysis. Polym. Chem. 4, 1610–1616. https://doi.org/10.1039/C2PY20793A.

Garcia, J.M., Robertson, M.L., 2017. The future of plastics recycling. Science 358, 870–872. https://doi.org/10.1126/science.aaq0324.

Gardea, F., Garcia, J.M., Boday, D.J., Bajjuri, K.M., Naraghi, M., Hedrick, J.L., 2014. Hybrid poly(aryl ether sulfone amide)s for advanced thermoplastic composites. Macromol. Chem. Phys. 215, 2260–2267. https://doi.org/10.1002/macp.201400267.

Garforth, A.A., Ali, S., Hernández-Martínez, J., Akah, A., 2004. Feedstock recycling of polymer wastes. Curr. Opin. Solid State Mater. Sci. 8, 419–425. https://doi.org/10.1016/j.cossms.2005.04.003.

Gondal, M.A., Siddiqui, M.N., 2007. Identification of different kinds of plastics using laser-induced breakdown spectroscopy for waste management. J. Environ. Sci. Health - Part A 42, 1989–1997. https://doi.org/10.1080/10934520701628973.

Goto, M., Koyamoto, H., Kodama, A., Hirose, T., Nagaoka, S., 2002. Depolymerization of polyethylene terephthalate in supercritical methanol. J. Phys. Condens. Matter 14, 11427. https://doi.org/10.1088/0953-8984/14/44/494.

Guddeti, R.R., Knight, R., Grossmann, E.D., 2000. Depolymerization of polypropylene in an induction-coupled plasma (ICP) reactor. Ind. Eng. Chem. Res. 39, 1171–1176. https://doi.org/10.1021/ie9906868.

Hahladakis, J.N., Velis, C.A., Weber, R., Iacovidou, E., Purnell, P., 2018. An overview of the chemical additives present in plastics: migration, release, fate and environmental impact during their use, disposal and recycling. J. Hazard Mater. 344, 179–199. https://doi.org/10.1016/j.jhazmat.2017.10.014.

Hatti-Kaul, R., Nilsson, L.J., Zhang, B., Rehnberg, N., Lundmark, S., 2020. Designing biobased recyclable polymers for plastics. Trends Biotechnol. 38, 50–67. https://doi.org/10.1016/j.tibtech.2019.04.011.

Hong, M., Chen, E.Y.X., 2017. Chemically recyclable polymers: a circular economy approach to sustainability. Green Chem. 19, 3692–3706. https://doi.org/10.1039/C7GC01496A.

Hu, B., Serranti, S., Fraunholcz, N., Di Maio, F., Bonifazi, G., 2013. Recycling-oriented characterization of polyolefin packaging waste. Waste Manag. 33, 574–584. https://doi.org/10.1016/j.wasman.2012.11.018.

Ignatyev, I.A., Thielemans, W., Vander Beke, B., 2014. Recycling of polymers: a review. Chem. Sustain. Energy Mater. 7, 1579–1593. https://doi.org/10.1002/cssc.201300898.

Jia, X., Qin, C., Friedberger, T., Guan, Z., Huang, Z., 2016. Efficient and selective degradation of polyethylenes into liquid fuels and waxes under mild conditions. Mater. Sci. 2, e1501591. https://doi.org/10.1126/sciadv.1501591.

Jiang, J., Deng, G., Chen, X., Gao, X., Guo, Q., Xu, C., Zhou, L., 2017. On the successful chemical recycling of carbon fiber/epoxy resin composites under the mild condition. Compos. Sci. Technol. 151, 243–251. https://doi.org/10.1016/j.compscitech.2017.08.007.

Jones, G.O., Yuen, A., Wojtecki, R.J., Hedrick, J.L., García, J.M., 2016. Computational and experimental investigations of one-step conversion of poly(carbonate)s into value-added poly(aryl ether sulfone)s. Proc. Natl. Acad. Sci. U.S.A. 113, 7722–7726. https://doi.org/10.1073/pnas.1600924113.

Kamimura, A., Shiramatsu, Y., Kawamoto, T., 2019. Depolymerization of polyamide 6 in hydrophilic ionic liquids. Green Energy Environ. 4, 166–170. https://doi.org/10.1016/j.gee.2019.01.002.

Keane, M.A., 2007. Catalytic conversion of waste plastics: focus on waste PVC. J. Chem. Technol. Biotechnol. 82, 787–795. https://doi.org/10.1002/jctb.1757.

La Mantia, F.P., 2004. Polymer mechanical recycling: downcycling or upcycling? Prog. Rubber Plast. Recycl. Technol. 20, 11–24. https://doi.org/10.1177/147776060402000102.

La Rosa, A.D., Banatao, D.R., Pastine, S.J., Latteri, A., Cicala, G., 2016. Recycling treatment of carbon fibre/epoxy composites: materials recovery and characterization and environmental impacts through life cycle assessment. Compos. B Eng. 104, 17–25. https://doi.org/10.1016/j.compositesb.2016.08.015.

Maio, F.D., Rem, P., Hu, B., Serranti, S., Bonifazi, G., 2010. The W2Plastics project: exploring the limits of polymer separation. Open Waste Manag. J. 3, 90–98. https://doi.org/10.2174/1876400201003010090.

Marczewski, M., Kamińska, E., Marczewska, H., Godek, M., Rokicki, G., Sokołowski, J., 2013. Catalytic decomposition of polystyrene. The role of acid and basic active centers. Appl. Catal. B Environ. 129, 236–246. https://doi.org/10.1016/j.apcatb.2012.09.027.

Merrington, A., 2017. Recycling of plastics. In: Kutz, M. (Ed.), Applied Plastics Engineering Handbook: Processing, Materials, and Applications, second ed. Elsevier. United States, pp. 167–187. https://doi.org/10.1016/B978-0-323-39040-8.00009-2.

Miskolczi, N., Bartha, L., Deák, G., Jóver, B., 2004. Thermal degradation of municipal plastic waste for production of fuel-like hydrocarbons. Polym. Degrad. Stabil. 86, 357–366. https://doi.org/10.1016/j.polymdegradstab.2004.04.025.

Morris, J., 1996. Recycling versus incineration: an energy conservation analysis. J. Hazard Mater. 47, 277–293. https://doi.org/10.1016/0304-3894(95)00116-6.

Nishida, H., 2011. Development of materials and technologies for control of polymer recycling. Polym. J. 43, 435–447. https://doi.org/10.1038/pj.2011.16.

North, E.J., 2013. Halden RU. Plastics and environmental health: the road ahead. Rev. Environ. Health 28, 1–8. https://doi.org/10.1515/reveh-2012-0030.

Özkan, K., Ergin, S., Işık, S., Işıklı, İ., 2015. A new classification scheme of plastic wastes based upon recycling labels. Waste Manag. 35, 29–35. https://doi.org/10.1016/j.wasman.2014.09.030.

Peacock, A., 2000. Handbook of Polyethylene. CRC Press, Boca Raton, pp. 1–544. https://doi.org/10.1201/9781482295467.

Ragaert, K., Delva, L., Van Geem, K., 2017. Mechanical and chemical recycling of solid plastic waste. Waste Manag. 69, 24–58. https://doi.org/10.1016/j.wasman.2017.07.044.

Rahimi, A., García, J.M., 2017. Chemical recycling of waste plastics for new materials production. Nature Reviews Chemistry 1, 1–11. https://doi.org/10.1038/s41570-017-0046.

Rajendran, K.M., Chintala, V., Sharma, A., Pal, S., Pandey, J.K., Ghodke, P., 2020. Review of catalyst materials in achieving the liquid hydrocarbon fuels from municipal mixed plastic waste (MMPW). Mater. Today Commun. 24, 100982. https://doi.org/10.1016/j.mtcomm.2020.100982.

Ran, F., 2015. Polyethersulfone (PES). In: Drioli, E., Giorno, L. (Eds.), Encyclopedia of Membranes. Springer, Berlin, pp. 1–2. https://doi.org/10.1007/978-3-642-40872-4_1909-1.

Rochman, C.M., Browne, M.A., Halpern, B.S., Hentschel, B.T., Hoh, E., Karapanagioti, H.K., Rios-Mendoza, L.M., Takada, H., Teh, S., Thompson, R.C., 2013. Classify plastic waste as hazardous. Nature 494, 169–171. https://doi.org/10.1038/494169a.

Ronca, S., 2017. Polyethylene. In: Gilbert, M. (Ed.), Brydson's Plastics Materials, eighth ed. Butterworth-Heinemann, Oxford, pp. 247–278. https://doi.org/10.1016/B978-0-323-35824-8.00010-4.

Sadat-Shojai, M., Bakhshandeh, G.R., 2011. Recycling of PVC wastes. Polym. Degrad. Stabil. 96, 404–415. https://doi.org/10.1016/j.polymdegradstab.2010.12.001.

Sakata, Y., Bhaskar, T., Uddin, M.A., Muto, A., Matsui, T., 2003. Development of a catalytic dehalogenation (Cl, Br) process for municipal waste plastic-derived oil. J. Mater. Cycles Waste Manag. 5, 113–124. https://doi.org/10.1007/s10163-003-0092-8.

Shah, J., Jan, M.R., 2015. Adnan. Tertiary recycling of waste polystyrene using magnesium impregnated catalysts into valuable products. J. Anal. Appl. Pyrol. 114, 163–171. https://doi.org/10.1016/j.jaap.2015.05.009.

Sharma, B.K., Moser, B.R., Vermillion, K.E., Doll, K.M., Rajagopalan, N., 2014. Production, characterization and fuel properties of alternative diesel fuel from pyrolysis of waste plastic grocery bags. Fuel Process. Technol. 122, 79–90. https://doi.org/10.1016/j.fuproc.2014.01.019.

Sheel, A., Pant, D., 2018. Recycling of Polyurethane Foams. In: Chemical Depolymerization of Polyurethane Foams via Glycolysis and Hydrolysis, pp. 67–75. https://doi.org/10.1016/b978-0-323-51133-9.00006-1.

Shin, S.R., Kim, H.N., Liang, J.Y., Lee, S.H., Lee, D.S., 2019. Sustainable rigid polyurethane foams based on recycled polyols from chemical recycling of waste polyurethane foams. J. Appl. Polym. Sci. 136, 47916. https://doi.org/10.1002/app.47916.

Singh, N., Hui, D., Singh, R., Ahuja, I.P.S., Feo, L., Fraternali, F., 2017. Recycling of plastic solid waste: a state of art review and future applications. Compos. B Eng. 115, 409–422. https://doi.org/10.1016/j.compositesb.2016.09.013.

Singh, R.K., Ruj, B., Sadhukhan, A.K., Gupta, P., 2020. Catalytic and non-catalytic thermolysis of waste polystyrene for recovery of fuel grade products and their characterization. In: Ghosh, S. (Ed.), Energy Recovery Processes from Wastes. Springer, Singapore, pp. 23–32. https://doi.org/10.1007/978-981-32-9228-4_3.

Solis, M., Silveira, S., 2020. Technologies for chemical recycling of household plastics – a technical review and TRL assessment. Waste Manag. 105, 128–138. https://doi.org/10.1016/j.wasman.2020.01.038.

Suresh, S.S., Mohanty, S., Nayak, S.K., 2017. Investigation into the mechanical and thermal properties of poly(methyl methacrylate) recovered from light guidance panels with a focus on future remanufacturing and sustainable waste management. J. Remanuf. 7, 217–233. https://doi.org/10.1007/s13243-017-0041-7.

Takahashi, A., Ohishi, T., Goseki, R., Otsuka, H., 2016. Degradable epoxy resins prepared from diepoxide monomer with dynamic covalent disulfide linkage. Polymer 82, 319–326. https://doi.org/10.1016/j.polymer.2015.11.057.

Thompson, R.C., Moore, C.J., vom Saal, F.S., Swan, S.H., 2009. Plastics, the environment and human health: current consensus and future trends. Phil. Trans. Biol. Sci. 364, 2153–2166. https://doi.org/10.1098/rstb.2009.0053.

Tilmatine, A., Medles, K., Bendimerad, S.E., Boukholda, F., Dascalescu, L., 2009. Electrostatic separators of particles: application to plastic/metal, metal/metal and plastic/plastic mixtures. Waste Manag. 29, 228–232. https://doi.org/10.1016/j.wasman.2008.06.008.

Tsujii, A., Namba, M., Okamura, H., Matsumoto, A., 2014. Radical alternating copolymerization of twisted 1,3-butadienes with maleic anhydride as a new approach for degradable thermosetting resin. Macromolecules 47, 6619–6626. https://doi.org/10.1021/ma501555n.

Ugarte, L., Calvo-Correas, T., Gonzalez-Gurrutxaga, I., Peña-Rodriguez, C., Etxeberria, O., Corcuera, M.A., Eceiza, A., 2018. Towards circular economy: different strategies for polyurethane waste recycling and the obtaining of new products. Proceedings 2, 1490. https://doi.org/10.3390/proceedings2231490.

UNEP, 2018. Single-Use Plastics: A Roadmap for Sustainability. UNEP., pp 1–104

Westhues, S., Idel, J., Klankermayer, J., 2018. Molecular catalyst systems as key enablers for tailored polyesters and polycarbonate recycling concepts. Sci. Adv. 4 https://doi.org/10.1126/sciadv.aat9669 eaat9669.

Xu, M.M., Huang, G.Y., Feng, S.S., McShane, G.J., Stronge, W.J., 2016. Static and dynamic properties of semi-crystalline polyethylene. Polymers 8, 77. https://doi.org/10.3390/polym8040077.

Yang, Y., Yang, J., Wu, W.M., Zhao, J., Song, Y., Gao, L., Yang, R., Jiang, L., 2015. Biodegradation and mineralization of polystyrene by plastic-eating mealworms: Part 1. Chemical and physical characterization and isotopic tests. Environ. Sci. Technol. 49, 12080–12086. https://doi.org/10.1021/acs.est.5b02661.

Yu, J., Sun, L., Ma, C., Qiao, Y., Yao, H., 2016. Thermal degradation of PVC: a review. Waste Manag. 48, 300–314. https://doi.org/10.1016/j.wasman.2015.11.041.

Zheng, Y., Chen, T., Sun, Y., Deng, R., 2016. Method for Preparing Pure Nylon-6 from Nylon-6 Waste as Raw Material. Chinese Patent CN105367818A.

Zhu, P., Liu, X., Wang, Y., Guan, C., Yang, Y., Zhu, J., Li, X., Qian, G., Frost, R.L., 2017. Production and characterization of recycled polycarbonate based composite material containing recycled glass fibers. J. Environ. Chem. Eng 5, 3439–3446. https://doi.org/10.1016/j.jece.2017.06.050.

Chapter 4

Biodegradable polymer chemistry: structural, physicochemical aspects, uses, and transformation technologies

4.1 Environmental overview of biodegradable functional polymers

As previously stated, synthetic petroleum-derived polymers revolutionized humanity's quality of life due to their optimal performance, low cost, and great versatility to be conveniently applied as structure materials, packaging supplies, and other basic products (Panchal and Vasava, 2020; Yin and Yang, 2020). Poly(ethylene), poly(propylene), poly(styrene), and poly(ethylene terephthalate) are the most commonly used fossil resource-based polymers, and have multiple applications in the development of bags, packaging, films, coatings, pipes, etc. (Marlina and Sato, 2020). For a long time, these materials were considered a gift from science to humanity because they improved people's quality of life, including the development of new technologies, and their production generated a positive economic impact in many countries. In 2014, the manufacture of these materials reached a total of 311 million tons, with the tendency to double in 20 years and quadruple by 2050 (Ellen MacArthur Foundation, 2017). However, over the past two decades, it has been shown that the production and processing of these polymers leads to an increase in greenhouse gas emissions that aggravate global warming. These materials also are accumulated in the environment due to their high stability, optimal mechanical properties, and wear resistance. It is estimated that a PET bottle, for example, has a useful life of 27–93 years (Luckachan and Pillai, 2011; Zeng et al., 2016; Haider et al., 2019; Mangaraj et al., 2019). Furthermore, these polymers can be fragmented to generate "microplastics," migrating from rivers to oceans (Ivleva et al., 2016; Haider et al., 2019). As if that were not enough, the management of these wastes is based on ecologically

Eco-friendly Functional Polymers
https://doi.org/10.1016/B978-0-12-821842-6.00018-X
Copyright © 2021 Elsevier Inc. All rights reserved.

unsustainable methods such as incineration, landfill waste, chemical degradation, etc. (Zeng et al., 2016; Yin and Yang, 2020). Fortunately, society has become increasingly aware that our natural resources are finite, and that we must preserve them, and so multiple sustainable alternatives have been proposed to prevent the increasing pollution generated by these materials. In this sense, the recycling culture emerged as a solution to this problem, however, Degnan and Shinde established that only 13% of these polymers are recycled annually (Degnan and Shinde, 2019), which is an insufficient alternative to remedy this environmental problem. In developed countries, governments and some NGOs choose to promote the development of research using a mechanism based on granting, hoping to find one or more alternatives to help to remedy this problem (Ibrahim et al., 2018). The European Union in 2018 suggested that, in order to achieve a sustainable plastics industry by 2030, fossil-based plastics need to be replaced with biodegradable polymers, as the solution to avoid an environmentally deteriorated future (European Union, 2018).

In the literature there are multiple definitions for establishing the meaning of biodegradable polymers. The International Union of Pure and Applied Chemistry (IUPAC) specifies that these are materials susceptible to degradation by biological activity, resulting in a decrease in their molecular mass (Haider et al., 2019). ASTM D883 states that a polymer is biodegradable when its deterioration occurs by microorganisms such as fungi, bacteria, and algae (Laycock et al., 2017). However, these definitions do not consider that the biodegradation process is also promoted by oxidation reactions, photodegradation, hydrolysis, etc., and therefore, a biodegradable polymer should be understood as a material experiencing a deterioration of physical and chemical properties due to the cleavage of their structure through the effect of environmental factors (Laycock et al., 2017; Feig et al., 2018; Agarwal, 2020; Yin and Yang, 2020). International certifications such as ASTM6400, EN13432, AS 4736, ISO 17088, and GreenPla, state that a material can be certified as biodegradable when it decomposes by 90% in 6 months (Elvers et al., 2016). These materials have been developed since the 1980s, exhibiting a rapid development because, in addition to contributing to the preservation of fossil resources, they present biodegradable properties at the end of their useful life, avoiding their accumulation in the environment and contributing to the development of ecological sustainability. European Plastic determined that biodegradable polymer production increased from 1.4 million tons in 2011 to 6.0 million tons in 2017, with promising growth in the future, and with the potential to replace conventional non-biodegradable materials (Luckachan and Pillai, 2011; Zeng et al., 2016).

Biodegradable polymers are distinguished, according to their origin, in natural materials, better known as biopolymers, and synthetic materials. Biopolymers are the product of an evolutionary course in nature and include polysaccharides such as cellulose or chitin, proteins such as collagen or albumin, and even lipids and nucleic acids. These present additional properties such as biocompatibility and non-toxicity, being attractive materials for application in the pharmaceutical, cosmetic, food, and biomedical industries (Shchipunov, 2012; Marlina and Sato, 2020). Synthetic biodegradable polymers, on the other hand, have been developed to obtain materials with specific morphological and mechanical properties, in conjunction with optimal reproducibility. Examples of synthetic biodegradable polymers are poly(lactic acid) (PLA), poly(ϵ-caprolactone) (PCL), poly(glycolic acid) (PGA), poly(butylene succinate) (PBS), polyhydroxyalkanoates (PHAs), etc. These materials are widely applied in medicine, pharmacology, agriculture, packaging, and even in the field of electronics (Zeng et al., 2016; Panchal and Vasava, 2020). It is accepted that, in most cases, the degradation of these materials occurs in two steps. The first consists of the excision of the polymer to produce molecules of low molecular weight through abiotic processes (e.g., hydrolysis, oxidation, photodegradation, etc.) or biotic processes (i.e., by the enzymatic action of microorganisms). In the second step, the bio-assimilation of the fragments generated for their subsequent mineralization in simple molecules such as CH_4, CO_2, H_2O, and biomass occurs (Fig. 4.1) (Vroman and Tighzert, 2009; Agarwal, 2020). A particular case is related with the accumulation of persistent organic molecules resulting in the biodegradation of biomass by soil microorganisms (Berrio et al., 2018; García et al., 2018; Mora et al., 2018).

Although the acceleration of biodegradable polymer production is promising, their development accounts for less than 2% of non-degradable polymer production, as these eco-friendly materials are generally more expensive, even today, and so they are not an economically viable option for industries and consumers in many sectors. In recent years, research into the development of inexpensive biodegradable polymers with optimal mechanical properties are still at an early stage, however, it is a topic of growing interest due to its

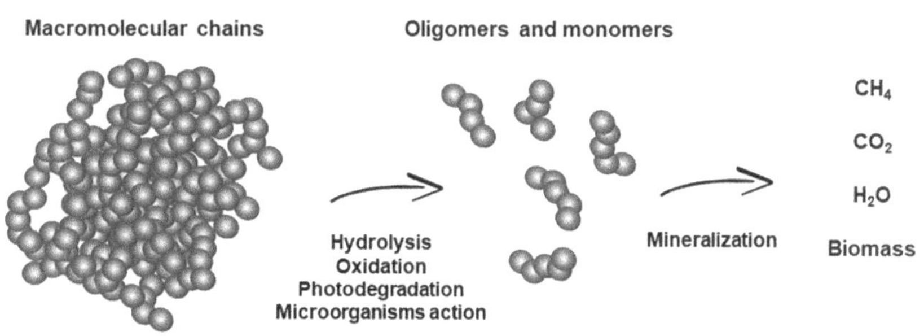

FIGURE 4.1 Degradation mechanism of biodegradable polymers. The first step consists of the cleavage of the macromolecular chains by the action of biotic or abiotic factors. The second consists of bio-assimilation and subsequent mineralization.

ecological benefits. Therefore, understanding its structure, morphology, and physicochemical properties is an important step to identify its degradation mechanism and its inclusion in many systems and technologies that today are dominated by non-biodegradable synthetic polymers.

4.2 Structural and morphological characteristics

The biodegradability of polymers depends, in addition to environmental factors, on the structural and morphological characteristics of the material since the presence of certain functional groups, bonds, aggregation structure, etc., will determine the ease with which fragmentation can be enabled. Thus, a material exhibits greater biodegradability when cleavage occurs easily, therefore it is important to know the structural and morphological characteristics that have a relevant impact on this property such as the presence of heteroatoms in the main chain of polymer, the presence of easily oxidizable groups, and side chains consisting of bulky groups that provide less chain packing. Fig. 4.2 shows the chemical structure of some biodegradable polymers used in the industry.

The molecular structure is the main characteristic controlling the capacity and degradation mechanism of a material. Zeng and collaborators, in a review into decomposition mechanisms, established that the biodegradation of a polymer descended in the following order of functional groups: aliphatic ester, peptide bond > carbamate > aliphatic ether > methylene (Zeng et al., 2016). This makes sense, since hydrolyzable groups in the chain, such as ester, urea, urethane, amide, and enamine bonds are susceptible to degradation by hydrolytic enzymes. In addition, it is congruent with the fact that biopolymers, such as proteins and polysaccharides, are degraded into biological systems

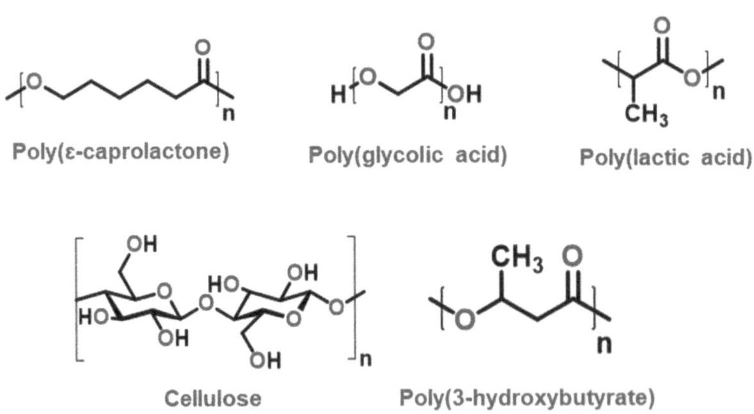

FIGURE 4.2 Chemical structure of biodegradable materials: PCL, PGA, PLA, PHB, and cellulose.

(Chen et al., 2020; Ghanbarzadeh and Almasi., 2013). Formerly, conjugated systems, such as aromatic groups, were considered to be more biodegradable than their aliphatic counterparts due to their showing a low reactivity, while conjugated systems would cleavage their C-H bonds when they are exposed to sunlight or ultraviolet. However, aromatic groups provide high rigidity and radical elimination properties to the material, and so subsequently this theory lost its validity. It is currently known that the incorporation of aliphatic esters into polymeric structures improves the biodegradability of polymer materials (Filiciotto and Rothenberg, 2020; Gewert et al., 2015). The structural characteristics of some biodegradable polymers are described in Table 4.1.

The hydrophobic and hydrophilic nature of polymeric materials, given by their structure, also affects their degree of biodegradability. Although enzyme-catalyzed reactions are commonly carried out in aqueous media, structures with equivalent hydrophobic and hydrophilic segments have been shown to be more biodegradable. This was described by DiBenedetto and collaborators in 1988, and recompiled years later, in 2013, by Ghanbarzadeh and Almasi in a chapter on biodegradable polymers. These researchers documented one of the first experiments to evaluate the effect of the hydrophobic and hydrophilic character of a polymer on its biodegradability, which consisted of inoculating the fungal species *Aspergillus niger* with a series of poly(alkylene tartrates). The results showed that the derivatives of hexamethylenediol and octamethylenediol were more degraded and assimilated by the fungus than the hydrophilic (ethylenediol or butylenediol) and hydrophobic polymers (decamethylenediol and dodecamethylene) (DiBenedetto et al., 1988; Ghanbarzadeh and Almasi., 2013). Another good example of the structural characteristics of biopolymers, as well as their functionality, was described by Nottelet et al. (2010). They synthesized amphiphilic diblock copolymers from a combination of poly(malic acid) and PLA with hydrophobic and hydrophilic functional segments, in order to develop a micellar drug-delivery system. This polymeric system exhibited high degradability under conditions of tumor tissue (pH from 5 to 37°C) (Nottelet et al., 2010; Zeng et al., 2016).

On the other hand, polymer morphology is also a relevant factor in the biodegradability of biopolymers. For example, the shape of the polymer has been shown to be a characteristic that affects biodegradability since enzymes degrade linear polymers easier than branched polymers. The flexibility of the material is another important factor to promote degradation by enzymatic catalysis since this property will allow it to be more accessible to the active site of enzymes. This explains the fact that an aromatic polyester, such as PET, is not biodegradable due to its structure that is composed of rigid aromatic rings, while aliphatic polyesters are highly biodegradable (Cameron and Shaver, 2011; Zeng et al., 2016). Polymer flexibility is related to the degree of crystallization of the macromolecule and in consequence to their molecular ordering. Heteropolymers, such as proteins, are less likely to crystallize due to irregularity in their repetitive units, being more biodegradable. This is one of

TABLE 4.1 Structural and morphological characteristics of some biodegradable polymers.

Polymer type	Polymer	Characteristics	Morphology	References
Aliphatic polyesters	PBS	Linear and hydrophobic polymer	Semicrystalline polymer	Luyt and Malik (2019)
	PCL	Linear and hydrophobic polymer	Semicrystalline polymer	Vroman and Tighzert (2009)
	PGA	Rigid thermoplastic polymer and high sensitivity to hydrolysis	Highly crystalline	Pascault et al. (2012)
	PLA	Hydrophobic polymer by effect of its side groups $-CH_3$. The rate of degradation will depend on the morphology	PLLA and PDLA are semicrystalline and PDLLA is amorphous	Luckachan and Pillai (2011), Vroman and Tighzert (2009)
Biopolymers	Cellulose	Linear polymer without branches consisting of glucose units linked together by $\beta(1 \rightarrow 4)$ glycosidic bonds. High tendency to form intra- and intermolecular hydrogen bonds by hydroxyl groups of glucose units	Microfibrils with crystalline and amorphous regions	Ghanbarzadeh and Almasi (2013)
	Chitosan	Linear polysaccharide composed mainly of (1–4)-linked 2-amino-2-deoxy-β-D-glucose monomers. High tendency to form intra- and intermolecular hydrogen bonds	Rigid crystalline	Priyadarshi and Rhim (2020)

TABLE 4.1 Structural and morphological characteristics of some biodegradable polymers.—cont'd

Polymer type	Polymer	Characteristics	Morphology	References
	Collagen	Triple helix structure with three polypeptide alpha chains rolled up	Fibril-shaped proteins	Balaji et al. (2018)
	Starch	Polysaccharide composed of anhydroglucose units linked together through $\alpha(1 \rightarrow 4)$ glycosidic bonds. It is fragile and hydrophilic	The most native starches are semicrystalline	Jiang et al. (2020); Zhong et al. (2020)
PHAs	PHB	Linear polyester composed of 3-hydroxybutanoic acid units	Highly crystalline	Mathuriya and Yakhmi (2017)
	PHBV	Linear polyester composed of 3-hydroxybutanoic acid and 3-hydroxypentanoic acid	Less crystalline than PHB	Philip et al. (2007)

the main challenges in the synthesis of biodegradable polymers because these are usually constituted by short and uniform repetitive units that contribute to the crystallization, negatively affecting their biodegradability due to hydrolyzable groups that are inaccessible to hydrolytic enzymes. Semicrystalline and amorphous polymers, on the other hand, exhibit greater biodegradability than crystalline ones. However, by breaking down the amorphous and semicrystalline regions first, a crystalline material is generated, preventing the enzymatic degradation (Ghanbarzadeh and Almasi, 2013; Laycock et al., 2017; Delamarche et al., 2020).

It has been shown that the number of chiral centers increases the specificity of enzymes and, consequently, the biodegradability of polymers. Therefore, during the design and synthesis of biodegradable polymeric materials, it is common to incorporate hydroxyl, benzyl, methyl, carboxyl, and other substituents, with the aim of generating chiral carbons close to the hydrolyzable regions of the polymer chain, increasing their biodegradability

(Ghanbarzadeh and Almasi, 2013). Tacticity also affects polymer morphology. For example, lactic acid-derived polymers, are presented in isotactic [poly(L-lactic acid) (PLLA) and poly(D-lactic acid) (PDLA)], and syndiotactic [poly(DL-lactic acid) (PDLLA)] stereochemical arrangements. This syndiotactic polymer, with alternating D and L units, constitutes a less crystalline material, with low tensile strength and, consequently, it is more biodegradable than semicrystalline materials based on PLLA and PDLA. Due to their optimal biodegradability, therefore, PDLLA have been attractive materials in multiple biomedical applications (Kricheldorf, 2001; Luckachan and Pillai, 2011). Finally, it is important to note that molecular weight is also an important aspect in decomposition. Polymers with lower molecular weight degrade faster. Biodegradation also occurs on the surface of the polymer, therefore, the larger surface area of a material, the greater its biodegradability (Vroman and Tighzert, 2009; Elvers et al., 2016).

4.3 Physicochemical properties

Biodegradable polymers are characterized by certain physical and chemical properties associated with the structural features set out above. These properties include the glass transition temperature (T_g), melting temperature (T_m), density (D), Young's modulus (E), tensile strength (TS), etc. (Guo et al., 2012). Some physicochemical properties for several polymers are set out in Table 4.2. The mechanical properties of biodegradable polymers depend on the

TABLE 4.2 Physicochemical properties of some biodegradable polymers.

Polymer	D (g cm^{-3})	T_m (°C)	T_g (°C)	E (GPa)	TS (MPa)	References
PBS	1.26	90	−32	0.6	35	Künkel et al. (2016)
PCL	1.11–1.15	60	50–60	0.4	20.7–42	Velasco et al. (2015)
PDLLA	1.25–1.27	165–180	40–69	1.9	29–150	Mangaraj et al. (2019)
PLLA	1.24–1.30	170–200	55–65	4.8	28–2300	Mangaraj et al. (2019)
PGA	1.50	220	35–40	12.5	60–99.7	Vroman and Tighzert (2009)
PHB	1.19	175	5–75	3.50	40	Rydz et al. (2018); Mangaraj et al. (2019)

morphology of the material. The most crystalline and dense materials will have better tensile properties. Tensile strength, in conjunction with E, are parameters used as mechanical failure criteria, so they are fundamental properties for polymers that will be applied in the construction field (Maraveas et al., 2020). Laycock and collaborators, in a recent review, established that when tensile strength has decreased by 5% of its initial value, the polymer has lost its strength. This property is affected by the forces present between the chains and the crystallinity of the material: the greater the crystalline regions in the polymer structure, the stiffer it will be. This can be a disadvantage, since rigid materials present localized tension that, in some cases, is much higher than the tensile strength, resulting in crack formation (Laycock et al., 2017). The functional groups in the polymeric structure influence the crystallinity and, therefore, the mechanical properties of the material because if hydrogen bonding and polar interactions are favored, then crystal packing is also favored (Ciardelli et al., 2020). A high hydrophilicity, resulting in the functional groups of the chains, will influence the mechanical properties because the polymer will tend to absorb water, producing swelling that can cause a loss of material strength (Van de Velde and Kiekens, 2001). Density is also closely related to mechanical resistance; thus, a higher density material is related to greater mechanical strength (Maraveas et al., 2020).

T_m is a parameter of interest because, at this temperature, thermal degradation of biodegradable thermoplastic materials take place, since the mobility of the chains occurs, reducing the rigidity of the material and making the bulk of the material more accessible to external factors. Therefore, polymers with lower T_m are more susceptible to degradation (Elvers et al., 2016; Laycock et al., 2017). Thermal degradation may limit the application of these materials in different areas. For example, PCL is a polyester exhibiting optimal thermal processability, however, its useful life is limited by its low T_m, as a consequence, PCL is mixed with other polymers, such as cellulose, to increase the scope of its applications (Navarro-Baena et al., 2016; Zhong et al., 2020).

On the other hand, biodegradable polymers that present low T_g are characterized as having poor mechanical properties, because, at room temperature, the flexibility of the chains is reduced, avoiding their movement, as they are fragile to traction or tension. T_g has also been considered as a limit at which, at higher temperatures t, the polymer degrades drastically (Laycock et al., 2017).

T_g and T_m can be estimated by difference scanning calorimetry (DSC) and thermogravimeter analysis (TGA). For the applications of these materials in the packaging area, the thermal properties are evaluated at room temperature (25°C), storage temperature (around 23°C), and at high temperatures (close to 30°C) as the polymer must be stable to these conditions (Din et al., 2020). The importance of thermal properties is that they are related to the structural characteristics of the material and, consequently, its ability to be biodegraded. Aliphatic polyesters, for example, have low T_g, while their T_m is high, contributing to their biodegradability, however at the same time, exhibiting poor performance and limiting their applications (Chen et al., 2008). Starch, on the other hand, has high T_g and low T_m, with the addition of plasticizers, such as glycerin, sorbitol, or urea, being necessary in conjunction with the application of shear forces and high temperatures, in order to improve their plasticity and to produce a thermoplastic material (Khan et al., 2016; Narancic et al., 2020).

It is important to note that all general physicochemical properties of biodegradable polymers depend on their structural nature. In the case of aliphatic polyesters, there is a wide variety of monomers with which these materials can be synthesized, and hence their physicochemical properties depend on the chosen monomer (see Table 4.2). Typically, these polymers are characterized by low T_m and flexibility (Künkel et al., 2016). However, these have low mechanical resistance, so their application is restricted. To this end, resins based on a combination of aliphatic and aromatic polyesters have been developed, which have the characteristic resistance of aromatic polyesters, preserving the biodegradability of aliphatic polyesters. BASF, in the late 1990s, introduced a material synthesized from terephthalic acid monomers, 1,4-butanodiol, and adipic acid. This polyester is known on the market as Ecoflex and was characterized by having mechanical properties similar to PET, being applied as a material for bags of organic waste or shopping and adhesive films (Gross and Karla, 2002; Künkel et al., 2016). PHAs, for their part, generally exhibit thermoplastic properties similar to fossil-based polymers, however, their other physicochemical properties depend on the chain length of the monomer that constitutes the material. For example, PHAs with short-chain monomers (four or five carbons in length) have mechanical properties similar to PP due to their tendency to crystallize, while their counterparts with chains that are 6–14 carbons in length exhibit elastic properties (Gopi et al., 2018; Narancic et al., 2020).

4.4 Transformation technologies

As discussed above, biodegradable polymers can be understood as materials that are degraded by environmental factors. A polymer is certified as a biodegradable material when it qualifies with the requirements of the international standard EN 13432. For these materials to meet the necessary requirements, several factors have been studied to develop transformation technologies that contribute to their biodegradation. In general terms, a polymer can undergo physical transformations promoted by mechanical, light, thermal factors, or chemical transformations controlled by enzymatic catalysis and by acids or bases (Haider et al., 2019; Rujnić-Sokele and Pilipović, 2017).

Enzyme-catalyzed degradation has also been extensively studied. In this technology, biodegradation occurs by erosion of the material surface due to enzymes which are usually larger than the degradable regions of the polymer (Feig et al., 2018). Enzymes produced by microorganisms may have hydrolytic or oxidative activity, causing material degradation. Narancic and O'Connor reported in a recent review, that the enzymes involved are α/β-hydrolase enzymes such as cutinases, lipases, and oxidases such as laccases (Narancic and O'Connor, 2019). Yang et al., for example, studied the α/β-hydrolase structures of oxidized poly(vinyl alcohol) hydrolase of the organisms *Sphingopyxis* sp. 113P3 and *Pseudomonas* sp. VM15C, which decompose poly(vinyl alcohol) by microbial oxidation and eventual enzymatic hydrolysis, obtaining products like acetic acid (Yang et al., 2014; Feig et al., 2018). Lipases and esterases, on the other hand, have been studied as enzymes that degrade aliphatic polyesters such as PCL (Tokiwa et al., 2009). This degradation technology is commonly referred to as biodeterioration, and the methods used to determine material biodegradation are based on the assessment of surface roughness disturbances, color changes, cracks, or hole formation, etc. (Lucas et al., 2008).

Although enzymatic decomposition mediated by microorganisms is the most studied transformation technology, it is not the only biofragmentation mechanism. It has been reported that microorganisms form communities in an extracellular matrix composed of biomolecules, such as polysaccharides or proteins, called biofilm, which alters the porous structure, and modifies the moisture levels and thermal transfer of the material, causing a weakening in the mechanical properties of the polymer (Cappitelli et al., 2006; Lucas et al., 2008; Espinosa et al., 2020). In addition, some bacteria, such as *Nitrobacter* spp., *Azotobacter* ssp., or *Thiobacillus* spp. promote chemical degradation by excreting nitric and sulfuric acid, or other low-molecular-weight compounds, which react with material components, causing surface erosion (Lugauskas et al., 2003; Rubio et al., 2006; Garcés et al., 2017; Lerma et al., 2020).

Another widely documented process of chemical transformation is abiotic hydrolysis technology by acid or base action (Tokiwa et al., 2009). To enable a polymer to be fragmented by water action, it must have hydrolyzable groups such as ester, amide, anhydride, urea, etc. This degradation process depends on the pH of the medium. Thus, for instance, biodegradable polymers such as PLA and PCL shown better degradation properties under acidic conditions (e.g., H_2SO_4 and HCl), while, in basic media (e.g., NaOH), cleavage is seen to be progressive. Degradation from hydrolysis is analyzed through structural changes in the material using techniques such as infrared spectroscopy (FTIR) or unclear magnetic resonance (NMR). However, a more effective method is to determine the presence of oligomers as products of the cleavage of polymer chains through high-performance liquid chromatography (HPLC) or gel permeation chromatography (GPC) (Yu et al., 2005; Lucas et al., 2008).

Abiotic degradation can also be improved through the incorporation of prooxidant additives, combined with antioxidants, which allow decomposition by oxidation. Prooxidants are commonly stearates of some metal that are regulated with phosphite or phenolic antioxidants (Filiciotto and Rothenberg, 2020).

Materials can also be photooxidized by adding antioxidants to the polymer chains that react by photolysis with ultraviolet radiation, producing radicals and promoting the cleavage of polymer chains. The photodegradation process is the most effective decomposition that takes place under environmental conditions (Ghanbarzadeh and Almasi, 2013). In addition, photodegradation can be carried out from prooxidant additives based on metal combinations such as Mn^{2+}/Mn^{3+}, degrading the material by cleavage mediated by free radical reactions. In this way, oxidation products of lower molecular weight are generated and can be degraded by microorganisms (Vroman and Tighzert, 2009). Photodegradative processes are easily studied from their outdoor exposure under real-world conditions. At the laboratory level, studies are based on exposing the material to UV radiation and determining the weight of the fragments generated from GPC. Although gravimetry could be considered as a simpler and more effective method, loss of polymer weight is often negligible, so a more sophisticated technique is usually needed (Lucas et al., 2008).

As regards physical methods, the effects of mechanical forces and thermodegradation are widely studied technologies. Mechanical forces are based on the application of pressure and stress to the material, contributing to polymer degradation due to residual stress resulting from manufacturing. Mechanical forces can be caused, in field conditions, by water or air turbulence, limitations in the installation of the material, or damage by surrounding organisms (Laycock et al., 2017). This degradation is not noticeable at the macroscopic level because decomposition by mechanical forces does not predominate in the biodegradation of the material, however, the application of these forces influences the speed at which the material decomposes (Briassoulis, 2005, 2007). For thermal degradation it is important to take into account the value of T_m (see Table 4.2). Above this temperature, materials are more susceptible to degradation because the chains are disorganized, being more accessible to chemical decomposition promoted by microorganisms. As in chemical methods, thermodegradation is studied through DSC, TGA, and structural changes, in conjunction with the fragments generated in decomposition, which are determined by GPC, FT-IR, and NMR (Laycock et al., 2017).

4.5 Uses of biodegradable polymers and their potential applications

Biodegradable polymers are materials used to replace materials based on synthetic petroleum-derived polymers. This chapter will set out in detail their applications. However, some specific applications are shown in Chapters 20–29.

In the area of medicine and biomedical applications, materials must be, in many cases, in addition to biodegradable, biocompatible and bioadsorbent. In this way, biopolymers such as polysaccharides and proteins have been attractive since they usually present these characteristics. Particularly, proteins are constituents of tissues in various organisms, such as skin, vascular, pulmonary, and musculoskeletal tissues, among others, so they are used as materials in drug-delivery systems, scaffolding in the area of tissue engineering, and even as materials for suture development (Zeng et al., 2016; Chen et al., 2020). Polysaccharides, such as chitin and cellulose, have also been arranged as materials for biomedical and pharmaceutical applications (Palencia et al., 2019), which are shown in Table 4.3. Within the group of biodegradable synthetic polymers, polyesters have also been optimal candidates due to their optimal mechanical properties and the possibility of controlling their biodegradation

TABLE 4.3 Summary of the most relevant applications of some biodegradable polymers.

Polymer	Application	References
Cellulose	Controlled release of pesticides, nutrients, and fertilizers. Material for the development of controlled drug-delivery systems and scaffolding for the regeneration of tissues	Vroman and Tighzert (2009); Palencia et al. (2019)
Chitin and chitosan	Products with antioxidant activity and radical elimination. Carriers of drugs, antitumor products, anticoagulants, antibacterial agents, and immune-adjuvants	Nair and Laurencin (2007); Priyadarshi and Rhim (2020)
Collagen	Wound dressings, bone graft, and tissue repair	Vroman and Tighzert (2009)
PBS	Medical supplies, packaging, bottles, and biomedical applications	Zhao et al. (2020)
PCL	Material for the development of controlled drug-delivery systems, scaffolding for the regeneration of tissues and support material for agricultural planting	Vroman and Tighzert (2009); Zeng et al. (2016)
PGA	Internal fixation devices for bones, absorbable sutures, and scaffolding for tissue regeneration	Yamane et al. (2014)
PHB	Disposable products, food packaging, and medical scaffolding like screws	Das et al. (2020); Zhao et al. (2020)
PLA	Waste bags, packaging systems, development of agricultural mulch films and fertilizer bags	Vroman and Tighzert (2009); Zhao et al. (2020)
Starch	Food packaging, development of agricultural mulch films	Zhong et al. (2020)

rate. Polyurethanes also exhibit optimal biodegradability and biocompatibility, so they have been promising candidates for the development of medical implants and materials for biomedical applications in contact with tissue and biological fluids (Palencia et al., 2020). PHAs are also attractive biodegradable materials due to their optimal mechanical properties, and so they have been used as medical scaffolding such as screws, pins, etc. (Vroman and Tighzert, 2009; Das et al., 2020; Yin and Yang, 2020).

Packaging materials, such as containers, are one of the most consuming plastic industries, accounting for 40.1% of the total production of these materials. They must have optimal sealed and air permeability, ease of processing and shaping, low cost for commonly used applications, and safety when in contact with food or other products. In particular, poly(ethylene) is the most widely used polymer to pack all variety of liquids, powders, and solids. Biodegradable polymers have been intended to replace this material over time with the aim of reducing the residues of conventional polymers (Zeng et al., 2016). In contrast, PLA has acceptable levels of oxygen permeability and water vapor, so it has wide application in storage systems; however, its cost is relatively higher than poly(ethylene). It is estimated that 86.9% of PET-based packaging is replaced by PLA materials (Dilkes et al., 2019; Yano et al., 2014). Also, PHA such as PHB, have exhibited optimal mechanical properties, being used as packaging material (Vroman and Tighzert, 2009). PHA have been proposed to replace PE and PP as packaging material; nevertheless, its application in food packaging is limited because it has been reported to have odor problems and the subsequent addition of additives (Meereboer et al., 2020).

In agriculture, the provision of biodegradable materials for plant protection is essential. The additional characteristics to biodegradability that these polymers must present include the ability to conserve moisture and increase soil temperature (Vroman and Tighzert, 2009). The materials used for this purpose are based on biopolymers such as cellulose, starch, and lignin, and synthetic polymers made from sorbitol, citric acid, glucamine, etc., which are degraded in soil in organic manure, contributing to silver growth and soil viability (Luckachan and Pillai, 2011; Zeng et al., 2016; Garcés et al., 2017; Palencia et al., 2017b). Biodegradable polyesters have also been attractive materials for agricultural applications (Palencia et al., 2017a; Yin and Yang, 2020). Particularly, PLA has been used as an agricultural mulch with the aim of replacing poly(ethylene), exhibiting optimal yield in crops, without harmful effects such as soil contamination (Yin and Yang, 2020). PHAs are also used as mulches because they have been reported to prevent weed growth and soil evaporation. In addition, these biodegradable polymers promote nitrogen fixation and act as an inoculant agent of bacteria (Das et al., 2020). Functional polyurethanes also have been designed for smart fertilization of crops activities (Palencia et al., 2016).

For the sake of humanity, and with the aim of an environmentally sustainable future, biodegradable polymers are expected to be the materials implemented in multiple previously exposed applications. Although the advent of biodegradable polymers has arisen with the aim of replacing conventional non-degradable materials, they present potential applications that can revolutionize the fields of medicine, technology, etc. For example, nanomedicine has a promising future that could improve the quality of life of many people, and in that way, biodegradable polymers have been evaluated as drug nanocarrier systems. To this end, in addition to in vivo toxicity studies, the physicochemical properties of these materials such as polymeric composition, tacticity, molecular weight, colloidal stability, hydrophilic/hydrophobic ratio, etc., are continuously explored and evaluated (Kakkar et al., 2017; Narancic et al., 2020). Narancic et al., in a recent review, highlighted current research in this field, where PHAs such as PHB are the protagonist, exhibiting optimal results of cell adhesion and good immune response (Narancic et al., 2020). Although promising, this field of medicine is emerging and continues to be studied today. Other biodegradable polymers such as alginate, quitosane, and PLA have potential application in the development of non-toxic carrier systems, contributing to the advancement of oral vaccines to obtain the desired immune response without the need for a booster dose (Jana et al., 2020).

In the area of tissue engineering, research is still in its early stages, with a few studies carried out in vivo, however, biodegradable biopolymers are attractive materials because they could replace materials that induce immune responses. In this way, polyesters such as PLA are materials widely studied for their optimal properties. However, they generate acid residues from hydrolysis, which could negatively impact cellular physiology (Narancic et al., 2020; Yin and Yang, 2020).

On the other hand, the electronic revolution has led to the generation of devices that, while they have improved communication and access to information, also are generators of pollution. Consequently, efforts have been made in the development of environmentally sustainable electronics. One of the most important works is that developed by Gao et al., who built an electret nanogenerator from cellulose and PLA, providing new knowledge for the development of ecological energy (Gao et al., 2016; Yin and Yang, 2020). Highly ion-conductive polymeric electrolytes are also attractive for use in ion batteries, supercapacitors, solar cells, etc., due to their high ion conductivity, mechanical resistance, and easy manufacture (Aziz and Chang, 2018; Li et al., 2018). Thus, PCL has been proposed as an electrode material due to its renewal capacity and biodegradability, contributing to the generation of sustainable energies (Bharti et al., 2020). Other biodegradable polymers, such as cellulose, have also been proposed as a material for the development of light-emitting diodes, solar cells, and field-effect transistors (Barr et al., 2011; Cao and Uhrich, 2019). However, the application of biodegradable polymers in the field of electronics is still under investigation, as the electrical and mechanical properties of these materials need to be improved.

In addition to ecologically preserving the planet, biodegradable polymers also have the potential to reduce or eliminate toxic waste that has been discharged into the environment as a product of industrialization (Hasan et al., 2020). For example, PCLs have been proposed as absorber of endosulfan, a toxic pesticide contained in groundwater. The results showed a good retention capacity, with it being a promising material for application in the removal of contaminants (Mourabit and Boulaid, 2019).

Biodegradable polymers have also been used for the development of nanocomposites with extremely novel applications. Nanocomposites based on starch and PLA with clay had excellent mechanical properties with little water penetration, contributing to the advancement of packaging technology. In the field of medicine, PLA nanofiber biocomposites with electro-threaded polyethylene glycol have the potential to be used as bone regeneration scaffolding since this material showed excellent mechanical properties, like polyethylene glycol, in conjunction with the biodegradability characteristics of PLA (Shi et al., 2018; Sun et al., 2018). Bone fillers based on biodegradable polymers such as soybeans had better tissue regeneration, compared with commercial bone grafts (Giavaresi et al., 2010). Bacterial nanocellulose tubes have also been suggested as biomedical prostheses that include artificial blood vessels, urethra, and esophagus (Tang et al., 2016).

4.6 Conclusion

In this chapter, the definition of biodegradable polymer was discussed, and it was explained how the presence of certain functional groups, the hydrophilic and hydrophobic nature, and even the morphology, were factors that influence material biodegradability. In addition, the main physicochemical properties were outlined, including how they influence the biodegradability of the polymer. The arrival of biodegradable polymers is revolutionizing the areas of medicine, packaging, and agriculture, and evidence potential applications that could improve the quality of life of humanity. The importance of these materials is reflected in their leading participation in the market, being desired by many countries to replace conventional non-biodegradable materials. However, despite all these advantages, these polymers continue to be more expensive than their non-biodegradable petroleum-based counterparts. This reflects the need for research to focus on methods that reduce the cost of obtaining, polymerization, and monomers. Only in this way, can these materials be widely accessed, avoiding an ecologically devastated future.

References

Agarwal, S., 2020. Biodegradable polymers: present opportunities and challenges in providing a microplastic-free environment. Macromol. Chem. Phys. 2000017. https://doi.org/10.1002/macp.202000017.

Aziz, S., Chang, S.H., 2018. Smart-fabric Sensor Composed of Single-Walled Carbon Nanotubes Containing Binary Polymer Composites for Health Monitoring, vol. 163, pp. 1−9. https://doi.org/10.1016/j.compscitech.2018.05.012.

Balaji, A.B., Pakalapati, H., Khalid, M., Walvekar, R., Siddiqui, H., 2018. Natural and synthetic biocompatible and biodegradable polymers. In: Biodegradable and Biocompatible Polymer Composites - Processing, Properties and Applications. Elsevier, Amsterdam, pp. 3−32. https://doi.org/10.1016/B978-0-08-100970-3.00001-8.

Barr, M.C., Rowehl, J.A., Lunt, R.R., Xu, J., Wang, A., Boyce, C.M., Im, S.G., Bulović, V., Gleason, K.K., 2011. Direct monolithic integration of organic photovoltaic circuits on unmodified paper. Adv. Mater. 23, 3499−3505. https://doi.org/10.1002/adma.201101263.

Berrio, M.E., Palencia, S.L., Lerma, T.A., Mora, M.A., 2018. Bacterial colonization modelling on soil particles: effect of humic acids on the formation of nitrogen-fixing bacteria biofilms. J. Sci. Technol. Appl. 5, 33−44. https://doi.org/10.34294/j.jsta.18.5.33.

Bharti, V., Singh, P.K., Sharma, J.P., 2020. Development of polymer electrolyte membranes based on biodegradable polymer. Mater. Today Proc. 1−7. https://doi.org/10.1016/j.matpr.2020.06.463.

Briassoulis, D., 2005. The effects of tensile stress and the agrochemical Vapam on the ageing of low density polyethylene (LDPE) agricultural films. Part I. Mechanical behaviour. Polymer Degradation Stability 88, 489−503. https://doi.org/10.1016/j.polymdegradstab.2004.11.021.

Briassoulis, D., 2007. Analysis of the mechanical and degradation performances of optimised agricultural biodegradable films. Polymer Degradation Stability 92, 1115−1132. https://doi.org/10.1016/j.polymdegradstab.2007.01.024.

Cameron, D.J.A., Shaver, M.P., 2011. Aliphatic polyester polymer stars: synthesis, properties and applications in biomedicine and nanotechnology. Chem. Soc. Rev. 40, 1761−1776. https://doi.org/10.1039/C0CS00091D.

Cao, Y., Uhrich, K.E., 2019. Biodegradable and biocompatible polymers for electronic applications: a review. J. Bioactive Compatible Polym. 34, 3−15. https://doi.org/10.1177/0883911518818075.

Cappitelli, F., Principi, P., Sorlini, C., 2006. Biodeterioration of modern materials in contemporary collections: can biotechnology help? Trends Biotechnol. 24, 350−354. https://doi.org/10.1016/j.tibtech.2006.06.001.

Chen, C.K., Huang, P.K., Law, W.C., Chu, C.H., Chen, N.T., Lo, L.W., 2020. Biodegradable polymers for gene-delivery applications. Int. J. Nanomed. 15, 2131−2150. https://doi.org/10.2147/IJN.S222419.

Chen, Y., Tan, L., Chen, L., Yang, Y., Wang, X., 2008. Study on biodegradable aromatic/aliphatic copolyesters. Brazilian J. Chem. Eng. 25, 321−335. https://doi.org/10.1590/S0104-66322008000200011.

Ciardelli, F., Bertoldo, M., Bronco, S., Passaglia, E., 2020. Molecular structure requirements. In: Polymers from Fossil and Renewable Resources. Springer, Cham, pp. 65−71. https://doi.org/10.1007/978-3-319-94434-0_3.

Das, D., Mulchandani, N., Kumar, A., Katiyar, V., 2020. Fabrication of stimuli-responsive polymers and their composites: candidates for resorbable sutures. In: Katiyar, V., Kumar, A., Mulchandani, N. (Eds.), Advances in Sustainable Polymers. Materials Horizons: From Nature to Nanomaterials. Springer, Singapore, pp. 121−144. https://doi.org/10.1007/978-981-15-1251-3_6.

Degnan, T., Shinde, S., 2019. Waste-plastic processing provides global challenges and opportunities. MRS Bull. 44, 436−437. https://doi.org/10.1557/mrs.2019.133.

Delamarche, E., Mattlet, A., Livi, S., Gérard, J.F., Bayard, R., Massardier, V., 2020. Tailoring biodegradability of poly(butylene succinate)/poly(lactic acid) blends with a deep eutectic solvent. Front. Mater. 7, 1–13. https://doi.org/10.3389/fmats.2020.00007.

DiBenedetto, L.J., Cameron, J.A., Huang, S., 1988. J. Biodegradation of hydroxylated polymers. In: Gebelein, C.G., Carraher, C.E., Foster, V.R. (Eds.), Applied Bioactive Polymeric Materials. Polymer Science and Technology, vol. 38. Springer, Boston, pp. 61–74. https://doi.org/10.1007/978-1-4684-5610-3_4.

Dilkes, L.S., Pratt, S., Laycock, L.B., 2019. The role of biodegradable plastic in solving plastic solid waste accumulation. In: Al-Salem, S.M. (Ed.), Plastics to Energy Fuel, Chemicals, and Sustainability Implications. William Andrew, Norwich, pp. 469–505. https://doi.org/10.1016/B978-0-12-813140-4.00019-4.

Din, M.I., Ghaffar, T., Najeeb, J., Hussain, Z., Khalid, R., Zahid, H., 2020. Potential perspectives of biodegradable plastics for food packaging application-review of properties and recent developments. Food Additiv. Contamin. A 37, 665–680. https://doi.org/10.1080/19440049.2020.1718219.

Ellen MacArthur Foundation, 2017. The New Plastics Economy: Rethinking the Future of Plastics and Catalysing Action, pp. 10–18. https://www.ellenmacarthurfoundation.org/publications/the-new-plastics-economy-rethinking-the-future-of-plastics-catalysing-action.

Elvers, D., Song, C.H., Steinbüchel, A., Leker, J., 2016. Technology trends in biodegradable polymers: evidence from patent analysis. Polym. Rev. 56, 584–606. https://doi.org/10.1080/15583724.2015.1125918.

Espinosa, A., Palencia, S.L., García, A., Palencia, M., 2020. Polymicrobial Biofilms: fundamentals, diseases, and treatments - a review. J. Sci. Technolo. Appl. 8, 4–54. https://doi.org/10.34294/j.jsta.20.8.54.

European Union, 2018. A European Strategy for Plastics in a Circular Economy, pp. 5–23. https://ec.europa.eu/environment/circular-economy/pdf/plastics-strategy-brochure.pdf.

Feig, V.R., Tran, H., Bao, Z., 2018. Biodegradable polymeric materials in degradable electronic devices. ACS Central Sci. 4, 337–348. https://doi.org/10.1021/acscentsci.7b00595.

Filiciotto, L., Rothenberg, G., 2020. Biodegradable plastics: standards, policies, and impacts. ChemSusChem 13, 1–18. https://doi.org/10.1002/cssc.202002044.

Garcés, V., Palencia, M., Combatt, E., 2017. Development of bacterial inoculums based on biodegradable hydrogels for agricultural applications. J. Sci. Technol. Appl. 2, 13–23. https://doi.org/10.34294/j.jsta.17.2.11.

Garcia, A., Combatt, E., Palencia, M., 2018. Structural study of humin and its interaction with humic acids by Fourier-transform mid-infrared spectroscopy. J. Sci. Technol. Appl. 4, 28–39. https://doi.org/10.34294/j.jsta.18.4.28.

Gao, X., Huang, L., Wang, B., Xu, D., Zhong, J., Hu, Z., Zhang, L., Zhou, J., 2016. Natural materials assembled, biodegradable, and transparent paper-based electret nanogenerator. ACS Appl. Mater. Interfaces 8, 35587–35592. https://doi.org/10.1021/acsami.6b12913.

Gewert, B., Plassmann, M.M., MacLeod, M., 2015. Pathways for degradation of plastic polymers floating in the marine environment. Environ. Sci. Process. Impacts 17, 1513–1521. https://doi.org/10.1039/C5EM00207A.

Ghanbarzadeh, B., Almasi, H., 2013. Biodegradable polymers. In: Chamy, R., Rosenkranz, F. (Eds.), Biodegradation – Life of Science. IntechOpen, London, pp. 141–185. https://doi.org/10.5772/56230.

Giavaresi, G., Fini, M., Salvage, J., Aldini, N.N., Giardino, R., Ambrosio, L., Nicolais, L., Santin, M., 2010. Bone regeneration potential of a soybean-based filler: experimental study in a rabbit cancellous bone defects. J. Mater. Sci. Mater. Med. 21, 615–626. https://doi.org/10.1007/s10856-009-3870-6.

Gopi, S., Kontopoulou, M., Ramsay, B.A., Ramsay, J.A., 2018. Manipulating the structure of medium-chain-length polyhydroxyalkanoate (MCL-PHA) to enhance thermal properties and crystallization kinetics. Int. J. Biol. Macromol. 119, 1248–1255. https://doi.org/10.1016/j.ijbiomac.2018.08.016.

Gross, R.A., Kalra, B., 2002. Biodegradable polymers for the environment. Science 297, 803–807. https://doi.org/10.1126/science.297.5582.803.

Guo, W., Tao, J., Yang, C., Song, C., Geng, W., Li, Q., Wang, Y., Kong, M., Wang, S., 2012. Introduction of environmentally degradable parameters to evaluate the biodegradability of biodegradable polymers. PLoS One 7, e38341. https://doi.org/10.1371/journal.pone.0038341.

Haider, T.P., Völker, C., Kramm, J., Landfester, K., Wurm, F., 2019. Plastics of the future? The impact of biodegradable polymers on the environment and on society. Angew. Chem. Int. Ed. 58, 50–62. https://doi.org/10.1002/anie.201805766.

Hasan, M.M., Shenashen, M.A., Hasan, M.N., Znad, H., Salman, M.S., Awual, M.R., 2020. Natural biodegradable polymeric bioadsorbents for efficient cationic dye encapsulation from wastewater. J. Mol. Liquids 114587. https://doi.org/10.1016/j.molliq.2020.114587.

Ibrahim, M.S., Sani, N., Adamu, M., Abubakar, M.K., 2018. Biodegradable polymers for sustainable environmental and economic development. MOJ Bioorgan. Organ. Chem. 2, 192–194. https://doi.org/10.15406/mojboc.2018.02.00080.

Ivleva, N.P., Wiesheu, A.C., Niessner, R., 2016. Microplastic in aquatic ecosystems. Angew. Chem. Int. Ed. 56, 1720–1739. https://doi.org/10.1002/anie.201606957.

Jana, P., Shyam, M., Singh, S., Jayaprakash, V., Dev, A., 2020. Biodegradable polymers in drug delivery and oral vaccination. Eur. Polymer J. 142, 110155. https://doi.org/10.1016/j.eurpolymj.2020.110155.

Jiang, T., Duan, Q., Zhu, J., Liu, H., Yu, L., 2020. Adv. Ind. Eng. Polymer Res. 3, 8–18. https://doi.org/10.1016/j.aiepr.2019.11.003.

Kakkar, A., Traverso, G., Farokhzad, O.C., Weissleder, R., Langer, R., 2017. Evolution of macromolecular complexity in drug delivery systems. Nat. Rev. Chem. 1, 0063. https://doi.org/10.1038/s41570-017-0063.

Khan, B., Niazi, M.B.K., Samin, G., Jahan, Z., 2016. Thermoplastic starch: a possible biodegradable food packaging material - a review. J. Food Process Eng. 40, e12447. https://doi.org/10.1111/jfpe.12447.

Kricheldorf, H.R., 2001. Syntheses and application of polylactides. Chemosphere 43, 49–54. https://doi.org/10.1016/S0045-6535(00)00323-4.

Künkel, A., Becker, J., Börger, L., Hamprecht, J., Koltzenburg, S., Loos, R., Schick, M.B., Schlegel, K., Sinkel, C., Skupin, G., Yamamoto, M., 2016. Polymers, biodegradable. Ullmann's Encyclopedia Ind. Chem. 1–29. https://doi.org/10.1002/14356007.n21_n01.pub2.

Laycock, B., Nikolić, M., Colwell, J.M., Gauthier, E., Halley, P., Bottle, S., George, G., 2017. Lifetime prediction of biodegradable polymers. Prog. Polymer Sci. 71, 144–189. https://doi.org/10.1016/j.progpolymsci.2017.02.004.

Lerma, T.A., Garcés, V., Palencia, M., 2020. Novel multi- and bio-functional hybrid polymer hydrogels based on bentonite-poly(acrylic acid) composites and sorbitol polyesters: structural and functional characterization. Eur. Polymer J. 128, 109627. https://doi.org/10.1016/j.eurpolymj.2020.109627.

Li, Y.J., Fan, C.Y., Zhang, J.P., Wu, X.L., 2018. A promising PMHS/PEO blend polymer electrolyte for all-solid-state lithium ion batteries. Dalton Trans. 47, 14932–14937. https://doi.org/10.1039/C8DT02904K.

Lucas, N., Bienaime, C., Belloy, C., Queneudec, M., Silvestre, F., Nava-Saucedo, J.E., 2008. Polymer biodegradation: mechanisms and estimation techniques — a review. Chemosphere 73, 429—442. https://doi.org/10.1016/j.chemosphere.2008.06.064.

Luckachan, G.E., Pillai, C.K.S., 2011. Biodegradable polymers - a review on recent trends and emerging perspectives. J. Polymers Environ. 19, 637—676. https://doi.org/10.1007/s10924-011-0317-1.

Lugauskas, A., Levinskaitė, L., Pečiulyte, D., 2003. Micromycetes as deterioration agents of polymeric materials. Int. Biodeterioration Biodegrad. 52, 233—242. https://doi.org/10.1016/S0964-8305(03)00110-0.

Luyt, A.S., Malik, S.S., 2019. Can biodegradable plastics solve plastic solid waste accumulation? In: Al-Salem, S.M. (Ed.), Plastics to Energy Fuel, Chemicals, and Sustainability Implications. Elsevier, Amsterdam, pp. 403—423. https://doi.org/10.1016/B978-0-12-813140-4.00016-9.

Mangaraj, S., Yadav, A., Bal, L.M., Dash, S.K., Mahanti, N., 2019. Application of biodegradable polymers in food packaging industry: a comprehensive review. J. Packaging Technol. Res. 3, 77—96. https://doi.org/10.1007/s41783-018-0049-y.

Maraveas, C., 2020. Production of sustainable and biodegradable polymers from agricultural waste. Polymers 12, 1127. https://doi.org/10.3390/polym12051127.

Marlina, D., Sato, H., 2020. Study on the higher-order structure and hydrogen bonding of biodegradable polymer by low-frequency vibrational spectroscopy. IOP Conference Series Mater. Sci. Eng. 835, 012004. https://doi.org/10.1088/1757-899X/835/1/012004.

Mathuriya, A.S., Yakhmi, J.V., 2017. Polyhydroxyalkanoates: biodegradable plastics and their applications. In: Martínez, L., Kharissova, O., Kharisov, B. (Eds.), Handbook of Ecomaterials. Springer, Cham, pp. 1—29. https://doi.org/10.1007/978-3-319-48281-1_84-1.

Meereboer, K.W., Misra, M., Mohanty, A.K., 2020. Review of recent advances in the biodegradability of polyhydroxyalkanoate (PHA) bioplastics and their composites. Green Chem. 22, 5519—5559. https://doi.org/10.1039/D0GC01647K.

Mora, M.A., Palencia, M., Combatt, E., 2018. Effect of dissolved organic matter and humic substances on transport of ions and low molecular weight molecules by liquid membranes. J. Sci. Technol. Appl. 4, 40—48. https://doi.org/10.34294/j.jsta.18.4.29.

Mourabit, F., Boulaid, M., 2019. Pesticide removal in drinking water treatment using biodegradable polymers. Mater. Today Proc. 13, 1033—1038. https://doi.org/10.1016/j.matpr.2019.04.068.

Narancic, T., Cerrone, F., Beagan, N., O'Connor, K.E., 2020. Recent advances in bioplastics: application and biodegradation. Polymers 12, 920. https://doi.org/10.3390/polym12040920.

Narancic, T., O'Connor, K.E., 2019. Plastic waste as a global challenge: are biodegradable plastics the answer to the plastic waste problem? Microbiology 165, 129—137. https://doi.org/10.1016/10.1099/mic.0.000749.

Navarro-Baena, I., Sessini, V., Dominici, F., Torre, L., Kenny, J.M., Peponi, L., 2016. Design of biodegradable blends based on PLA and PCL: from morphological, thermal and mechanical studies to shape memory behavior. Polymer Degrad. Stability 132, 97—108. https://doi.org/10.1016/j.polymdegradstab.2016.03.037.

Nair, L.S., Laurencin, C.T., 2007. Biodegradable polymers as biomaterials. Prog. Polymer Sci. 32, 762—798. https://doi.org/10.1016/j.progpolymsci.2007.05.017.

Nottelet, B., Tommaso, C.D., Mondon, K., Gurny, R., Möller, M., 2010. Fully biodegradable polymeric micelles based on hydrophobic- and hydrophilic-functionalized poly(lactide) block copolymers. J. Polymer Sci. A Polymer Chem. 48, 3244—3254. https://doi.org/10.1002/pola.24100.

Panchal, S.S., Vasava, D.V., 2020. Biodegradable polymeric materials: synthetic approach. ACS Omega 5, 4370–4379. https://doi.org/10.1021/acsomega.9b04422.

Palencia, M., Berrio, M.E., García-Quintero, A., Lerma, T., 2019. Cellulose-based nanocomposites for pharmaceutical and biomedical applications. In: Kumar Giri, T., Ghosh, B. (Eds.), Polysaccharide-based Nano-Biocarrier in Drug Delivery. CRC Press, pp. 39–61.

Palencia, M., Mora, M., Palencia, S.L., 2017b. Biodegradable polymer hydrogels based in sorbitol and citric acid for controlled release of bioactive substances from plants (polyphenols). Curr. Chem. Biol. 11, 36–43. https://doi.org/10.2174/2212796810666161028114432.

Palencia, M., Lerma, T.A., Arrieta, A., 2020. Antibacterial ans non-hemolytic cationic polyurethanes with N-carboxymethyl-N,N,N-triethylammonium groups for bacteremia-control in biomedical-using materials. Mater. Today Commun. 22, 100708. https://doi.org/10.1016/j.mtcomm.2019.100708.

Palencia, M., Lerma, T., Combatt, E., 2017a. Hydrogels based in cassava starch with antibacterial activity for controlled release of cysteamine-silver nanostructured agents. Curr. Chem. Biol. 11, 28–35. https://doi.org/10.2174/2212796810666161108152319.

Palencia, M., Lerma, T., Córdoba, A., 2016. Polyurethanes with boron retention properties for the development of agricultural fertilization smart systems. J. Sci. Technol. Appl. 1, 39–52. https://doi.org/10.34294/j.jsta.16.1.4.

Pascault, J.P., Höfer, R., Fuertes, P., 2012. Mono-, di-, and oligosaccharides as precursors for polymer synthesis. In: Matyjaszewski, K., Möller, M. (Eds.), Polymer Science: A Comprehensive Reference, vol. 10. Elsevier, Amsterdam, pp. 59–82. https://doi.org/10.1016/B978-0-444-53349-4.00254-5.

Philip, S., Keshavarz, T., Roy, I., 2007. Polyhydroxyalkanoates: biodegradable polymers with a range of applications. J. Chem. Technol. Biotechnol. 82, 233–247. https://doi.org/10.1002/jctb.1667.

Priyadarshi, R., Rhim, J.W., 2020. Chitosan-based biodegradable functional films for food packaging applications. Innov. Food Sci. Emerg. Technol. 62, 102346. https://doi.org/10.1016/j.ifset.2020.102346.

Rubio, C., Ott, C., Amiel, C., Dupont-Moral, I., Travert, J., Mariey, L., 2006. Sulfato/thiosulfato reducing bacteria characterization by FT-IR spectroscopy: a new approach to biocorrosion control. J. Microbiol. Methods 64, 287–296. https://doi.org/10.1016/j.mimet.2005.05.013.

Rujnić-Sokele, M., Pilipović, A., 2017. Challenges and opportunities of biodegradable plastics: a mini review. Waste Manag. Res. J. Sustain. Circ. Econ. 35, 132–140. https://doi.org/10.1177/0734242X16683272.

Rydz, J., Musioł, M., Zawidlak-Wegrzyńska, B., Sikorska, W., 2018. Present and future of biodegradable polymers for food packaging applications. In: Grumezescu, A.M., Holban, A.A. (Eds.), Biopolymers for Food Design - Handbook of Food Bioengineering. Academic Press, Cambridge, pp. 431–467. https://doi.org/10.1016/B978-0-12-811449-0.00014-1.

Shchipunov, Y., 2012. Bionanocomposites: green sustainable materials for the near future. Pure Appl. Chem. 84, 2579–2607. https://doi.org/10.1351/PAC-CON-12-05-04.

Shi, M., Yang, R., Li, Q., Lv, K., Miron, R.J., Sun, J., Li, M., Zhang, Y., 2018. Inorganic self-assembled bioactive artificial proto-osteocells inducing bone regeneration. ACS Appl. Mater. Interfaces 10, 10718–10728. https://doi.org/10.1021/acsami.8b00385.

Sun, J., Shen, J., Chen, S., Cooper, M.A., Fu, H., Wu, D., Yang, Z., 2018. Nanofiller reinforced biodegradable PLA/PHA composites: current status and future trends. Polymers 10, 505. https://doi.org/10.3390/polym10050505.

Tang, J., Li, X., Bao, L., Chen, L., Hong, F.F., 2016. Comparison of two types of bioreactors for synthesis of bacterial nanocellulose tubes as potential medical prostheses including artificial blood vessels. J. Chem. Technol. Biotechnol. 92, 1–31. https://doi.org/10.1002/jctb.5111.

Tokiwa, Y., Calabia, B.P., Ugwu, C.U., Aiba, S., 2009. Biodegradability of plastics. Int. J. Mol. Sci. 10, 3711–3742. https://doi.org/10.3390/ijms10093722.

Van de Velde, K., Kiekens, P., 2001. Biopolymers: overview of several properties and consequences on their applications. Polymer Testing 21, 433–442. https://doi.org/10.1016/S0142-9418(01)00107-6.

Velasco, M.A., Narváez-Tovar, C.A., Garzón-Alvarado, D.A., 2015. Design, materials, and mechanobiology of biodegradable scaffolds for bone tissue engineering. BioMed Res. Int. 2015, 1–22. https://doi.org/10.1155/2015/729076.

Vroman, I., Tighzert, L., 2009. Biodegradable polymers. Materials 2, 307–344. https://doi.org/10.3390/ma2020307.

Yamane, K., Sato, H., Ichikawa, Y., Sunagawa, K., Shigaki, Y., 2014. Development of an industrial production technology for high-molecular-weight polyglycolic acid. Polymer J. 46, 769–775. https://doi.org/10.1038/pj.2014.69.

Yang, Y., Ko, T.P., Liu, L., Li, J., Huang, C.H., Chan, H.C., Ren, F., Jia, D., Wang, A.H.J., Guo, R.T., Chen, J., Du, G., 2014. Structural insights into enzymatic degradation of oxidized polyvinyl alcohol. ChemBioChem 1–5. https://doi.org/10.1002/cbic.201402166.

Yano, J., Hirai, Y., Sakai, S.I., Tsubota, J., 2014. Greenhouse gas emissions from the treatment of household plastic containers and packaging: replacement with biomass-based materials. Waste Manag. Res. J. Sustain. Circ. Eco. 32, 304–316. https://doi.org/10.1177/0734242X14525820.

Yin, G.Z., Yang, X.M., 2020. Biodegradable polymers: a cure for the planet, but a long way to go. J. Polymer Res. 27, 1–14. https://doi.org/10.1007/s10965-020-2004-1.

Yu, J., Plackett, D., Chen, L.X.L., 2005. Kinetics and mechanism of the monomeric products from abiotic hydrolysis of poly[(R)-3-hydroxybutyrate] under acidic and alkaline conditions. Polymer Degrad. Stability 89, 289–299. https://doi.org/10.1016/j.polymdegradstab.2004.12.026.

Zeng, S.H., Duan, P.P., Shen, M.X., Xue, Y.J., Wang, Z.Y., 2016. Preparation and degradation mechanisms of biodegradable polymer: a review. IOP Conf. Ser. Mater. Sci. Eng. 137, 012003. https://doi.org/10.1088/1757-899X/137/1/012003.

Zhao, L., Rong, L., Zhao, L., Yang, J., Wang, L., Sun, H., 2020. Plastics of the future? The impact of biodegradable polymers on the environment. In: He, D., Luo, Y. (Eds.), Microplastics in Terrestrial Environments. The Handbook of Environmental Chemistry, vol. 95. Springer, Cham, pp. 423–445. https://doi.org/10.1007/698_2020_462.

Zhong, Y., Godwin, P., Jin, Y., Xiao, H., 2020. Biodegradable polymers and green-based antimicrobial packaging materials: a mini-review. Adv. Ind. Eng. Polym. Res. 3, 27–35. https://doi.org/10.1016/j.aiepr.2019.11.002.

Chapter 5

Market, economic impact, and perspectives of eco-friendly polymers: biological, biodegradable, recyclable, and reusable polymers

5.1 An overview of the polymer market

The era of plastics emerged in the 1960s, revolutionizing the industry of materials, the market of multiple everyday products, and the global economy, due to the profitable price of oil and their multiple applications. In Chapter 4 it was stated that the development of polymeric materials was considered a gift to society, which dominated the market and presented an unprecedented success that has led to an annual growth of 250,000 kilotons per year, and it has been estimated that, at the end of this century, this value will increase to 1 million kilotons. However, society has been faced with rapid economic growth, the positioning of polymers as raw materials, the practical and technological developments derived from them, and improvements in the quality of life in many areas, but has ignored the environmental consequences that these polymers would generate half a century later, and even today, due to the accumulation of residues of these materials, their excessive consumption, and the bad practices of their use and disposal, which inevitably have led to serious problems of contamination of the oceans, aquifers, cities, and fields. It was not until 2 decades ago that growing social awareness of the preservation of petroleum resources, in conjunction with environmental concerns, generated the global production of biological and biodegradable polymers, presenting its significant participation in global markets, competing with conventional polymers, and positively impacting the economies of many countries (Babu et al., 2013; Kukoyi, 2016; Mangaraj et al., 2019; Prieto, 2016; Badia et al., 2017; Fagnani et al., 2021).

Biological materials, and those obtained from biological sources, have the advantage that they replace fossil resources, eliminating the reliance on limited

Eco-friendly Functional Polymers
https://doi.org/10.1016/B978-0-12-821842-6.00014-2
Copyright © 2021 Elsevier Inc. All rights reserved.

supplies such as oil and natural gas, and contributing to the development of a sustainable industry. Biodegradable polymers, on the other hand, solve another serious ecological problem: waste accumulation in the environment (CEBR, 2015; Nova Institute, 2020). The recycling of polymers, such as poly(ethylene terephthalate) (PET), has also emerged as a possible solution to these ecological problems, so that their production and market share have also increased in recent years (Geng et al., 2019; Markets and Markets, 2018).

Biopolymer global production has shown gradual growth in recent years. In 2013, the Centre for Economics and Business Research (CEBR) reported that global production of these biomaterials was 1.6 million tons (CEBR, 2015). In 2017, biopolymer production had reached 2.1 million tons (Niaounakis, 2019). Nova–Institute reported, in 2019, that the total production of polymers developed from biological precursors was 3.8 million tons, corresponding to a growth of 3% compared to 2018. It is important to note that Asia leads the research and development of biological and biodegradable polymers, producing 46% of all biopolymers worldwide (European Bioplastics and Nova Institute, 2020). Despite this promising growth, this value is equivalent to just 1% of total production of petroleum-based materials, therefore, biological polymers do not currently represent an effective solution to conserve oil resources (Babu et al., 2013; Nova Institute, 2020). The obtaining of biological polymers involves, in some cases, the large-scale production of monomers as raw material for the development of these materials, and in other cases, the extraction and purification of biomaterials. In 2018, the production rate of bio-monomers was 120,000 tons per year. The most used precursors are based on biogenic by-products, such as glycerol in the production of biodiesel, which is used to produce epoxy resins. Biomass, which is composed of biopolymers such as starch, cellulose, and fatty acids, is also used in the development of these materials (Nova Institute, 2020). In 2020, European Bioplastics reported that approximately 0.7 million hectares were earmarked for the cultivation of raw materials, such as plants, cereals, and tubers, from which biopolymers of market interest, such as starch and cellulose, are extracted (Ilyas and Sapuan, 2020; Zhong et al., 2020).

Nonbiodegradable biopolymers, such as poly(ethylene) (PE) and PET, obtained from biological sources, account for approximately 40% of global production of biological materials (European Bioplastics and Nova Institute, 2020). Although the participation of bio-PE and bio-PET in the global market enables the preservation of fossil resources, it promotes pollution because solid wastes are accumulated in the environment. In order to solve this problem, some governments have promoted a recycling culture, however, few materials are used after their first use. In 2009, up to 25 million tons of post-consumption waste was collected in Europe, increasing the environmental crisis by accumulating plastics in the environment (Ignatyev et al., 2014). In 2016, the Ellen MacArthur Foundation published a reporting stating that only 14% of polymers arranged as packaging material were recycled (Ellen

MacArthur Foundation, 2017). For this reason, it is currently of interest to produce biodegradable materials from biological sources. The Institute of Bioplastics and Biocomposites established that, between 2013 and 2015, the production of biodegradable polymers increased production from 1.6 to 2.0 million tons. In 2019, the production of these materials was 2.1 million tons (Filiciotto and Rothenberg, 2020). Thus, despite the growth in the production of these polymers, their implementation is limited by poor political support, mainly in underdeveloped countries with petroleum-dependent economies. If the promotion of biological and biodegradable polymers were equivalent to those of their fossil-based counterparts, their annual growth rate would be in the range of 10%–20% (Nova Institute, 2020).

This chapter aims to present details on the markets, economic impact, and prospects of biological, biodegradable, and recyclable polymers, with the objective of motivating future researchers and entrepreneurs to produce and use these materials, as an economically and ecologically sustainable alternative.

5.2 Ecofriendly polymer market

5.2.1 Biologic polymer market

It was previously stated that the production of polymers based on biological precursors accounted for only 1% of the total polymer and plastics market, which is integrated by brands aspiring to offer to consumers several sustainable eco-friendly alternatives. The biological polymer market has been driven by factors internal and external to these polymers. Internal factors are based on advances in the production of biological and biodegradable polymers, obtaining functional materials with optimal mechanical properties from renewable raw material; whereas external factors are caused by governments of developed countries, such as Italy, France, Spain, China, United States, etc., that have prohibited the use of nondegradable fossil-based materials in one or more sectors of their economies (Nova Institute, 2020). In Italy, for example, the materials industry based on biological polymers is a success due to policies and regulations against polymers from petroleum sources, which caused the growth of up to 150 companies that produce these biomaterials for use as shopping bags, coatings, and packaging, with total annual sales of 370 million euros. Some cities and states in the United States, such as San Francisco and California, also promoted biodegradable polymer markets by introducing legislation regulating and prohibiting nonbiodegradable plastic and transport bags (CEBR, 2015). These regulations, together with different initiatives such as the Lead Market Initiative in the European Union, and BioPreferred in the United States, have increased the demand for renewable resources to produce polymers to replace materials based on petroleum-derived precursors (Babu et al., 2013).

The main markets for biological polymers correspond to the fields of packaging, agriculture, medicine, electronics, textiles, etc. (European Bioplastics

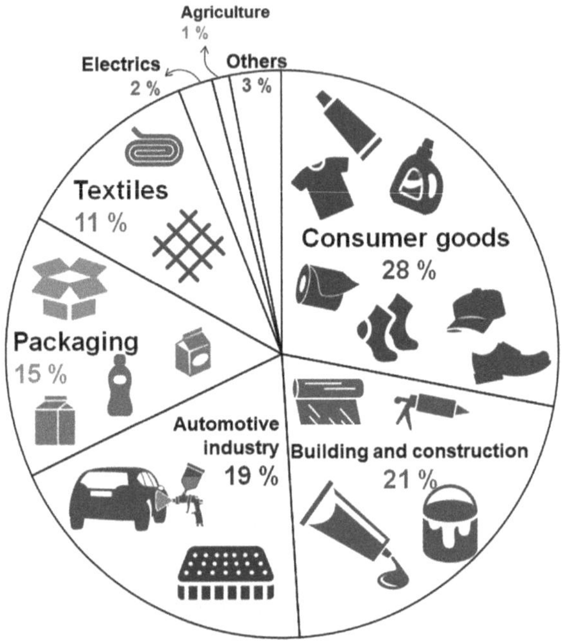

FIGURE 5.1 Main markets for biological polymers.

and Nova Institute, 2020; Kukoyi, 2016). The consumer goods market was demonstrated, in 2018, to be the largest market consumer of biological polymers (28%), including biological polymers such as cellulose, starches, polyamides, polyurethanes, etc. The construction sector, meanwhile, accounted for 21% of the market to be applied in construction adhesives, coatings, and paints, while the automotive industry accounted for 19% to be used as seat foams and automotive coatings. Biomaterials based on poly(lactic acid) (PLA) and starch are widely used in the packaging industry, which represents 15% of the market. North America and Europe account for 62% of the demand for polymers based on biological sources for use in the packaging sector (CEBR, 2015; Nova Institute, 2018). Fig. 5.1 illustrates the distribution of the main sectors of the market that consume biological polymers.

Some world market-leading producers of biological polymers include Ashland Inc. (United States), Economy Polymers and Chemicals (United States), NatureWorks LLC (United States), Novamont (Italy), GreenBio (China), and Kaneka Corporation (Japan) (Kukoyi, 2016). Novamont produces more than 120,000 tons per year of starch- and polyester-based biomaterials (Byun and Kim, 2014). NatureWorks LLC, for its part, focuses on PLA production for the packaging and bag market (Grujić et al., 2017). GreenBio built a production plant that produces 10,000 tons of polyhydroxyalkanoates (PHAs) per year for use in biocoatings, bags, films, among others (Modi et al., 2011). Coca-Cola is an international company that demands biological polymers such as bio-PET and PLA for its products (Byun and Kim, 2014; Grujić et al., 2017). In fact, Coca-Cola launched the Plantbottle, which is a bio-PET-based bottle, being an outstanding biopolymer on the market, followed by PLA, cellulose, and starch (Hahladakis et al., 2020; Hottle et al., 2017).

5.2.2 Biodegradable polymer markets

Biodegradable polymers also made a great contribution to the market. In 2010, the global market for biodegradable polymers reported a total of 206 million pounds, reaching an annual growth rate of 12.6% (Mangaraj et al., 2019; Prieto, 2016). It is estimated that in 2025 the global market for biodegradable polymers will reach USD 6.73 billion (Narancic et al., 2020). Asia leads the production of these materials, contributing 63.1%, followed by North America with 13.5%, Europe with 13.0%, and finally South America with 10.0%. The most biodegradable polymers desired on the market are PLA, aliphatic polyesters, starch, and PHA, whose production percentages are around 10.9%, 10.8%, 9.4%, and 3.6%, respectively. Their main market is the packaging industries, contributing 50% of the consumption of biodegradable polymers (Dilkes-Hoffman et al., 2019; Mangaraj et al., 2019). The impact of these polymers in the world economic environment is such that it has led to organizations such as the Bio-based Industries (BBI), in conjunction with the European Union, investing 3.7 billion into research projects promoting technologies to produce biodegradable polymers (RameshKumar et al., 2020).

A report by European Bioplastics, in collaboration with Nova–Institute, informed that the market for biodegradable polymers, such as PLA and PHA, has shown high growth in their production due to their commercial demand to replace conventional polymers such as PE, PET, and PP (European Bioplastics and Nova Institute, 2020; RameshKumar et al., 2020). In 2014, production of these materials reached around 200,000 tons for the packaging market to produce bottles and bags (Castro-Aguirre et al., 2016; Prieto, 2016), generating a global demand of up to 360,000 tons (Fortman et al., 2018; RameshKumar et al., 2020). In 2018, industrial PLA production was 218 kilotons (Zhao et al., 2020). Starch is another biodegradable polymer that has gained great popularity in the market for the development of materials in agriculture, packaging, and even in new markets such as the production of coffee capsules (Prieto, 2016). On the other hand, the current global production of PHA has reached 25,3000 tons, demonstrating the great interest in these polymers in the international market due to their optimal properties (RameshKumar et al., 2020). These are used in the agricultural, medical, and packaging industries. There are other emerging fields, such as the pharmaceutical area, which do not have a significant impact on the market because the application of biodegradable polymers in this area is still under investigation, however, the demand for these biomaterials in the pharmaceutical industry is expected to grow in the coming years (Kukoyi, 2016).

5.2.3 Recyclable polymer markets

Despite the above, the market for biological and biodegradable materials remains challenging due to poor political support in many developing countries, along with low and attractive oil prices (Nova Institute, 2020). In this sense, international companies such as Coca-Cola, Walmart, PepsiCo, Amcor Werner, etc., have chosen to use recyclable polymers such as PET. These companies are committed to the Plastic Recyclers Association in the United States, increasing the market for these materials by using 100% reusable packaging. Prestigious companies in the electrical and electronics industry such as HP, Dell, Samsung, Lenovo, and Sony have also chosen to use recyclable materials (Markets and Markets, 2018). However, polymers that have already been used are only employed again in limited applications, as there remains a safety concern about their contaminants, therefore even though these materials are reusable, they are rarely applied after their first use, so they accumulate in the environment, generating the same problem as their nonreusable counterparts (Markets and Markets, 2018). According to a report from Markets and Markets, there are few markets worldwide focused on the use of post-consumption polymers, such as Waste Management Inc., Republic Services Inc., SUEZ, etc. (Markets and Markets, 2020b).

5.3 Economic impact

In the global economy, plastics production positively affects gross domestic product, because people commonly use them in their daily lives. Despite the global economic crisis caused by the COVID-19 pandemic, markets of major powers are expected to have a positive increase in purchasing power parity ($ PPP). For example, in the US market, it is expected to increase from $ 44,000 PPP to $ 81,000 PPP, while China, India, and Indonesia estimate increases by a factor of five–six of a base of $14,000 PPP (Filiciotto and Rothenberg, 2020). This economic growth will inevitably increase the demand for plastics from petroleum waste, and so significant replacement of these materials with biodegradable polymers is not expected in the coming decade. In order to avoid a possible environmental crisis, the Scientific Advisory Council of the European Academies (EASAC) indicated that a possible solution would be to increase the price of materials based on conventional polymers. Yin and Yang established, in a recent review, that only the cost control of these polymers will be able to replace conventional polymers. To this end, it is important that there is no significant expense in the cost of monomers, and polymerization is optimized in terms of simplifying the process and reducing energy consumption (Filiciotto and Rothenberg, 2020; Yin and Yang, 2020).

Fortunately, pollution concerns have raised alarms in many countries. In the United States, prices of biological biodegradable polymers have fallen from USD 1000 per kilogram to a few dollars over the past 20 years, and so these are already at a price close to conventional materials (Haider et al., 2019). In this way, people can choose to use biological and biodegradable materials without being economically affected.

Global production of biological and biodegradable polymers, in conjunction with the growing market due to high demand, has generated a positive economic impact in many countries (see Table 5.1). In 2012, a report by Transparency Market Research described that demand for these polymers for that year was USD 4.95 billion (Kukoyi, 2016). CEBR reported that the production of biological polymers contributed USD 369 billion to the US economy. In addition, the industries of these materials generate four million jobs (CEBR, 2015). Other countries have also benefited economically from the polymer industry, such as Brazil, where the bio-based polymer market reached USD 1.7 billion in 2015, with a compound annual rate of 19.5% between 2010 and 2015, contributing to the country's economic growth (Kukoyi, 2016). The socioeconomic impact also has the potential to be positive in the coming decades. In 2013, the biomaterials industry created a total of 23,000 jobs. The European company, EuropaBio, developed an analysis in 2016 setting that, by 2030, this industrial sector could generate up to 300,000 highly skilled jobs (European Bioplastics and Nova-Institute, 2018).

TABLE 5.1 Estimated global market value of some biological and biodegradable polymers (EGM is estimated global market).

Polymer	Value of EGM (USD Million)	Years	References
Alginate	409.2	2019	Liu et al. (2019)
Chitosan	1205	2015	Brasselet et al. (2019), Chen et al. (2017)
CMC	1600	2017	Candido and Gonçalves (2019)
Collagen	599	2020	Markets and Markets (2020a)
PCL	113.9	2015	Leluk et al. (2018)
PGA	182.0–240.0	2014–2017	Samantaray et al. (2020)
PHA	73.6–215.0	2016–2020	Kourmentza et al. (2017), Morya et al. (2021)
PLA	304.9–698.0	2014–2017	Jabbar et al. (2020), Liu and Matuana (2019)

Although the medical and biomedical markets are not the largest, they are the sectors generating the most revenue. For example, the market for biological polymers for application in the orthopedic area generated revenues of USD 236.5 million in the United States. In the field of regenerative medicine, biomaterials such as chitosan, starch, collagen, etc., have generated sales rates such as USD 240 million per year (Kukoyi, 2016). Packaging has also generated million-dollar revenue, and it was estimated that, in 2013, the market in this field reached USD 597 billion (CEBR, 2015). The recyclable polymer market, on the other hand, was estimated at USD 36.9 million in 2017, with an estimated annual growth rate of 6.4%. To market-generated income, revenue from initiatives that promote the use of these materials should be added. For example, in the United Kingdom high taxes have been established on landfills, collecting large revenues that are subsequently earmarked for innovation projects for the development of biodegradable and reusable materials (Markets and Markets, 2018).

Despite the aspects set out in this section, the market for biological, biodegradable, and recyclable polymers remains limited because, in most applications, durability prevails more than biodegradability. If materials could adjust and control their biodegradability over time, they could replace approximately 50% of total fossil resource-based polymer production (Filiciotto and Rothenberg, 2020). Furthermore, the high cost of these materials restricts their production for disposal in the above applications. Ciriminna and Pagliaro collected a comment from an entrepreneur in India, who established that it was inconceivable to sell a fish worth Rs 20 in a biodegradable and biological bag that costs Rs 25 (Ciriminna and Pagliaro, 2019). This comment summarizes why, despite the multiple advantages of the biological and biodegradable polymers set out herein, they do not yet exceed their petroleum-derived counterparts due to their high cost of production and high cost to the consumer.

5.4 Perspectives

As laws emerge to regulate the use of petroleum resources, and society becomes more environmentally aware, the demand for biological and biodegradable polymers gradually increases. Biopolymer production in 2022 is expected to reach 2.4 million tons (Niaounakis, 2019; RameshKumar et al., 2020). Nova–Institute estimates that the production of biological polymers will have an annual growth rate of 4% until 2023. However, this rate is equivalent to the growth of polymers based on fossil sources, so the biopolymer market would remain at 2% (Nova Institute, 2018). Asia is expected to continue leading the production of biological polymers, followed by Europe. North America, for its part, is estimated to increase its share of global biopolymer production by 18% by 2025 (European Bioplastics and Nova-Institute, 2018). Regarding the market, CEBR estimates that, by 2023,

TABLE 5.2 Estimated global market perspective values of some biological and biodegradable polymers (EGM is estimated global market).

Polymer	Value of EGM (USD Million)	Projection year	References
Alginate	923.8	2025	Anderson et al. (2018), Ihua et al. (2020)
Chitosan	2550.0	2022	Brasselet et al. (2019)
CMC	2250.0	2024	Candido and Gonçalves (2019)
Collagen	795.0	2025	Markets and Markets (2020a)
PCL	214.5	2021	Leluk et al. (2018)
PGA	470.0	2024	Samantaray et al. (2020)
PHA	119.2	2024	Martínez-Martínez et al. (2019)
PLA	6500	2025	Shojaeiarani et al. (2020)

the value of the global packaging market will increase to USD 828 billion (CEBR, 2015). Table 5.2 shows the projections of some biological and biodegradable polymers.

The production pattern of biological polymers is also expected to change. Currently, approximately 57.1% of biological polymers are not biodegradable, meaning that 42.9% are biodegradable (Niaounakis, 2019). Considering the growing market for biodegradable polymers, European Bioplastics estimates that the production of nonbiodegradable biopolymers will decrease by 37% by 2025 (European Bioplastics and Nova-Institute, 2018), contributing to solving the problem of plastic accumulation. The global market for recyclable polymers also presents optimal expectations, estimating achieving USD 50.4 million in 2022, where PET is the fastest growing polymer, while the post-consumer materials market in 2025 is expected to be USD 18.8 million (Markets and Markets, 2018, 2020b).

5.5 Conclusions

Society is facing an ecological crisis generated by the exploitation of oil resources for the generation of materials, and the subsequent accumulation of these in the environment. In this sense, it is necessary to find alternatives to help solve this problem. A first solution consists in the replacement of these nondegradable materials from fossil sources with biological, biodegradable, and recyclable polymers. The environmental benefits they offer have led to an increase in demand around the world. This high demand is reflected in its high production, and its prominence in the world's largest markets.

On the other hand, the high production of biological and biodegradable polymers for eventual application in fields such as packaging, medicine, agriculture, etc., allows the gradual replacement of their nonbiodegradable counterparts based on fossil resources. The market, widely promoted by laws prohibiting conventional plastics, is a great mechanism that wants to offer consumers environmentally friendly products. As an additional benefit, the production of bioplastics has a positive impact on the developing countries to the extent that they enter this production line, the reason behind the high income generated by the high potential demand, which will undoubtedly contribute to the establishment of a sustainable economy. However, the main limitation of these materials is their extremely high cost, mainly in underdeveloped countries with oil-dependent economies. Therefore, a large part of society is unwilling to invest more money for more expensive materials, even if that has a positive environmental impact. Research should therefore focus on developing more economical and efficient polymerization methods, obtaining monomers and biopolymers, with the aim of reducing material costs. In this way, it will be possible to replace conventional polymers with polymers that will help prevent an environmentally deteriorated future.

Certainly, all the perspectives raised in this chapter are subject to change due to the crisis over the COVID-19 pandemic, as the social distancing and blockade of several countries around the world has prevented the growth of the plastics industry in general, affecting the market for biological, biodegradable, and recyclable polymers.

References

Anderson, L.A., Islam, M.A., Prather, K.L.J., 2018. Synthetic biology strategies for improving microbial synthesis of "green" biopolymers. J. Biol. Chem. 293, 5053–5061. https://doi.org/10.1074/jbc.TM117.000368.

Babu, R.P., O'Connor, K., Seeram, R., 2013. Current progress on bio-based polymers and their future trends. Prog. Biomater. 2, 1–16. https://doi.org/10.1186/2194-0517-2-8.

Badia, J.D., Gil-Castell, O., Ribes-Greus, A., 2017. Long-term Properties and End-of-Life of Polymers from Renewable Resources, vol. 137, pp. 35–57. https://doi.org/10.1016/j.polymdegradstab.2017.01.002.

Brasselet, C., Pierre, G., Dubessay, P., Dols-Lafargue, M., Coulon, J., Maupeu, J., Vallet-Courbin, A., de Baynast, H., Doco, T., Michaud, P., Delattre, C., 2019. Modification of chitosan for the generation of functional derivatives. Appl. Sci. 9, 1321. https://doi.org/10.3390/app9071321.

Byun, Y., Kim, Y.T., 2014. Utilization of bioplastics for food packaging industry. In: Han, J.H. (Ed.), Innovations in Food Packaging, second ed., pp. 369–390. https://doi.org/10.1016/B978-0-12-394601-0.00015-1.

Candido, R.G., Gonçalves, A.R., 2019. Evaluation of two different applications for cellulose isolated from sugarcane bagasse in a biorefinery concept. Ind. Crop. Prod. 142, 111616. https://doi.org/10.1016/j.indcrop.2019.111616.

Castro-Aguirre, E., Iñiguez-Franco, F., Samsudin, H., Fang, X., Auras, R., 2016. Poly(lactic acid)—mass production, processing, industrial applications, and end of life. Adv. Drug Deliv. Rev. 107, 333–366. https://doi.org/10.1016/j.addr.2016.03.010.

CEBR, 2015. The Future Potential Economic Impacts of a Bio-Plastics Industry in the UK, pp. 1–51. https://bbia.org.uk/wp-content/uploads/2015/11/BBIA-CEBR-Report.compressed.pdf.

Chen, X., Yang, H., Zhong, Z., Yan, N., 2017. Base-catalyzed, one-step mechanochemical conversion of chitin and shrimp shells into low molecular weight chitosan. Green Chem. 19, 2783–2792. https://doi.org/10.1039/C7GC00089H.

Ciriminna, R., Pagliaro, M., 2019. Biodegradable and compostable plastics: a critical perspective on the dawn of their global adoption. Chem. Open 9, 8–13. https://doi.org/10.1002/open.201900272.

Dilkes-Hoffman, L.S., Pratt, S., Laycock, L.B., 2019. The role of biodegradable plastic in solving plastic solid waste accumulation. In: Al-Salem, S.M. (Ed.), Plastics to Energy Fuel, Chemicals, and Sustainability Implications. William Andrew, Norwich, pp. 469–505. https://doi.org/10.1016/B978-0-12-813140-4.00019-4.

Ellen MacArthur Foundation, 2017. Plastics and the Circular Economy. https://www.ellenmacarthurfoundation.org/explore/plastics-and-the-circular-economy.

European Bioplastics, Nova Institute, 2020. Bioplastics Market Data. https://www.european-bioplastics.org/market/.

Fagnani, D.E., Tami, J.L., Copley, G., Clemons, M.N., Getzler, Y.D.Y.L., McNeil, A.J., 2021. 100th anniversary of macromolecular science viewpoint: redefining sustainable polymers. ACS Macro Lett. 10, 41–53. https://doi.org/10.1021/acsmacrolett.0c00789.

Filiciotto, L., Rothenberg, G., 2020. Biodegradable plastics: standards, policies, and impacts. ChemSusChem 13, 1–18. https://doi.org/10.1002/cssc.202002044.

Fortman, D.J., Brutman, J.P., De Hoe, G.X., Snyder, R.L., Dichtel, W.R., Hillmyer, M.A., 2018. Approaches to sustainable and continually recyclable cross-linked polymers. ACS Sustain. Chem. Eng. 6, 11145–11159. https://doi.org/10.1021/acssuschemeng.8b02355.

Geng, Y., Sarkis, J., Bleischwitz, R., 2019. How to globalize the circular economy. Nature 565, 153–155. https://doi.org/10.1038/d41586-019-00017-z.

Grujić, R., Vujadinović, D., Savanović, D., 2017. Biopolymers as food packaging materials. In: Pellicer, E., Nikolic, D., Sort, J., Baró, M., Zivic, F., Grujovic, N., Grujic, R., Pelemis, S. (Eds.), Advances in Applications of Industrial Biomaterials. Springer, Cham, pp. 139–160. https://doi.org/10.1007/978-3-319-62767-0_8.

Hahladakis, J.N., Iacovidou, E., Gerassimidou, S., 2020. Plastic waste in a circular economy. In: Letcher, T.M. (Ed.), Plastic Waste and Recycling - Environmental Impact, Societal Issues, Prevention, and Solutions, pp. 481–512. https://doi.org/10.1016/B978-0-12-817880-5.00019-0.

Haider, T.P., Völker, C., Kramm, J., Landfester, K., Wurm, F., 2019. Plastics of the future? The impact of biodegradable polymers on the environment and on society. Angew. Chem. Int. Ed. 58, 50–62. https://doi.org/10.1002/anie.201805766.

Hottle, T.A., Bilec, M.M., Landis, A.E., 2017. Biopolymer production and end of life comparisons using life cycle assessment. Resour. Conserv. Recycl. 122, 295–306. https://doi.org/10.1016/j.resconrec.2017.03.002.

Ignatyev, I.A., Thielemans, W., Beke, B.V., 2014. Recycling of polymers: a review. ChemSusChem 7, 1579–1593. https://doi.org/10.1002/cssc.201300898.

Ihua, M.W., FitzGerald, J.A., Guihéneuf, F., Jackson, S.A., Claesson, M.J., Stengel, D.B., Dobson, A.D.W., 2020. Diversity of bacteria populations associated with different thallus regions of the brown alga *Laminaria digitata*. PLoS ONE 15 (11). https://doi.org/10.1371/journal.pone.0242675. e0242675.

Ilyas, R.A., Sapuan, S.M., 2020. Biopolymers and biocomposites: chemistry and technology. Curr. Anal. Chem. 16, 500–503. https://doi.org/10.2174/157341101605200603095311.

Jabbar, A., Tausif, M., Tahir, H.R., Basit, A., Bhatti, M.R.A., Abbas, G., 2020. Polylactic acid/lyocell fibre as an eco-friendly alternative to polyethylene terephthalate/cotton fibre blended yarns and knitted fabrics. J. Textil. Inst. 111, 129–138. https://doi.org/10.1080/00405000.2019.1624070.

Kourmentza, C., Costa, J., Azevedo, Z., Servin, C., Grandfils, C., De Freitas, V., Reis, M.A.M., 2017. *Burkholderia thailandensis* as a microbial cell factory for the bioconversion of used cooking oil to polyhydroxyalkanoates and rhamnolipids. Bioresour. Technol. 247, 829–837. https://doi.org/10.1016/j.biortech.2017.09.138.

Kukoyi, A.R., 2016. Economic impacts of natural polymers. In: Olatunji, O. (Ed.), Natural Polymers. Springer, Cham, pp. 339–362. https://doi.org/10.1007/978-3-319-26414-1_11.

Leluk, K., Sobczyk, K., Borowczak, M., 2018. Polycaprolactone – new insight into classic material. E3S Web Conf. 44, 1–6. https://doi.org/10.1051/e3sconf/20184400094.

Liu, J., Yang, S., Li, X., Yan, Q., Reaney, M.J.T., Jiang, Z., 2019. Alginate oligosaccharides: production, biological activities, and potential applications. Compr. Rev. Food Sci. Food Saf. 18, 1–23. https://doi.org/10.1111/1541-4337.12494.

Liu, Y., Matuana, L.M., 2019. Surface texture and barrier performance of poly(lactic acid)–cellulose nanocrystal extruded-cast films. J. Appl. Polym. Sci. 136, 47594. https://doi.org/10.1002/app.47594.

Mangaraj, S., Yadav, A., Bal, L.M., Dash, S.K., Mahanti, N., 2019. Application of biodegradable polymers in food packaging industry: a comprehensive review. J. Packag. Technol. Res. 3, 77–96. https://doi.org/10.1007/s41783-018-0049-y.

Markets and Markets, 2020a. Collagen Peptides Market by Source (Bovine, Porcine, and Marine & Poultry), Form (Dry and Liquid), Application (Nutritional Products, Food & Beverages, Pharmaceuticals, and Cosmetics & Personal Care Products) and Region – Global Forecast to 2025. https://www.marketsandmarkets.com/Market-Reports/collagen-peptide-market-248590980.html.

Markets and Markets, 2020b. Post-consumer Recycled Plastics Market by Polymer Type, Service, Processing Type (Mechanical, Chemical, Biological), End-Use Application (Packaging, Building & Construction, Automotive, Electronics, Furniture, Others), and Region – Global Forecast to 2025. https://www.marketsandmarkets.com/Market-Reports/post-consumer-recycled-plastic-market-222347130.html.

Markets and Markets, 2018. Recycled Plastics Market by Source (Bottles, Films, Fibers, Foams), Type (PET, PE, PP, PVC, PS), End-Use Industry (Packaging, Building & Construction, Textiles, Automotive, Electrical & Electronics) and Region – Global Forecast to 2022. https://www.marketsandmarkets.com/Market-Reports/recycled-plastic-market-115486722.html.

Martínez-Martínez, M.D., González-Pedrajo, B., Dreyfus, G., Soto-Urzúa, L., Martínez-Morales, L.J., 2019. Phasin PhaP1 is involved in polyhydroxybutyrate granules morphology and in controlling early biopolymer accumulation in *Azospirillum brasilense* Sp7. Amb. Express 9, 1–15. https://doi.org/10.1186/s13568-019-0876-4.

Modi, S., Koelling, K., Vodovotz, Y., 2011. Assessment of PHB with varying hydroxyvalerate content for potential packaging applications. Eur. Polym. J. 47, 179–186. https://doi.org/10.1016/j.eurpolymj.2010.11.010.

Morya, R., Sharma, A., Kumar, M., Tyagi, B., Singh, S.S., Thakur, I.S., 2021. Polyhydroxyalkanoate Synthesis and Characterization: A Proteogenomic and Process Optimization Study for Biovalorization of Industrial Lignin, vol. 320, p. 124439. https://doi.org/10.1016/j.biortech.2020.124439.

Narancic, T., Cerrone, F., Beagan, N., O'Connor, K.E., 2020. Recent Adv. Bioplast. Appl. Biodegrad. 12, 920. https://doi.org/10.3390/polym12040920.

Niaounakis, M., 2019. Recycling of biopolymers – the patent perspective. Eur. Polym. J. 114, 464–475. https://doi.org/10.1016/j.eurpolymj.2019.02.027.

Nova Institute, 2018. Bio-based Building Blocks and Polymers – Global Capacities, Production and Trends 2018–2023, pp. 1–16. http://bio-based.eu/downloads/bio-based-building-blocks-and-polymers-global-capacities-and-trends-2016-2022/.

Nova Institute, 2020. The Global Bio-Based Polymer Market 2019 – A Revised View on a Turbulent and Growing Market, pp. 1–5. http://nova-institute.eu/press/?id=164.

Prieto, A., 2016. To be, or not to be biodegradable... that is the question for the bio-based plastics. Microb. Biotechnol. 9, 1–6. https://doi.org/10.1111/1751-7915.12393.

RameshKumar, S., Shaiju, P., O'Connor, K.E., Babu, P.R., 2020. Bio-based and biodegradable polymers – state-of-the-art, challenges and emerging trends. Curr. Opin. Green Sustain. Chem. 21, 75–81. https://doi.org/10.1016/j.cogsc.2019.12.005.

Samantaray, P.K., Little, A., Haddleton, D.M., McNally, T., Tan, B., Sun, Z., Huang, W., Ji, Y., Wan, C., 2020. Poly(glycolic acid) (PGA): a versatile building block expanding high performance and sustainable bioplastic applications. Green Chem. 22, 4055–4081. https://doi.org/10.1039/D0GC01394C.

Shojaeiarani, J., Bajwa, D.S., Stark, N.M., Bergholz, T.M., Kraft, A.L., 2020. Spin coating method improved the performance characteristics of films obtained from poly(lactic acid) and cellulose nanocrystals. Sustain. Mater. Technol. 26, e00212. https://doi.org/10.1016/j.susmat.2020.e00212.

Yin, G.Z., Yang, X.M., 2020. Biodegradable polymers: a cure for the planet, but a long way to go. J. Polym. Res. 27, 1–14. https://doi.org/10.1007/s10965-020-2004-1.

Zhao, L., Rong, L., Zhao, L., Yang, J., Wang, L., Sun, H., 2020. Plastics of the future? The impact of biodegradable polymers on the environment. In: He, D., Luo, Y. (Eds.), Microplastics in Terrestrial Environments. The Handbook of Environmental Chemistry, vol. 95. Springer, Cham, pp. 423–445. https://doi.org/10.1007/698_2020_462.

Zhong, Y., Godwin, P., Jin, Y., Xiao, H., 2020. Biodegradable polymers and green-based anti-microbial packaging materials: a mini-review. Adv. Ind. Eng. Polym. Res. 3, 27–35. https://doi.org/10.1016/j.aiepr.2019.11.002.

Part II

Theoretical and experimental aspects in the design of eco-friendly polymers

Chapter 6

Polymer biosynthesis and biotransformations

6.1 Biosynthesis and biotransformations for obtaining bio-based monomers and polymers

Biosynthesis is the chemical transformation of one or more substances in others through biological mechanisms. In general, biosynthesis is a process associated with the metabolism of living entities including animals, plants, fungi, and bacteria. As a consequence, biosynthetic processes imply an intercellular synthesis. On the other hand, a biotransformation like biosynthesis is a chemical transformation of one or more substances in others through biological mechanisms, but its nature is extracellular. Obviously, a biosynthetic process is a biotransformation occurring intracellularly, but not all biotransformations are biosynthetic processes. For instance, the release of a biomolecule which is a precursor or intermediate in a chemical reaction in a membrane bioreactor where microorganisms are in the feed and their excreted biomolecules are able to pass through the membrane are collected and used to carry out some chemical reaction in the permeate (see Fig. 6.1A) (Koller et al., 2010; Carlini et al., 2016).

On the other hand, bio-based monomers and polymers are biomolecules obtained through the action of living organisms or materials derived from them. As a consequence, a biopolymer is a bio-based polymer since it can be obtained from bio-based molecules provided that these are polymerizable or can be easily transformed into a polymerizable molecule; but also, they can be obtained directly from living organisms (Carlini et al., 2016; Andreeben and Steinbuchel et al., 2019; Jha and Kumar, 2019; Cotted et al., 2020; Liu et al., 2020; George et al., 2020). Thus, in principle, bio-based monomers and polymers are eco-friendly in respect of their origin, although this does not mean that they are eco-friendly when their uses or wastes are considered. The chemical structures of some biopolymers are shown in Fig. 6.1B.

Although different classifications of biopolymers are found in the literature, many are limited to polymers biosynthesized by microorganisms as a result of the metabolic processes inherent in them, while others focus on their origin and the type of living source from which they are derived. The drawback of the latter

Eco-friendly Functional Polymers
https://doi.org/10.1016/B978-0-12-821842-6.00029-4
Copyright © 2021 Elsevier Inc. All rights reserved.

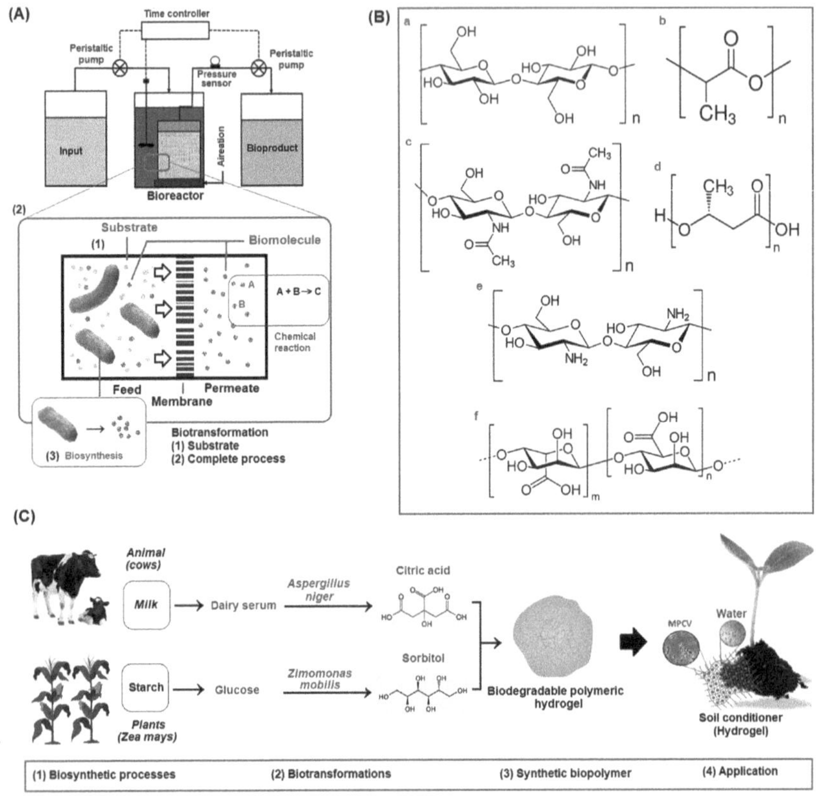

FIGURE 6.1 (A) Illustration of a membrane bioreactor: (1) biotransformation of substrate inside the reactor; (2) combination of biosynthesis and biotransformation in the complete processes; and (3) biosynthesis of precursor by bacterial metabolic activity. (B) Examples of biopolymers: (a) cellulose; (b) poly(lactic acid); (c) chitin; (d) polyhydroxyalcanoate; (e) chitosan; and (f) alginic acid. (C) Illustration of the complete route for obtaining biodegradable synthetic biopolymer: (1) biosynthetic processes from animals and plants; (2) biotransformations by fungi and bacteria to obtain biomolecules; (3) polymerization for making biodegradable hydrogels; and (4) application as a soil conditioner type hydrogel and bacterial inoculum.

approach is that the same biopolymers can be produced by completely different organisms. Biopolymers can be (1) biological biopolymers, which are biosynthesized by living organisms, (2) synthetic biopolymers, also described as bio-obtained, which are synthesized from biological precursors of low or high molecular weight, and (3) biotechnologically mediated biopolymers, which are a type of biopolymer based on extracellular biotransformations of one or more substrates, or whose bioproducts are the result of chemical modification of substrate by the action of living organisms and not as a direct result of their internal metabolism (see Fig. 6.1B and C; Younes, 2015; Pugazhendhi et al., 2018; Hassan et al., 2019; Acquavia et al., 2021).

A biological biopolymer is the result of a direct biosynthetic process carried out by a living organism, and is associated with a "natural" mechanism for establishing its internal components. This definition includes biopolymers synthesized inside living organisms that are later excreted as a result of biological processes. As a consequence, proteins, carbohydrates, and nucleic acids are typical biological polymers in animals, plants, fungi, insects, and bacteria. In the first case, proteins are biosynthesized biomolecules with multiple functions; they are present in all organisms and, although their functionality and structure are very varied and complex, all are the result of the linking of amino acids through amide-type bonds. Widely known examples are collagen, enzymes like urease, keratin, etc. (Gupta and Nayak, 2015; Nagarajan et al., 2019). In the specialized literature, "protein biopolymers" are usually referred to in very specific contexts, for example, collagen in fish characterization studies (Anaya et al., 2019), urease in applied microbiology studies due to its biological function in specific microorganisms (Palencia et al., 2016; Stave et al., 2017), unspecific proteins in microorganisms like adhesins (Palencia et al., 2016), and keratin for biomedical applications (Saha et al., 2019).

Carbohydrates, for their part, are biomolecules characterized by glycosidic bonds of carbohydrates. Their structural and functional variety are also very great, and they are also very common in nature. Some examples include starches, cellulose, and chitin, among others (Neri-Badang and Chakraborty, 2019; Kaur and Singh, 2020; Khademian et al., 2020). Finally, there are the nucleic acids, intra- and extracellular DNA and RNA. In this case, they are characterized for being biopolymers constituted from nucleotides, which in turn are made up of a nitrogenous base (e.g., cytosine, thymine, adenine, guanine, and uracil), a sugar, and a phosphodiester group. Furthermore, it is a supramolecular structure with two chains forming a helical system (Neffe et al., 2013; Zhang et al., 2020). In addition, other biopolymers should be included also: geo-biopolymers including humic and fulvic acids, which are described as being topic specific in specialized studies (García et al., 2018; Mora et al., 2018; Restrepo et al., 2018; Berrio et al., 2018), and condensation biopolymers with phenolic groups (Shavani et al., 2018). In addition, biological biopolymers can be intracellular (structural proteins, polyhydroxyalkanoates or PHA, etc.) (Anjum et al., 2016; Winnacker 2019) or extracellular (e.g., microbial polysaccharides and polyesters generated during the formation of biofilms, extracellular bacterial proteins, and DNA, etc.) (Espinosa et al., 2020).

Synthetic biopolymers can be the result of polymerization reactions of biomolecules (e.g., fatty acids, amino acids, amino alcohols, etc.) and bioprecursors (i.e., molecules without a direct biological function but which can be obtained as by-products of metabolic processes). In the first case, for example, both sorbitol and citric acid are biomolecules that can be polymerized to obtain biodegradable polyesters (Garcés et al., 2017). Note that sorbitol can be obtained from plants (e.g., *Sorbus aucuparia*), fungi (e.g., *Zymomonas mobilis*), and also by catalytic hydrogenation of glucose (Silveira and Jonas, 2002). Other microorganisms able to produce sorbitol are *Candida boidinii*, *Saccharomyces cerevisiae*, *Candida famata*, and *Z. mobilis* (Silveira and Jonas, 2002). Glucose is in turn obtained directly from plants and microorganisms by biotransformation of biopolymers such as cellulose or starches. On the other hand, citric acid can be extracted from citric fruits, but at an industrial level, it is usually produced by *Aspergillus niger* (Max et al., 2010). In the second case, cellulose, which is a biopolymer, can be acetylated to obtain cellulose acetate (Suzuki et al., 2020), while chitosan can be obtained from chitin through deacetylation reactions (Rizeq et al., 2019).

On the other hand, biotechnologically mediated biopolymers are obtained from bioprocesses designed for the control and use of enzymatic machinery; these are based on their action on specific substrates under specific conditions and not as a direct result of internal metabolism. Behind them there is usually extensive and rigorous optimization work on biotransformations. An example is the production of poly(lactic acid) by biofermentation of starches, other polysaccharides and biowastes (Lasprilla et al., 2012; Pleissner et al., 2017). It is important to indicate that, through biotechnological processes, beside biopolymers their precursors can be obtained. For example, the biocatalytic transformation of adipic acid into 1,6-hexamethylenediamine (HEDA), but also, microorganism-mediated production of HEDA, succinic acid, and others, which are used as chemical precursors in the synthesis of bio-based biopolymers (Nakajima et al., 2017). Other examples previously indicated include the production of citric acid and sorbitol by microorganisms (see Fig. 6.1C).

It should be kept in mind that, although biopolymers are a promising technology, the explosion of research into biomaterials converges with the exponential increase in research into nanomaterials, in particular, nanocomposites of polymers and biopolymers (Kalia et al., 2011). Thus, new research is focused toward the improvement of properties and functionality of polymers and biopolymers by inclusion of nanomaterials in the polymer phase using different methods (Kalia et al., 2011; Palencia et al., 2016; Hernandez et al., 2019; Villalobos et al., 2019).

6.2 Biosynthesis of biopolymers

Although they show generalities, biosynthetic processes are characterized by having a high degree of specificity in terms of the kingdom, gender, class, and species of organism. Even for microorganisms of the same species it is possible to identify significant biosynthetic differences, this being the case of bacteria. Therefore, it is not possible to show details of all the mechanisms, and to a certain extent, it would be meaningless given the enormous volume of specialized literature focused on specific details. In the case of animals, there are relatively few examples of biopolymers produced on a large scale, however, some examples of "protein biopolymer" are wool and collagen. On the other hand, at the plant level, the main biopolymer is cellulose, followed by xylan in terms of abundance. In addition, starches, natural rubber, and suberin are other typical examples. Unlike animals, genetic changes are more widely accepted in plants. Therefore, the genetic manipulation of plants for the genetically directed production of biopolymers, in particular polyhydroxyalkanoates, has been studied in some species such as cotton (*Gossypium* sp.), flax (*Linum usitatissimum*), poplar tree (*Populus* sp.), African oil palm (*Elaeis guineensis*), camelina (*Camelina sativa*), maize (*Zea mays*), switchgrass (*Panicum virgatum*), and sugarcane (*Saccharum officinarum*) (Dobrogojski et al., 2018).

In the case of insects, perhaps the most relevant biopolymer is chitin and its main derivative, chitosan: whose origin includes crustaceans, plants, animals and other species. However, fungi and bacteria are the main synthesizers of this biopolymer and other bioprecursors due to their ease of control, adaptability and powerful metabolic machinery and therefore this material is low cost and the time associated with its production is short (Younes, 2015; Pugazhendhi et al., 2018; Hassan et al., 2019; Acquavia et al., 2021).

In order to illustrate some aspects related to the biosynthesis of biopolymers with industrial uses, some important biological biopolymers are described below.

6.2.1 Cellulose biosynthesis

At the biosynthetic level, to achieve an accurate description of cellulose biosynthesis is required to identify the functions of numerous genes and their corresponding expressions, including biosynthesis of enzymes, proteins, and other biomolecules that are required. Thus, the main catalytic subunit for the polymerization of cellulose is cellulose synthase, which consists of two glycosyltransferases, which is characterized because it remains strongly linked to its polymer product, and it has been identified in plants, fungi, and bacteria (Brown and Saxena, 2000; McNamara et al., 2015; Polko and Kieber, 2019). However, although this enzyme is found in all living organisms capable of biosynthesizing cellulose, it is believed that, due to the evolutionary and adaptive differences of organisms, biosynthetic pathways can change significantly between different organisms and even between organisms of the same kingdom. Evidence of this is that regulatory proteins and the underlying mechanisms for cellulose synthesis show great differences between different types of organisms (Li et al., 2014). Thus, for example, in plants, cellulose biosynthesis is characterized by a strict process of control and regulation of the biosynthetic process. In this way, the amount of cellulose, the presence of other polymers, the degree of polymerization, the crystallinity, and the orientation of the cellulose microfibrils show a very high degree of reproducibility among individuals of the same species (Brown and Saxena, 2000; Li et al., 2014).

The catalytic mechanism for cellulose biosynthesis has been described as being governed by a SN_2-type substitution reaction, or bimolecular nucleophilic substitution, in which the nonreducing C4 hydroxyl group of the polymer (acceptor) acts as nucleophilic, and attacks the anomeric C1 carbon of the donor glucose. This reaction occurs more easily through the simultaneous deprotonation of the nucleophile and in the presence of Mg^{2+} and Mn^{2+}, which are responsible for stabilization of the pyrophosphate group of the substrate. In addition, during this process, the anomeric carbon configuration of the donor sugar is reversed (i.e., from α to β) and generates uridine diphosphate (UDP) as a leaving group (see Fig. 6.2A; McNamara et al., 2015).

On the other hand, bacterial cellulose can be produced by oxidative fermentation using microorganisms such as *Gluconacetobacter* and *Agrobacterium*. Although the molecular structure of bacterial cellulose is identical to that of cellulose from the originated plant, it is characterized by a higher degree of purity, a higher degree of polymerization, crystallinity, absorption and retention of water, and adaptability; and therefore, it is associated with greater biocompatibility and biodegradability (Costa et al., 2017). The biosynthetic pathway for bacterial cellulose production involves at least three stages: polymerization of glucose residues in the β1−4-glucan chain, the extracellular secretion of the chains, and their subsequent crystallization (i.e., formation of microfibrils through hydrogen bonds). In addition, multiple enzymes participate in this process, such as glucokinase, phosphoglucomutase, glucose-1-phosphate uridylyltransferase, and cellulose synthase (Lustri et al., 2015). However, it is important to reiterate that different routes are possible for the same microorganism, including exo- and endogenous sources (Lustri et al., 2015; Polko and Kieber, 2019).

6.2.2 Biosynthesis of starches

Starch is one of the most important energy reserves in plants, and one of the most promising biomaterials. This biopolymer is active in the regulation of plant growth, since it is a long-term carbon reserve which is translocated to support different phases of growth (Cai et al., 2014; Cornejo-Ramirez et al., 2018; Fan and Picchioni, 2020). Its biosynthetic mechanism has been extensively studied, and it is well known in comparison to the pathways of other biopolymers. This mechanism involves units of glucose-1-phosphate (Glu-1-P), adenosine triphosphate (ATP), adenosine diphosphate (ADP), pyrophosphate (PP) and enzymes ADP-Glc pyrophosphorylase (AGPase), starch synthases (SS) of which five classes have been described (SS I−IV are soluble and associated with amylopectin synthesis, whereas granule-bound SS is associated with the synthesis of amylose), isoamylase and pullulanase which are debranching enzymes and act by introducing branching points, and glucose (Wang et al., 2014; Geigenberger, 2011). The biosynthetic process begins with the conversion of Glu-1-P and ATP into ADP-Glu and PP, which is catalyzed by AGPase. The resulting ADP-Glu acts as a glucosyl donor for SS in order to

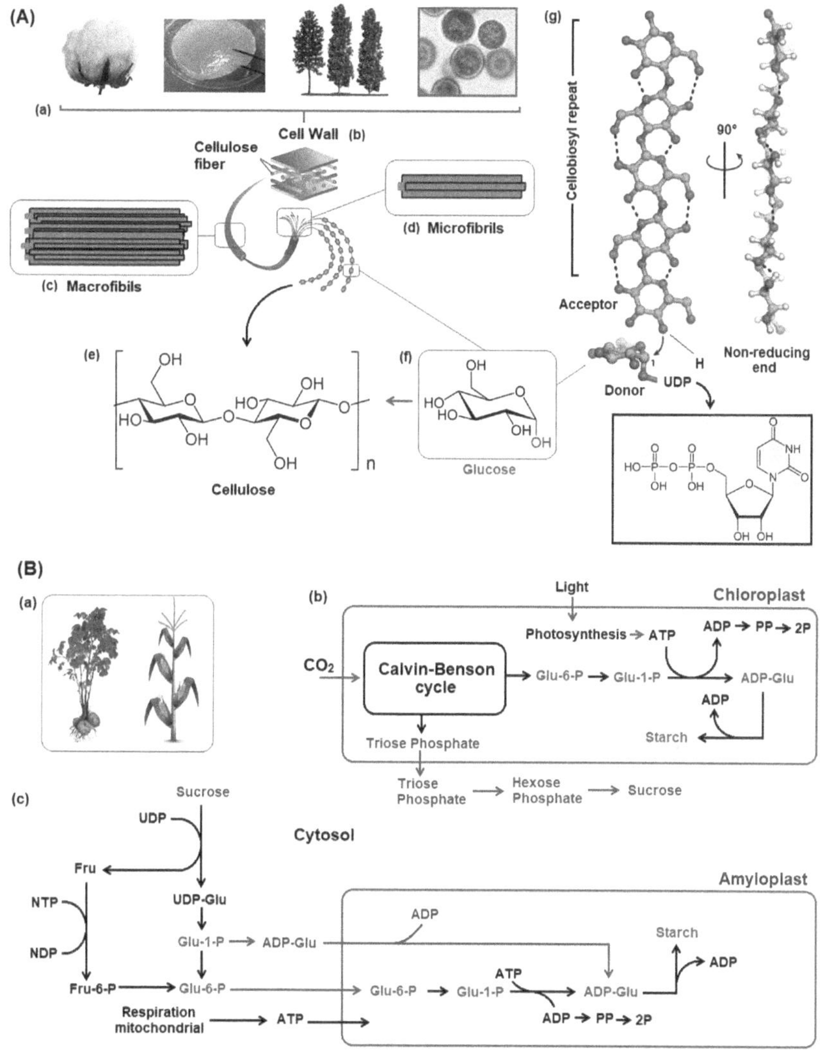

FIGURE 6.2 (A) Illustration of composition levels of cellulose from biological raw materials (a), cell wall (b), macrofibrils (c) microfibrils, (d) cellulose, and (f) monomer (glucose). Illustration of main steps of polymerization stage is shown in (g). (B) Illustration of starch biosynthetic pathways: (a) example of biological sources, (b) starch biosynthetic pathway in chloroplasts, and (c) starch biosynthetic pathway in amyloplasts.

cause elongation of glucan chains with α-1,4 bonds (Geigenberger, 2011). As in cellulose, different biosynthetic pathways are possible depending on the type of plant.

As for cellulose, different metabolic pathways have been described for starch biosynthesis, in particular, these depend on whether the biosynthetic process takes place in amyloplasts or chloroplasts. If starch is synthesized in chloroplasts then it is characterized by short-term storage and used during the day (photosynthetic pathways), whereas long-term storage is located in the amyloplasts (nonphotosynthetic pathways) (Orzechowski, 2008; Li et al., 2019). In the case of chloroplasts, starch biosynthesis is the end product of the metabolic pathway that is linked to the Calvin–Benson cycle for the production of Glu-6-P, its subsequent conversion to Glu-1-P, followed of ADP-Glu, and finally starch (see Fig. 6.2B); meanwhile, starch biosynthesis within amyloplasts is an alternate pathway based on the conversion of sucrose to starch where AGPase is the only source to obtain ADP-Glu (Bahaji et al., 2014; Subasinghe et al., 2014; Garcés and Chate-Galvis, 2020). Thus, outside amyloplasts, UDP-Glu and fructose (Fru) are obtained from sucrose. From UDP-Glu Glu-1-P is obtained, followed of Glu-6-P, which is biotransformed to Glu-1-P inside the amyloplasts, and subsequently it is transformed to ADP-Glu and finally to starch. In a parallel process, Fru is transformed to Fru-6-P by consuming nucleoside triphosphate (NDP), which is converted to nucleoside diphosphate (NDP). Fru-6-P is transformed to Glu-6-P in the cytosol entering amyloplasts and is transformed by Glu-6-P → Glu-1-P → ADP-Glu → starch (see Fig. 6.2B; Geigenberger, 2011; Garcés and Chate-Galvis, 2020).

6.2.3 Chitin biosynthesis

Chitin is a linear biopolymer which can be obtained from insects, crustaceans, mollusks, fungi, bacteria, and algae (Merzendorfer, 2006; Zargar et al., 2015). The structure of chitin is composed of (1–4)-linked 2-acetamido-2-deoxy-b-D-glucopyranose units and its modification by deacetylation produces chitosan. Biosynthetic mechanisms change in terms of biological entity associated with its production, however, generalized pathways can be defined for this biopolymer (Zargar et al., 2015). In insects and fungi, chitin biosynthesis is catalyzed by chitin synthases which are a family of glycosyltransferases (class I to class VII) and their function is directly related to the formation of a glycosidic bond from sugar moieties. In addition, other enzymes participate in the biosynthetic process such as cellulose synthases and hyaluronan synthases (Merzendorfer, 2006, 2011; Lenardon et al., 2010; Zargar et al., 2015).

The chitin biosynthetic pathway in insects begins with glycogen, which is a multibranched polysaccharide based on glucose. Through the action of glycogen phosphorylase Glu-1-P is produced. At this point, two pathways are possible, the first is directed toward the formation of Glu-6-P, while the second advances toward the formation of trehalose-6-phosphate, trehalose, and glucose in that order, and at this point, hexokinase acts on glucose to convert it to Glu-6-P, which is a convergence point between both pathways. Thus, the conversion from Glu-1-P to Glu-6-P is catalyzed by phosphoglucomutase and from Glu-6-P to Fru-6-P by Glu-6-P isomerase. The participation of glutamine, which is transformed into glutamic acid, mediates the production of glucosamine-6-phosphate

(GluAm-6-P). The corresponding GluAm-6-P N-acetyltransferase controls the step from GluAm-6-P to N-acetylglucosamine-6-P (N-AcGluAm-6-P). N-AcGluAm-6-P can also be obtained by recycling of chitin with the participation of chitinolytic enzymes, la formation of N-AcGluAm, and finally N-AcGluAm-6-P which is transformed to N-AcGluAm-1-P. At this point, UTP is consumed and PP is produced resulting in UDP-N-AcGluAm from N-AcGluAm-6-P and by action of chitin synthase chitin is produced (Merzendorfer, 2011; Muthukrishnan et al., 2012; Zargar et al., 2015).

6.2.4 Biosynthesis of polyhydroxyalkanoates (PHAs)

PHAs are biopolyesters from hydroxyalkanoic acids intracellularly synthesized by microorganisms such as bacteria, archaea and algae. Their functions are as carbon and energy sources. In general, the biosynthesis of PHA proceeds by condensation of the carboxyl group in the hydroxyalkanoic acid, the hydroxyl group of another molecule, for instance, a carbohydrate, to form an ester-type bond through the action of polymerase and depolymerase enzymes (Chen et al., 2016; Tarrahi et al., 2020; Garcés and Chate-Galvis, 2020).

More than 14 different metabolic pathways have been described for the synthesis of PHAs, however, three are the most relevant. The first begins with sugar transformation into acetyl-CoA, followed by bimolecular condensation of two acetyl-CoA molecules and formation of acetoacetyl-CoA by action of β-ketothiolase. Later, acetoacetyl-CoA is reduced to 3-hydroxybutyryl-CoA and finally the respective PHAs. In these steps, acetoacetyl-CoA reductase and PHA synthase act, respectively. The second pathway begins with fatty acids which are transformed into acetyl-CoA and used as a substrate for the β-oxidation cycle where three intermediate molecules are generated: 2-trans-enoylCoA, (S)-hydr-CoA, and ketoacyl-CoA. During these stages three enzymes act: 3-ketoacyl-ACP reductase, 3-hydroxyacyl-CoA epimerase, and enoyl-CoA hydratase, respectively. Finally, the third pathway uses any type of carbohydrate as substrate which is transformed to acetyl-CoA, and this molecule enters through a route known as "novo of fatty acids" (Sudesh et al., 2000; Agnew et al., 2013; Meng et al., 2014).

6.3 Biotransformations of biopolymers

In general, biotransformations can be subdivided into two groups or main types: (1) fermentations, which use natural carbon sources as substrates and enzymes for the synthesis of new products, such as alcohols, ketones, vitamins, antibiotics, or amino acids; and (2) synthetic biotransformations where the substrate is biochemically modified to obtain a new product, without this implying a degradative process (Rizzarelli and Degli, 2020). Note that, whereas fermentations can be defined to be biotechnological processes, synthetic biotransformations are bioengineering synthetic processes.

Examples of biotransformations can be very varied. As a consequence, depending on the purpose, biotransformations require a design in stages including pretreatment, enzymatic digestibility, fermentation, and other specific aspects of each process (Bilal et al., 2017). Typical pretreatments include: mechanical processing such as gridding or milling, thermal pretreatment at relatively high temperatures or by addition of some hot phase as hot water, chemical pretreatment usually directed toward acid or basic hydrolysis using alkali, inorganic acids, organic solvents, biological pretreatment usually mediated by fungi and bacteria, and enzymatic pretreatment which is based on enzymes such as purified and semipurified ligninolytic enzymes and fungal laccase (Bilal et al., 2017). Some examples of biotransformations are described below.

The biotransformation strategy called BioSwitch is directed toward the biotransformation of peptides. This method is based on enzymatic dephosphorylation of phosphothreonine residues by phosphatases. Using this approach, biotransformation of poly(ethylene oxide)−blockpeptide copolymers have been described (Kühnle and Börner, 2009). Enzymatic biotransformation has been used as an alternative to substitute chemical treatments during the modification of synthetic and natural fibers. Modifications of natural fibers have been described to be carried out through a fermentation bioreactor using *Acetobacter xylinum* (Pommet et al., 2008; Silva et al., 2010). Solid-state fermentation uses a solid substrate with a low moisture content for the production of microbial metabolites, biopolymers as poly(hydroxyalkanoates), and enzymes as cellulases, among others (Sindhu et al., 2015; Behera and Ray, 2016; Srivastava et al., 2019). In the field of energy production, biotransformation of lignocellulosic biomass has been carried out with the use of lignocellulolytic enzymes and involved the hydrolysis of polysaccharide biomass. Enzymes used include endoglucanases, cellobiohydrolases, and β-glucosidases (Adegboye et al., 2021). In addition, some microorganisms used for biofuel production are *Corynebacterium glutamicum*, *E. coli*, *Saccharomyces cerevisiae*, *Klebsiella pneumoniae*, *C. celluvorans*, *C. cellulolyticum*, and others (Naik et al., 2010; Adegboye et al., 2021). For this application, usually biopolymers are transformed in molecules with low molecular weight, such as 3-hydroxypropic acid, ethanol, isobutanol, 1,3-propanediol, and fatty alcohol, (Pollegioni et al., 2015; Adegboye et al., 2021).

6.4 Conclusions and remarks

Biosynthetic processes and biotransformations offer a strategy for carrying out chemical synthesis and modifications through biological processes. Consequently, the synthetic and industrial potential is enormous, being a research field with growing and marked development. In addition, due to the greater degree of environmental awareness, the existing problems at the levels of waste and pollution, the tendency to design eco-friendlier processes, and the relatively low cost offered by biotechnological methods, allow us to infer that growth in this field will continue to increase.

Another important aspect to highlight is the fact that the inclusion of new techniques and disciplines further strengthens the area of biosynthesis and biotransformation. New fields in this regard are genetic engineering, nanotechnology, and supramolecular chemistry. Perhaps, the main disadvantage of these strategies for the bio-obtaining of raw materials and specific products is the control of bioprocesses, and the adequate description and standardization of the conditions prior to their escalation. It must be taken into account that these conditions are both biotic and abiotic, and therefore, a relatively higher degree of difficulty is to be expected, although this does not mean that it is a limiting difficulty of these processes.

References

Acquavia, M.A., Pascale, R., Martelli, G., Bondoni, M., Bianco, G., 2021. Natural polymeric materials: a solution to plastic pollution from the agro-food sector. Polymers 13, 158. https://doi.org/10.3390/polym13010158.

Adegboye, M.F., Ojue, O.B., Talia, P.M., Oluranti, O., 2021. Bioprospecting of microbial strains for biofuel production: metabolic engineering, applications, and challenges. Biotechnol. Biofuels 14, 5. https://doi.org/10.1186/s13068-020-01853-2.

Agnew, D.E., Pfleger, B.F., 2013. Synthetic biology strategies for synthesizing polyhydroxyalkanoates from unrelated carbon sources. Chem. Eng. Sci. 103, 58–67. https://doi.org/10.1016/j.ces.2012.12.023.

Anaya, L.R., Libreros, K.H., Palencia, V.J., Atencio, V.J., Palencia, M., 2019. Mid-infrared spectral characterization of fish scales: "Bocachico" (Prochilodus magdalenae) by functionally-enhanced derivative spectroscopy (FEDS) - a methodological approach. J. Sci. Technol. Appl. 6, 28–39. https://doi.org/10.34294/j.jsta.19.6.39.

Andreeben, C., Steinbuchel, A., 2019. Recent developments in non-biodegradable biopolymers: precursors, production processes, and future perspectives. Appl. Microbiol. Biotechnol. 103, 143–157. https://doi.org/10.1007/s00253-018-9483-6.

Anjum, A., Zuber, M., Mahmood, K., Noreen, A., Anjum, N., Tabasum, S., 2016. Microbial production of polyhydroxyalkanoates (PHAs) and its copolymers: a review of recent advancements. Int. J. Biol. Macromol. 89, 161–174. https://doi.org/10.1016/j.ijbiomac.2016.04.069.

Bahaji, A., et al., 2014. Starch biosynthesis, its regulation and biotechnological approaches to improve crop yields. Biotechnol. Adv. 32 (1), 87–106. https://doi.org/10.1016/j.biotechadv.2013.06.006.

Behera, S.S., Ray, R.C., 2016. Solid state fermentation for production of microbial cellulases: recent advances and improvement strategies. Int. J. Biol. Macromol. 86, 656–669. https://doi.org/10.1016/j.ijbiomac.2015.10.090.

Berrio, M.E., Palencia, S.L., Lerma, T.A., Mora, M.A., 2018. Bacterial colonization modelling on soil particles: effect of humic acids on the formation of nitrogen-fixing bacteria biofilms. J. Sci. Technol. Appl. 5, 33–44. https://doi.org/10.34294/j.jsta.18.5.33.

Bilal, M., Asgher, M., Iqbal, H., Hu, H., Zhang, X., 2017. Biotransformation of lignocellulosic materials into value-added products—a review. Int. J. Biol. Macromol. 98, 447–458. https://doi.org/10.1016/j.ijbiomac.2017.01.133.

Brown, R.M., Saxena, I.M., 2000. Cellulose biosynthesis: a model for understanding the assembly of biopolymers. Plant Physiol. Biochem. 38, 57–67. https://doi.org/10.1016/S0981-9428(00)00168-6.

Cai, C., Zhao, L., Huang, J., Chen, Y., Wei, C., 2014. Morphology, structure and gelatinization properties of heterogeneous starch granules from high-amylose maize. Carbohydr. Polym. 102, 606–614. https://doi.org/10.1016/j.carbpol.2013.12.010.

Carlini, A.S., Adamiak, L., Gianneschi, N.C., 2016. Biosynthetic polymers as functional materials. Macromolecules 49, 4379–4394. https://doi.org/10.1021/acs.macromol.6b00439.

Chen, G.Q., Jiang, X.R., Guo, Y., 2016. Synthetic biology of microbes synthesizing polyhydroxyalkanoates (PHA). Synthet. Syst. Biotechnol. 1, 236–242. https://doi.org/10.1016/j.synbio.2016.09.006.

Cornejo-Ramirez, Y.I., Martínez-Cruz, O., Del Toro-Sánchez, C.L., Wong-Corral, F.J., Borboa-Flores, J., Cinco-Moroyaqui, F.J., 2018. The Structural Characteristics of Starches and Their Functional Properties, vol. 16, pp. 1003–1017. https://doi.org/10.1080/19476337.2018.1518343.

Costa, A., Almeida, F., Vinhas, G., Sarubbo, L., 2017. Production of bacterial cellulose by gluconacetobacter hansenii using corn steep liquor as nutrient sources. Front. Microbiol. 8 https://doi.org/10.3389/fmicb.2017.02027.

Cottet, C., Ramirez-Tapias, Y.A., Delgado, J.F., de la Osa, O., Salvay, A.G., Peltzer, M., 2020. Biobased materials from microbial biomass and its derivatives. Materials 13, 1263. https://doi.org/10.3390/ma13061263.

Dobrogojski, J., Spychalski, M., Lucinski, R., Borek, S., 2018. Transgenic plants as a source of polyhydroxyalkanoates. Acta Physiol. Plant. 40, 162. https://doi.org/10.1007/s11738-018-2742-4.

Espinosa, A., Palencia, S.L., García, A., Palencia, M., 2020. Polymicrobial Biofilms: fundamentals, diseases, and treatments - a review. J. Sci. Technol. Appl. 8, 4–54. https://doi.org/10.34294/j.jsta.20.8.54.

Fan, Y., Picchioni, F., 2020. Modification of starch: a review on the application of "green" solvents and controlled functionalization. Carbohydr. Polym. 241, 116350. https://doi.org/10.1016/j.carbpol.2020.116350.

Garcés, V., 2020. Chate-galvis N. Biosynthesis and biotransformations of biopolymers: starch and poly(hydroxyalkanoates). J. Sci. Technol. Appl. 9, 4–9. https://doi.org/10.34294/j.jsta.20.9.60.

Garcés, V., Palencia, M., Combatt, E., 2017. Development of bacterial inoculums based on biodegradable hydrogels for agricultural applications. J. Sci. Technol. Appl. 2, 13–23. https://doi.org/10.34294/j.jsta.17.2.11.

Garcia, A., Combatt, E., Palencia, M., 2018. Structural study of humin and its interaction with humic acids by Fourier-transform mid-infrared spectroscopy. J. Sci. Technol. Appl. 4, 28–39. https://doi.org/10.34294/j.jsta.18.4.28.

Geigenberger, P., 2011. Regulation of starch biosynthesis in response to a fluctuating environment. Plant Physiol. 155, 1566–1577. https://doi.org/10.1104/pp.110.170399.

George, A., Sanjay, M.R., Srisuk, R., Parameswaranpillai, J., Siengchin, S., 2020. A comprehensive review on chemical properties and applications of biopolymers and their composites. Int. J. Biol. Macromol. 154, 329–338. https://doi.org/10.1016/j.ijbiomac.2020.03.120.

Gupta, P., Nayak, K., 2015. Characteristics of protein-based biopolymer and its application. Polym. Eng. Sci. 55, 485–498. https://doi.org/10.1002/pen.23928.

Hassan, M., Bai, J., Dou, D.Q., 2019. Biopolymers; definition, classification and applications. Egypt. J. Chem. 62, 1725–1737. https://doi.org/10.21608/EJCHEM.2019.6967.1580.

Hernandez, F.E., Jaramillo, A.F., Medina, C., Flores, P., 2019. Effect of temperature on the mechanical and structural properties of vulcanized natural rubber: estimation of long-term polymer degradation. J. Sci. Technol. Appl. 6, 16–27. https://doi.org/10.34294/j.jsta.19.6.38.

Jha, A., Kumar, A., 2019. Biobased technologies for the efficient extraction of biopolymers from waste biomass. Bioproc. Biosyst. Eng. 42, 1893–1901. https://doi.org/10.1007/s00449-019-02199-2.

Kaur, T., Singh, R.P., 2020. Carbohydrate biopolymers: diversity, applications, and challenges. In: Singh, J., Meshram, V., Gupta, M. (Eds.), Bioactive Natural Products in Drug Discovery. Springer, pp. 517–545. https://doi.org/10.1007/978-981-15-1394-7_18.

Kalia, S., Dufresne, A., Mathew, B., Cherian, B.M., Kaith, B.S., Averous, L., Njuguna, J., Nassiopoulos, E., 2011. Cellulose-based bio- and nanocomposites: a review. Int. J. Polym. Sci. 837875. https://doi.org/10.1155/2011/837875.

Khademian, E., Saleih, E., Saneepur, H., Galiano, F., Figoli, A., 2020. A systematic review on carbohydrate biopolymers for adsorptive remediation of copper ions from aqueous environments-part A: classification and modification strategies. Sci. Total Environ. 738, 139829. https://doi.org/10.1016/j.scitotenv.2020.139829.

Koller, M., Salerno, A., De Sousa, M.M., Reiterer, A., Braunegg, G., 2010. Modern biotechnological polymer synthesis: a review. Food Technol. Biotechnol. 48, 255–269.

Kühnle, H., Börner, H., 2009. Biotransformation on polymer–peptide conjugates: a versatile tool to trigger microstructure formation. Angew. Chem. 48, 6431–6434. https://doi.org/10.1002/anie.200805768.

Lasprilla, A.J.R., Martínez, G.A.R., Lunelli, B.H., Jardini, A.L., Filho, R.M., 2012. Poly-lactic acid synthesis for application in biomedical devices — a review. Biotechnol. Adv. 30, 321–328. https://doi.org/10.1016/j.biotechadv.2011.06.019.

Lenardon, M., Munro, C.A., Gow, N.A., 2010. Chitin synthesis and fungal pathogenesis. Curr. Opin. Microbiol. 13, 416–423. https://doi.org/10.1016/j.mib.2010.05.002.

Li, S., Bashline, L., Lei, L., Gu, Y., 2014. Cellulose synthesis and its regulation. Arabidopsis Book 12, e0169. https://doi.org/10.1199/tab.0169.

Liu, C., Luan, P., Li, Q., Cheng, Z., Xiang, P., Liu, D., Hou, D., et al., 2020. Biopolymers derived from trees as sustainable multifunctional materials: a review. Adv. Mater. 2001654. https://doi.org/10.1002/adma.202001654.

Li, H., et al., 2019. Starch branching enzymes contributing to amylose and amylopectin fine structure in wheat. Carbohydr. Polym. 224, 115185. https://doi.org/10.1016/j.carbpol.2019.115185.

Lustri, W.R., Gomes de Oliveira, H., da Silva, H., Peres, M., Gutierrez, J., Tercjak, A., et al., 2015. Microbial Cellulose — Biosynthesis Mechanisms and Medical Applications. In: Poletto, M., Ornaghi, H.L. (Eds.), Cellulose - Fundamental Aspects and Current Trends. IntechOpen. https://doi.org/10.5772/61797.

Max, B., Salgado, J.M., Rodríguez, N., Cortés, S., Converti, A., Domínguez, J.M., 2010. Biotechnological production of citric acid. Braz. J. Microbiol. 41, 862–875. https://doi.org/10.1590/S1517-83822010000400005.

McNamara, J.T., Morgan, J.L.W., Zimmer, J., 2015. A molecular description of cellulose biosynthesis. Annu. Rev. Biochem. 84, 895–921. https://doi.org/10.1146/annurev-biochem-060614-033930.

Meng, D.C., Shen, R., Yao, H., Chen, J.C., Wu, Q., Chen, G.Q., 2014. Engineering the diversity of polyesters. Curr. Opin. Biotechnol. 29, 24–33. https://doi.org/10.1016/j.copbio.2014.02.013.

Merzendorfer, H., 2006. Insect chitin synthases: a review. J. Comp. Physiol. B 176, 1–15. https://doi.org/10.1007/s00360-005-0005-3.

Merzendorfer, H., 2011. The cellular basis of chitin synthesis in fungi and insects: common principles and differences. Eur. J. Cell Biol. 90, 759–769. https://doi.org/10.1016/j.ejcb.2011.04.014.

Mora, M., Palencia, M., Combatt, E., 2018. Effect of dissolved organic matter and humic substances on transport of ions and low molecular weight molecules by liquid membranes. J. Sci. Technol. Appl. 4, 40–48. https://doi.org/10.34294/j.jsta.18.4.29.

Muthukrishnan, S., Merzandorfer, H., Arakane, Y., Kramer, K.J., 2012. Chitin metabolism in insects. In: Gilbert, L.I. (Ed.), Insect Molecular Biology and Biochemistry, pp. 193–225. https://doi.org/10.1016/B978-0-12-384747-8.10007-8.

Nagarajan, S., Radhakrishnan, S., Narayana, S., Balme, S., Miele, P., Bechelany, M., 2019. Overview of protein-based biopolymers for biomedical application. Macromol. Chem. Phys. 220, 1900126. https://doi.org/10.1002/macp.201900126.

Naik, S.N., Goud, V.V., Rout, P.K., Dalai, A.K., 2010. Production of first and second generation biofuels: a comprehensive review. Renew. Sustain. Energy Rev. 14, 578–597. https://doi.org/10.1016/j.rser.2009.10.003.

Nakajima, H., Dijstra, P., Loos, K., 2017. The recent developments in biobased polymers toward general and engineering applications: polymers that are upgraded from biodegradable polymers, analogous to petroleum-derived polymers, and newly developed. Polymers 9, 523. https://doi.org/10.3390/polym9100523.

Neffe, A.T., Wischke, C., Racheva, M., Lendlein, A., 2013. Progress in biopolymer-based biomaterials and their application in controlled drug delivery. Expet Rev. Med. Dev. 10, 813–833. https://doi.org/10.1586/17434440.2013.839209.

Neri-Badang, M.C., Chakraborty, S., 2019. Carbohydrate polymers as controlled release devices for pesticides. J. Carbohydr. Chem. 38, 67–85. https://doi.org/10.1080/07328303.2019.1568449.

Orzechowski, S., 2008. Starch metabolism in leaves. Acta Biochim. Polonica 55 (3), 435–445.

Palencia, M., Berrio, M.E., Melendrez, M.F., 2016. Nanostructured polymer composites with potential applications into the storage of blood and hemoderivates. J. Sci. Technol. Appl. 1, 4–14. https://doi.org/10.34294/j.jsta.16.1.1.

Palencia, S.L., Lagos, G., García, A., 2016. Fungi-bacterium interactions: helycobacter pyroli - Candida albicans. J. Sci. Technol. Appl. 1, 53–69. https://doi.org/10.34294/j.jsta.16.1.5.

Pleissner, D., Semichelis, F., Mariano, S., Fiore, S., Michelle, I., Gutierrez, N., Schneider, R., Venus, J., 2017. Direct production of lactic acid based on simultaneous saccharification and fermentation of mixed restaurant food waste. J. Clean. Prod. 143, 615–623. https://doi.org/10.1016/j.jclepro.2016.12.065.

Polko, J.K., Kieber, J.J., 2019. The regulation of cellulose biosynthesis in plants. Plant Cell 31 (2), 282–296. https://doi.org/10.1105/tpc.18.00760.

Pollegioni, L., Tonin, F., Rosini, E., 2015. Lignin-degrading enzymes. FEBS J. 282, 1190–1213. https://doi.org/10.1111/febs.13224.

Pommet, M., Juntaro, J., Hang, J., Mantalaris, A., Lee, A.F., et al., 2008. Surface modification of natural fibers using bacteria: depositing bacterial cellulose onto natural fibers to create hierarchical fiber reinforced nanocomposites. Biomacromolecules 9, 1643–1651. https://doi.org/10.1021/bm800169g.

Pugazhendhi, A., Indira, K., Jacob, J.M., Mukesh, M., Kumar, G., 2018. Biopolymers: classification and applications. In: Ahmed, S., Kanchi, S., Kumar, G. (Eds.), Handbook of Biopolymers: Advances and Multifaceted Applications, pp. 1–22. https://doi.org/10.1201/9780429024757.

Restrepo, D., Palencia, M., Palencia, V.J., 2018. Study by attenuated total reflectance spectroscopy of structural changes of humified organic matter by chemical perturbations via alkaline dissolution. J. Sci. Technol. Appl. 4, 49–59. https://doi.org/10.34294/j.jsta.18.4.30.

Rizzarelli, P., Degli, F., 2020. Biodegradation of green polymer composites: laboratory procedures and standard test methods. In: Advanced Applications of Bio-Degradable Green Composites. Materials Research Forum LLC, pp. 1–44. https://doi.org/10.21741/9781644900659-1.

Rizeq, B.R., Younes, N.N., Rasool, K., Nasrallah, G.K., 2019. Synthesis, bioapplications, and toxicity evaluation of chitosan-based nanoparticles. Int. J. Mol. Sci. 20, 5776–5800. https://doi.org/10.3390/ijms20225776.

Saha, S., Arshad, M., Zubair, M., Ullah, A., 2019. Keratin as a biopolymer. In: Sharma, S., Kumar, A. (Eds.), Keratin as a Protein Biopolymer. Springer. https://doi.org/10.1007/978-3-030-02901-2_6163.

Shavani, a., Ahmed, A.E.D., Saeedi, P., Izafar, Z., Bekuit, A.A., Khademhosseini, A., 2018. Polyphenol uses in biomaterials engineering. Biomaterials 167, 91–106. https://doi.org/10.1016/j.biomaterials.2018.03.018.

Silva, C., Matama, T., Cavaco-Paulo, A., 2010. Biotransformation of synthetic fibers. In: Encyclopedia of Industrial Biotechnology: Bioprocess, Bioseparation, and Cell Technology. John Wiley & Sons. https://doi.org/10.1002/9780470054581.eib630.

Silveira, M.M., Jonas, R., 2002. The biotechnological production of sorbitol. Appl. Microbiol. Biotechnol. 59, 400–408. https://doi.org/10.1007/s00253-002-1046-0.

Sindhu, R., Pandey, A., Binod, P., 2015. Solid-state fermentation for the production of poly(-hydroxyalkanoates). Chem. Biochem. Eng. 29, 173–181. https://doi.org/10.15255/CABEQ.2014.2256.

Srivastava, N., Srivastava, M., Ramteke, P.W., Mishra, P.K., 2019. Solid-state fermentation strategy for microbial metabolites production: an overview. In: Kumar, V., Pandey, A. (Eds.), New and Future Developments in Microbial Biotechnology and Bioengineering, pp. 345–354. https://doi.org/10.1016/B978-0-444-63504-4.00023-2.

Stave, K., Berrio, M.E., Palencia, S.L., 2017. Methods of diagnosis of *Helicobacter pylori* in children: current status and prospects. J. Sci. Technol. Appl. 2, 24–43. https://doi.org/10.34294/j.jsta.17.2.12.

Subasinghe, R.M., et al., 2014. Multimeric states of starch phosphorylase determine protein–protein interactions with starch biosynthetic enzymes in amyloplasts. Plant Physiol. Biochem. 83, 168–179. https://doi.org/10.1016/j.plaphy.2014.07.016.

Sudesh, K., Abe, H., Doi, Y., 2000. Synthesis, structure and properties of polyhydroxyalkanoates: biological polyesters. Prog. Polym. Sci. 25 (10), 1503–1555. https://doi.org/10.1016/S0079-6700(00)00035-6.

Suzuki, S., Yada, R., Hamamo, Y., Wada, N., Takahashi, K., 2020. Green synthesis and fractionation of cellulose acetate by controlling the reactivity of polysaccharides in Sugarcane bagasse. ACS Sustain. Chem. Eng. 8, 9002–9008. https://doi.org/10.1021/acssuschemeng.0c01639.

Tarrahi, R., Fathi, Z., Ozgur, M., Doustkhah, E., Khataee, A., 2020. Polyhydroxyalkanoates (PHA): from production to nanoarchitecture. Int. J. Biol. Macromol. 146, 596–619. https://doi.org/10.1016/j.ijbiomac.2019.12.181.

Villalobos, R.J., Medina, C., Canales, C., Melendrez, M., Flores, P., 2019. Development of nanocomposite materials with thermostable matrix from nanoreinforcement of titanium dioxide (TNPS and TNTS) and zinc oxide (ZNPS and ZNBS) on epoxy resin matrix. J. Sci. Technol. Appl. 6, 65–78. https://doi.org/10.34294/j.jsta.19.6.42.

Wang, K., Henry, R.J., Gilbert, R.G., 2014. Causal relations among starch biosynthesis, structure, and properties. Springer Sci. Rev. 2, 15–33. https://doi.org/10.1007/s40362-014-0016-0.

Winnacker, M., 2019. Polyhydroxyalkanoates: recent advances in their synthesis and applications. Eur. J. Lipid Sci. Technol. 121, 1900101. https://doi.org/10.1002/ejlt.201900101.

Younes, B., 2015. Classification, characterization, and the production processes of biopolymers used in the textiles industry. J. Textil. Inst. 108, 674–682. https://doi.org/10.1080/00405000.2016.1180731.

Zagar, V., Asghari, M., Dashti, A., 2015. A review on chitin and chitosan polymers: structure, chemistry, solubility, derivatives, and applications. ChemBioEng Rev. 2, 204–226. https://doi.org/10.1002/cben.201400025.

Zhang, L., Yang, Y., Tan, J., Yuan, Q., 2020. Chemically modified nucleic acid biopolymers used in biosensing. Mater. Chem. Front. 4, 1315–1327. https://doi.org/10.1039/D0QM00026D.

Chapter 7

Eco-friendly composites and nanocomposites

7.1 Composites and nanocomposites

Over the past century, thanks to progressive technological developments, there has been a steady improvement in the quality of life for most of humanity. Inherently, this development requires the design of new materials that make up for the deficiencies or overcome the limitations of those conventional materials; this innovation in advanced materials technology includes metallic alloys, ceramics, polymeric materials, nanomaterials, composite materials, etc. Thus, for instance, the aeronautical industry is immersed in a constant search for new and novel structural materials with low densities, and high mechanical and chemical resistance, or in other words, light, strong, and corrosion-resistant materials (Sapuan, 2017). In this way, all those industries characterized as carrying out technological developments in their activities such as aerospace, transportation, construction, maritime transport, sports, and infrastructure, generally have as a common denominator the constant development of new materials. One of the most important advances in the development of new materials arises from the synergy resulting from the combination of two or more materials. The materialization of this premise has given rise to so-called composite materials, or simply composites. Thus, today, composites have replaced many conventional materials, thanks to the excellent combinations of properties in a single material. Therefore, it is now common to observe composite materials in everyday life (Kumar, 2019).

A composite material is defined as a multicomponent material, that is, a material made up of two or more chemically different materials, producing a new material with unique, new, or improved properties and characteristics compared with those exhibited separately by its constituent materials (Cardarelli, 2018). It is important to note that, among the community in the area of composite materials, it is common to use the term "phase" instead of "component" This is due to the term "phase" being used to define those regions of a material's space where physicochemical properties are homogeneous. Therefore, the use of the term "phase" does not imply that they must have a different chemical nature, and so their use to refer to composites is

Eco-friendly Functional Polymers
https://doi.org/10.1016/B978-0-12-821842-6.00019-1
Copyright © 2021 Elsevier Inc. All rights reserved.

incorrect (Farikov, 2015). Consequently, the appropriate term to refer to composite materials is "component" which allows them to be differentiated from those materials called "homocomposites" or "single-phase composite materials" These multiphase materials are characterized by being single-component materials, that is, materials in which their constituents have the same chemical nature, however, they differ in the phase that they are in. An example of the above is single-polymer materials, in which their constituents differ from each other based on their polymorphic phases (Matabola et al., 2009; Liu et al., 2012; Karger-kocsis and Barany, 2014).

In general, composites are made up of two or more materials with different chemical compositions, which are combined and processed according to their proportions (concentration), size, or shape (particles, fibers, sheets, etc.). Commonly, for the description of composites, the terms colloid chemistry, continuous phase, and disperse phase, have been extrapolated to refer to the components of the material. In this way, the term continuous phase is associated with the "matrix" or component in a greater proportion in the material and the term dispersed phase is associated with the "reinforcement phase" that is, those components in lower proportions in the material (Agrawal et al., 2011). The use of the term "reinforcement phase" appears due to the fact that, for the making of composites, a small-size dispersed phase is used for the reinforcement of the matrix. This improving is usually based on the increase of mechical and thermal properties; however, it is important to note that these are not the only characteristics improved in the material. Thus, for example, the purpose of investigations in functional composite materials is the development of materials that not only have excellent improved mechanical properties, but also other types of improved properties, such as, electrical, thermal, magnetic, optical, etc. (Chung, 2019). The continuous phase or matrix is also called the binder material, as it surrounds and binds the reinforcing material or filling material, which is presented in the form of sheets, particles, and/or fibers, of natural or synthetic origin. Among the materials that are most commonly used as matrices for the development of composites are polymers, metals, and ceramics (Borodulin et al., 2019).

In particular, when the reinforcing material in a composite is processed and incorporated on a nanometric scale, that is, it has at least one, two, or three of its dimensions in the nanometer range (<100 nm), it is known as "nanocomposite" In this way, the main advantage of the nanometric size presented by the filler material in nanocomposites is improvements in the mechanical and optical properties exhibited by nanocomposites compared to their conventional counterparts (Safdari and Al-Haik, 2018). However, reducing the dimensions of the reinforcement material to a nanometric level leads to an improvement in the compatibilization between the components of the system, improving their interactions, and the properties of materials. These new characteristics are a consequence of the high surface area/volume ratio of the reinforcing nanoparticles used during the preparation of the nanocomposites.

In addition, it is important to note that not only the size, but also the shape of nanometric systems, for example, nanoparticles, nanosheets, nanotubes, etc., which are incorporated into nanocomposites is directly involved in understanding their structure−property relationships (Omanović-Mikličanin et al., 2020). Among these nanocomposites are those of the polymeric matrix that are composed of a polymer and a nanoscale reinforcing material; among the most commonly used reinforcing materials are clays, carbon nanofibers, nanosilica, and aluminum oxide and titanium nanoparticles. Thus, research into the development of nanocomposites based on clays and polymers has recently been through rapid growth, thanks to the naturally nanometric sizes and high natural abundance of clays, which allow hybrid nanocomposites to be obtained from widely available raw materials (Lerma, 2018; Mukhopadhyay et al., 2020).

7.2 Classification of composites

Composites can be classified according to the origin, type of matrix, or the characteristics, geometry, and size of the reinforcement material (Fig. 7.1). The most common classification criteria are described below.

7.2.1 Classification based on the type of matrix

Polymeric matrix composites: Polymeric matrix compounds are characterized by being composed of thermosetting or thermoplastic polymers with dispersed reinforcing materials, for example, glass, carbon, metallic fibers, natural fibers, metallic particles, etc. (Sathishkumar et al., 2014). Thermoset polymers, based on polyester, epoxy, and phenolic resins, have been widely used in the development of composites thanks to their excellent mechanical and thermal properties, as a result of the high degree of cross-linking of their polymeric chains (Hamidi and Altan, 2018; Raquez et al., 2010). Examples of thermoset polymers used as matrices in composites are epoxy and phenolic resins, unsaturated polyesters, polyimides, polyesters, polyurethane, silicones, and vinyl esters. Some examples of thermoplastic polymers used as matrices are polystyrene, polycarbonate, polyacetal, polyamides, polyacrylates, polyethylene terephthalate, polybutylene terephthalate, polyethylene, poly(vinyl chloride), and polypropylene (Supian et al., 2018; Deng et al., 2015; Chaudhary and Ahmad, 2020).

Ceramic matrix composites: Ceramic matrix composites are heterogeneous materials consisting of a ceramic matrix commonly reinforced with carbon fibers, silicon carbide, aluminum oxide, or silicon nitride. These ceramic matrix composites are developed with the purpose of obtaining materials with excellent properties such as low density, high mechanical, electrical and thermal resistance, as well as high chemical resistance (May et al., 2020). The properties of ceramic matrix composites depend on several factors, including

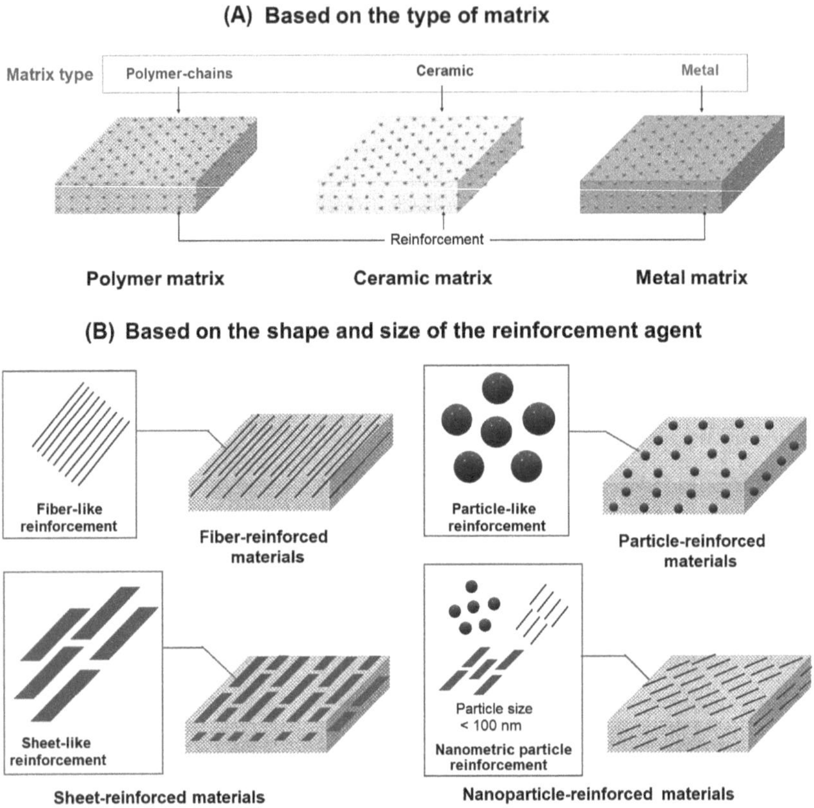

FIGURE 7.1 Illustration of composites according to matrix type (A) and type of reinforcement agent (B).

the type of manufacturing method. Commonly, chemical vapor infiltration, polymer infiltration, and pyrolysis methods are widely used for the manufacture of long-fiber-reinforced ceramic matrix composites, in which the ceramic matrix is formed from a fluid infiltrated into the structure of the fibers. The infiltration techniques differ from each other in the types of fluids and the processes of conversion of the fluid into ceramic, which influences the properties of the composite material obtained (Kopeliovich, 2018).

Metal matrix composites: Metal matrix compounds are characterized by having a metal matrix, for example, aluminum, magnesium, copper, or titanium, reinforced with other metals, ceramics, or other material. Given their excellent properties, these materials are widely used in the space, aeronautical, and transportation industries. Thanks to their low densities, high rigidity, adjustable thermal expansion coefficients, and relatively low production costs, the use of metal matrix composites is attractive in a number of application fields (Chawla, 2012). Ceramic matrix composites, depending on the shape of the reinforcing material, can be classified as continuously reinforced composites, to refer to those in which the reinforcing material is in form of fibers, and discontinuously reinforced composites, to refer to those composites of metal matrix reinforced with particles. In this way, the form of reinforcement to be used will depend on the type of material and application to be obtained, since in these systems the use of discontinuous or continuous reinforcements will favor the isotropy or anisotropy of the material, respectively. Thus, fiber-reinforced composite materials are isotropic materials characterized by having high specific strength, high rigidity, and excellent fatigue properties. Meanwhile, particle-reinforced metal matrix composites are isotropic materials with high rigidity (May et al., 2020).

7.2.2 Classification according to the shape and size of the reinforcement agent

Fiber-reinforced composite materials: In fiber-reinforced materials, their dispersed or reinforcing phase has a structure in the form of fibers, commonly of synthetic origin, for example, glass, carbon, Kevlar, etc. However, at present, it is natural fibers such as flax, hemp, jute, sisal, etc. that dominate the industry, thanks to their low densities, excellent mechanical properties, low cost, and renewable natural sources (Fuqua et al., 2012). However, one of the main disadvantages of the use of natural fibers as reinforcing agents for composite materials is the low phases compatibilization between the fiber and the polymer matrix. In addition, the high hydrophilicity presented by this type of fiber causes materials with a relatively high rate of moisture absorption. Therefore, it is necessary to carry out physical and chemical modifications to the natural fibers that generate a change in their surface properties in order to improve the adherence between the different matrices (Faruk et al., 2014).

Particle-reinforced composites: The inclusion of small inorganic objects to reinforce organic or inorganic matrices has been practiced for decades, with one of the earliest examples representing a composite material with particles being concrete. Concrete is a material widely used for construction, and is a composite material made up of a reinforcing or filling material, with coarse rock or gravel being commonly used, which is incorporated and dispersed in a cement matrix. Here, the reinforcing material provides rigidity and strength, while the cement matrix acts as a binder that holds the structure together (Yang et al., 2017). In this way, currently, in response to the limitations presented by conventional materials, for example, low rigidity and low resistance, the micro- and nanoparticles of silicon oxide, glass, aluminum oxide, crystalline cellulose, chitosan, etc., are widely used as reinforcement materials that modify and improve the physical and mechanical properties of the materials (Casati and Vedani, 2014). In particular, in particle-reinforced composites, it is important to note that both the geometry and size of the particles play important roles in determining the properties of the material. Likewise, the degree of dispersion and the distribution of the reinforcement particles in the matrix determine and modulate its properties, for example, mechanical strength, wear resistance, and thermal stability. In addition, among the main advantages of using particle-reinforced composites are their low cost and their easy procurement and composite formation (Fu et al., 2019).

Sheet-reinforced composite: In the field of composite materials, the term "sheet" is commonly used to refer to composite materials reinforced with glass and carbon fibers molded into sheets. These laminated composite materials have reinforcing material fibers with specific orientations in order to obtain lightweight and high-performance structural materials, however, they do not satisfy the condition of being composites in which the reinforcement material has a geometry in the form of sheets (Liu et al., 2012; Kumar, 2019). A typical example of a sheet-reinforced composite is clay–polymer hybrid composites. The use of clays has gained particular interest due to their intrinsic geometry in the form of sheets that have dimensions on the nanometric scales, and natural abundance thanks to being one of the main constituents of soil. These composites consist of a dispersed phase of clay mineral sheets, commonly obtained through physical exfoliation processes, which are incorporated into polymeric matrices in order to improve their mechanical, thermal, and functional properties (Palencia et al., 2017; Buruga et al., 2019). Likewise, polymeric nanocomposites reinforced with graphite nanosheets have been widely investigated and used as structural materials due to their markedly superior mechanical and thermal properties compared to conventional structural composites. As in clay-based composites, the most important stage in the preparation of these composites is the cleavage of coarse graphite crystals into stacks of finer, exfoliated graphene sheets (Shan et al., 2015; Szabó et al., 2005).

Nanocomposites: The term nanocomposite is used to refer to composite materials of two or more components where at least one of these has one, two, or three of its dimensions in a nanometric scale, that is, the reinforcing material is at the nanoscale (<100 nm). The geometries of nanomaterials commonly used in nanocomposite development include nanoparticles, nanofibers, nanosheets, etc. (Omanović-Mikličanin et al., 2020; Palencia et al., 2018; Villalobos et al.,2019a, 2019b). Nanocomposites show unique properties due to the effect of their nanometric size compared to conventional compounds, even with a low filler content. In fact, studies have also shown that nanofillers have strong reinforcing effects, producing a positive impact on mechanical, surface, and barrier properties. Consequently, polymer nanocomposites offer significant potential in the development of advanced materials for numerous applications. These new materials benefit from the synergy between filler particles and polymer chains that have similar length scales and a large interfacial area relative to the material volume (Chen et al., 2021; Hassan and Hu, 2020).

In general, the available synthesis methods of nanocomposites can be classified as ex situ or in situ methods. Ex situ methods comprise all techniques by which the reinforcing nanoparticulate material is formed outside the matrix, that is, the nanoparticles are prefabricated and incorporated directly into the matrix of the composite. Therefore, the ex situ manufacture of nanoparticulate material allows control of its size, shape, and density. However, one of the main challenges and limitations of this method is the achievement of adequate reinforcing material dispersion through the matrix and the compatibility between the phases to avoid their segregation (Zhao et al., 2011). In contrast, in situ methods, as the name implies, involve the formation and growth of nanoparticulate material within the matrix of the composite. These in situ methods stand out mainly for being simple techniques, and in many cases, they are described as one-pot synthesis, which allows the growth of the nanoparticles within the matrix, minimizing their possible agglomeration and maintaining a good spatial distribution. However, in in situ methods, reaction compounds or by-products generated remain in the material, compromising the properties and applications of the nanocomposite prepared when these residues cannot be easily removed (Umoren and Solomon, 2019).

7.3 Eco-friendly composites and nanocomposites

Over the last few decades, composites and nanocomposites have widely replaced conventional materials, and their markets are growing continuously. In particular, it is widely known that the use of a polymer and one (or more) reinforcing materials makes it possible to obtain novel materials with new properties and, in particular, the ability to modulate and combine the constituent materials' properties. Among the main composites and nanocomposites are those based on polymers, that is, a composite material in which the matrix or continuous phase is a natural or synthetic polymer (La Mantia and Morreale, 2011). However, given the structural and compositional complexity of some of these composite materials, their reuse and recycling once they have complete their life cycle may be a complicated task, resulting in some cases their requiring direct disposal or incineration, which implies high energetic costs and a negative environmental impact (Utekar et al., 2021). In addition, many polymeric composite systems are based on synthetic and nonbiodegradable polymer matrices derived from fossil fuels (nonrenewable sources), which increases their environmental impact as a consequence of their low degradability and high persistence (Ahmed et al., 2018).

Due to the above, researchers have focused their studies on finding materials from sustainable and ecological sources. Thus, nature has become the focus of attention as an abundant source of renewable and biodegradable materials, which comply with the guidelines of sustainability, eco-efficiency, and green chemistry for the development of novel materials, products, and processes (Abdul et al., 2012). Some examples of applications of these materials include in food packaging (Di Filipo et al., 2021), biomedical devices

(Palencia et al., 2016; Subhedar et al., 2021), for the making of drug-delivery systems (Pourmoazzen et al., 2020), environmental applications (Lerma et al., 2020), making of structural materials and coatings (Villalobos et al., 2019a), etc. In this way, the term "eco-friendly" has been coined to refer to those composite and nanocomposite materials in which all the constituent materials of the system, matrix, and reinforcement are obtained from renewable, sustainable, and biodegradable natural sources, thus giving rise to eco-friendly composites and nanocomposites (Singh et al., 2020). It is important not to confuse the term "biodegradable" with those systems consisting only of at least one component of natural origin, for example, polyethylene and nanocrystalline cellulose nanocomposites (Sapkota et al., 302AD). From a biodegradability approach, to consider a compound as eco-friendly is expected that all its constituents are characterized to be completely biodegradable and, ideally, they can be obtained from renewable sources. However, they also can be defined as eco-friendly materials from a functional or sustainable approach; by instance, when materials are used in environmental applications, or are characterized to show recyclability, low energy consumption, come from a sustainable processes, which consider their manufacture, use, and even their final disposal.

7.4 Natural polymer matrices and reinforcements used in eco-friendly composites

At present, as a consequence of the accumulation of plastics, the depletion of nonrenewable resources such as petroleum, and the growing concern about environmental problems associated with the use of unsustainable processes and materials, the development of environmentally friendly composite materials has received widespread attention (Yin et al., 2020). In this way, for the development of eco-friendly composites and nanocomposites, nanoparticles or fibers with natural origins are commonly used as reinforcing agents for bio-based polymer matrices, that is, polymeric matrices that are obtained naturally or synthetically from monomers of natural origin or polymers directly extracted from natural sources (Zini and Scandola, 2011). An example of the development of eco-friendly compounds is observed in adobe constructions, a brick made from clay and straw, which is dried in the sun and used for the construction of houses, a process which dates back thousands of years (Yu et al., 2019).

Currently, bio-based polymers have emerged as a renewable alternative to potential substitutes for conventional polymers for the development of eco-friendly composites due to their biodegradability and improved performance after reinforcement. Among the different bio-based polymers that are used commercially are poly(hydroxyalkanoates), poly(lactic acid), poly(ethylene) obtained from ethanol, alginate, cellulose, starch, chitosan, polyesters obtain from bioprecursors, proteins, polyether, and epoxy resins. Generally, biopolymers used as a matrix in the field of composites are also known as resin, as a consequence of the typical use of natural resins as a matrix in the development processes of eco-friendly composites (Sharma et al., 2018).

In eco-friendly composites, regarding the selection of reinforcing material based on natural origin, fibers, or particles, the criteria are generally defined by the stiffness and resistance values of the compound of interest. Therefore, it is expected that the intrinsic properties of each reinforcement agent or matrix component will influence the mechanical properties of the composite material and depend on the geometry, concentration, and orientation of the reinforcement, and on the adhesion at the reinforcement–matrix interface. Several types of biological reinforcements such as cellulose, cotton, flax, hemp, starch, clays, etc., have also begun to be used (La Mantia and Morreale, 2011; Singh et al., 2020).

7.5 Chemical and physical treatments for the development of eco-friendly composites

In general, many compounds of natural origin used as reinforcements in the development of composite materials have high hydrophilicity, which results in compatibilization between the reinforcement phases and the essentially hydrophobic polymeric matrix not being promoted. One of the main challenges in its processing is the adequate compatibilization, stabilization, and dispersion of the reinforcement phase in the polymeric matrix, in order to avoid nonuniform distribution in the matrix and the formation of voids (Zini and Scandola, 2011). The use of different types of techniques, both physical and chemical, has been reported in order to improve the compatibility between the reinforcement–matrix phases. These techniques have in common that their general objective is modification of the surface properties of the reinforcement or matrix phase, that is, at the molecular level they seek the modification or introduction of surface functional groups that promote specific intermolecular interactions and, therefore, the adhesion forces between the phases of the composite are increased. However, both in chemical and physical techniques, the modification generated should only occur at the surface level of the reinforcement or matrix in order to avoid altering its mechanical properties. Thus, among the chemical techniques used for phase compatibilization are alkaline treatment, treatment with silane, acetylation, peroxidation, benzoylation, treatment with potassium permanganate, and treatment with stearic acid, etc. Regarding physical techniques, the use of plasma stands out (Mukesh and Godara, 2019; Ramesh and Deepa, 2019). For example, esterification, etherification, and treatment with silanes or isocyanates, have been reported as methods of chemical surface modification of fibers and natural particles with polar hydroxyl groups on their surface to incorporate nonpolar groups through specific chemical reactions and hence improve the mechanical properties of eco-friendly compounds (Saha et al., 2016).

Alkaline treatment: Alkaline treatment is one of the most effective and low-cost methods used for the chemical treatment of materials of natural origin, specifically those of lignocellulosic origin. In general, alkaline treatment seeks surface modification by breaking internal hydrogen bonds of materials, which changes the surface topography, crystallinity, unit cell structure, functional groups, and possible orientation of the material, to improve its surface properties (Ramesh and Deepa, 2019). In general, there are two different objectives for which it is sought to apply alkaline pretreatment in lignocellulosic materials, the first is to improve the tensile strength of the fibers by eliminating amorphous materials such as hemicelluloses, lignin, and pectin, and the second is to expose the cellulose chains for better functionalization, that is, as a pretreatment for the subsequent application of another modification technique (Saha et al., 2016).

Esterification: In general, the esterification process involves the reaction between the hydroxyl groups on the surface of the organic material with organic compounds such as acetic anhydride, benzoyl chloride, fatty acids, etc., with the purpose of replacing the polar hydroxyl groups with functional groups of nonpolar nature, or related to the matrix to be used, and therefore, to modify the surface properties of the material and improve the compatibilization between phases (Saha et al., 2016).

Etherification: Etherification, like esterification, seeks the introduction of nonpolar groups on the surface that increases the hydrophobic nature of materials and resistance to degradation. Thus, benzyl chloride or alkyl chlorides, for example, ethyl iodide, in the presence of alkaline catalysts, have been used to modify the hydrophilicity of fibers and particles (Ramesh and Deepa, 2019).

Treatment with isocyanates: Thanks to the fact that isocyanates are widely used on industrial and daily levels, their industrialized production makes these compounds economically viable to be used in the surface chemical modification of materials. Furthermore, the high reactivity of isocyanate groups with nucleophilic groups, for example, hydroxyl groups, carboxylic acids, and amines, allows them to be used in the modification of a wide range of natural matrices. Thus, the isocyanate group reacts with the surface nucleophilic groups of the reinforcing material, improving the adhesion of the interface with the matrix. These treatments are characterized by being carried out at low temperatures of around 50°C thanks to the use of catalysts (Ramesh and Deepa, 2019; Saha et al., 2016).

Treatment with peroxides and oxidizing agents: Treatments with peroxide agents are used in the modification and compatibilization of phases by generating free radicals on the surface of materials. Thus, organic peroxides react with hydrogen atoms on the surface, in order to activate and be able to react with the surfaces of the other constituents of the composite material, improving its compatibility and adherence. These peroxide treatments are similar to the free radical polymerization initiation step, which induces adhesion between polymeric compounds reinforced with natural compounds.

On the other hand, treatment with permanganate induces the oxidation of the surface groups of the materials in order to reduce their hydrophilicity and improve the compatibilization between hydrophobic phases (Ahmad et al., 2019).

Treatment by silane: Organosilane compounds are also used in surface modification and phase compatibilization for the development of eco-friendly composites. In general, these compounds are characterized by having at least one alkyl group attached to their central silicon atom, which also has hydroxyl or halogen groups (X: Cl, Br, or I), which can be hydrolyzed and bind to the hydroxyl groups in the polymer chains through covalent bonds, for example, cellulose or hemicelluloses. Thus, when silanes react with hydroxyl groups of the fibers and particles, they decrease their swelling capacity in the presence of aqueous solvents due to the increase in their hydrophobicity, and in addition, increase their chemical and tensile resistance (Shokoohi et al., 2008; Xie et al., 2010).

Graft polymers: Chemically or photo-catalytically catalyzed copolymerization has also been used to generate graft polymers with the purpose of making the reinforcement and matrix phases compatible in eco-friendly composites. Thus, the reinforcement acts as the support for chains of a polymer, which is properly selected depending on the type of interactions to be promoted or improved between the reinforcement–matrix. The hydrophilic or hydrophobic graft polymer chains act as an interfacial agent and improve the adhesion between the reinforcement and continuous phase, generating an increase in the tensile, thermal, and chemical resistance, and in the hydrophobicity of the treated materials due to the substitution of hydroxyl groups by polymer chains (Gao et al., 2012; Thakur et al., 2014; Saha et al., 2016).

Mechanical and solvent extraction methods: The physical methods for the treatment of reinforcement materials and matrices used in the development of eco-friendly composites are based on the separation, purification, and structural and surface modification of natural materials, in order to improve the affinity between the phases without compromising their chemical composition. Mechanical methods are based on grinding, stretching, lamination, and pressing of materials in order to modify their geometry, increase surface areas and, therefore, improve the union between the matrix and the reinforcement. On the other hand, solvent extraction is perhaps the simplest method to generate fractionation of the components of materials of natural origin, causing an increase in their surface area after eliminating soluble impurities (Ahmad et al., 2019).

Plasma treatment: Plasma processing is a physical treatment that induces chemical changes on the surface of materials. This treatment is widely used as a replacement for conventional chemical modification techniques, thanks to the fact that it does not require chemical products to modify the surfaces and, therefore, does not generate waste and minimizes environmental pollution. In general, plasma can be described as a medium of unbound negative and positive particles, with a total electrical charge of approximately zero, which is influenced by variables such as gas type, flow, pressure, temperature, and concentration. It has been proposed that the plasma produces a physical modification in the surface through the roughness of the material due to the sputtering effect producing an expansion of the contact area that increases the friction between the matrix and the reinforcement (Mukhopadhyay and Fangueiro, 2009). Currently, nonthermal plasma processing is an emerging ecological technology with great potential for the treatment of thermolabile natural compounds, producing morphological changes and variation in the contact area that increase the friction between the reinforcement and the matrix in composites through the generation of free electrons in the discharge of plasma that is heated up to $10-100 \times 10^3$ K, while the gas is kept at moderate temperatures of between 25°C and 100°C (Gholampour and Ozbakkaaloglu, 2020; Zhu, 2017).

7.6 Conclusions and remarks

Despite intense academic and technological research, the large-scale commercialization of eco-friendly composites and nanocomposites is still in its initial stages compared to the large volume of composites containing petroleum-derived polymers. This is a consequence of the increase in the costs of obtaining and treating matrices and bio-based reinforcements for the compatibility of phases. Furthermore, the development and implementation of fast and efficient processing procedures for eco-friendly composites are key factors in the advance toward their industrialized production and commercialization. It is important to note that biodegradability is a property that is not always sought in eco-friendly compounds, consequently, in the automotive, aerospace, and construction industries, materials that are resistant and durable to the environment are sought. Therefore, the disposal at the end of the life cycle of eco-friendly composites and nanocomposites will depend on their biodegradability properties. Under these criteria, although an eco-friendly composite is made up of materials of natural or bio-based origin, it does not necessarily meet the biodegradability criteria. Thus, composting eco-friendly compounds with biodegradable matrices is an effective disposal strategy, contributing to a reduction of waste sent to municipal landfills and the generation of organic fertilizers. In contrast, when the eco-friendly composite matrix is not biodegradable, as a consequence of its structural complexity, which makes it difficult to separate and reuse its components using efficient and economical methods, its disposal options are limited to incineration or accumulation in a sanitary landfill, processes that by definition are not sustainable over the long term.

References

Abdul, H.P.S., Bhat, A.H., Ireana, A.F., 2012. Green composites from sustainable cellulose nanofibrils: a review. Carbohydr. Polym. 87 (2), 963–979. https://doi.org/10.1016/j.carbpol.2011.08.078.

Agrawal, M., Gupta, S., Stamm, M., 2011. Recent developments in fabrication and applications of colloid based composite particles. J. Mater. Chem. 21, 615–627. https://doi.org/10.1039/c0jm02631j.

Ahmad, R., Hamid, R., Osman, S.A., 2019. Physical and chemical modifications of plant fibres for reinforcement in cementitious composites. Adv. Civ. Eng. 5185806. https://doi.org/10.1155/2019/5185806.

Ahmed, T., Shahid, M., Azeem, F., Rasul, I., Shah, A.A., Noman, M., Hameed, A., Manzoor, N., Manzoor, I., Muhammad, S., 2018. Biodegradation of plastics: current scenario and future prospects for environmental safety. Environ. Sci. Pollut. Control Ser. 25, 7287–7298. https://doi.org/10.1007/s11356-018-1234-9.

Borodulin, A., Kalinnikov, A., Tereshkov, A., Kharaev, A., 2019. New polymeric binders for the production of composites. Mater. Today: Proceedings 11 (1), 139–143. https://doi.org/10.1016/j.matpr.2018.12.121.

Buruga, K., Song, H., Shang, J., Bolan, N., Jagannathan, T.K., Kim, K.H., 2019. A review on functional polymer-clay based nanocomposite membranes for treatment of water. J. Hazard Mater. 379, 120584. https://doi.org/10.1016/j.jhazmat.2019.04.067.

Cardarelli, F., 2018. Composite materials. In: Cardarelli, F. (Ed.), Materials Handbook. Springer, pp. 1497–1517. https://doi.org/10.1007/978-3-319-38925-7_18.

Casati, R., Vedani, M., 2014. Metal matrix composites reinforced by nano-particles—a review. Metals 4 (1), 65–83. https://doi.org/10.3390/met4010065.

Chaudhary, V., Ahmad, F., 2020. A review on plant fiber reinforced thermoset polymers for structural and frictional composites. Polym. Test. 91, 106792. https://doi.org/10.1016/j.polymertesting.2020.106792.

Chawla, K.K., 2012. Metal matrix composites. In: Chawala, K.K. (Ed.), Composite Materials. Springer, pp. 197–248. https://doi.org/10.1007/978-0-387-74365-3_6.

Chen, H., Pan, Z., Wang, W., Chen, Y., Xing, S., Cheng, Y., Ding, X., Liu, J., Zhai, J., Yu, J., 2021. Ultrahigh discharge efficiency and improved energy density in polymer-based nanocomposite for high-temperature capacitors application. Compos. Appl. Sci. Manuf. 142, 106266. https://doi.org/10.1016/j.compositesa.2020.106266.

Chung, D.D.L., 2019. A review of multifunctional polymer-matrix structural composites. Compos. B Eng. 160, 644–660. https://doi.org/10.1016/j.compositesb.2018.12.117.

Deng, S., Djukic, L., Paton, R., Ye, L., 2015. Thermoplastic–epoxy interactions and their potential applications in joining composite structures – a review. Compos. Appl. Sci. Manuf. 68, 121–132. https://doi.org/10.1016/j.compositesa.2014.09.027.

Di Filippo, M.F., Dolci, L.S., Liccardo, L., Bigi, A., Bonvicini, F., Gentilomi, G.A., Passerini, N., Panzavolta, S., Albertini, B., 2021. Cellulose derivatives-snail slime films: new disposable eco-friendly materials for food packaging. Food Hydrocoll. 111, 106247. https://doi.org/10.1016/j.foodhyd.2020.106247.

Farikov, S., 2015. Composite materials – is the use of proper definitions important? Mater. Today 18 (10), 528–529. https://doi.org/10.1016/j.mattod.2015.10.001.

Faruk, O., Bledzki, A.K., Fink, H.P., Sain, M., 2014. Progress report on natural fiber reinforced composites. Macromol. Mater. Eng. 299 (1), 9–26. https://doi.org/10.1002/mame.201300008.

Fu, S., Sun, Z., Huang, P., Li, Y., Hu, N., 2019. Some basic aspects of polymer nanocomposites: a critical review. Nano Mater. Sci. 1 (1), 2–30. https://doi.org/10.1016/j.nanoms.2019.02.006.

Fuqua, M.A., Huo, S., Ulven, C.A., 2012. Natural fiber reinforced composites. Polym. Rev. 52 (3), 259–320. https://doi.org/10.1080/15583724.2012.705409.

Gao, H., Xie, Y., Ou, R., Wang, Q., 2012. Grafting effects of polypropylene/polyethylene blends with maleic anhydride on the properties of the resulting wood–plastic composites. Compos. Appl. Sci. Manuf. 43 (1), 150–157. https://doi.org/10.1016/j.compositesa.2011.10.001.

Gholampour, A., Ozbakkaloglu, T., 2020. A review of natural fiber composites: properties, modification and processing techniques, characterization, applications. J. Mater. Sci. 55, 829–892. https://doi.org/10.1007/s10853-019-03990-y.

Hamidi, Y.K., Altan, C.M., 2018. 2.5 process-induced defects in resin transfer molded composites. In: Beaumont, P.W.R., Zweben, C.H. (Eds.), Comprehensive Composite Materials II. Elsevier, pp. 95–106. https://doi.org/10.1016/B978-0-12-803581-8.09902-1.

Hassan, Y.A., Hu, H., 2020. Current status of polymer nanocomposite dielectrics for high-temperature applications. Compos. Appl. Sci. Manuf. 138, 106064. https://doi.org/10.1016/j.compositesa.2020.106064.

Karger-Kocsis, J., Bárány, T., 2014. Single-polymer composites (SPCs): status and future trends. Compos. Sci. Technol. 92, 77–94. https://doi.org/10.1016/j.compscitech.2013.12.006.

Kopeliovich, D., 2018. 5 - advances in manufacture of ceramic matrix composites by infiltration techniques. In: Low, I.M. (Ed.), Woodhead Publishing Series in Composites Science and Engineering. Advances in Ceramic Matrix Composites, second ed. Woodhead Publishing, pp. 93–119. https://doi.org/10.1016/B978-0-08-102166-8.00005-0.

Kumar, D., Pagar, D.D., Kumar, R., Pruncu, C.I., 2019. Recent progress of reinforcement materials: a comprehensive overview of composite materials. J. Mater. Res. Technol. 8 (6), 6354–6374. https://doi.org/10.1016/j.jmrt.2019.09.068.

La Mantia, F.P., Morreale, M., 2011. Green composites: a brief review. Compos. Appl. Sci. Manuf. 42 (6), 579–588. https://doi.org/10.1016/j.compositesa.2011.01.017.

Lerma, T.A., Garces, V., Palencia, M., 2020. Novel multi- and bio-functional hybrid polymer hydrogels based on bentonite-poly(acrylic acid) composites and sorbitol polyesters: structural and functional characterization. Eur. Polym. J. 128, 109627. https://doi.org/10.1016/j.eurpolymj.2020.109627.

Lerma, T.A., Palencia, M., Combatt, E.M., 2018. Soil polymer conditioner based on montmorillonite-poly(acrylic acid) composites. J. Appl. Polym. Sci. 135, 46211. https://doi.org/10.1002/app.46211.

Liu, P., Lu, R., Huang, T., Cong, P., Jiang, S., Li, T., 2012. Tensile and tribological properties of polytetrafluroethylene homocomposites. Wear 289, 65–72. https://doi.org/10.1016/j.wear.2012.04.013.

Matabola, K.P., De Vries, A.R., Moolman, F.S., Luyt, A.S., 2009. Single polymer composites: a review. J. Mater. Sci. 44, 6213–6222. https://doi.org/10.1007/s10853-009-3792-1.

May, M., Rupakula, G.D., Matura, P., 2020. Non-polymer-matrix composite materials for space applications. Comp. Part C: Open Access 3, 100057. https://doi.org/10.1016/j.jcomc.2020.100057.

Mukesh, G.S.S., 2019. Effect of chemical modification of fiber surface on natural fiber composites: a review. Mater. Today Proc. 18 (7), 3428–3434. https://doi.org/10.1016/j.matpr.2019.07.270.

Mukhopadhyay, R., Bhaduri, D., Sarkar, B., Rusmin, R., Hou, D., Khanam, R., Sarkar, S., Biswas, J.K., Vithanage, M., Bhatnagar, A., Ok, Y.S., 2020. Clay–polymer nanocomposites: progress and challenges for use in sustainable water treatment. J. Hazard Mater. 383, 121125. https://doi.org/10.1016/j.jhazmat.2019.121125.

Mukhopadhyay, S., Fangueiro, R., 2009. Physical modification of natural fibers and thermoplastic films for composites — a review. J. Thermoplast. Compos. Mater. 22 (2), 135–162. https://doi.org/10.1177/0892705708091860.

Omanović-Mikličanin, E., Badnjević, A., Kazlagić, A., Hajlovac, M., 2020. Nanocomposites: a brief review. Health Technol. 10, 51–59. https://doi.org/10.1007/s12553-019-00380-x.

Palencia, M., Berrio, M., Melendrez, M., 2016. Nanostructured polymer composites with potential applications into the storage of blood and hemoderivates. J. Sci. Technol. Appl. 1, 4–14. https://doi.org/10.34294/j.jsta.16.1.1.

Palencia, M., Berrío, M.E., García-Quintero, A., Lerma, T.A., 2018. Cellulose-based nanocomposites for pharmaceutical and biomedical applications. In: Kumar, T., Ghosh, B. (Eds.), Polysaccharide Based Nano-Biocarrier in Drug Delivery, pp. 39–61. https://doi.org/10.1201/9780429449437-3.

Palencia, M., Lerma, T., Arrieta, A., 2017. Desarrollo de compositos hidrofóbicos híbridos basados en bentonita y poli(vinilestearato). Revista Latinoamericana de Metalurgia y Materiales 37 (1), 11–18. http://ve.scielo.org/scielo.php?script=sci_arttext&pid=S0255-69522017000100003.

Pourmoazzen, Z., Sadeghifar, H., Yang, G., Lucia, L., 2020. Cholesterol-modified lignin: a new avenue for green nanoparticles, meltable materials, and drug delivery. Colloids Surf. B Biointerf. 186, 110685. https://doi.org/10.1016/j.colsurfb.2019.110685.

Ramesh, M., Deepa, C., 2019. Processing of green composites. In: Muthu, S. (Ed.), Green Composites. Textile Science and Clothing Technology. Springer, Singapore, pp. 47–72. https://doi.org/10.1007/978-981-13-1972-3_2.

Raquez, J.M., Deléglise, M., Lacrampe, M.F., Krawczak, P., 2010. Thermosetting (bio)materials derived from renewable resources: a critical review. Prog. Polym. Sci. 35 (4), 487–509. https://doi.org/10.1016/j.progpolymsci.2010.01.001.

Safdari, M., Al-Haik, M.S., 2018. 5 - a review on polymeric nanocomposites: effect of hybridization and synergy on electrical properties. In: Ismail, A.F., Goh, P.S. (Eds.), Carbon-Based Polymer Nanocomposites for Environmental and Energy Applications. Elsevier, pp. 113–146. https://doi.org/10.1016/B978-0-12-813574-7.00005-8.

Saha, P., Chowdhury, S., Roy, D., Adhikari, B., Kim, J.K., Thomas, S., 2016. A brief review on the chemical modifications of lignocellulosic fibers for durable engineering composites. Polym. Bull. 73, 587–620. https://doi.org/10.1007/s00289-015-1489-y.

Sapkota, J., Natterodt, J.C., Shirole, A., Foster, E.J., Weder, C., 302. Fabrication and properties of polyethylene/cellulose nanocrystal composites. Macromol. Mater. Eng. 1, 1600300. https://doi.org/10.1002/mame.201600300.

Sapuan, S.M., 2017. Chapter 3 - composite materials. In: Sapuan, S.M. (Ed.), Composite Materials: Concurrent Engineering Approach. Butterworth-Heinemann, pp. 57–93. https://doi.org/10.1016/B978-0-12-802507-9.00003-9.

Sathishkuman, T.P., Naveen, J., Satheeshkumar, S., 2014. Hybrid fiber reinforced polymer composites – a review. J. Reinforc. Plast. Compos. 33 (5 8), 454–471. https://doi.org/10.1177/0731684413516393.

Shan, L., Chen, X., Tian, X., Chen, J., Zhou, Z., Jiang, M., Xu, X., Hui, D., 2015. Fabrication of polypyrrole/nano-exfoliated graphite composites by in situ intercalation polymerization and their microwave absorption properties. Compos. B Eng. 73, 181–187. https://doi.org/10.1016/j.compositesb.2014.11.042.

Sharma, B., Malik, P., Jain, P., 2018. Biopolymer reinforced nanocomposites: a comprehensive review. Mater. Today Commun. 16, 353–363. https://doi.org/10.1016/j.mtcomm.2018.07.004.

Shokoohi, S., Arefazar, A., Khosrokhavar, R., 2008. Silane coupling agents in polymer-based reinforced composites: a review. J. Reinf. Plas. Comp. 27 (5), 473–485. https://doi.org/10.1177/0731684407081391.

Singh, G., Pal, L., Kumar, P., Singh, S., 2020. Green composites: a review of processing technologies and recent applications. J. Thermoplast. Compos. Mater. 33 (8), 1145–1171. https://doi.org/10.1177/0892705718816354.

Subhedar, A., Bhadauria, S., Ahankari, S., Kargarzadeh, H., 2021. Nanocellulose in biomedical and biosensing applications: a review. Int. J. Biol. Macromol. 166, 587–600. https://doi.org/10.1016/j.ijbiomac.2020.10.217.

Supian, A.B.M., Sapuan, S.M., Zuhri, M.Y.M., Zainudin, E.S., Ya, H.H., 2018. Hybrid reinforced thermoset polymer composite in energy absorption tube application: a review. Defence Technol. 14 (4), 291–305. https://doi.org/10.1016/j.dt.2018.04.004.

Szabó, T., Szeri, A., Dékány, I., 2005. Composite graphitic nanolayers prepared by self-assembly between finely dispersed graphite oxide and a cationic polymer. Carbon 43 (1), 87–94. https://doi.org/10.1016/j.carbon.2004.08.025.

Thakur, V., Thakur, M., Gupta, R., 2014. Graft copolymers of natural fibers for green composites. Carbohydr. Polym. 104, 87–93. https://doi.org/10.1016/j.carbpol.2014.01.016.

Umoren, S.A., Solomon, M.M., 2019. Protective polymeric films for industrial substrates: a critical review on past and recent applications with conducting polymers and polymer composites/nanocomposites. Prog. Mater. Sci. 104, 380–450. https://doi.org/10.1016/j.pmatsci.2019.04.002.

Utekar, S.V.K.S., More, N., Rao, A., 2021. Comprehensive study of recycling of thermosetting polymer composites – driving force, challenges and methods. Compos. B Eng. 207, 108596. https://doi.org/10.1016/j.compositesb.2020.108596.

Villalobos, R., Medina, C., Canales, C., Melendrez, M., Flores, P., 2019a. Development of nanocomposite materials with thermostable matrix from nanoreinforcement of titanium dioxide (TNPS and TNTS) and zinc oxide (ZNPS and ZNBS) on epoxy resin matrix. J. Sci. Technol. Appl. 6, 65–78. https://doi.org/10.34294/j.jsta.19.6.42.

Villalobos, R., Rojas, D., Diaz, A., Ramirez, J., Melendrez, M., Flores, P., 2019b. Synthesis and characterization of 1d metal oxide nano-structures: titanium dioxide nanotubes (TNTS) and zinc oxide nanobars (ZNBS). J. Sci. Technol. Appl. 6, 79–95. https://doi.org/10.34294/j.jsta.19.6.43.

Xie, Y., Hill, C.A.S., Xiao, Z., Militz, H., Mai, C., 2010. Silane coupling agents used for natural fiber/polymer composites: a review. Compos. Appl. Sci. Manuf. 41 (7), 806–819. https://doi.org/10.1016/j.compositesa.2010.03.005.

Yang, H., Cui, H., Tang, W., Li, Z., Han, N., Xing, F., 2017. A critical review on research progress of graphene/cement based composites. Compos. Appl. Sci. Manuf. 102, 273–296. https://doi.org/10.1016/j.compositesa.2017.07.019.

Yin, H., Zheng, P., Zhang, E., Rao, J., Lin, Q., Fan, M., Zhu, Z., Zeng, Q., Chen, N., 2020. Improved wet shear strength in eco-friendly starch-cellulosic adhesives for woody composites. Carbohydr. Polym. 250, 116884. https://doi.org/10.1016/j.carbpol.2020.116884.

Yu, W.H., Sing, S.L., Chua, C.K., Kuo, C.N., Tian, X.L., 2019. Particle-reinforced metal matrix nanocomposites fabricated by selective laser melting: a state of the art review. Prog. Mater. Sci. 104, 330–379. https://doi.org/10.1016/j.pmatsci.2019.04.006.

Zhao, X., Lv, L., Pan, B., Zhang, W., Zhang, S., Zhang, Q., 2011. Polymer-supported nanocomposites for environmental application: a review. Chem. Eng. J. 170 (2–3), 381–394. https://doi.org/10.1016/j.cej.2011.02.071.

Zhu, F., 2017. Plasma modification of starch. Food Chem. 232, 476–486. https://doi.org/10.1016/j.foodchem.2017.04.024.

Zini, E., Scandola, M., 2011. Green composites: an overview. Polym. Compos. 32, 1905–1915. https://doi.org/10.1002/pc.21224.

Chapter 8

Eco-friendly supramolecular systems: polymer geo- and biomimicry

8.1 Supramolecular systems: a fundamental approach and critical analysis

8.1.1 Description of matter organization levels

Matter, along with energy, are the great constituents of the universe. In particular, matter can be described, in a simplified way, to be interacting particles forming a fundamental system called an "atom." Depending on the number of particles and their organization, there are multiple types of atoms, with hydrogen being the simplest. At present, it is accepted that atoms are the simplest organization level with the intrinsic properties of what is understood by matter. As they interact, atoms give rise to "molecules," and sets of molecules give rise to what it is known as "phase." Depending on the interactions existing and the energetic state, phases can differ from each other without changing their nature. A set of phases gives rise to mixtures, which can be homogeneous or heterogeneous, and in this way, it is possible to arrive at more complex structures in terms of the degree of organization of matter. As a result of interactions between molecules, it is possible to define levels of organization mediated by noncovalent bonds, for example, the quaternary structure of proteins (Yu et al., 2006), micellar systems resulting in surfactant dissolution when their concentration is higher than critical micellar concentration (Wu et al., 2020), simple emulsions and double emulsions stabilized by surfactants (Palencia and Rivas, 2011), inorganic nanoparticles stabilized by adsorption of organic molecules in their interface (Palencia et al., 2016a,b), polyelectrolyte complexes between polyanions and polycations (Meka et al., 2017), carbon nanotubes occluded in a polymeric phase (Villalobos et al., 2019), etc. However, the intermolecular interactions that give rise to systems with a degree of organization above the molecules can also be mediated by covalent and coordination bonds. For example, nanoparticles stabilized with organic molecules through thiols (Battocchino et al., 2012), polymer–clay multilayer

Eco-friendly Functional Polymers
https://doi.org/10.1016/B978-0-12-821842-6.00010-5
Copyright © 2021 Elsevier Inc. All rights reserved.

systems obtained by covalent anchoring of the polymer on the clay surface (Garcés et al., 2018; Lerma et al., 2018a,b), the natural humin–clay system (Garcia et al., 2018), interpenetrating polymer networks, which although their linking is by physical restriction, the impediment of its mechanical separation is the result of covalent interactions in superimposed three-dimensional structures (Palencia et al., 2016a,b; Lerma et al., 2017), and polymer–membrane hybrid systems, resulting from the insertion of polymer chains on the surface of a porous material are also of polymeric origin (Arbeláez et al., 2017).

Traditionally, according to the IUPAC the term "supramolecular" is used to refer to "any system formed from two or more molecular entities held together and organized by means of intermolecular binding interactions" (IUPAC, 1997). However, this definition is limited to systems where their interactions are noncovalent. Likewise, they define supramolecular chemistry as "a field of chemistry related to species of greater complexity than molecules, that are held together and organized by means of intermolecular interactions" (IUPAC, 1997); a definition that would not cover the study of those systems of two or more molecular units and organized through covalent interactions. In this way, in this chapter the concept of a "supramolecular system" will be taken from a broader point of view, breaking with the traditional vision and limited to noncovalent interactions. The above classifies consequently with the matter organization levels, those systems that are found above the molecules, but linked together by covalent interactions. As previously indicated, matter is organized from simple to complex systems, following a progressive organization level, that is, atoms, molecules, supramolecular systems, mixtures (e.g., homo- and heterogeneous), compartmentalized systems (e.g., the cell), specialized systems (e.g., tissues), multicomponent biofunctional systems (e.g., organs), individual entities (e.g., plants and animals), and this progression could even be extended to populations, communities, and ecosystems. In consequence, the reevaluation of the supramolecular concept is proposed here, given the impossibility of classifying, using the classical approach, those systems based on molecular associations of a covalent nature. This classical approach could be the result of an interpretation from some specializations of knowledge focused on the study of systems based on noncovalent interactions, and not from chemistry and the analysis of the organization of matter.

The development and study of supramolecular systems has a particular interest, thanks to the fact that their properties and applications will depend on the characteristics of the construction units, and the type of interaction that occurs between the different molecular entities. Therefore, the structures and properties of supramolecular assemblies are controlled by properly choosing the construction materials and type of interaction. Each type of interaction (ionic, covalent, coordination or, noncovalent) has a unique bonding force and characteristics, obtaining systems with excellent and modulable properties. Thus, supramolecular systems based on "weak" interactions, for example, the spontaneous intermolecular association of noncovalent interactions such as hydrogen bonding, $\pi-\pi$ stacking, metal–ligand coordination, ionic interaction, or hydrophobic interaction, have allowed the design and development of complex and well-defined chemical systems in the field of molecular engineering, which are kinetically or thermodynamically stable and susceptible to react to stimuli or external factors such as, pH, temperature, ionic strength, etc., allowing the modulating of its properties and advancing in the development of functional systems (Lin and Grinstaff, 2013). In contrast, for example, those supramolecular systems with covalent interactions will present greater structural stability against the indicated factors. Thus, supramolecular systems have been widely used in applications such as molecular recognition, catalysis, bio- and geomimetic systems, intelligent substance transporters, functional materials, etc. (Nitschke, 2014; Lerma et al., 2020).

In conclusion, supramolecular systems can be classified according to the type of interaction they present in (1) ionic supramolecular systems, (2) covalent supramolecular systems, (3) coordination supramolecular systems, and (4) noncovalent or van der Waals supramolecular systems. The properties of these systems are determined by the forces of the bonds between the components of the supramolecular assembly. Thus, covalent bonds are generally considered irreversible with bond strengths greater than 60 kcal/mol, ionic and metal–ligand coordination interactions are classified as medium intensity with bond strengths between 20 and 60 kcal/mol, and noncovalent interactions are weak interactions with forces less than 20 kcal/mol (Lin and Grinstaff, 2013).

8.2 Eco-friendly supramolecular polymeric systems

At present, many technological developments require the constant search for new sources and methods of obtaining materials. In particular, polymeric materials have been and are one of the most widely used materials in countless applications, and can be found in clothing, household appliances, kitchen utensils, the aeronautical industry, the marine industry, etc., thanks to their excellent physical, chemical, mechanical, and functional properties, low production costs, etc. However, the use of polymers from the oil industry, a nonrenewable natural source, has been so extensive that they currently represent a major source of environmental pollution as a result of the low biodegradability of most of these materials (Lambert and Wagner, 2018; Toussaint et al., 2019). That is why polymers of natural origin and sustainable sources emerge as promising materials that can supplement and improve the characteristics of conventional polymers. In addition, thanks to the high diversity of functional groups that these polymers have, it is possible to use them as building blocks for the development of complex supramolecular systems with adaptable properties. Thus, for example, in contrast to the few examples of synthetic polymers that can be transiently cross-linked through noncovalent interactions, there is a great variety of biopolymers that can interact through hydrogen bonds and form supramolecular gels, among these, polysaccharides such as cellulose, starch, agarose, and dextran stand out, whose cross-linking can occur through the hydrogen bonding of their hydroxyl groups (Rossow and Seiffert, 2015). In this way, the main characteristics that eco-friendly supramolecular polymeric systems possess are the ability to spontaneously and reversibly self-assemble complex and ordered systems, the ability to respond to external stimuli such as pH, pressure, temperature, radiation, etc., and finally, self-repair capacity (Brunsveld et al., 2001; Fan et al., 2020). In general, supramolecular polymeric systems can be classified according to the number of components in "supermolecules" to refer to those well-defined oligomolecular systems that are the origin of the interaction of a few components and in "polymolecular assemblies" to refer to those complex structures supramolecular or to specific phases composed of a large number of components and with defined microscopic organization and macroscopic characteristics depending on their nature (films, layers, membranes, vesicles, micelles, mesophases, surfaces, solids, etc.) (Lehn, 2002).

It should be noted that the appropriate selection and design of the monomers, or polymers of natural origin, which will be used as building blocks and, especially, the selection and modulation of the type of interaction between the different components of the supramolecular polymeric systems, allow the incorporation of specific characteristics. Thus, in addition to being the supramolecular polymeric systems one of the main alternatives for the substitution of conventional polymers, their unique functional properties promote their use in the design of intelligent systems for various applications, for example, in biomaterials, drug transporters, energy storage systems, biomimetic systems, etc. (Lin and Grinstaff, 2013). Generally, electrostatic interactions between polyelectrolyte layers are traditionally the most used to build supramolecular systems through the layer-by-layer deposition technique (Wang et al., 2018). However, at present, researchers have focused their study interest toward the development of multilayer systems of molecular self-assembly where noncovalent interactions, such as hydrogen bonding, eliminates the need to use polyelectrolytes and allows the use of neutral polymers with different functional groups, for example, -OH, -COOH, $-NH_2$, -SH, etc., which can act as donors and acceptors of hydrogen bonds in the supramolecular structure (Kharlampieva and Sukhishvili, 2006). As a result of the use of intermolecular forces of lower intensity, supramolecular systems with a multilayer structure with noncovalent interactions such as hydrogen bonding, will have a greater sensitivity and response to external stimuli such as temperature, pH, redox reactions, and ionic strength, converting them into intelligent systems of technological interest in fields of medicine, sensor development, etc. (Zhang et al., 2007).

Likewise, in the development of multilayer supramolecular polymeric systems, strong and stable interactions like covalent interactions can be used.

Hence, these types of interactions produce supramolecular assembly with greater stability and structural strength. As a consequence of the high stability exhibited by covalent-type interactions in supramolecular polymeric systems, the functionality of the polymers is suitably selected based on the desired characteristics in order to be obtained in the supramolecular system. Consequently, in these covalent-type systems, the response to external stimuli will not be directly linked to the stability and modulation of the interactions that make up the supramolecular system, but to the functional response of each of the polymeric layers (Stupp et al., 2020). In addition, hydrophobic interactions occurring between nonpolar molecules or compounds are also used in the assembly of multilayer systems, where compatibility between them plays a crucial role in their stability in order to reduce a possible phase segregation (Ma and Tian, 2014). Likewise, bio-specific interactions are used as a technique for the construction of supramolecular polymeric systems thanks to the fact that the bio-specific interactions that some biomaterials are capable of exhibiting with specific components through different molecular interactions are used or imitated, for example, electrostatic and hydrophobic interactions (Yan et al., 2014). Finally, the coordination complexes between a great diversity of metal ions and organic ligand present in polymers can generate strong and stable coordination complexes that also allow the construction of eco-friendly supramolecular polymeric systems with potential photochemical, electrochemical, and photophysical applications (Weck, 2007).

8.3 Biomimicry: biomimetic supramolecular systems

Nature throughout millions of years of evolution has developed endless compounds, materials, supramolecular systems, chemical processes, etc., with exceptional properties and functionalities. These are the result of an extensive process of adaptation, modification, and improvement, emerging as a source of study and inspiration for the realization of new and innovative applications. Bio-inspired or biomimetic materials are defined as materials whose manufacture and/or design are inspired by the structure and/or functions of natural systems, in order to imitate or improve the excellent properties of materials found in nature. Thanks to the advance in the compression of the molecular structure of organisms, such as cells, surfaces, and materials, advances in the knowledge of the functions of molecules, molecular engineering and, in the development of high-resolution analytical techniques, it has been possible to understand the various biological structures and processes at molecular and supramolecular levels, allowing the development of biomimetic systems which imitate and improve the properties of natural systems (Shen et al., 2014).

The development of biomimetic systems requires the initial simplification of the biological model in order to achieve the proper design of an artificial system, capable of appropriately replicating a function or a property of the model existing in nature. Thus, biomimetic design is based on the study and understanding of natural processes and systems in order to develop systems with the same or better characteristics or functionalities. The concept of biomimicry is universal, that is, we can find it in various fields of study, for example, architecture, computer science, physical sciences, materials sciences, etc. Therefore, understanding how complex molecules, materials, and systems contribute to the functions in natural systems has become the study objective of scientists working in different fields and using design inspired by nature. Thus, the interpretation of the mechanisms behind each process or phenomenon is not only of biological interest, but also allows the design of materials and systems replicating the natural functions, or that serve as a design basis for analogous synthetic systems. In particular, each one of the natural materials and systems exhibits a characteristic behavior, either in response to an external or internal factor of the organism, which gives it the ability to modulate its physicochemical properties, for example, the hardening of sea cucumbers against a threat or the change of color in the chameleon skin (Salmon et al., 2019; Shoffstal and Capadona, 2018).

8.3.1 Approaches to biomimetic design

The accomplishment of biomimetic designs has generally followed two design routes: (1) problem–bioinspiration–solution, this approach is based on the premise of defining as the first stage of the design process a problem or need to be solved, in order to find in nature, organisms or ecosystems to solve the problem and, (2) bioinspiration–solution–problem, this approach to biomimetic design focuses in the first instance on studying those characteristics, behaviors, or particular functions of organisms or ecosystems to design processes or materials that can be used as design alternatives to those established or as a solution to an emerging problem (Aziz and El sherif, 2016). Both approaches are described below.

Problem–bioinspiration–solution: this approach is one of the most commonly used by designers who seek inspiration from nature to solve their investigation problems. In general, six crucial stages can be indicated for the application of this approach: (1) identifying and adequately analyze the critical points of the designs or problems to be solved, (2) searching and identifying analogous biological systems that contribute to solving the problem proposed, (3) carrying out a detailed study of the biological system, (4) designing the biomimetic system or material based on the study results of the natural system, (5) evaluating and analyzing results and providing feedback on the designed bio-inspired systems, and (6) designing the system or prototype material that solves the problem initially raised.

Bioinspiration–solution–problem: In contrast to the traditional approach, this approach is based on the study and prior knowledge of the functioning and characteristics of biological systems in order to design biomimetic systems that do not necessarily intend to solve an immediate problem. Likewise, six stages are identified for its application which consist of: (1) studying biological systems, (2) identifying and understanding biological mechanisms with particular properties or characteristics, (3) understanding the principles of phenomena observed in biological systems, (4) designing the biomimetic system or material based on the results of the study of the natural system, (5) evaluating and implement the desired system, and (6) searching the applicability of the systems.

8.3.2 Polymer-based biomimetic supramolecular systems

Polymeric materials can be considered as a cornerstone of modern industry and high technology, which since their discovery, thanks to their excellent properties, have allowed them to be used in various applications. Commonly used polymeric materials come mostly from unsustainable fossil sources and are classified as thermosetting and thermoplastic. This is, in the same order, materials being actionable and recyclable, and materials having a highly cross-linked molecular structure, high chemical and solvent resistance, higher glass transition temperature, and thermal stability (Luzi et al., 2019). In particular, biomimetic materials are designed with the aim of mimicking the properties and functions of compounds of biological origin. Thus, in the field of biomimetic polymeric systems development, new materials of natural origin are constantly being sought, for example, proteins and carbohydrates, and the production of eco-friendly materials that contribute to the resolution of environmental problems associated with the accumulation of waste (Silverman and Roberto, 2007). In addition, with the advancement of supramolecular studies, it is possible to modulate the interactions between the polymeric components of the systems, through electrostatic, covalent, coordination, or noncovalent interactions, to obtain materials with unique properties and functionalities. The most widely used interactions in the field of biomimicry, for the design of complex supramolecular systems, are noncovalent, electrostatic, and coordination interactions, thanks to their high sensitivity exhibited to factors such as temperature and pH. In this sense, biomimetic-supramolecular polymeric materials have been based on the development of supramolecular polymeric systems with noncovalent interactions. These systems, thanks to the fact that they present weak interactions, have properties that can be influenced in specific and predictable ways when they are exposed to a predefined stimulus (Zhou et al., 2017). Thus, materials and systems that are mechanically transformed or adapted have emerged as a topic of interest, both in the field of biology, where the emphasis of the research is on developing a better understanding of the way in which natural mechanisms allow the survival of organisms that take advantage of mechanical transformations for key functions, as well as, in the materials field, where the main objective is their potential use in applications ranging from active damping systems to soft robotics (Montero et al., 2017). In addition, it is highlighted that, as a solution to the high stability and low response to stimuli presented by supramolecular covalent polymeric systems, supramolecular assemblies with dynamic main or side chain covalent bonds have been designed, with the aim that they may possess characteristics of repeatable processing and self-healing (Huang et al., 2020).

A case of particular interest is the biomimetization of cellular systems. Cellular systems have served as inspiration for the development of supramolecular biofunctional materials, thanks to the way in which cellular systems are held together mainly by noncovalent intermolecular interactions between small molecules, the ability of molecular self-assembly and, to the high suprastructural order that natural cellular systems possess. Furthermore, the study of the different cellular processes, for example, enzymatic reactions, ligand–receptor binding, or electron transfer, have provided the fundamental bases for the bio-inspired development of processes that can respond to multiple stimuli or compounds (Zhou et al., 2017). In particular, in relation to the capacity of biological systems that can spontaneously self-repair and adapt to environmental or external stimuli, most synthetic materials do not show self-repair after mechanical deformation or dynamic adaptability (Liu et al., 2017). In this addressing, noncovalent interactions, that is, hydrogen bonding, play a crucial role in many biological processes such as DNA base pairing, receptor–ligand binding, enzymatic catalysis, and protein folding, etc. These have become the most widely used interactions for the synthesis of supramolecular polymers with self-curing properties and are sensitive to stimuli, thanks to the low binding force that these bonds present and their high directionality and versatility (Rossow and Seiffert, 2015). One interesting example is the biomimicry of the marine mussel. The outstanding mechanical and adhesive properties of the marine mussel, which allow these mollusks to adhere to underwater rocks and against the high forces caused by tides, have inspired several works that focus on improvement of the mechanical properties of polymers through similar supramolecular interactions. The marine mussel has a material composed mainly of proteins that contain in high concentration the amino acid 3,4-dihydroxyphenylalanine (L-DOPA), which is believed to participate in both adhesion processes and processes of cross-linking, through the formation of metal–ligand complexes (Silverman and Roberto, 2007). Therefore, taking as a source of inspiration the adhesion mechanisms of the mussel, several research studies have been described as being inspired by this operating principle, from which materials showing deformation hardening or shape memory have been developed. As in the biological model, the interactions that have been widely used to mimic mechanical reinforcement are based on the formation of complexes between metal ions and properly designed ligands in polymeric matrices (Torres et al., 2012).

8.4 Geomimicry: a bioinspired approach to the development of artificial soils

8.4.1 Soil degradation as a current global problem

The increase in the global population, together with the accelerated development of science over the last century, have fuelled an increase in the consumption of natural resources. Focusing on soil, it is estimated that only 11% of the total solid surface of the Earth is classified as agricultural soil, of which around 80% suffers moderate to severe erosion, while 10% experiences mild erosion (Pimentel and Burgess, 2013). The rest of the surface has characteristics either due to wetness, superficiality, chemical imbalance, or that is constantly frozen, which makes it unsuitable for agricultural use. In addition, agricultural activities is not the only ones carried out on this resource, but one of a great variety of uses that man has developed on the Earth's surface, such as livestock, mineral extraction, support for construction, disposal waste, spaces for leisure activities, etc. Therefore, soil degradation is not only related to natural events but also to anthropogenic processes (Nearing et al., 2017).

Soil degradation is understood as the partial or total loss of its productivity, resulting in processes such as water or wind erosion, salinization, waterlogging, nutrient depletion, deterioration of the soil structure, and pollution. Soil degradation can be classified into three categories: chemical, physical, and biological degradation. Chemical degradation is understood as that which occurs when acidification, salinization, sodification, loss of nutrients, and toxicity take place, either by release or concentration, of different chemical elements in the soil, for example, heavy metals and metalloids (Bhattacharyya et al., 2015). On the other hand, physical degradation occurs with the loss of structure, decrease in permeability, and water retention of soils. Finally, biological degradation refers to the loss of plant cover, a decrease in the content of organic matter, and a reduction in macro- and microfauna (Paz-Ferrreiro and Fu, 2016). The causes of soil degradation can be varied, for example, in soils used for cultivation, its degradation may be associated with transformations in irrigation, intensive agriculture, burning, livestock operations, etc. The use of pesticides and chemical fertilizers, essential in current agricultural production, are also a source of deterioration, since they can transmit toxic products to the food chain, as well as generate the gradual loss of soil biota, replacing its biological functions, and as a consequence increase the dependence of soils on external inputs to maintain their productivity (Tetteh, 2015). Industrial activity is also a source of degradation, and is directly associated with the inadequate disposal of waste products generated by various manufacturing or mining processes, for example, waste rich in lead, copper, etc. However, it can also be associated with accidents that can generate degradation and/or irreversible loss of soil (Yasurani et al., 2011; Yao et al., 2012).

8.4.2 A brief description of soil composition and interactions

Soil can be defined as the unconsolidated mineral material on the surface of the Earth, which serves as a natural medium for the growth of plants, and that is subject to and influenced by various factors of its origin and the environment, which are: the parent material, the climate (humidity and temperature effects), organisms, and topography (van Es, 2017). Its constituents can be divided into four fractions: (1) inorganic fraction, (2) organic fraction, (3) soil solution, and (4) gaseous phase, with the first two being those that originate the solid phase of the soil, which is responsible for its biogeochemical behavior (Glinski, 2011).

The inorganic fraction of the soil is made up of minerals, oxides, and salts, with its composition depending mainly on the type of rock from which it was formed. This fraction can be classified according to its size into gravel (>2 mm), sands (2 mm < 60 μm), silts (60 μm < 2 μm), and clays (<2 μm); being the proportion of each of them a determining factor of agricultural characteristics of the soils (Blume et al., 2016). In particular, clay minerals are laminar-type aluminosilicates, the product of primary mineral weathering processes, which have a particle diameter of less than 2 μm, forming the colloidal fraction of the soil and giving it important properties, for example, water retention, ion exchange, and structure (Kome et al., 2019).

On the other hand, the organic fraction of soils is made up of living organic matter, for example, microorganisms, fresh organic matter, and humic substances. This fraction in the soils is one of the components that mainly acts on the balances of soil–plant activity, its effects on soil properties can include indirect effects influencing and modifying the properties determining soil fertility, or direct effects, which act on the physiological and/or metabolic mechanisms of plants (Lehmann and Kleber, 2015). In particular, among living organic matter there are different communities of bacteria, which act as vital components of soils, since they are involved in various biotic activities of the ecosystem so that it is rich in nutrients and sustainable for the production of crops. In addition, bacteria stimulate plant growth through the mobilization of nutrients in soils, producing numerous plant growth regulators, protecting plants against phytopathogens by controlling or inhibiting them, and improving soil structure (Zaidi et al., 2017). This is the case of rhizobacteria, which are housed around the roots of plants, facilitating the transformation, mobilization, and solubility of the nutrients present in soil (Vurukonda et al., 2016). On the other hand, humified matter (or humic substances) is a heterogeneous mixture of natural organic substances widely distributed in soils, the product of complex biogeochemical processes. Humic substances are classified based on their solubility in acidic and alkaline solutions in fulvic acids, soluble in all pH ranges, in humic acids, soluble at alkaline pH, and insoluble at acidic pH, and in humins, insoluble in all ranges pH. The chemical composition of these substances varies from the fulvic acids to humin, which has a higher degree of polymerization, molecular weight, and carbon content. In particular, these substances give the soil buffer capacity, ion exchange, retention of nutritional elements (Fe, Cu, Zn), texture, water retention, and coloration, therefore, a decrease in these fractions, clay−organic matter, causes the inevitable loss of soil properties and, consequently, its deterioration and a reduction of its fertility, translating into low agricultural production (Filip and Demnerova, 2009; Trevisan et al., 2010).

In soil, humified organic matter and mineral components, particularly clays and silts, are closely related. The interaction mechanisms of these organo-mineral complexes can be described by van der Waal's forces, ion exchange, hydrogen bonding, and ligand exchange. These mechanisms can occur simultaneously depending on the loading characteristics of the materials (Cornejo and Hermosin, 1996; Smernik and Kookana, 2015). In this sense, the charge characteristics of humic substances are derived mainly from ionized carboxyl groups, and to a lesser extent, from phenolic units present in their structure, which exhibit polar and nonpolar characteristics depending on the proportion of hydroxyl groups, aliphatic units, aldehydes, and quinones (Gomes et al., 2016). On the other hand, the charging properties of clays depend on the factors of chemical composition, nature of the atoms on their surface (mainly oxygen and hydrogen), the extent and type of their defective sites, and the class of exchangeable cations. Such properties confer the ability to interact with each other and form stable aggregates, to such an extent that, in soils rich in humic substances, all the clay can be covered by these (Han et al., 2016; Kloster and Avena, 2015).

8.4.3 Clay−polymer supramolecular systems

In recent decades, the development of hybrid supramolecular systems based on clays and polymers has seen considerable growth, due to the improvements in the properties exhibited by these materials, for example, mechanical properties, flammability, etc., compared to their predecessors (Fu et al., 2019; Wypych et al., 2018). Clays acquire particular interest due to their geometry in the form of sheets with nanometric dimensions, natural abundance, and especially, due to functions as main constituents of soil. In addition to the above, clay surface modification with polymers is one of the most effective methods for obtaining polymeric hybrid composites, thanks to the great functional variety of polymers and the improvements of material properties resulting from the addition of clays (Kango et al., 2013). However, to obtain these structures, the properties of the different inorganic−organic phases are important, with the hydrophilicity being a decisive factor in the making of these materials. Thus, it is of vital importance that some type of pretreatment of the clay surface is carried out in order to modify its surface properties, and promote a better interface interaction. In this way, the clay surface modification with species able to change the properties of silicates emerges as the best route toward obtaining these hybrid systems (Gul et al., 2016; Pavlidou and Papaspyrides, 2008).

The methods of surface modification of clays can be classified, according to the type of interaction between the inorganic and organic phase, as physical adsorption or chemical anchoring. Although physical adsorption has the advantage of not altering the structure of the mineral, its interactions, mainly by van der Waals or electrostatic mechanisms, is weak and easily overcome, therefore these characteristics are transferred to the resulting material (Lerma et al., 2018a,b). In contrast, the chemical anchoring of precursor polymers or monomers does not have such limitations, allowing the surface insertion of organic groups through the appropriate selection of coupling agents, for example, vinyl, amino, epoxides, etc., which act as posterior modification points (He et al., 2013; Yang et al., 2017). The chemical anchor (interaction mediated by covalent bonds) is formed usually through organosilane coupling agents, these coupling agents are effectively used for organic modification of the clay surface through a covalent bond, because clay minerals are characterized by organophilic properties. Organosilanes generally contain R-Si-X groups (X = OR, Cl) and are capable of reacting with the silanol groups on the mineral surface of clays, causing stable siloxane bridges (Abeywardena et al., 2017). Surface modification by chemical anchoring is an effective method for changing the hydrophilic properties of silicates, giving them organophilic domains capable of interacting with polymers, thus composites can be easily obtained by different preparation techniques. Also, it is possible for the coupling of organosilanes containing unsaturated groups or easily modifiable groups in their structure, which gives rise to hybrid structures capable of carrying out polymerization reactions and the subsequent formation of hybrid polymeric composites (Jauhari et al., 2017; Lerma et al., 2018a,b; Mansoori et al., 2010).

8.4.4 Geomimicry: mimetic systems of the structure and/or function of soils

Geomimicry arises as an alternative for the achievement of new materials and supramolecular assemblies whose design and function are based on systems analogous to those existing in nature, having as its central axis the development of materials of geological origin resulting from complex biogeochemical processes, for example, soil. These natural systems are the result of an extensive process of adaptation, modification, and improvement, allowing them to be ideal models for the achievement of new and innovative applications. Thus, taking soil as a source of inspiration, at the level of its structure and functionality, is an ideal model for making artificial materials for monitoring, improving, and rehabilitation of natural and productive ecosystems (Lerma et al., 2020).

Starting from the premise that the mimetic systems must be similar to the real ones. both in structure and in functionality, it is clear that an important strategy to define the precursors to be used must be based on an analysis of the real system, but also, as a starting point, the "guide" system must be simplified in terms of the desired properties. For example, as previously indicated, organic matter can be divided into two large groups, that is, humified organic matter and living organic matter, however, in addition, clay minerals give soils different and important properties that influence the exchange capacity, cationic, water absorption, plasticity, etc. In this way, a soil particle can in a first step be simplified as a core–shell system formed by a nucleus of clay covered with an organic layer (humin), which in turn, forms a supramolecular structure with an outermost envelope of organic compounds (humic and fulvic acids). In this way, the external envelope gives the system properties such as moisture retention, buffer capacity, cation exchange capacity, interparticle cementation, etc. Consequently, the surface modification of clay particles and the construction of organic coatings through the use of functional polymers is a first approach for the construction of geomimetic systems. Due to the enormous structural variety and properties of polymers, a large number of potential systems for different applications can be designed (Lerma et al., 2018a,b; Lerma et al., 2020) (see Fig. 8.1). Thus, synthetic polymers such as poly(acrylic acid), poly(methacrylic acid), poly(sodium 4-styrene sulfonate), poly(vinyl pyrrolidone), among others, and natural polymers such as lignin, polyphenols, polysaccharides, etc., are an important and promising alternative for the retention, release, and immobilization of different substrates such as

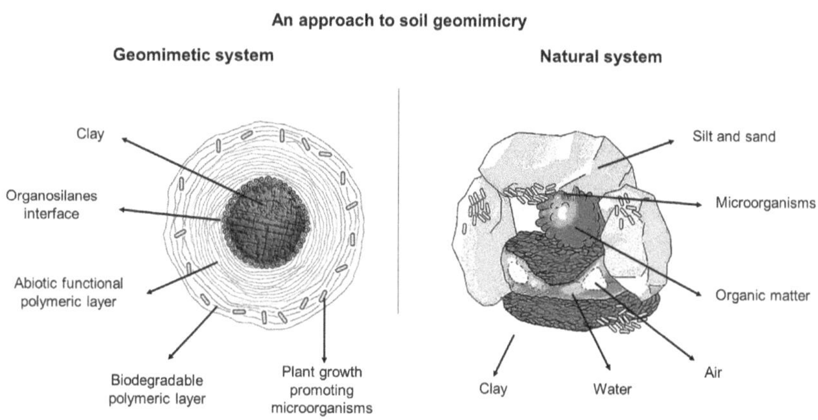

FIGURE 8.1 Comparison of the structure of the soil-inspired geomimetic system, based on covalent supramolecular hybrid systems and their natural counterparts.

macromolecules, proteins, pollutants, and micro- and macronutrients, etc. The use of this class of polymers has a great number of advantages, including that they can be anchored in a variety of functional surfaces by means of different synthesis methodologies, selecting a great variety of functional groups, for example, -COOH, -NH_2, SH, terminal double or triple bonds, etc., with susceptibility of chemical modifications in a second stage (Palencia et al., 2017; Garces et al., 2017; Zhang et al., 2017; Bee et al., 2018; Lerma et al., 2018a,b; Palencia et al., 2019; Gu et al., 2020; Lerma et al., 2020).

Finally, geomimetic systems with a high degree of structural and functional similarity of soils have been developed. In particular, the introduction of the biotic component and its functional activation is one of the great achievements in this type of materials. For the above, multilayer systems consisting of a clay core covered with a first layer, structures similar to humin-humic acids, were covered with a second polymeric layer based on sorbitol polyesters. In this way, the second layer was evaluated as a growth medium for plant growth-promoting bacteria, for which, these were inactivated by drying in order to bring them to their dormant state, and later they were activated by hydrating the system. Biodegradation of the outer envelope then allowed the release of the metabolite, a longer adaptation time, and growth of the microorganism population, with the release of water through a bio-assisted erosion mechanism, and conservation of the geomimetic particles (Garces et al., 2017; Lerma et al., 2018a,b; Lerma et al., 2020).

8.5 Conclusions and remarks

Supramolecular systems must be understood as a level of organization of matter characterized by being located above the molecular organization. These systems can be the result of covalent and noncovalent interactions, the latter being the most studied due to their importance in biological systems, and consequently, the initial and dominant referent when conceptualizing this type of system. An analysis of the levels of organization of matter allows identification of the supramolecular systems mediated through covalent and coordination interactions. Regardless of the conceptual discussion, it is clear that supramolecular systems, in the broad sense of the term, are a field with enormous technological potential.

On the other hand, the constant advances in the understanding of how molecular systems contribute to the functionality of biological systems has allowed the application of this knowledge for the design and development of materials and improved biomimetic processes, that is, systems are designed to incorporate properties that exceed the characteristics of the natural systems that were used as a model. However, due to the high complexity of natural systems, bio- and geomimicry have as their starting point the initial simplification of the model system to achieve the proper design of an artificial system. In particular, from the study of the structure and composition of geological systems such as soil, geomimicry arises as an alternative for the achievement of new materials and supramolecular assemblages capable of monitoring, supplying, improving, and rehabilitating the properties of soils and contributing to mitigation of its deterioration.

References

Abeywardena, S.B.Y., Perera, S., de Silva, N.K.M., Tissera, N.P., 2017. A facile method to modify bentonite nanoclay with silane. Int. Nano Lett. 7, 237–241. https://doi.org/10.1007/s40089-017-0214-2.

Arbeláez, N., Lerma, T., Córdoba, A., 2017. Modification of membranes by insertion of short-chain alcohols on reactive porous substrates: effect of chain length on the surface free energy. J. Sci. Technol. Appl. 2, 75–83. https://doi.org/10.34294/j.jsta.17.2.16.

Aziz, M.S., El sheriff, A.Y., 2016. Biomimicry as an approach for bio-inspired structure with the aid of computation. Alex. Eng. J. 55 (1), 707–714. https://doi.org/10.1016/j.aej.2015.10.015.

Battocchino, C., Meneghin, C., Fratoddi, I., Venditti, L., Russo, M.V., et al., 2012. Silver nanoparticles stabilized with thiols: a close look at the local chemistry and chemical structure. J. Phys. Chem. 116, 19571–19578. https://doi.org/10.1021/jp305748a.

Bee, S.L., Abdullah, M.A.A., Bee, S.T., Sin, L.T., Rahmat, A.R., 2018. Polymer nanocomposites based on silylated-montmorillonite: a review. Prog. Polym. Sci. 85, 57–82. https://doi.org/10.1016/j.progpolymsci.2018.07.003.

Bhattacharyya, R., Ghosh, B.N., Mishra, P.K., Mandal, B., Rao, C.S., Sarkar, D., Das, K., Anil, K.S., Lalitha, M., Hati, K.M., Franzluebbers, A.J., 2015. Soil degradation in India: challenges and potential solutions. Sustainability 7 (4), 3528–3570. https://doi.org/10.3390/su7043528.

Blume, H.P., Brümmer, G.W., Feige, H., Horn, R., Kandeler, E., Kögel-Knabner, I., Kretzschmar, R., Stahr, K., Wilke, B.M., 2016. Inorganic soil components—minerals and rocks. In: Scheffer/SchachtschabelSoil Science. Springer, Berlin, Heidelberg, pp. 7–53. https://doi.org/10.1007/978-3-642-30942-7_2.

Brunsveld, L., Folmer, B.J.B., Meijer, E.W., Sijbesma, R.P., 2001. Supramolecular polymers. Chem. Rev. 101 (12), 4071–4098. https://doi.org/10.1021/cr990125q.

Cornejo, J., Hermosín, M.C., 1996. Chapter 15 - interaction of humic substances and soil clays. In: Piccolo, A. (Ed.), Humic Substances in Terrestrial Ecosystems. Elsevier Science B.V., pp. 595–624. https://doi.org/10.1016/B978-044481516-3/50016-5

Fan, W., Jin, Y., Shi, L., Du, W., Zhou, R., 2020. Transparent, eco-friendly, super-tough "living" supramolecular polymers with fast room-temperature self-healability and reprocessability under visible light. Polymer 190, 122199. https://doi.org/10.1016/j.polymer.2020.122199.

Filip, Z., Demnerova, K., 2009. Humic substances as natural interacting agents: their characteristics and possible fate in environment. In: Simeonov, L.I., Hassanien, M.A. (Eds.), Exposure and Risk Assessment of Chemical Pollution - Contemporary Methodology. Springer, Netherlands, pp. 247–257. https://doi.org/10.1007/978-90-481-2335-3_17.

Fu, S., Sun, Z., Huang, P., Li, Y., Hu, N., 2019. Some basic aspects of polymer nanocomposites: a critical review. Nano Mater. Sci. 1 (1), 2–30. https://doi.org/10.1016/j.nanoms.2019.02.006.

Garcés, V., Palencia, M., Combatt, E., 2017. Development of bacterial inoculums based on biodegradable hydrogels for agricultural applications. J. Sci. Technol. Appl. 2, 13–23. https://doi.org/10.34294/j.jsta.17.2.11.

Garcés, V., Lerma, T., Palencia, M., Combatt, E., 2018. Building a probe-type passive sampler to study the contents of heavy metals in the aqueous phase in the soil. J. Sci. Technol. Appl. 4, 4–16. https://doi.org/10.34294/j.jsta.18.4.26.

Garcia, A., Combatt, E., Palencia, M., 2018. Structural study of humin and its interaction with humic acids by Fourier-transform mid-infrared spectroscopy. J. Sci. Technol. Appl. 4, 28–39. https://doi.org/10.34294/j.jsta.18.4.28.

Gliński, J., 2011. Soil phases. In: Gliński, J., Horabik, J., Lipiec, J. (Eds.), Encyclopedia of Agrophysics. Encyclopedia of Earth Sciences Series. Springer, Dordrecht, pp. 760–761. https://doi.org/10.1007/978-90-481-3585-1_151.

Gomes, B.A., Lopes, F., Andrade, M.H., 2016. Humic acids: structural properties and multiple functionalities for novel technological developments. Mater. Sci. Eng. C 62, 967–974. https://doi.org/10.1016/j.msec.2015.12.001.

Gu, Y., Dong, K., Geisen, S., Yang, W., Yan, Y., Gu, D., Liu, N., Borisjuk, N., Luo, Y., Friman, V.P., 2020. The effect of microbial inoculant origin on the rhizosphere bacterial community composition and plant growth-promotion. Plant Soil 452, 105–117. https://doi.org/10.1007/s11104-020-04545-w.

Gul, S., Kausar, A., Muhammad, B., Jabeen, S., 2016. Research progress on properties and applications of polymer/clay nanocomposite. Polym. Plast. Technol. Eng. 55 (7), 684–703. https://doi.org/10.1080/03602559.2015.1098699.

Han, L., Sun, K., Jin, J., Xing, B., 2016. Some concepts of soil organic carbon characteristics and mineral interaction from a review of literature. Soil Biol. Biochem. 94, 107–121. https://doi.org/10.1016/j.soilbio.2015.11.023.

He, H., Tao, Q., Zhu, J., Yuan, P., Shen, W., Yang, S., 2013. Silylation of clay mineral surfaces. Appl. Clay Sci. 71, 15–20. https://doi.org/10.1016/j.clay.2012.09.028.

Huang, S., Kong, X., Xiong, Y., Zhang, X., Chen, H., Jiang, W., Niu, Y., Xu, W., Ren, C., 2020. An overview of dynamic covalent bonds in polymer material and their applications. Eur. Polym. J. 141, 110094. https://doi.org/10.1016/j.eurpolymj.2020.110094.

IUPAC, 1997. Compendium of Chemical Terminology. In: (the "Gold Book"). Compiled by A. D. McNaught and A. Wilkinson, second ed. Blackwell Scientific Publications, Oxford. https://doi.org/10.1351/goldbook.ST07111.

Jauhari, J., Liao, C.Y., Chiou, J.Y., Chang, S.J., Tsai, Y.T., Lin, J.J., 2017. Functionalizing and molecular bonding nanoscale silicate-polymer composites of epoxies and Polyacrylates. J. Polym. Res. 24, 6. https://doi.org/10.1007/s10965-016-1158-3.

Kango, S., Kalia, S., Celli, A., Njuguna, J., Habibi, Y., Kumar, R., 2013. Surface modification of inorganic nanoparticles for development of organic–inorganic nanocomposites—a review. Prog. Polym. Sci. 38 (8), 1232–1261. https://doi.org/10.1016/j.progpolymsci.2013.02.003.

Kharlampieva, E., Sukhishvili, S.A., 2006. Hydrogen-bonded layer-by-layer polymer films. J. Macromol. Sci., Part C. 46 (4), 377–395. https://doi.org/10.1080/15583720600945386.

Kloster, N., Avena, M., 2015. Interaction of humic acids with soil minerals: adsorption and surface aggregation induced by Ca2+. Environ. Chem. 12, 731–738. https://doi.org/10.1071/EN14157.

Kome, G., Enang, R., Tabi, F., Yerima, B., 2019. Influence of clay minerals on some soil fertility attributes: a review. Open J. Soil Sci. 9 (9), 155–188. https://doi.org/10.4236/ojss.2019.99010.

Lambert, S., Wagner, M., 2018. Microplastics are contaminants of emerging concern in freshwater environments: an overview. In: Wagner, M., Lambert, S. (Eds.), Freshwater Microplastics : Emerging Environmental Contaminants? Springer International Publishing, Cham, pp. 1–23. https://doi.org/10.1007/978-3-319-61615-5_1.

Lehmann, J., Kleber, M., 2015. The contentious nature of soil organic matter. Nature 528, 60–68. https://doi.org/10.1038/nature16069.

Lehn, J.-M., 2002. Supramolecular polymer chemistry—scope and perspectives. Polym. Int. 51 (10), 825–839. https://doi.org/10.1002/pi.852.

Lerma, T., Martinez, G., Palencia, M., 2017. Generation of thiolated porous surfaces by interpenetrating polymeric networks: study of their surface properties. J. Sci. Technol. Appl. 3, 56–65. https://doi.org/10.34294/j.jsta.17.3.24.

Lerma, T., Combatt, E., Palencia, M., 2018a. Soil-mimicking hybrid composites based on clay, polymers and nitrogen-fixing bacteria for the development of remediation systems of degraded soil. J. Sci. Technol. Appl. 4, 17–27. https://doi.org/10.34294/j.jsta.18.4.27.

Lerma, T.A., Palencia, M., Combatt, E.M., 2018b. Soil polymer conditioner based on montmorillonite-poly(acrylic acid) composites. J. Appl. Polym. Sci. 135, 46211. https://doi.org/10.1002/app.46211.

Lerma, T.A., Garcés, V., Palencia, M., 2020. Novel multi- and bio-functional hybrid polymer hydrogels based on bentonite-poly(acrylic acid) composites and sorbitol polyesters: structural and functional characterization. Eur. Polym. J. 128, 109627. https://doi.org/10.1016/j.eurpolymj.2020.109627.

Lin, X., Grinstaff, M.W., 2013. Ionic supramolecular assemblies. J. Chem. 53 (8), 498–510. https://doi.org/10.1002/ijch.201300034.

Liu, J., Tan, C.S.Y., Yu, Z., Lan, Y., Abell, C., Scherman, O.A., 2017. Biomimetic supramolecular polymer networks exhibiting both toughness and self-recovery. Adv. Mater. 29 (14), 1604951. https://doi.org/10.1002/adma.201604951.

Luzi, F., Torre, L., Kenny, J.M., Puglia, D., 2019. Bio- and fossil-based polymeric blends and nanocomposites for packaging: structure‑Property relationship. Materials 12 (3), 471. https://doi.org/10.3390/ma12030471.

Ma, X., Tian, H., 2014. Stimuli-responsive supramolecular polymers in aqueous solution. Acc. Chem. Res. 47 (7), 1971–1981. https://doi.org/10.1021/ar500033n.

Mansoori, Y., Atghia, S.V., Zamanloo, M.R., Imanzadeh, G., Sirousazar, M., 2010. Polymer–clay nanocomposites: free-radical grafting of polyacrylamide onto organophlic montmorillonite. Eur. Polym. J. 46 (9), 1844–1853. https://doi.org/10.1016/j.eurpolymj.2010.07.006.

Meka, V.S., Sing, M.K.G., Pichika, M.R., Nali, S.R., Kolapalli, V.R.M., Kesharwani, P., 2017. A comprehensive review on polyelectrolyte complexes. Drug Discov. Today 22, 1697–1706. https://doi.org/10.1016/j.drudis.2017.06.008.

Montero, L., Meesorn, W., Moatsou, D., Weder, C., 2017. Bioinspired polymer systems with stimuli-responsive mechanical properties. Chem. Rev. 117 (20), 12851–12892. https://doi.org/10.1021/acs.chemrev.7b00168.

Nearing, M.A., Xie, Y., Liu, B., Ye, Y., 2017. Natural and anthropogenic rates of soil erosion. Int. Soil Water Conser. Res. 5 (2), 77–84. https://doi.org/10.1016/j.iswcr.2017.04.001.

Nitschke, J.R., 2014. Supramolecular and dynamic covalent reactivity. Chem. Soc. Rev. 43 (6), 1798–1799. https://doi.org/10.1039/C4CS90006E.

Palencia, M., Rivas, B.L., 2011. Metal-ion retention properties of water-soluble amphiphilic block copolymer in double emulsion systems (w/o/w) stabilized by non-ionic surfactants. J. Colloid Interface Sci. 363, 682–689. https://doi.org/10.1016/j.jcis.2011.05.089.

Palencia, M., Berrio, M., Melendrez, M., 2016a. Nanostructured polymer composites with potential applications into the storage of blood and hemoderivates. J. Sci. Technol. Appl. 1, 4–14. https://doi.org/10.34294/j.jsta.16.1.1.

Palencia, M., Restrepo, D., Combatt, E., 2016b. Functional polymer from high molecular weight linear polyols and polyurethane-based crosslinking units: synthesis, characterization, and boron retention properties. J. Appl. Polym. Sci. 133, 43895. https://doi.org/10.1002/app.43895.

Palencia, M., Córdoba, A., Arrieta, Á., 2017. Stimuli-sensitive nanostructured poly(sodium 4-styrene sulfonate): synthesis, characterization, and study of metal ion retention properties. J. Appl. Polym. Sci. 135 (11), 46001. https://doi.org/10.1002/app.46001.

Palencia, M., Córdoba, A., Melendrez, M.F., 2019. Nanocomposites based on cationic polyelectrolytes and silver nanoparticles: synthesis, characterization, molybdate retention and antimicrobial activity. Arab. J. Chem. 12 (6), 825–834. https://doi.org/10.1016/j.arabjc.2017.11.020.

Pavlidou, S., Papaspyrides, C.D., 2008. A review on polymer–layered silicate nanocomposites. Prog. Polym. Sci. 33 (12), 1119–1198. https://doi.org/10.1016/j.progpolymsci.2008.07.008.

Paz-Ferreiro, J., Fu, S., 2016. Biological indices for soil quality evaluation: perspectives and limitations. Land Degrad. Dev. 27 (1), 14–25. https://doi.org/10.1002/ldr.2262.

Pimentel, D., Burgess, M., 2013. Soil erosion threatens food production. Agriculture 3 (3), 443–463. https://www.mdpi.com/2077-0472/3/3/443.

Rossow, T., Seiffert, S., 2015. Supramolecular polymer networks: preparation, properties, and potential. In: Seiffert, S. (Ed.), Supramolecular Polymer Networks and Gels. Advances in Polymer Science, vol. 268. Springer, Cham, pp. 1–46. https://doi.org/10.1007/978-3-319-15404-6_1.

Salmon, A.R., Cormier, S., Wang, W., Abell, C., Baumberg, J., 2019. Motile artificial chromatophores: light-triggered nanoparticles for microdroplet locomotion and color change. Adv. Opt. Mat. 1900951. https://doi.org/10.1002/adom.201900951.

Shen, Y.X., Saboe, P.O., Sines, I.T., Erbakan, M., Kumar, M., 2014. Biomimetic membranes: a review. J. Membr. Sci. 454, 359–381. https://doi.org/10.1016/j.memsci.2013.12.019.

Shoffstall, A.J., Capadona, J.R., 2018. Bioinspired materials and systems for neural interfacing. Curr. Opin. Biomed. Eng. 6, 110–119. https://doi.org/10.1016/j.cobme.2018.05.002.

Silverman, H.G., Roberto, F.F., 2007. Understanding marine mussel adhesion. Mar. Biotechnol. 9, 661–681. https://doi.org/10.1007/s10126-007-9053-x.

Smernik, R.J., Kookana, R.S., 2015. The effects of organic matter–mineral interactions and organic matter chemistry on diuron sorption across a diverse range of soils. Chemosphere 119, 99–104. https://doi.org/10.1016/j.chemosphere.2014.05.066.

Stupp, S.I., Clemons, T.D., Carrow, J.K., Sai, H., Palmer, L.C., 2020. Supramolecular and hybrid bonding polymers. J. Chem. 60, 124–131. https://doi.org/10.1002/ijch.202000005.

Tetteh, R.N., 2015. Chemical soil degradation as a result of contamination: a review. J. Soil Sci. Environ. Manag. 6 (11), 301–308. https://doi.org/10.5897/JSSEM15.0499.

Torres, F.G., Troncoso, O.P., Torres, C.E., 2012. Chapter 14 mussel byssus fibres: a tough biopolymer. In: John, M.J., Thomas, S. (Eds.), Natural Polymers: Volume 1: Composites, vol. 1. The Royal Society of Chemistry, pp. 305–329. https://doi.org/10.1039/9781849735193-00305.

Toussaint, B., Raffael, B., Angers-Loustau, A., Gilliland, D., Kestens, V., Petrillo, M., Rio-Echevarria, I.M., Eede, G.V., 2019. Review of micro- and nanoplastic contamination in the food chain. Food Addit. Contam. 36 (5), 639–673. https://doi.org/10.1080/19440049.2019.1583381.

Trevisan, S., Francioso, O., Quaggiotti, S., Nardi, S., 2010. Humic substances biological activity at the plant-soil interface: from environmental aspects to molecular factors. Plant Sig. Behav. 5 (6), 635–643. https://doi.org/10.4161/psb.5.6.11211.

van Es, H., 2017. A new definition of soil. CSA News 62 (10), 20–21. https://doi.org/10.2134/csa2017.62.1016.

Villalobos, R., Medina, C., Canales, C., Melendrez, M., Flores, P., 2019. Development of nanocomposite materials with thermostable matrix from nanoreinforcement of titanium dioxide (TNPS and TNTS) and zinc oxide (ZNPS and ZNBS) on epoxy resin matrix. J. Sci. Technol. Appl. 6, 65–78. https://doi.org/10.34294/j.jsta.19.6.42.

Vurukonda, S.S., Vardharajula, S., Shrivastava, M., SkZ, A., 2016. Enhancement of drought stress tolerance in crops by plant growth promoting rhizobacteria. Microbiol. Res. 184, 13–24. https://doi.org/10.1016/j.micres.2015.12.003.

Wang, S., Yang, L., Wang, Q., Fan, Y., Shang, J., Qiu, S., Li, J., Zhang, W., Wu, X., 2018. Supramolecular self-assembly of layer-by-layer graphene film driven by the synergism of $\pi-\pi$ and hydrogen bonding interaction. J. Photochem. Photobiol. Chem. 355, 249–255. https://doi.org/10.1016/j.jphotochem.2017.09.023.

Weck, M., 2007. Side-chain functionalized supramolecular polymers. Polym. Int. 56 (4), 453–460. https://doi.org/10.1002/pi.2200.

Wu, S., Liang, F., Hu, D., Li, H., Yang, W., Zhu, Q., 2020. Determining the critical micelle concentration of surfactants by a simple and fast titration method. Anal. Chem. 92, 4259–4265. https://doi.org/10.1021/acs.analchem.9b04638.

Wypych, F., Bergaya, F., Schoonheydt, R.A., 2018. 10 - from polymers to clay polymer nanocomposites. In: Schoonheydt, R., Johnston, C.T., Bergaya, F. (Eds.), Developments in Clay Science, vol. 9. Elsevier, pp. 331–359. https://doi.org/10.1016/B978-0-08-102432-4.00010-X.

Yan, Y., Björnmalm, M., Caruso, F., 2014. Assembly of layer-by-layer particles and their interactions with biological systems. Chem. Mater. 26 (1), 452–460. https://doi.org/10.1021/cm402126n.

Yang, Y., Chen, Y., Leng, F., Huang, L., Wang, Z., Tian, W., 2017. Recent advances on surface modification of halloysite nanotubes for multifunctional applications. Appl. Sci. 7 (12), 1215. https://doi.org/10.3390/app7121215.

Yao, Z., Li, J., Xie, H., Yu, C., 2012. Review on remediation technologies of soil contaminated by heavy metals. Proc. Environ. Sci. 16, 722–729. https://doi.org/10.1016/j.proenv.2012.10.099.

Yasunari, T.J., Stohl, A., Hayano, R.S., Burkhart, J.F., Eckhardt, S., Yasurani, T., 2011. Cesium-137 deposition and contamination of Japanese soils due to the Fukushima nuclear accident. Proc. Natl. Acad. Sci. U.S.A. 108 (49), 19530–19534. https://doi.org/10.1073/pnas.1112058108.

Yu, X., Wang, C., Li, Y., 2006. Classification of protein quaternary structure by functional domain composition. BMC Bioinf. 7, 187. https://doi.org/10.1186/1471-2105-7-187.

Zaidi, A., Khan, M.S., Saif, S., Rizvi, A., Ahmed, B., Shahid, M., 2017. Role of nitrogen-fixing plant growth-promoting rhizobacteria in sustainable production of vegetables: current perspective. In: Zaidi, A., Khan, M. (Eds.), Microbial Strategies for Vegetable Production. Springer, Cham, pp. 49–79. https://doi.org/10.1007/978-3-319-54401-4_3.

Zhang, X., Chen, H., Zhang, H., 2007. Layer-by-layer assembly: from conventional to unconventional methods. Chem. Commun. 14, 1395–1405. https://doi.org/10.1039/b615590a.

Zhang, X., Peng, X., Zhang, S.W., 2017. 7 - synthetic biodegradable medical polymers: polymer blends. In: Zhang, X. (Ed.), Science and Principles of Biodegradable and Bioresorbable Medical Polymers. Woodhead Publishing., pp. 217–254. https://doi.org/10.1016/B978-0-08-100372-5.00007-6

Zhou, J., Li, J., Du, X., Xu, B., 2017. Supramolecular biofunctional materials. Biomaterials 129, 1–27. https://doi.org/10.1016/j.biomaterials.2017.03.014.

Chapter 9

Eco-friendly hydrogels

9.1 Hydrogels: concepts and characteristics

A hydrogel is a special type of material characterized as a highly hydrophilic substance, conferring its ability to absorb a large amount of water. Their high hydrophilicity is a result of polar, ionic, or ionizable functional groups on the structure of molecules conforming them; however, their water absorption capacity is directly related to their swelling capacity, which is produced by the internal flexibility of material and is a result of intermolecular interactions, both physical interactions (e.g., steric hindrance between polymer chains, strong noncovalent interactions, mechanical impossibility of reorganization of its internal structure, mechanical hindrance, and little free volume for internal molecular reorganization) and chemical interactions (e.g., cross-linking points by covalent or coordination bonds) (Sannino et al., 2009; Chen, 2020). For specific examples see Table 9.1.

In the usual terminology of hydrogel science and technology, the term "colloidal gel" is used, since when hydrogel is dispersed in water it shows viscoelastic properties due to the intra- and intermolecular interactions present in the polymer chains and the solvents incorporated in the system (Ahmed et al., 2013). Similarly, a dehydrated hydrogel is called a "xerogel" which begins to hydrate, when left in contact with water, absorbing the liquid as it swells until the thermodynamic force of swelling is fully counteracted by the elastic and retraction force exerted by the crossover points (Nayak and Das, 2018). Another commonly used adjective is "superabsorbent" which is a qualitative description that refers to the fact that the material absorbs several times its equivalent mass in water. However, there is no standard criterion for defining when a hydrogel should be described as superabsorbent, or simply as absorbent.

Thanks to their properties, to date there has been a growing demand for the use of hydrogels in different areas, including products for personal care and hygiene (diapers and sanitary napkins), pharmaceutical products (transport system and controlled drug release), healthcare products (scaffolds in tissue engineering, wound care dressings), agriculture (retention and controlled release of water and fertilizers), and even in continuous research and development activities, where materials have been developed for smart devices in

Eco-friendly Functional Polymers
https://doi.org/10.1016/B978-0-12-821842-6.00015-4
Copyright © 2021 Elsevier Inc. All rights reserved.

TABLE 9.1 Examples of different types of interactions in the internal phase of polymers.

Interaction type	Interaction	Type of material	Example	References
Physical	Steric hindrance	Hydrogel	Chitosan-PAA	Jing et al. (2000)
	Steric hindrance	Hydrogel	Hyperbranched PEG	Postnova et al. (2013)
	Mechanical	IPN	PP-PU	Lerma et al. (2016)
	Mechanical	IPN	PP-fiberglass	Lerma et al. (2017)
	Mechanical	IPN	PP- poly(VBNMDG)	Palencia et al. (2017)
Chemical	Covalent	Hydrogel	Poly(sorbitol citrate)	Garcés et al. (2017)
	Covalent	Hydrogel	Starch-Dextrose-Agar	Garcés et al. (2017)
	Covalent	Hybrid hydrogel	Clay-PAA	Lerma et al. (2018)
	Coordinating	Hydrogel	Cathecol-Fe polymers	Picchioni and Muljana (2018)
	Coordinating	Hydrogel	Disulfide-acylhydrazone	Picchioni and Muljana (2018)
	Covalent	Hydrogel	Polyureas	Otálora et al. (2019)

IPN, interpenetrating polymer network; *PAA*, poly(acrylic acid); *PEG*, poly(ethylene glycol); *PP*, poly(propylene); *PU*, poly(urethane); *VBNMDG*, (4-vinylbenzyl)-N-methyl-D-glucamine.

biomedical areas (e.g., sensors, microfluidic systems, nano-reactors, and interactive surfaces) (Ahmed, 2015; Caló and Khutoryanskiy, 2015; Chen et al., 2019; Wang et al., 2020).

The terms "eco-friendly hydrogels" or "ecological hydrogels" refer to a hydrogel whose components are mostly of natural origin and are degraded in the soil into nontoxic products for humans, animals, and plants. Therefore, its most striking characteristics are its biocompatibility and high biodegradability without the formation of toxic species after degradation (see Fig. 9.1) (Wang and Wang, 2009; Caló and Khutoryanskiy, 2015). These eco-friendly

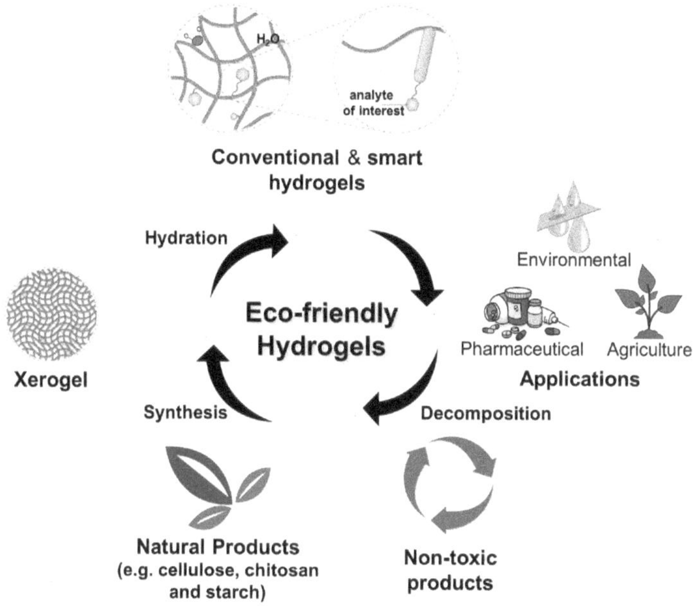

FIGURE 9.1 Representation of the life cycle of eco-friendly hydrogels.

hydrogels have begun to replace traditional hydrogels synthesized from petroleum-derived polymers, such as polyacrylamide and poly(acrylic acid), since they are not completely biodegradable, and eco-toxic and carcinogenic residues can be produced from their wastes, which is why they are considered potential pollutants of soil, and surface and groundwater, and consequently a health risk (Chadwick et al., 2015).

9.2 Classification of hydrogels

Hydrogels can be classified according to their origin, composition, nature of the cross-linking, nature of the monomer, functionality, and nature of the external stimulus capable of producing an effect on them (Ullah et al., 2015). This classification is described below.

9.2.1 Hydrogels according to the origin

Hydrogels can be classified into three categories based on their origin, including synthetic polymers, natural polymers, and hybrid polymers. Synthetic polymers for structural applications, that is, plastics, are usually hydrophobic or have very low polarity, in addition, they can be generalized to show good mechanical and thermal properties and be chemically more resistant than natural polymers, which usually are more susceptible to chemical oxidation. However, synthetic polymers used in making hydrogels do not have the same relationship in terms of properties as their analog hydrogels. The main condition of hydrogel is the absorption and retention of water, as a consequence, the polymer molecules that make up the hydrogel must be largely hydrophilic. This does not exclude hydrophobic molecules from being part of the hydrogel, since the requirement of its nature as a hydrogel is functional and not the chemical nature of its components. Examples of polymers used for typical synthetic hydrogels are poly(acrylic acid), poly(-styrene sulfonate), poly(ethylene glycol), etc.˙ On the other hand, natural hydrogels are obtained from natural polymers, or from precursors obtained from natural sources. In comparison with synthetic hydrogels, natural hydrogels are safer and, in many cases, they are biocompatible and biodegradable. Examples of natural hydrogels include protein-based hydrogels (e.g., gelatin, collagen), DNA-based hydrogels, hydrogels based on polyesters obtained from low-molecular-weight biomolecules, and polysaccharide-based hydrogels (e.g., starch, hyaluronic acid, chitosan, alginate, and cellulose) (Liu et al., 2008; Sannino et al., 2009; Palencia et al., 2017a, 2017b; Park et al., 2018; Notodihardjo et al., 2019; Seidlits et al., 2019). In contrast, hybrid hydrogels are defined using two different approaches. In the first, these hydrogels are prepared by adding some synthetic parts to natural substrates, for example, the graft copolymerization of vinyl monomers on polysaccharides (Kandile et al., 2019; Peng et al., 2019) or grafted starch-acrylic acid polymers (Pei et al., 2013; Peng et al., 2016; Ye et al., 2017). Note that this definition includes two different types of materials in terms of synthetic or natural origin. In this way, blends of polymers meeting the above condition also should be conceptualized as hybrid hydrogels. However, the concept "hybrid hydrogels" also refers to those hydrogels formed from hydrophilic polymers and a non-organic component, for example, clays or metals. Some examples are geomimetic hydrogels obtained from montmorillonite and poly(acrylic acid) (Lerma et al., 2018), bentonite and poly(acrylic acid) functionalized with sorbitol polyesters (Lerma et al., 2020), and hydrogels based on starch and silver nanoparticles (Palencia et al., 2017a).

9.2.2 Hydrogels according to the polymer composition

Regarding polymer composition, hydrogels can be classified as homopolymers or copolymers, depending on whether the polymer is composed of one, two, or more types of monomer. In the case of hydrogels composed of two or more different monomer types, at least one of the components must be hydrophilic in order to give the desired functionality (Batista et al., 2019). This classification is an extension of the conventional classification of polymers. As a consequence, it is reasonable to assume that classification can be easily expanded to describe the different types of copolymers: random, alternating, graft, and block copolymers.

9.2.3 Hydrogels according to the nature of cross-linking

Hydrogels can be classified from the nature of the cross-linking, and thus, depending on whether their cross-linking is physical or chemical. In the first case, physical processes such as hydrophobic association, the aggregation, entanglement of chains, hydrogen bonds, and forming of interpenetrating polymer networks could exist in the polymer matrix. Meanwhile, in chemical cross-linking, covalent interactions, which can be purely covalent or by co-ordination bonds, are responsible for holding the polymer chains together, forming a single highly hydrophilic three-dimensional network. These hydrogels are characterized as stronger, and more resistant and stable than physical-cross-linking hydrogels, and, therefore, they are the most commonly used hydrogels in conditions of excess water content (Pei et al., 2013; Ye et al., 2017; Peng et al., 2016). In contrast, and as a special case, hydrogels based on interpenetrating polymeric networks are made of two or more physically cross-linked polymers, which are part of a double network. This type of network can be completely interpenetrated or semi-interpenetrated, in the first case, the interconnection and mechanical restriction of chains forming each of the networks is total, whereas in a semi-interpenetrated network, one polymer is a cross-linked polymer and the other is a noncross-linked polymer (Ahmed, 2015).

9.2.4 Hydrogels according to the nature of the monomer

Monomers are the basic units that make up the polymer chain. The physical and chemical properties of the macromolecules will be directly related to their nature. Strictly speaking, it is expected that in a polymer most of the monomers present are hydrophilic, however, in the case of copolymers, non-hydrophilic monomers may be forming the polymer chain to a lesser extent. Therefore, interest in the nature of monomers focuses almost exclusively on those of a hydrophilic nature. These can be classified as neutral monomers with polar groups, monomers with ionic groups (e.g., R_4N^+, N-SO_3^-), with ionizable groups (e.g., R-COOH, RNH_2), and zwitterionic monomers (Ahmed, 2015).

As can be concluded from the different types of polymers that make up the hydrogel, and from the different types of hydrogel, the combinations of different types of monomer in the same network make their difficult classification. Consequently, the practical utility of this classification can be strongly limited. Thus, for example, chitosan-based hydrogels are made up of monomeric units containing ionizable groups (i.e., R_2-NH) and neutral, polar, and nonionizable groups (i.e., R-OH). The same chitosan can be quaternized, and, depending on the degree of quaternization achieved, the groups present in this new hydrogel will be ionizable, ionic, and polar neutral groups (i.e., R-OH, R_2-NH, and R_4N^+). In addition, the chain inserted for quaternization can contain other functional groups than those initially present (e.g., R_1-OCO-R_2 and R-COOH) (Palacio et al., 2019).

9.2.5 Hydrogels according to the function and nature of the external stimulus

It is also possible to classify hydrogels as conventional or intelligent, depending on whether their function is to retain and release water, or if, in addition to this, they retain and release different analytes of interest by a controlled mechanism, such as fertilizers, phytohormones, proteins, drugs, etc. This is in response to small changes in the environment, such as pH or ionic strength. Furthermore, these changes are reversible; therefore, hydrogels are able to return to their initial state after a reaction as soon as desired. For all the above, smart hydrogels are currently the subject of considerable scientific research due to their potential in high-tech applications in the biomedical, pharmaceutical, biotechnological, bioseparation, biosensor, agriculture, and cosmetic fields (Ullah et al., 2015). Smart hydrogels can respond to external stimuli, either biochemically (antigens, enzymes, ligands), chemically (pH, solvent composition, ionic strength, molecular species), or physically (temperature, electric or magnetic field, light, pressure, sound) (Cui et al., 2019; Zhang et al., 2020a,b). This stimulus-sensitive behavior makes hydrogels attractive for the design of "smart" devices such as ionic skin, tissue engineering, wound dressing, healing, etc. (Sun et al., 2014; Yegappan et al., 2019; Craciun et al., 2019; Martínez-Martínez et al., 2019; Zhang et al., 2019a,b; Hsu et al., 2019). Details of the applications are given in Chapters 12–19.

9.3 Synthesis of hydrogels

In general, hydrogels are synthesized from one or more polar monomers, with the chemical properties of the monomer being the main factor determining the presence of an initiator and/or cross-linking agent. When synthesis of eco-friendly is carried out, it is expected that all components should be nontoxic and, preferibly, thay should be able to hydrolyze their bonds in the presence of degrading agents. In addition, eco-friendly hydrogels also can be obtained from natural sources such as cellulose, starch, or various natural polymers that have suitable functional groups or have been functionalized with polymerizable groups that promote cross-linking (Shantha, 2002).

During synthesis reactions, in order to control the heat of polymerization and the final properties of the hydrogels, it is possible to use diluents, such as water or other aqueous solutions. Finally, the hydrogel must be washed to remove any impurities left over from the preparation process. These include unreacted monomers, initiators, cross-linkers, and by-products. The wide variety of starting materials allows the preparation of hydrogels with the desired physical properties for a given application (Ahmed, 2015). Considering that a hydrogel is simply a cross-linked hydrophilic polymeric network, any technique that can be used to create a cross-linked polymer can be used to produce a hydrogel, either by bulk, solution, suspension, or reverse suspension polymerization, by grafting or by irradiation (Pei et al., 2013; Peng et al., 2016 Ye et al., 2017). Each of these polymerization techniques is described and summarized in Table 9.2.

TABLE 9.2 Summary and description of polymerization techniques useful for making hydrogels.

Preparation technique	Description	References
Bulk polymerization	The polymerization reaction is usually started with ultraviolet radiation, an increase in temperature, or chemical catalysts. It is carried out in the absence of solvents	Suda (2007)
Polymerization in solution	The polymerization is initiated thermally by ultraviolet irradiation, temperature, or by a redox initiator system. The presence of solvent acts as a heat sink. It is carried out in the presence of solvents	Ahmed (2015)
Suspension and reverse suspension polymerization	Monomers and the initiator are mixed and dispersed in a phase of different polarity. Continuous stirring and the addition of a surfactant agent are usually required. The viscosity of the monomer solution, stirring speed, rotor design, and dispersant type primarily govern the size and shape of the resin particles. Powder or microspheres of polymer are obtained	Watanabe et al. (1993); Ogata (2006)
Graft polymerization	This corresponds with a copolymerization reaction. A polymer is grown using functional groups of another polymer as an anchor point, or by reaction through functional groups, an already-formed polymer chain is chemically anchored to another polymer. The grafts can be low-molecular-weight molecules and oligomers	Talaat et al. (2008); Duquette et al. (2020)
Radiation polymerization	High-energy ionizing radiation is used to activate the unsaturated monomers. During the synthesis process, the formation of free radicals is generated both in the starting monomers, polymers, and in the water molecules. Recombination of these radicals results in the formation of a cross-linked structure	Wach et al. (2002); Ajji et al. (2008); Makuuchi (2010)

Continued

TABLE 9.2 Summary and description of polymerization techniques useful for making hydrogels.—cont'd

Preparation technique	Description	References
Polyelectrolyte complexing	The reaction requires at least two polymers with opposite charges, the interaction between the surface charges brings about the formation of the hydrogel. Noncovalent polymers are obtained	Palacio et al. (2019)
Interpenetrating polymer network (IPN)	IPN can be obtained by: (1) monomer polymerization and cross-linking in the inside of a cross-linked polymer, (2) simultaneous monomer polymerization, (3) monomer linear polymerization in the inside of a cross-linked polymer, and (4) mixing of two polymers and subsequent cross-linking of both, or one of them	Palencia et al. (2014a, 2016a)
Core–shell systems	The reaction is started by anchoring a polymer to an organic or inorganic core. Subsequently, through reactions of functional groups, a second hydrogel layer is built as a coating of the first. Polymer layers can be by grafting or by in situ synthesis of polymers	Lerma et al. (2020)
Cross-linking reaction via side chain	A polymer with reactive functional groups can be cross-linked through its functional groups intra- or intermolecularly. The cross-linking can be mediated by a low-molecular-weight cross-linker, an oligomer, or a polymer	Palencia et al. (2016b)
Physical mixtures	A physical mixture of hydrophilic polymer with high molecular weights or high-branched chains produces a polymer matrix physically prevented from reorganization and dissolution. Therefore, the addition of water makes these systems behave like physical hydrogels	Ahmed (2015)
Hydrodynamic hydrogels	Hydrogels are obtained in situ by the application of pressure to polymer dissolution in the inside of an ultrafiltration membrane system. A dead-end membrane module is used. Hydrogel is formed as a result of concentration polarization and the density can be changed by changes to pressure, concentration, and stirring rate	Palencia et al. (2014b, 2014c)
Polymer hydration	Hydration of high-molecular-weight hydrophilic polymers by the addition of water, or an aqueous solution, in an amount below that required for dissolution results in the formation of a hydrogel	Ahmed (2015)

9.4 Applications of hydrogels

In general, the application of hydrogels is focused on taking advantage of its water absorption capacity. They have multiple applications in different fields of science and technology. Details of these are shown in later chapters, as follows: applications in agriculture (see Chapter 12), in medicine and biomedicine (see Chapter 17), in the study of the environment and environmental remediation activities (see Chapter 13), in microbiology (see Chapters 27, 28, and 29), in the food industry (see Chapter 14), in the pharmaceutical industry (see Chapter 18), in the production of nanoparticles (see Chapter 26), etc.

9.5 Conclusions

Hydrogels are materials characterized by absorbing water as a result of their high hydrophilicity. This unique feature makes them versatile materials for multiple applications. Clearly their properties can be modulated by controlling the properties of the polymers that make them up, and by controlling the composition of the liquid absorbed by them. Although they are widely studied materials, with the exception of hydrodynamic polymers, there have been no major advances in the development of new hydrogels. This should not be confused with the development of specific hydrogels with multiple properties. The incursion of nanotechnology has been another point of progress in this type of materials. Of course, in the latter case it is the direct consequence of this incursion of the scientific world in the development of nanotechnological systems. Undoubtedly the greatest advances in hydrogels in the medium term will be directed at the construction of intelligent, multifunctional, nanostructured systems, and based on biodegradable polymers.

References

Ahmed, E., Aggor, F., Awad, A., El-Aref, A., 2013. An innovative method for preparation of nanometal hydroxide superabsorbent hydrogel. Carbohydr. Polym. 91, 693–698. https://doi.org/10.1016/j.carbpol.2012.08.056.

Ahmed, E., 2015. Hydrogel: preparation, characterization, and applications: a review. J. Adv. Res. 6, 105–121. https://doi.org/10.1016/j.jare.2013.07.006.

Ajji, Z., Mirjalili, G., Alkhatab, A., Dada, H., 2008. Use of electron beam for the production of hydrogel dressings. Radiat. Phys. Chem. 77, 200–202. http://doi./10.1016/j.radphyschem.2007.05.016.

Batista, R., Perez, P., Siqueira, J., Machado, M., Cerqueira, M., Teixeira, J., Cordeiro, J., 2019. Hydrogel as an alternative structure for food packaging systems. Carbohydr. Polym. 205, 106–116. https://doi.org/10.1016/j.carbpol.2018.10.006.

Caló, E., Khutoryanskiy, V., 2015. Biomedical applications of hydrogels: a review of patents and commercial products. Eur. Polym. J. 65, 252–267. https://doi.org/10.1016/j.eurpolymj.2014.11.024.

Chadwick, D., Wei, J., Yanan, T., Guanghui, Y., Qirong, S., Qing, C., 2015. Improving manure nutrient management towards sustainable agricultural intensification in China. Agric. Ecosyst. Environ. 209, 34–46. https://doi.org/10.1016/j.agee.2015.03.025.

Chen, Y., 2020. 1 - properties and development of hydrogels. In: Chen, Y. (Ed.), Hydrogels Based on Natural Polymers. Elsevier, pp. 3–16. https://doi.org/10.1016/B978-0-12-816421-1.00001-X.

Chen, Y., Binti, S., Yahya, M., Lee, H., 2019. 6 - novel superabsorbent cellulose-based hydrogels: present status, synthesis, characterization, and application prospects. In: Mondal, I. (Ed.), Cellulose-Based Superabsorbent Hydrogels, Polymers and Polymeric Composites: A Reference Series. Springer, pp. 156–189. https://doi.org/10.1007/978-3-319-77830-3_9.

Craciun, M., Tartau, M., Pinteala, M., Marin, L., 2019. Nitrosalicyl-imine-chitosan hydrogels based drug delivery systems for long term sustained release in local therapy. J. Colloid Interface Sci. 536, 196–207. https://doi.org/10.1016/j.jcis.2018.10.048.

Cui, X., Lee, J., Chen, W., 2019. Eco-friendly and biodegradable cellulose hydrogels produced from low cost okara: towards non-toxic flexible electronics. Sci. Rep. 9, 18166. https://doi.org/10.1038/s41598-019-54638-5.

Duquette, D., Nzediegwu, C., Portillo-Perez, G., Dumont, M., Prasher, S., 2020. Eco-friendly synthesis of hydrogels from starch, citric acid, and itaconic acid: swelling capacity and metal chelation properties. Starch - Stärke 72, 1900008. https://doi.org/10.1002/star.201900008.

Garcés, V., Palencia, M., Combatt, E., 2017. Development of bacterial inoculums based on biodegradable hydrogels for agricultural applications. J. Sci. Technol. Appl. 2, 13–23. https://doi.org/10.34294/j.jsta.17.2.11 (2017).

Hsu, L., Peng, B., Chen, M., Thalib, B., Ruslin, M., Xuan, T., Chou, H., Ou, K., 2019. The potential of the stem cells composite hydrogel wound dressings for promoting wound healing and skin regeneration: in vitro and in vivo evaluation. J. Biomed. Mater. Res. B Appl. Biomater. 107, 278–285. https://doi.org/10.1002/jbm.b.34118.

Jing, Y., Khutoryanskiy, V., Kan, V., Gabdulina, Y.R., Mun, G., et al., 2000. Interaction of chitosan with hydrogel of poly(acrylic acid) and preparation of encapsulated drugs. Euras. Chem. Technol. J. 3, 191–194. https://doi.org/10.18321/ectj565.

Kandile, N., Mohamed, H., 2019. Chitosan nanoparticle hydrogel based sebacoyl moiety with remarkable capability for metal ion removal from aqueous systems. Int. J. Biol. Macromol. 122, 578–586. https://doi.org/10.1016/j.ijbiomac.2018.10.198.

Lerma, T., Collazos, S., Córdoba, A., 2016. Effect of side chain length of carbamates on the surface properties of porous interpenetrating networks. J. Sci. Technol. Appl. 1, 30–38. https://doi.org/10.34294/j.jsta.16.1.3.

Lerma, T.A., Combatt, E.M., Palencia, M., 2018. Soil-mimicking hybrid composites based on clay, polymers and nitrogen-fixing bacteria for the development of remediation systems of degraded soil. J. Sci. Technol. Appl. 4, 17–27. https://doi.org/10.34294/j.jsta.18.4.27.

Lerma, T.A., Garcés, V., Palencia, M., 2020. Novel multi- and bio-functional hybrid polymer hydrogels based on bentonite-poly(acrylic acid) composites and sorbitol polyesters: structural and functional characterization. Eur. Polym. J. 128, 109627. https://doi.org/10.1016/j.eurpolymj.2020.109627.

Lerma, T., Martinez, G., Palencia, M., 2017. Generation of thiolated porous surfaces by interpenetrating polymeric networks: study of their surface properties. J. Sci. Technol. Appl. 3, 56–65. https://doi.org/10.34294/j.jsta.17.3.24.

Liu, W., Griffith, M., Fengfu, L., 2008. Alginate microsphere-collagen composite hydrogel for ocular drug delivery and implantation. J. Mater. Sci. Mater. Med. 19, 3365–3371. https://doi.org/10.1007/s10856-008-3486-2.

Makuuchi, K., 2010. Critical review of radiation processing of hydrogel and polysaccharide. Radiat. Phys. Chem. 79, 267–271. https://doi.org/10.1016/j.radphyschem.2009.10.011.

Martínez-Martínez, M., Rodríguez-Berna, G., Bermejo, M., Gonzalez-Alvarez, I., Gonzalez-Alvarez, M., Merinoc, V., 2019. Covalently crosslinked organophosphorous derivatives-chitosan hydrogel as a drug delivery system for oral administration of camptothecin. Eur. J. Pharm. Biopharm. 136, 174–183. https://doi.org/10.1016/j.ejpb.2019.01.009.

Nayak, A., Das, B., 2018. 1 - introduction to polymeric gels. In: Pal, K., Banerjee, I. (Eds.), Polymeric Gels. Woodhead Publishing Series in Biomaterials, pp. 3–27. https://doi.org/10.1016/B978-0-08-102179-8.00001-6.

Notodihardjo, S., Morimoto, N., Kakudo, N., Mitsui, T., Minh, T., Tabata, Y., Kusumoto, K., 2019. Efficacy of gelatin hydrogel impregnated with concentrated platelet lysate in murine wound healing. J. Surg. Res. 234, 190–201. https://doi.org/10.1016/j.jss.2018.09.037.

Ogata, T., Nagayoshi, K., Nagasako, T., Kurihara, S., Nonaka, T., 2006. Synthesis of hydrogel beads having phosphinic acid groups and its adsorption ability for lanthanide ions. React. Funct. Polym. 66, 625–633. https://doi.org/10.1016/j.reactfunctpolym.2005.10.015.

Otálora, A., Lerma, T.A., Palencia, M., 2019. Synthesis and characterization of polurea-based Hydrogels by Multicomponent Polycondensation of 1,6-Hexamethylenediisocyanate, sorbitol and cysteine. J. Sci. Technol. Appl. 7, 5–16. https://doi.org/10.34294/j.jsta.19.7.47.

Palacio, D., Urbano, B.F., Palencia, M., Rivas, B.L., 2019. Preparation of alkylated chitosan-based polyelectrolyte hydrogels: the effect of monomer charge on polymerization. Eur. Polym. J. 118, 551–560. https://doi.org/10.1016/j.eurpolymj.2019.06.024.

Palencia, M., Córdoba, A., Rivas, B.L., 2014c. Concentration–polarization effect of poly(sodium styrene sulfonate) on size distribution of colloidal silver nanoparticles during diafiltration experiments. Colloid Polym. Sci. 292, 619–626. https://doi.org/10.1007/s00396-013-3096-5.

Palencia, M., Lerma, T., Córdoba, A., 2016a. Polyurethanes with boron retention properties for the development of agricultural fertilization smart systems. J. Sci. Technol. Appl. 1, 39–52. https://doi.org/10.34294/j.jsta.16.1.4.

Palencia, M., Lerma, T., Berrio, M., 2017. Thin-film composite by in situ polymerization of 4-chloromethyl styrene functionalized with N-methyl-D-glucamine in pore-type micro-reactors. J. Sci. Technol. Appl. 3, 66–76. https://doi.org/10.34294/j.jsta.17.3.25.

Palencia, M., Lerma, T., Combatt, E., 2017a. Hydrogels based in cassava starch with antibacterial activity for controlled release of cysteamine-silver nanostructured agents. Curr. Chem. Biol. 11, 28–35. https://doi.org/10.2174/2212796810666161108152319.

Palencia, M., Mora, M., Palencia, S., 2017b. Biodegradable polymer hydrogels based in sorbitol and citric acid for controlled release of bioactive substances from plants (polyphenols). Curr. Chem. Biol. 11, 36–43. https://doi.org/10.2174/2212796810666161028114432.

Palencia, M., Restrepo, D.F., Combatt, E., 2016b. Functional polymer from high molecular weight linear polyols and polyurethane-based crosslinking units: synthesis, characterization, and boron retention properties. J. Appl. Polym. Sci. 133, 43895. https://doi.org/10.1002/app.43895.

Palencia, M., Rivas, B.L., Valle, H., 2014b. Size separation of silver nanoparticles by dead-end ultrafiltration: description of fouling mechanism by pore blocking model. J. Membr. Sci. 455, 7–14. https://doi.org/10.1016/j.memsci.2013.12.051.

Palencia, M., Vera, M., Combatt, E., 2014a. Polymer networks based in (4-vinylbenzyl)-N-methyl-D-glucamine supported on microporous polypropylene layers with retention boron capacity. J. Appl. Polym. Sci. 131, 40653. https://doi.org/10.1002/app.40653.

Park, S., Mun, S., Kim, Y., 2018. Effect of xanthan gum on lipid digestion and bioaccessibility of β-carotene-loaded rice starch-based filled hydrogels. Food Res. Int. 105, 440–445. https://doi.org/10.1016/j.foodres.2017.11.039.

Pei, Y., Wang, X., Huang, W., Liu, P., Zhang, L., 2013. Cellulose-based hydrogels with excellent microstructural replication ability and cytocompatibility for microfluidic devices. Cellulose 20, 1897–1909. https://doi.org/10.1007/s10570-013-9930-6.

Peng, H., Huang, Q., Yue, H., Li, Y., Wu, M., Liu, W., Zhang, G., Fu, S., Zhang, J., 2019. The antitumor effect of cisplatin-loaded thermosensitive chitosan hydrogel combined with radiotherapy on nasopharyngeal carcinoma. Int. J. Pharm. 556, 97–105. https://doi.org/10.1016/j.ijpharm.2018.11.068.

Peng, N., Wang, Y., Ye, Q., Liang, L., An, Y., Li, Q., Chang, C., 2016. Biocompatible cellulose-based superabsorbent hydrogels with antimicrobial activity. Carbohydr. Polym. 137, 59–64. https://doi.org/10.1016/j.carbpol.2015.10.057.

Picchioni, F., Muljana, H., 2018. Hydrogels based on dynamic covalent and non-covalent bonds: a chemistry perspective. Gels 4, 21. https://doi.org/10.3390/gels4010021.

Postnova, I., Silant'ev, V., Kim, M.H., Song, G.Y., Kim, I., et al., 2013. Hyperbranched polyglycerol hydrogels prepared through biomimetic mineralization. Colloids Surf. B Biointerfaces 103, 31–37. https://doi.org/10.1016/j.colsurfb.2012.10.026.

Sannino, A., Demitri, C., Madaghiele, M., 2009. Biodegradable cellulose-based hydrogels: design and applications. Materials 2, 353–373. https://doi.org/10.3390/ma2020353.

Seidlits, S., Liang, J., Bierman, R., Sohrabi, A., Karam, J., Holley, S., Cepeda, C., Peptide-Modified, W.C., 2019. Hyaluronic acid-based hydrogels as a 3D culture platform for neural stem/progenitor cell engineering. J. Biomed. Mater. Res. 107, 704–718. https://doi.org/10.1002/jbm.a.36603.

Shantha, K., Harding, D., 2002. Synthesis and evaluation of sucrose-containing polymeric hydrogels for oral drug delivery. J. Appl. Polym. Sci. 84, 2597. https://doi.org/10.1002/app.10378.

Suda, K., 2007. Superabsorbent polymers and superabsorbent polymer composites. Sci. Asia 33, 39–43. https://doi.org/10.2306/scienceasia1513-1874.2007.33(s1).039.

Sun, J., Keplinger, C., Whitesides, G., Suo, Z., 2014. Ionic skin. Adv. Mater. 26, 7608–7614. https://doi.org/10.1002/adma.201403441.

Talaat, H., Sorour, M., Aboulnour, A., Shaalan, H., Ahmed, E., Awad, A., Ahmed, M., 2008. Development of a multi-component fertilizing hydrogel with relevant techno-economic indicators. Am.-Eurasian J. Agric. Environ. Sci. 3, 764–770.

Ullah, F., BisyrulM, Othman, H., Javed, F., Ahmad, Z., 2015. Akil H. Classification, processing and application of hydrogels: a review. Mater. Sci. Eng. C 57, 414–433. https://doi.org/10.1016/j.msec.2015.07.053.

Wach, R., Mitomo, H., Yoshii, F., Kume, T., 2002. Hydrogel of radiation-induced cross-linked hydroxypropylcellulose. Macromol. Mater. Eng. 287, 285–295. https://doi.org/10.1002/1439-2054(20020401)287:4<285::AID-MAME285>3.0.CO;2-3.

Wang, W., Wang, A., 2009. Preparation, characterization and properties of superabsorbent nanocomposites based on natural guar gum and modified rectorite. Carbohydr. Polym. 77, 891–897. https://doi.org/10.1016/j.carbpol.2009.03.012.

Wang, Y., Cao, H., Wang, X., 2020. Synthesis and characterization of an injectable ε-polylysine/carboxymethyl chitosan hydrogel used in medical application. Mater. Chem. Phys. 248, 122902. https://doi.org/10.1016/j.matchemphys.2020.122902.

Watanabe, N., Hosoya, Y., Tamura, A., Kosuge, H., 1993. Characteristics of water-absorbent polymer emulsions. Polym. Int. 30, 525–531. https://doi.org/10.1002/pi.4990300417.

Ye, D., Cheng, Q., Zhang, Q., Wang, Y., Chang, C., Li, L., Peng, H., Zhang, L., 2017. Deformation drives alignment of nanofibers in framework for inducing anisotropic cellulose hydrogels with high toughness. ACS Appl. Mater. Interfaces 9, 43154–43162. https://doi.org/10.1021/acsami.7b14900.

Yegappan, R., Selvaprithiviraj, V., Amirthalingam, S., Mohandas, A., Hwang, N., Jayakumar, R., 2019. Injectable angiogenic and osteogenic carrageenan nanocomposite hydrogel for bone tissue engineering. Int. J. Biol. Macromol. 122, 320–328. https://doi.org/10.1016/j.ijbiomac.2018.10.182.

Zhang, Q., Yu, G., Zhou, Q., Li, J., Feng, Y., Wang, L., Tang, Y., Peng, Y., 2020a. Eco-friendly interpenetrating network hydrogels integrated with natural soil colloid as a green and sustainable modifier for slow release of agrochemicals. J. Clean. Prod. 269, 122060. https://doi.org/10.1016/j.jclepro.2020.122060.

Zhang, R., Lei, L., Song, Q., Li, X., 2019a. Calcium ion cross-linking alginate/dexamethasone sodium phosphate hybrid hydrogel for extended drug release. Colloids Surf. B Biointerfaces 175, 569–575. https://doi.org/10.1016/j.colsurfb.2018.11.083.

Zhang, X., Ma, X., Hou, T., Guo, K., Yin, J., Wang, Z., Shu, L., He, M., Yao, J., 2019b. Inorganic salts induce thermally reversible and anti-freezing cellulose hydrogels. Angew. Chem. Int. Ed. 131, 7444–7448. https://doi.org/10.1002/ange.201902578.

Zhang, X., Liu, Y., Lu, P., Zhang, M., 2020b. Preparation and properties of hydrogel based on sawdust cellulose for environmentally friendly slow release fertilizers. Green Process. Synth. 9, 139–152. https://doi.org/10.1515/gps-2020-0015.

Chapter 10

Eco-friendly chemical transformations: strategies and technologies based on green chemistry principles, click chemistry, and eco-sustainable development

10.1 Chemical transformations in the current environment

Chemistry and chemical transformations play a key role nowadays because they allow the control of matter, its properties, the design of new products, and transformation of raw materials in various sectors including pharmaceuticals, agrochemicals, textiles, plastics, coatings, etc. (Fortunak, 2010; Parveen et al., 2016). However, some transformations end up being relatively inefficient, for example, it is estimated that only about 10% of the resources obtained from the Earth by extractive industries, end up in the final product, with 90% ending up as waste in various forms (Glenn et al., 2015; Li, 2016a,b). In addition, some chemical reactions require multiple steps during the synthesis and purification processes, large volumes of organic solvents, high-energy consumption, and the use of toxic catalysts and reagents, making the processes costly and harmful to the environment. Therefore, it is sought to increase the efficiency of resources so that ecological processes can be implemented in accordance with the 12 principles of green chemistry and under the principles of sustainable development (see Chapter 1). The importance of this is illustrated by the Millennium Project, which recently listed sustainability as one of the 15 global challenges facing humanity (Vaidya and Mayer, 2016), and similarly, the Global Agreement of the United Nations, which focuses on sustainability as a goal for future global businesses (Izzo et al., 2020). In accordance with the above, this chapter addresses strategies and technologies based on green chemistry, click chemistry, and sustainable development.

Eco-friendly Functional Polymers
https://doi.org/10.1016/B978-0-12-821842-6.00027-0
Copyright © 2021 Elsevier Inc. All rights reserved.

10.2 Definitions of green chemistry, click chemistry, and eco-sustainable development

The term "green chemistry" presented by the scientific community in 1991, was aimed at the elimination or reduction of hazardous substances, in order to reduce the exposure of humans and the environment to chemicals during synthesis procedures in accordance with its 12 key principles (Anastas and Warner, 1998; Ribeiro et al., 2010). These principles consist of preventing unnecessary waste (e.g., by recycling solvents, catalysts, and reagents), economy of matter (i.e., by minimizing the loss of precursors and intermediate compounds), avoiding harmful synthetic methods (e.g., by using safe and low-toxicity chemicals), reducing the toxicity of the product (e.g., by using low-toxicity separation methods), using safe solvents or avoiding their use altogether, reducing energy consumption (e.g., by avoiding high temperatures and pressures), using renewable resources, reducing the number of steps in synthesis (e.g., by avoiding derivatization), using selective catalysts to increase performance, using a design for degradation, process monitoring (e.g., prevention of contamination through permanent reaction analysis), and finally, preventing negative impacts on human health and the environment (Marquez and Machado, 2014; Unterlass, 2016; Kharissova et al., 2019). Currently, these principles are used in specialized technologies dedicated to water purification, air purification, energy generation and storage, manufacture of electronic products, medicines, plastics and pesticides, polymers, etc. To give a perspective on the extent of the problem, in 2015 alone, approximately 322 million tons of plastics (synthetic polymers derived from petroleum) were produced, and consequently, plastics have become the most produced synthetic materials on Earth. This is in contrast to the fact that the most abundant biological material on Earth, cellulose, is also polymeric in nature (Worthington et al., 2017).

However, although the solution may be thought of as simple, that is, the problem is solved by replacing plastic-based products with biopolymer-based products, the truth is that it is not as simple when taking a broad approach. For example, if the millions of tons of petroleum-derived plastics were replaced by the equivalent millions of tons of biodegradable biopolymers, what would happen to the millions of tons of by-products that were previously used as inputs for the manufacture of plastics and that under this new context would no longer be used? Consequently, it is clear that a paradox arises, which we might call the "useful-unused waste paradox" (which refers to the fact that if a waste is useful, it cannot be considered a waste, but if it is useful and not used, then even if it is not a waste, it will become one, because it is its use and not its production which removes it from the category of waste). This apparent contradiction comes from the fact that replacing plastics only changes the context of the problem, but not the problem itself. The use of petroleum derivatives to produce plastic materials is in itself a solution to the waste problem of a global industry, that is highly profitable and "dirty" in terms of the high volume of waste it generates. In this context, plastic materials are a technology-based solution to the use of waste, giving it added value, and transforming it from waste into useful materials. Therefore, it is clear that the replacement of plastics by biopolymers, although these are biodegradable, does not reduce the volume of waste derived from oil, since they continue to be produced in the same quantity but in a different way. It is at this point where the paradox is configured, and from which the importance of the balance between the problem and the solution must be made, taking into account the context and related aspects. The conclusions that have been derived are interesting, and allow us to call attention to the importance of identifying the real problem, which is contextualized with the following premise, and could even be considered the basis of eco-sustainable development: It The requirement for clean energy sources for the efficient replacement of oil as an energy resource is urgent.

On the other hand, click chemistry was described by Huisgen in 1950. However, it was Sharpless, the winner of the Nobel Prize in Chemistry in 2001, who in 1999 coined and presented this term at the 217th annual meeting of the American Chemical Society. He used the term to refer to highly efficient synthetic reactions where molecular building blocks readily intervene under ambient conditions. As a result, click chemistry has become a very popular topic, as evidenced by the almost exponential growth in the number of related publications (Kolb et al., 2001; Hein et al., 2008). The term "click chemistry" refers to a group of reactions that are fast, easy to apply and purify, versatile, with a wide tolerance of functional groups, regiospecific, and with a high yield, whose by-products must be harmless and allow its elimination by nonchromatographic methods, but also, the reaction must be carried out under ambient conditions, preferably in water or without solvents. The general rule for this approach is that "... all synthetic explorations should be limited, as far as possible, to easily synthesized molecules" (Kolb et al., 2001; Hein et al., 2008).

Currently, transformation strategies and technologies are projected to be based on eco-sustainable development, which, as the name indicates, means that environmental characteristics and other sustainability characteristics must be met, that is, all efforts are directed toward achieving an adequate balance between the ecological, economic, social, cultural ,and political development of communities. It ensures satisfaction of the needs of the present without compromising the ability of future generations to meet their own needs. In this same context, it is convenient to highlight the importance of "producing more with less" which is the basis of the concept of "eco-efficiency" (Kirchhoff, 2005; Chouhan et al., 2013; Marques and Machado, 2014).

The three key challenges facing our society in terms of sustainability are the rapid depletion of natural resources, the increase in population, with the growing demand for food and resources to meet basic needs, including an ever-increasing demand for energy, and the increasing pressure on natural resources and ecosystems with consequent environmental deterioration (Glenn et al., 2015; Li, 2016a,b). To address these major challenges, it is necessary to explore new fundamental tools in the chemical sciences that can speed up the production of chemicals using sustainable and eco-friendly processes.

10.3 Strategies and technologies based on the principles of green chemistry

Polymers are among the most important mass-produced materials on the planet. However, this is largely a consequence of the continuous supply of petrochemicals. To guarantee the sustainable production of polymers and functional materials, there are two requirements, first, alternative raw materials such as by-products or industrial waste in order to generate added value, and second, synthesis methods with an approach toward green chemistry, as far as possible, assisted by physical or biological methods. In the first case, microwave irradiation, ultrasound, ultraviolet radiation, application of magnetic fields, etc., are alternatives. Meanwhile, for biological methods, bacteria, viruses, yeasts, plant extracts, fungi, or algae are used. However, if the use of chemical methods is necessary, ideally these should be based on the use of catalysts, which should be characterized as nontoxic and with a long useful life (Heise and Meulen, 2011).

An example of an industrial process based on green chemistry and biological methods is the production of bio-based polyethylene, which is obtained through catalytic conversion of bioethanol into ethylene, where bio-based alcohols can be obtained by fermentation of biomass such as starch, sucrose, and cellulose. In particular, methanol and dimethyl ether are obtained, polymerized by free-radical polymerization, and finally, transformed by physical methods in functional materials (Ying et al., 2015) (Wang et al., 2020). Companies that have implemented these processes include Braskem in Brazil, Dow in the United States, and Axens in France (Zhang and Yu, 2013).

Various eco-friendly polymers have also been developed from industrial wastes, for instance, those produced by the petrochemical industry. Thus, from low-cost alkenes and elemental sulfur, several polymers known as polysulfides and sulfur copolymers are obtained. The corresponding reactions are often solvent-free and based on principles of atomic economy (Worthington et al., 2017). In particular, the synthesis of sulfur copolymers proceeds through a reverse vulcanization process. To start the reaction, the elemental sulfur is first melted at a relatively high temperature close to 185°C using a thermostatic oil bath to ensure opening of the sulfur rings and the formation of long sulfur chains with radicals at both ends, which allows the reaction of sulfur with divinilic groups and the polymerization process. These high-sulfur materials are suitable for various applications, including lithium-sulfur batteries, mercury adsorbents, optical plastics, thermal and electrical insulators, mercury sensors, and fertilizers (Abbasi et al., 2020).

Taking into account that lignin is the third most abundant natural polymer in nature, the development of value-added products, such as lignin-based thermoplastics, has been encouraged. However, the main limitation of this polymer is its polydispersity and hyperbranched structure for obtaining biopolymers. To correct this, initially, the molecular weight of the lignin must be modified, either by fractionation with methanol, or by cross-linking with formaldehyde and later mixing with soft telechelic polybutadiene segments. The cross-linking process allows the obtaining of a material with a high molecular weight and the characteristics of thermoplastic rubbers. The use of higher molecular weight lignin improved the resulting shear modulus due to the formation of a network of telechelic polybutadiene bridges. This new method of synthesis of thermoplastics based on lignin will open a new pathway toward the use of biomass and a reduction of agrochemical waste (Saito et al., 2012). In the same context, wood components are used, mainly lignin, for the synthesis of polyamides from 4-propylcyclohexanol, which is one of the best known (and most abundant) hydrogenative depolymerization products of lignin (Kühlborn et al., 2017).

Other renewable materials used as starting materials for the production of polymers include lactic acid, itaconic anhydride, vegetable oils, and cardanol. As a green method, the enzyme lipase is used as a polymerization catalyst, as lipase is a naturally benign catalyst, which shows a specific catalysis and recyclable character. Its mode of action is by opening rings of cyclic esters (lactones) and obtaining reactive polyesters (Kobayashi, 2017). This type of biological catalysis, compared with traditional catalysis, is widely used as it has the potential to reduce chemical waste by 85% (Bérubé et al., 2017). In general, these and other advances allow energy-efficient processes to be used in order to prevent the formation of waste, providing savings in matter and energy, with the use of safer solvents and renewable chemicals, and producing less hazardous materials that are of lower toxicity (Kharissova et al., 2019).

10.4 Strategies and technologies based on the principles of click chemistry

Strategies and technologies based on the principles of click chemistry, as mentioned in Section 10.2, seek to shorten the extensive steps in the synthetic routes under mild reaction conditions, with high regio- and stereoselectivity, as well as high efficiency of the process (Qin et al., 2010; Li, 2016a,b). To date, four main classifications of click reactions have been identified. The first of these is the cycloadditions, which mainly refer to the 1,3-dipolar cycloaddition (HDC) of Huisgen. These are achieved from azides and terminal alkynes in the presence of a copper catalyst to form 1,2,3-triazoles. These compounds are mainly applied in the pharmaceutical industry in the synthesis of new polymeric administration systems, dendrimers, and the modification of surfaces (Kolb et al., 2001; Hein et al., 2008). Their high application due to their tolerance of typical biological conditions, high aqueous solubility, and biological activity such as antimicrobial, antioxidant, anticancer, antiviral, anti-HIV, and antituberculosis activities. However, their main limitation has been the need to use copper catalysts, since a high level of copper in the body can be harmful to health, and even fatal. For this reason, new click reactions have now been developed to obtain azides from primary amines, which are one of the most abundant functional groups (see Fig. 10.1) (Arslan and Tesdelen, 2019; Shi et al., 2017).

The second classification includes opening nucleophile rings such as aziridines, epoxides, cyclic sulfates, aziridinium ions, episulfonium ions, etc.; while the third of these reactions is related to the chemistry of the carbonyl of the aldol type, not as formations of ureas, thioureas, hydrazones, oxime ethers, amides, aromatic heterocycles, etc. In addition, there are reactions that include additions to multiple carbon–carbon bonds such as epoxidations, aziridinations, dihydroxylations, sulfenyl halide additions, nitrosyl halide additions, and certain Michael additions (Kolb et al., 2001; Hein et al., 2008).

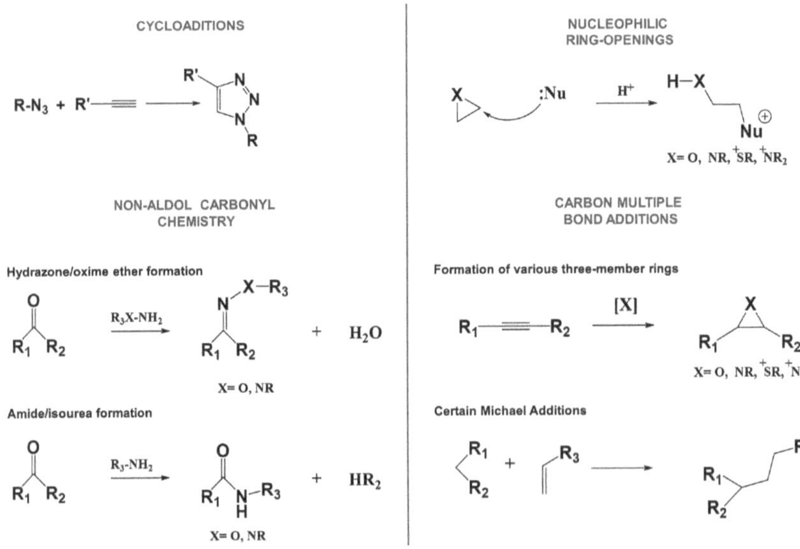

FIGURE 10.1 Classification of click chemistry reactions.

The efficiency of click reactions have allowed them to be extended to polymer science, in such a way that these materials can be practically and quantitatively synthesized. A polymer obtained from click reactions and with promising biomedical applications is the biodegradable citrate-based elastomer, such as poly(1,8-octanediol citrate), polymerized from functional azide and alkyne diols under a cross-linking mechanism, unique thermal synchronous binary (TSB), which allows the esterification of carboxylic acid and hydroxyl groups. Simultaneously, thermal click reactions are carried out between the alkyne and azide groups to finally allow the formation of polymeric films. Residual azide groups on the surface of polymer films have allowed for easy and efficient surface conjugation of bioactive molecules (such as peptides, proteins, and even viruses and cells) through strain-promoted alkyne–azide cycloaddition in an aqueous medium and at room temperature. This increased the mechanical strength of the material, demonstrating that it is possible to carry out two types of copper-free click reactions (thermal and strain-promoted click reaction), which are useful as a new cross-linking method and conjugation pathway for an efficient surface. Furthermore, it was shown that when a collagen mimetic peptide is employed, it is possible to effectively promote endothelial cell adhesion and proliferation (Droumaguet and Velonia, 2008; Guo et al., 2014).

In this same area, polymeric nanoparticles have been developed that contain biodegradable azide in their formulation, allowing these particles to be functionalized with bio-orthogonal click reactions to easily modify their surface and their ability to interact with biological systems and respond to external stimuli. These properties give these materials excellent clinical potential (Olejniczak et al., 2017). As such, there are many investigations carried out to obtain cross-linked polymers through click reactions, such as hydrogels based on biopolymers (collagen, cellulose, alginate, dextran, chitosan, and even DNA) that are subsequently chemically and rapidly cross-linked in the presence of, for example, biological or physical cross-linkers. These biopolymers are renewable and nontoxic, and so they have potential applications in regenerative medicine (for the regeneration of skin, bone, spinal cord, cartilage, and cornea) (Li et al., 2020).

Various types of click polymerizations have been developed already. Among them are polymerizations based on triple-bond building blocks, such as Cu (I) catalyzed azide-alkyne and thiol-yne polymerizations. It has been established as a powerful and selective method for polymer synthesis, where the Cu (I) catalyst is obtained by reducing Cu (II) ions in situ with sodium ascorbate or metallic copper (Cu) (Cao et al., 2016; Dong et al., 2016). These reactions can also be assisted by microwave and ultrasonic irradiation (direct reaction), as external stimuli offer additional reaction energy to polymerization, leading to higher efficiency and a shorter reaction time (Hou et al., 2015). In this case, ultraviolet radiation triggered an intramolecular charge transfer from the π system of a ligand to the metal ion, which resulted in the reduction of species from Cu (II) to Cu (I), and therefore a ligand stabilizer is necessary, making the indirect method the most efficient. However, as mentioned above, the complete removal of transition metal residues, even when supported on another material, is difficult as these metal residues could coordinate with the nitrogen atoms of the triazole rings formed (Shi et al., 2017).

In this regard, progress continues in the search for new synthesis routes, for example, organocatalysts of tetramethylammonium hydroxide (NMe_4OH) have been used at room temperature to carry out polycycloaddition of aromatic alkynes and azides. The results reflect the obtaining of 1,2,3-triazole of high molecular weight and high yields (up to 96%) (Liu et al., 2015). Another attractive click polymerization is that based on thiol, which is essentially a new polymerization activated either by free radicals, amines, or transition metals, which is, of course, accompanied by physical methods as a source of irradiation. For example, photoinitiated and heat-induced thiol-ene click polymerizations of thiolin and thiol-epoxy click polymerizations have been reported, from which functional polymers are easily prepared (Li and Cheng, 2013; Shi et al., 2017).

10.5 Conclusions

It is essential to introduce the concepts of green chemistry and click chemistry in chemical transformations in order to achieve a methodological outlook about the concept eco-sustainability. The implementation of green chemistry allows the design of chemical products and processes reducing or eliminating the use and generation of toxic substances based on its 12 principles, which, in short, seek to use alternative raw materials to generate new materials with added value. In addition, green synthesis methods should be able to be assisted by physical or biological methods.

On the other hand, click chemistry seeks to shorten the extensive steps in the synthetic routes under mild reaction conditions, with high regio- and stereoselectivity, as well as high efficiency of the process. This approach of synthetic chemistry is based on four types of reactions, where the most widely used to date has been cycloadditions from azides and alkynes catalyzed with copper. However, its main limitation for biomedical applications is the possibility of contamination by copper. For this reason, the development of environmentally friendly catalysts that can be implemented in the reaction is booming, as is the expansion of research toward new synthetic routes catalyzed by physical or biological methods. Finally, the application of these strategies in the synthesis of eco-friendly polymers has made it possible to obtain biodegradable and easily functionalizable materials, which have potential applications in various areas of science, mainly in medicine, thanks to their high biocompatibility and solubility in aqueous media. However, the improvement of the mechanical properties of these materials continues.

References

Abbasi, A., Nasef, M., YahyaWan, Z., 2020. Sulfur-based polymers by inverse vulcanization: a novel path to foster green chemistry. Green Mater. 8, 172–180. https://doi.org/10.1680/jgrma.19.00053.

Anastas, P., Warner, J., 1998. Principles of green chemistry. In: Anastas, P., Warner, J. (Eds.), Green Chemistry: Theory and Practice. Oxford University Press, Oxford, UK, pp. 29–56.

Arslan, M., Tasdelen, M., 2019. Click chemistry in macromolecular design: complex architectures from functional polymers. Chem. Africa 2, 195–214. https://doi.org/10.1007/s42250-018-0030-8.

Bérubé, C., Barbeau, X., Lague, P., Voyer, N., 2017. Revisiting the Juliá–Colonna enantioselective epoxidation: supramolecular catalysis in water. Chem. Commun. 53, 5099–5102. https://doi.org/10.1039/C7CC01168G.

Cao, X., Shi, Y., Wang, X., Graff, R., Gao, H., 2016. Design a highly reactive trifunctional core molecule to obtain hyperbranched polymers with over a million molecular weight in one-pot click polymerization. Macromolecules 49, 760–766. https://doi.org/10.1021/acs.macromol.5b02678.

Chouhan, N., Kumar, A., Sharma, A., Ameta, R., 2013. 3. Eco-friendly products. In: Ameta, S., Ameta, R. (Eds.), Green Chemistry: Fundamentals and Applications, vol. 1. Apple Academics Press CRC Press, pp. 41–84. https://doi.org/10.1155/2012/626759.

Dong, W., Wu, H., Chen, M., Shi, Y., Sun, J., Qin, A., Tang, B., 2016. Anionic conjugated polytriazole: direct preparation, aggregation-enhanced emission, and highly efficient Al3+ sensing. Polym. Chem. 7, 5835–5839. https://doi.org/10.1039/C6PY01202G.

Droumaguet, B., Velonia, K., 2008. Click chemistry: a powerful tool to create polymer-based macromolecular chimeras. Macromol. Rapid Commun. 29, 1073–1089. https://doi.org/10.1002/marc.200800155.

Fortunak, J., 2010. Eco-friendly synthesis of fine chemicals. J. Am. Chem. Soc. 132, 2846. https://doi-org.bd.univalle.edu.co/10.1021/ja1000483.

Glenn, J., Florescu, E., Millennium project Team, 2015. Executive Summary. State Fut. 19.

Guo, J., Xie, Z., Tran, R., Xie, D., Jin, D., Bai, X., Yang, J., 2014. Click chemistry plays a dual role in biodegradable polymer design. Adv. Mater. 26, 1906–1911. https://doi.org/10.1002/adma.201305162.

Hein, C., Liu, X., Wang, D., 2008. Click chemistry, a powerful tool for pharmaceutical sciences. Pharmaceut. Res. 25, 2216–2230. https://doi.org/10.1007/s11095-008-9616-1.

Heise, A., Meulen, I., 2011. Enzymatic polymer synthesis in green chemistry. In: Mathers, R., Meiers, M. (Eds.), Green Polymerization Methods: Renewable Starting Materials, Catalysis and Waste Reduction. Wiley, pp. 291–311. https://doi.org/10.1002/9783527636167.ch13.

Hou, Y., Cao, S., Wang, L., Pei, Y., Zhang, G., Zhang, S., Pei, Z., 2015. Morphology-controlled dual clickable nanoparticles via ultrasonic-assisted click polymerization. Polym. Chem. 6, 223–227. https://doi.org/10.1039/C4PY01045K.

Izzo, M., Ciaburri, M., Tiscini, R., 2020. The challenge of sustainable development goal reporting: the first evidence from Italian listed companies. Sustainability 12, 3494. https://doi.org/10.3390/su12083494.

Kharissova, O., Kharisov, B., González, C., Méndez, Y., López, I., 2019. Greener Synthesis of Chemical Compounds and Materials, vol. 6. Royal Society open science, p. 191378. https://doi.org/10.1098/rsos.191378.

Kirchhoff, M., 2005. Promoting sustainability through green chemistry. Resour. Conserv. Recycl. 44, 237–243. https://doi.org/10.1016/j.resconrec.2005.01.003.

Kolb, H., Finn, M., Sharpless, K., 2001. Click chemistry: diverse chemical function from a few good reactions. Angew. Chem. Int. Ed. 40, 2004–2021. https://doi.org/10.1002/1521-3773(20010601)40:11<2004::aid-anie2004>3.3.co;2-x.

Kobayashi, S., 2017. Green polymer chemistry: new methods of polymer synthesis using renewable starting materials. Struct. Chem. 28, 461–474. https://doi.org/10.1007/s11224-016-0861-3.

Kühlborn, J., Danner, A., Frey, H., Iyer, R., Arduengo, A., Opatz, T., 2017. Examples of xylochemistry: colorants and polymers. Green Chem. 19, 3780–3786. https://doi.org/10.1039/C7GC01244F.

Li, C., 2016a. Reflection and perspective on green chemistry development for chemical synthesis—daoist insights. Green Chem. 18, 1836–1838. https://doi.org/10.1039/C6GC90029A.

Li, C., 2016b. Exploration of new chemical reactivities for sustainable molecular transformations. Chemistry 1, 423–437. https://doi.org/10.1016/j.chempr.2016.08.007.

Li, Y., Cheng, B., 2013. One-pot synthesis of precise polyisoxazoles by click polymerization: copper (I)-catalyzed 1,3-dipolar cycloaddition of nitrile oxides with alkynes. J. Polym. Sci. Polym. Chem. 51, 1645–1650. https://doi.org/10.1002/pola.26537.

Liu, Y., Wang, J., Huang, D., Zhang, J., Guo, S., Hu, R., Zhao, Z., Qin, A., Tang, B., 2015. Synthesis of 1,5-regioregular polytriazoles by efficient NMe4OH-mediated azide–alkyne click polymerization. Polym. Chem. 6, 5545–5549. https://doi.org/10.1039/C5PY00186B.

Li, Y., Wang, X., Han, Y., Sun, H., Hilborn, J., Shi, L., 2020. Click chemistry based biopolymeric hydrogels for regenerative medicine. Biomed. Mater. 16 (2), 1–10, 022003. https://doi.org/10.1088/1748-605X/abc0b3.

Marquez, C., Machado, A., 2014. Environmental sustainability: implications and limitations to green chemistry. Found. Chem. 16, 125–147. https://doi.org/10.1007/s10698-013-9189-x.

Olejniczak, J., Collet, G., Huu, V., Chan, M., Lee, S., Almutairi, A., 2017. Biorthogonal click chemistry on poly(lactic-co-glycolic acid)-polymeric particles. Biomat. Sci. 5, 211–215. https://doi.org/10.1039/C6BM00721J.

Parveen, K., Banse, V., Ledwani, L., 2016. Green synthesis of nanoparticles: their advantages and disadvantages. AIP Conf. Proc. 1724, 20048. https://doi.org/10.1063/1.4945168.

Qin, A., Lam, J., Tang, B., 2010. Click polymerization: progresses, challenges, and opportunities. Macromolecules 43, 8693–8702. https://doi.org/10.1021/ma101064u.

Ribeiro, M., Costa, D., Machado, A., 2010. Green chemistry letters and reviews 'Green Star': a holistic green chemistry metric for evaluation of teaching laboratory experiments. Green Chem. Lett. Rev. 3, 149–159. https://doi.org/10.1080/17518251003623376.

Saito, T., Brown, R., Hunt, M., Pickel, D., Pickel, J., Messman, J., Baker, F., Kellerd, M., Naska, A., 2012. Turning renewable resources into value-added polymer: development of lignin-based thermoplastic. Green Chem. 14, 3295–3303. https://doi.org/10.1039/C2GC35933B.

Shi, Y., Sun, J., Qin, A., 2017. Click polymerization: the aurora of polymer synthetic methodology. J. Polym. Sci. Polym. Chem. 55, 616–621. https://doi.org/10.1002/pola.28419.

Unterlass, M., 2016. Green synthesis of inorganic–organic hybrid materials: state of the art and future perspectives. Eur. J. Inorg. Chem. 1135–1156. https://doi.org/10.1002/ejic.201501130.

Vaidya, A., Mayer, A.L., 2016. Critical review of the Millennium project in Nepal. Sustainability 8, 1043. https://doi.org/10.3390/su8101043.

Wang, Z., Ganewatta, M., Tang, C., 2020. Sustainable polymers from biomass: bridging chemistry with materials and processing. Prog. Polym. Sci. 101, 101197. https://doi.org/10.1016/j.progpolymsci.2019.101197.

Worthington, M., Kuceraa, R., Chalker, J., 2017. Green chemistry and polymers made from sulfur. Green Chem. 19, 2748–2761. https://doi.org/10.1039/C7GC00014F.

Ying, L., Yuan, X., Ye, M., Cheng, Y., Li, X., Liu, Z., 2015. A seven lumped kinetic model for industrial catalyst in DMTO process. Chem. Eng. Res. Des. 100, 179–191. https://doi.org/10.1016/j.cherd.2015.05.024.

Zhang, M., Yu, Y., 2013. Dehydration of ethanol to ethylene. Ind. Eng. Chem. Res. 52, 9505–9514. https://doi.org/10.1021/ie401157c.

Chapter 11

Eco-friendly particles, fibers, and surfaces

11.1 Particles, fibers, and surfaces: an overview

The morphological characteristics of these materials are related to their applications, and are the result of the macroscopic manifestation of properties at a molecular level. The morphology determines the extension of the boundary delimiting the constitutive phase of material systems. In this way, a system with a specific volume and mass can be described by its density, dimensions, and mass, as well as by other inherent properties associated with its nature, such as the melting and boiling temperatures. However, for the same material, this can have different shapes without changing its density, mass, or volume. The change in shape implies a change in the contact surface and therefore a change in the interaction surface with the environment. In general, from a geometric point of view, irregular particles will have a greater contact surface area than the equivalent sphere that circumscribes it. Similar generalizations can be made from other geometric shapes, for example, an equivalent cube or an equivalent ellipse. In a 2D projection, an analogous inference can be made by changing the volume for area, and the surface for perimeter (Gun'ko, 2018; Popelka et al., 2020).

Among the most common morphologies in materials science and technology are particles, fibers, and surfaces.

In a broad sense and from a morphological point of view, a particle is a fraction of a material which can be separated as an individual entity which is morphologically characterized by its spatial dimensions, longitudinal, altitudinal, and latitudinal, that are approximately equal. Examples of particles in multiple applications are commonly identified, including: polymer capsules (Larrañaga et al., 2017), polymer nanoparticles (Zielinska et al., 2020), metallic nanoparticles (Meléndrez et al., 2011; Charvali and Nikolova, 2019; Sanchez et al., 2019), emulsion droplets (Rivas and Palencia, 2011), etc. A particle can be regular or irregular, however, it is usually described by its diameter, volume, or equivalent sphere surface. However, in some cases, the particle size analysis is done using 2D projections of, for example, a photograph, and consequently the aforementioned descriptors are reduced to their

Eco-friendly Functional Polymers
https://doi.org/10.1016/B978-0-12-821842-6.00001-4
Copyright © 2021 Elsevier Inc. All rights reserved.

corresponding 2D descriptors (Shihavudding et al., 2017). In general, the properties of particles depend on their size and shape, which reduce the population of molecules or atoms in their interface as the size is decreased. The most popular example is nanoparticles, for which the effects of size and shape strongly impact the nanometric properties (Palencia et al., 2011). Particulate catalysts used in photocatalytic processes are another common example in which particulates show a greater surface activity resulting from the smaller size (Wang and Dommen, 2020). However, other properties are closely related to the morphological characteristics of the particles, such as hydrodynamic properties (e.g., diffusion, colloidal behavior, susceptibility to transport when they are released to the environment, bioavailability of liquid formulations, etc.) (Martchenko et al., 2011; Monaghan, 2012; Wagner et al., 2017). Details about specific applications of polymer and inorganic particles are given in Chapters 12, 14, 18, 23, and 26.

On the other hand, fiber-type materials, or fibers, are materials characterized as having a length-to-diameter ratio much greater than 1 (i.e., materials significantly longer than they are wide), for which some authors have suggested that this value can be intuitively defined as being greater than 100 (Hawkins, 2014). These can be natural or synthetic materials, and can be differentiated clearly from particles by morphological criteria as basically they are materials with extended morphology. Some examples include glass fiber (Thomason, 2019), polymer fiber (Sanjay et al., 2018), wool (Hearle, 2000), carbon fiber (Newcomb, 2016), and cotton (Pinho and Soares, 2018). Strictly speaking, there is no clear and specific conceptualization of the fiber concept. At a macroscopic level, the material resulting from the spinning of multiple fibers is usually classified as fiber, but also as yarn, and therefore, it is evident that there is an overlap in these terms. At the microscopic level, some materials can be included in the concept of fiber, but they are often referred to by other names, for example, nanotube, fibril, monofilament, and fiber yarn (Hawkins, 2014). In a more specific context, for example, researchers refer to carbon nanofibers (Feng et al., 2014; Yadav et al., 2020), carbon nanotubes (Gupta et al., 2019; Aznar et al., 2020), carbon nanofibrils (Ali and Al-Asmari, 2012; Chen et al., 2020), and carbon nanowires (Li et al., 2005; Kumar et al., 2007). Details of the specific applications of polymer-based fibers are given in Chapters 14, 17, and 19.

Finally, a surface is the interface between a condensed phase, that is, a solid and liquid phase, and a gaseous phase. In some cases, the solid phase is included, forming a solid–liquid interface, usually when the target phase is the solid phase. Surfaces are the support for many processes, and can act as catalysts for chemical reactions (Wang, 2017; Boukhavolov et al., 2021), chemical reaction substrates (Jiang and Guo, 2019; Boukhavolov et al., 2021), functional materials for a wide variety of applications, from sensors to antimicrobial agents (Sanchez et al., 2018; Palencia et al., 2016a,b,c, 2020a,b), biomarkers (Anaya et al., 2019), adsorbents in pollutant removal and monitoring applications

(Palencia et al., 2016a,b,c, 2017; Martínez et al., 2017; Garcés et al., 2018), porous media for mediated separation by membranes (Palencia et al., 2017), biointerfaces (Palencia et al., 2016a,b,c; Berrio et al., 2018; Espinosa et al., 2020), etc. Morphologically they are characterized by being a fraction of the material not delimited by its dimensions, but by its location, that is, by definition a surface will differ from the bulk material in terms of the intermolecular forces acting on it, which can be defined as the bulk or cohesive forces, and the forces of the adjacent phase in contact, that is, the other phase constituting the interface. In general, a surface can be homogeneous or heterogeneous, isotropic or anisotropic, passive or active, and synthetic or biological. In addition, they will present macroscopic properties describing it as a whole, and microscopic properties associated with its composition and local characteristics at the molecular level (Palencia et al., 2016a,b,c, 2020a,b; Martínez et al., 2017; Wang, 2017; Berrio et al., 2018; Garcés et al., 2018; Sanchez et al., 2018; Anaya et al., 2019; Jiang and Guo, 2019; Espinosa et al., 2020; Boukhavolov et al., 2021). Details of the specific applications of polymer-based surfaces are given in Chapters 16, 20, 23, and 24.

The enormous variety of applications of synthetic polymers, of biological origin, and mixed, added to the great variety of possibilities that derive from the modification of this type of material, allow it to be concluded that particle systems, fibers, and surfaces have great potential from an eco-friendly point of view. The foregoing makes them a technological ally for scientific goals, but they are also of a sustainable nature such as those defined in the sustainable development goals defined by the United Nations (Rajeswari et al., 2020).

11.2 Polymer-based particles, fibers, and surfaces

In relation to "pure" polymeric particles, that is to say, those where the polymer is not acting as a coating of an inorganic particle, these are usually produced by various methods, including: (1) polymerization in a heterogeneous phase, in which the polymer chain forms precipitates in a nonsolvent; (2) emulsion polymerization, which is a type of polymerization in a heterogeneous phase characterized by being a stabilized system by surfactants; (3) by phase exchange, in which a polymer is started, which precipitates when passing from a solvent to a nonsolvent, injection techniques are commonly used to guarantee a homogeneous morphology; and (4) through resin formation, which is characterized by the presence of the cross-linker to obtain a continuous material, and by requiring grinding by mechanical means of the resulting material to obtain particles of irregular morphology (Gokmen and Du, 2012; Nasrabadi et al., 2019). The incorporation of a porogenic agent allows the particles to be characterized by high porosity, and consequently, a greater surface area (Gokmen and Du, 2012).

On the other hand, "no-pure" or "hybrid" polymer particles can be obtained by the coating of inorganic particles with a polymer. Usually, metal and oxide nanoparticles, and clays are used. Of the many materials that are found naturally, clays are the most abundant, and they are particularly interesting due to their different geometric shapes, molecular plate structure, cation exchange capacity, swelling, porosity, and high stability. The fundamental units are multiple layers of silicate stacked with divalent ionic species. The clay surface can be modified in different ways to develop clay composite materials, with the most common method being modification based on the adsorption of an organic component on the clay surface either by electrostatic interaction or nonionic adsorption. Due to the negative charge of clays, many of their surface modifications are made through the interaction of the negatively charged surface with cationic molecules. In addition, thanks to their intrinsic properties, clays are considered materials with high potential for the elimination of pollutants in low-cost water sources, and are more stable, safer, ecological, and efficient, and so can be used for the elimination of pollutants from water in discontinuous processes. For example, the elimination of heavy metals copper and cadmium was described being carried out from functionalization of bentonite with alginate, surface anchoring of functionalized montmorillonite with polyelectrolytes, polymeric coatings of montmorillonite particles, etc. (Garcés et al., 2018; Lerma et al., 2018; Liu et al., 2019a,b; Garmia, 2018; Li et al., 2018; Pawar et al., 2020). The biocompatibility of clays and the biodegradability properties of biopolymers make their combination a successful approach for the development of bionanocomposites that can be used for environmental remediation, with biopolymers, polysaccharides, and polypeptides being considered the best candidates due to their economic and nontoxic properties (Lerma et al., 2018; Liu et al., 2019a,b). In particular, sodium alginate is a natural polymer extracted mainly from brown algae, which has a high affinity for toxic divalent cations and other dangerous pollutants. As particulate materials, composite polymer particles have high porosity, high surface area, high affinity in charged molecules, high stability in organic solvents, adaptability to extreme experimental conditions, easy biodegradation, and nontoxicity (Pawar et al., 2020) (see Fig. 11.1A). On the other hand, bentonite–polyelectrolyte systems can be used for sampling of soils (Garcés et al., 2018; see Fig. 11.1B), and bentonite-poly(acrylic acid) has been described as a geomimetic system for soil conditioning (Lerma et al., 2018).

On the other hand, the fibers can be obtained from synthetic methods or natural sources. Fibers are a field of work that is in constant expansion, since not only are the applications increasing regularly, but also because their applications are becoming increasingly specialized, technological, and demanding from the point of view of the desired areas of use. In general, the trend in the design and application of fibers follows the same design pathway observed in polymers for advanced applications: biopolymeric nature, multifunctionality, and smart materials (Keya et al., 2019).

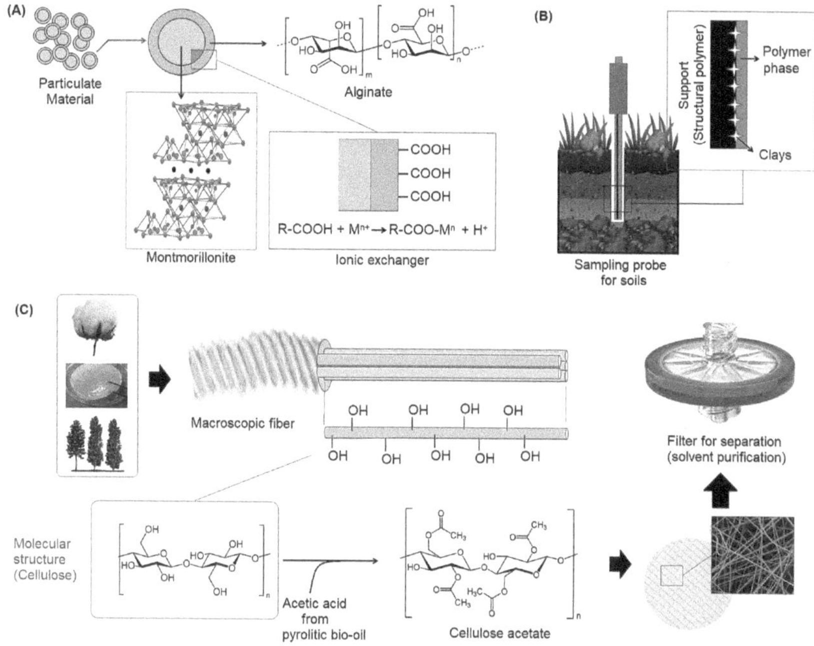

FIGURE 11.1 (A) Illustration of clay–alginate hybrid particles with core–shell morphology; (B) illustration of particles as the anchoring point of polymer chains and their application as a sampling probe for soils; and (C) illustration of the modification of natural fibers (cellulose); in this figure the extraction sources, macroscopic structure of fibers, modification by esterification using pyrolytic acetic acid, and illustration of technological applications of cellulose acetate fibers in membrane-based separation systems are exemplified.

Natural fibers are considered a promising alternative to synthetic fibers. In this way, significant advances in the conversion of lignocellulose fibers into materials of great technological importance have been obtained. They are used in engineering applications, such as the reinforcement of polymers and bioplastics, films, production of biofuels, etc. In particular, fiber-reinforced biocomposites have some advantages, thanks to their renewable origin, biodegradability, low cost, and low density; however, the low thermal stability of natural fibers limits their use in composite applications (Keya et al., 2019). Vegetable fibers are mainly composed of cellulose, lignin, and hemicelluloses, with a large amount of hydroxyl and carboxyl groups in their chemical structure, providing lignocellulosic fibers with high polarity and hydrophobicity. In particular, the mechanical properties of the fibers are affected by the low mechanical performances and low physicochemical interactions at the fiber–matrix interface, and therefore, they exhibit low interfacial adhesion and reactive functional groups. Other parameters affecting the application of fibers during the making of more complex materials include the presence of humidity and their intrinsic porosity, which weaken the microstructure and consequently the fiber–matrix interface. Among the most used fibers are banana, coconut, jute, cotton hemp jute, sisal, bamboo, wood fibers, etc. From an experimental design point of view, surface modifications and surface chemical functionalization are among the strategies used for fiber modification and subsequent use in more complex materials (Bian et al., 2020).

The techniques used to carry out the surface modification of natural fibers include treatment with alkalis or mercerization (Zuluaga et al., 2009), peroxide (Abdul et al., 2014), silane (Lin et al., 2020a,b; Asim et al., 2016), plasma (Katouah and El-Metwaly, 2021), benzoylation (Lu et al., 2015), and a direct reaction with surface functional groups (Palencia et al., 2020a,b, see Fig. 11.1C). Among the different types of treatment, one of the most widely used techniques is alkaline modification. This technique consists of the depolymerization of cellulose by the addition of an alkali, for example, sodium hydroxide, which breaks down the hydrogen bonds of the hydroxyl groups on the structure of the cell wall, which is covered by lignin, cellulose, hemicellulose, wax, and oils. Some investigations suggest that the alkaline treatment acts mainly on the rigidity and resistance of the fiber as a consequence of an increase in the amount of exposed cellulose. Recent studies have shown the potential of jute fibers with subsequent alkaline pretreatment, in the manufacture of fiber cement by means of the extruded products technique whose advantage is the use of simpler machines for continuous production. It has been recommended that the alkaline treatment of jute fibers increases the surface area, reactivity, and roughness of the fiber surface, and consequently, facilitates the production of nanofibrils. In addition, the presence of extracts on the surface of the fibers influences the exothermic hydration characteristic of Portland cement, increasing the curing time due to the inhibitory effect on the hydration of the cement (Fonseca et al., 2019). Another treatment for the surface modification of fiber is based on the use of silane to obtain surface anchoring points (Sabarinathan et al., 2020; Liu et al., 2019a,b).

Other techniques are based on plasma treatment, which consists of the breaking of different chemical bonds and the formation of species such as free radicals, ions, etc. (Putra et al., 2020; Macedo et al., 2020). For example, in the research work of Da Silva et al. the surface of the sisal fiber was modified using cold plasma (a mixture of argon and carbon dioxide gases) assisted by microwaves for the removal of the methylene blue dye. The mechanisms of interaction between the CO_2 plasma are associated with radical and ionic oxygen species, which are capable of breaking the hydroxyl group of lignocellulosic compounds and forming unsaturated carbon bonds (da Silva et al., 2020).

With respect to polymer surfaces, more detailed information is provided in Chapter 16. Polymeric surfaces are a higher order concept than the concepts of particles and fibers. Both particles and fibers have surfaces that, using a certain approach, can be considered the object of study. In the previous examples the importance of the surfaces was shown, for example, in the generation of functional coatings of clays, and in modification of the properties of cellulose fibers.

Excluding the surfaces of particulate materials and fibers, the polymeric surfaces fulfill different functions according to their porosity. If the porosity is very low, the polymeric phase behaves like a dense material, and consequently, the surface will be a containment barrier to substances that cannot cross it. Furthermore, depending on the surface hydrophilic–lipophilic balance, the surface will be susceptible to being wetted by liquids for which it has a higher affinity. This can be easily assessed by contact angle measurements (Arbelaez et al., 2017; Palencia, 2017). In cases where the polymeric phase that makes up the surface is dense, it may still be the case that it is permeable to gases, which will be a function of the partition coefficient, the pressure, and the affinity of the gas molecules for the polymer phase. On the other hand, as the porosity of the surface increases, the permeability of substances through it increases, and for the special case of liquids, the permeability can be described by the Darcy equation that relates pressure with velocity flow per unit effective surface area (Palencia, 2019).

Films are a special type of surface; they are characterized by being dense with respect to the application medium. They can therefore fulfill applications analogous to coatings. A particular case is biofilms, which are biopolymeric films that are characterized by being anisotropic, heterogeneous, and of variable composition. Biofilms are the result of the combination of extracellular secretions of microorganisms, and therefore their composition is linked to the nature of the biological agents involved (Espinosa et al., 2020).

11.3 Conclusions and remarks

The morphology of polymeric materials is a determining aspect for their applications. The three main morphologies are particles, fibers, and surfaces. In all cases, their reactivity, and consequently the possibility of being modified, will be given by their surface characteristics, both from structural and physicochemical points of view. However, the surface characteristics also strongly influence the compatibility of the phases, which is an aspect of great relevance in the manufacture of composites and nanocomposites (see Chapter 7), in the barrier properties, the phenomena of adsorption, diffusion, wettability, etc. Undoubtedly, for practical applications of new materials, a niche with great development opportunities is focused on the evolution of these morphologies.

References

Abdul, N.I., Azowa, N., Zainnuddin, N., Rayung, M., Zuhainis, W., 2014. The influence of chemical surface modification of kenaf fiber using hydrogen peroxide on the mechanical properties of biodegradable kenaf fiber/poly(lactic acid) composites. Molecules 19, 2957–2968. https://doi.org/10.3390/molecules19032957.

Ali, A.A., Al-Asmari, A.K., 2012. Wet-electrospun CuNP/carbon nanofibril composites: potential application for micro surface-mounted components. Appl. Nanosci. 2, 55–61. https://doi.org/10.1007/s13204-011-0042-z.

Anaya, L., Libreros, K., Palencia, V., Atencio, V., Palencia, M., 2019. Mid-infrared spectral characterization of fish scales: "Bocachico" (*Prochilodus magdalenae*) by functionally-enhanced derivative spectroscopy (FEDS) - a methodological approach. J. Sci. Technol. Appl. 6, 28–39. https://doi.org/10.34294/j.jsta.19.6.39.

Arbeláez, N., Lerma, T., Córdoba, A., 2017. Modification of membranes by insertion of short-chain alcohols on reactive porous substrates: effect of chain length on the surface free energy. J. Sci. Technol. Appl. 2, 75–83. https://doi.org/10.34294/j.jsta.17.2.16.

Asim, M., Jawaid, M., Abdan, K., Ishak, M.R., 2016. Effect of alkali and silane treatments on mechanical and fibre-matrix bond strength of kenaf and pineapple leaf fibres. JBE 13 (3), 426–435. https://doi.org/10.1016/S1672-6529(16)60315-3.

Aznar, N., Hasan, R., Tyagi, M., Yadav, N., Narang, J., 2020. Carbon nanotube - a review on synthesis, properties and plethora of applications in the field of biomedical science. Sensors Int. 1, 100003. https://doi.org/10.1016/j.sintl.2020.100003.

Berrio, M., Palencia, S., Lerma, T., Mora, M., 2018. Bacterial colonization modelling on soil particles: effect of humic acids on the formation of nitrogen-fixing bacteria biofilms. J. Sci. Technol. Appl. 5, 33–44. https://doi.org/10.34294/j.jsta.18.5.33.

Bian, H., Chen, L., Dong, M., Wang, L., Wang, R., Zhou, X., Dai, H., 2020. Natural lignocellulosic nanofibril film with excellent ultraviolet blocking performance and robust environment resistance. Int. J. Biol. Macromol. 166, 1578–1585. https://doi.org/10.1016/j.ijbiomac.2020.11.037.

Boukhavolov, D.W., Paolucci, V., D'Olimpio, G., Cantalini, C., Politano, A., 2021. Chemical reactions on surfaces for applications in catalysis, gas sensing, adsorption-assisted desalination and Li-ion batteries: opportunities and challenges for surface science. Phys. Chem. Chem. Phys. 23, 7541–7552. https://doi.org/10.1039/D0CP03317K.

Charvali, M.S., Nikolova, M.P., 2019. Metal oxide nanoparticles and their applications in nanotechnology. SN Appl. Sci. 1, 607. https://doi.org/10.1007/s42452-019-0592-3.

Chen, J., Liu, Y., Liu, Z., Chen, Y., Zhang, C., et al., 2020. Carbon nanofibril composites with high sulfur loading fabricated from nanocellulose for high-performance lithium-sulfur batteries. Colloid. Surface. Physicochem. Eng. Aspect. 603, 125249. https://doi.org/10.1016/j.colsurfa.2020.125249.

da Silva, C.C., de Faria Lima, A., Moreto, J.A., Dantas, S., Henrique, M.A., Pasquini, D., Gelamo, R.V., 2020. Influence of plasma treatment on the physical and chemical properties of sisal fibers and environmental application in adsorption of methylene blue. Mater. Today Commun. 23, 101140. https://doi.org/10.1016/j.mtcomm.2020.101140.

Espinosa, A., Palencia, S., García, A., Palencia, M., 2020. Polymicrobial biofilms: fundamentals, diseases, and treatments - a review. J. Sci. Technol. Appl. 8, 4–54. https://doi.org/10.34294/j.jsta.20.8.54.

Feng, L., Xie, N., Zhong, J., 2014. Carbon nanofibers and their composites: a review of synthesizing, properties and applications. Materials 7, 3919–3945. https://doi.org/10.3390/ma7053919.

Fonseca, C.S., Silva, M.F., Mendes, R.F., Hein, P.R.G., Zangiacomo, A.L., Savastano Jr., H., Tonoli, G.H.D., 2019. Jute fibers and micro/nanofibrils as reinforcement in extruded fiber-cement composites. Construct. Build. Mater. 211, 517–527. https://doi.org/10.1016/j.conbuildmat.2019.03.236.

Garcés, V., Lerma, T., Palencia, M., Combatt, E., 2018. Building a probe-type passive sampler to study the contents of heavy metals in the aqueous phase in the soil. J. Sci. Technol. Appl. 4, 4−16. https://doi.org/10.34294/j.jsta.18.4.26.

Garmia, D., Zaghouane-Boudiaf, H., Ibbora, C.V., 2018. Preparation and characterization of new low cost adsorbent beads based on activated bentonite encapsulated with calcium alginate for removal of 2, 4-dichlorophenol from aqueous medium. Int. J. Biol. Macromol. 115, 257−265. https://doi.org/10.1016/j.ijbiomac.2018.04.064.

Gokmen, M.T., Du, F.E., 2012. Porous polymer particles—a comprehensive guide to synthesis, characterization, functionalization and applications. Prog. Polym. Sci. 37, 365−405. https://doi.org/10.1016/j.progpolymsci.2011.07.006.

Gun'ko, V., 2018. Various Methods to Describe the Morphological and Textural Characteristics of Various Materials, vol. 9, pp. 317−340. https://doi.org/10.15407/hftp09.04.317.

Gupta, N., Gupta, S.M., Sharma, S.K., 2019. Carbon nanotubes: synthesis, properties and engineering applications. Carbon Lett. 29, 419−447. https://doi.org/10.1007/s42823-019-00068-2.

Hawkins, S.C., 2014. Introduction to fiber materials. In: Schulz, M.J., Shanov, V., Yin, Z. (Eds.), Nanotube Superfiber Materials. Science Direct, pp. 1−32. https://doi.org/10.1016/B978-1-4557-7863-8.00001-3.

Hearle, J.W.S., 2000. A critical review of the structural mechanics of wool and hair fibres. Int. J. Biol. Macromol. 27, 123−138. https://doi.org/10.1016/S0141-8130(00)00116-1.

Jiang, B., Guo, H., 2019. Dynamics in reactions on metal surfaces: a theoretical perspective. J. Chem. Phys. 150, 180901. https://doi.org/10.1063/1.5096869.

Katouah, H., El-Metwaly, N.M., 2021. Plasma treatment toward electrically conductive and superhydrophobic cotton fibers by in situ preparation of polypyrrole and silver nanoparticles. React. & Funct. Polym. 104810. https://doi.org/10.1016/j.reactfunctpolym.2021.104810.

Keya, K.N., Kona, N.A., Koly, F.A., Maraz, K.M., Islam, M.N., Khan, R.A., 2019. Natural fiber reinforced polymer composites: history, types, advantages and applications. Mater. Eng. Res. 1, 69−85. https://doi.org/10.25082/MER.2019.02.006.

Kumar, A., Avasthi, D., Tripathi, A., Dragos, L., Carrey, J.D., Pivin, J., 2007. Formation and characterization of carbon nanowires. J. Appl. Phys. 102, 04430. https://doi.org/10.1063/1.2767227.

Larrañaga, A., Lomora, M., Sarasua, J.R., Palivan, C.G., Pandit, A., 2017. Polymer capsules as micro-/nanoreactors for therapeutic applications: current strategies to control membrane permeability. Prog. Mater. Sci. 90, 325−357. https://doi.org/10.1016/j.pmatsci.2017.08.002.

Lerma, T.A., Palencia, M., Combatt, E.M., 2018. Soil polymer conditioner based on montmorillonite-poly(acrylic acid) composites. J. Appl. Polym. Sci. 135, 46211. https://doi.org/10.1002/app.46211.

Li, J., Ng, T., Chen, H., 2005. Carbon nanotubes and nanowires for biological sensing. In: Vo-Dinh, T. (Ed.), Protein Nanotechnology, Methods in Molecular Biology. Humana Press, pp. 191−223. https://doi.org/10.1385/1-59259-858-7:191.

Li, W., Ma, Q., Bai, Y., Xu, D., Wu, M., Ma, H., 2018. Facile fabrication of gelatin/bentonite composite beads for tunable removal of anionic and cationic dyes. Chem. Eng. Res. Des. 134, 336−346. https://doi.org/10.1016/j.cherd.2018.04.016.

Lin, D., Zheng, Y., Wang, X., Huang, Y., Ni, L., Chen, X., Wu, D., 2020a. Study on physicochemical properties, antioxidant and antimicrobial activity of okara soluble dietary fiber/sodium carboxymethyl cellulose/thyme essential oil active edible composite films incorporated with pectin. Int. J. Biol. Macromol. 165, 1241−1249. https://doi.org/10.1016/j.ijbiomac.2020.10.005.

Lin, G., Wang, H., Yu, B., Qu, G., Chen, S., Kuang, T., Liang, Z., 2020b. Combined treatments of fiber surface etching/silane-coupling for enhanced mechanical strength of aramid fiber-reinforced rubber blends. Mater. Chem. Phys. 255, 123486. https://doi.org/10.1016/j.matchemphys.2020.123486.

Liu, Y., Guo, L., Wang, W., Sun, Y., Wang, H., 2019a. Modifying wood veneer with silane coupling agent for decorating wood fiber/high-density polyethylene composite. Construct. Build. Mater. 224, 691–699. https://doi.org/10.1016/j.conbuildmat.2019.07.090.

Liu, Y., Lv, X., Bao, J., Xie, J., Tang, X., Che, J., Tong, J., 2019b. Characterization of silane treated and untreated natural cellulosic fibre from corn stalk waste as potential reinforcement in polymer composites. Carbohydr. Polym. 218, 179–187. https://doi.org/10.1016/j.carbpol.2019.04.088.

Lu, N., Oza, S., Ghaempahah, M., 2015. Surface modification of natural fibers for reinforcement in polymeric composites. In: Kumar, V., Singh, A. (Eds.), Surface Modification of Biopolymers. Wiley, pp. 224–237. https://doi.org/10.1002/9781119044901.ch9.

Macedo, M.J., Silva, G.S., Feitor, M.C., Costa, T.H., Ito, E.N., Melo, J.D., 2020. Surface modification of kapok fibers by cold plasma surface treatment. J. Mater. Res. & Technol. 9 (2), 2467–2476. https://doi.org/10.1016/j.jmrt.2019.12.077.

Martchenko, I., Dietsch, H., Moitzi, C., Schurtenberger, P., 2011. Hydrodynamic properties of magnetic nanoparticles with tunable shape anisotropy: prediction and experimental verification. J. Phys. Chem. B 115, 14838–14845. https://doi.org/10.1021/jp2078264.

Martinez, J., Troches, E., Páez, M., 2017. Evaluation of poly(ethylene-co-vinylacetate) as polymeric adsorbent of polychlorinated biphenyls in water. J. Sci. Technol. Appl. 2, 94–107. https://doi.org/10.34294/j.jsta.17.2.18.

Meléndrez, M.F., Cárdenas, G., Palencia, M.S., 2011. Colloidal germanium nanoparticles and nanocubes synthesized by chemical liquid deposition. Res. Sci. Lett. 4, 526–535.

Monaghan, J.J., 2012. Smoothed particle hydrodynamics and its diverse applications. Annu. Rev. Fluid Mech. 44, 323–346. https://doi.org/10.1146/annurev-fluid-120710-101220.

Nasrabadi, M.N., Goli, S.A.H., Doost, A.S., Dewettinck, K., Van der Meeren, P., 2019. Bioparticles of flaxseed protein and mucilage enhance the physical and oxidative stability of flaxseed oil emulsions as a potential natural alternative for synthetic surfactants. Colloids Surf. B Biointerfaces 184, 110489. https://doi.org/10.1016/j.colsurfb.2019.110489.

Newcomb, B.A., 2016. Processing, structure, and properties of carbon fibers. Compos. Appl. Sci. Manuf. 91, 262–282. https://doi.org/10.1016/j.compositesa.2016.10.018.

Palencia, M., 2017. Surface free energy of solids by contact angle measurements. J. Sci. Technol. Appl. 2, 84–93. https://doi.org/10.34294/j.jsta.17.2.17.

Palencia, M., 2019. Fundamental and methodological aspects of porous membrane characterization by hydrodynamic permeability test - a review. J. Sci. Technol. Appl. 7, 17–25. https://doi.org/10.34294/j.jsta.19.7.48.

Palencia, S.L., Buelvas, A.M., Palencia, M.S., 2011. Interaction mechanisms of inorganic nanoparticles and biomolecular systems of microorganisms. Curr. Chem. Biol. 9, 11–23. https://doi.org/10.2174/2212796809666151022201811.

Palencia, M., Berrio, M., Melendrez, M., 2016a. Nanostructured polymer composites with potential applications into the storage of blood and hemoderivates. J. Sci. Technol. Appl. 1, 4–14. https://doi.org/10.34294/j.jsta.16.1.1.

Palencia, M., Lerma, T., Córdoba, A., 2016b. Polyurethanes with boron retention properties for the development of agricultural fertilization smart systems. J. Sci. Technol. Appl. 1, 39–52. https://doi.org/10.34294/j.jsta.16.1.4.

Palencia, S., Lagos, G., Garcia, A., 2016c. Fungi-bacterium interactions: *Helycobacter pyroli* - Candida Albicans. J. Sci. Technol. Appl. 1, 53–69. https://doi.org/10.34294/j.jsta.16.1.5.

Palencia, M., Martinez, J., Arrieta, A., 2017. Removal of acid blue 129 dye by polymer-enhanced ultrafiltration (PEUF). J. Sci. Technol. Appl. 2, 65–74. https://doi.org/10.34294/j.jsta.17.2.15.

Palencia, M., Lerma, T.A., Arrieta, A.A., 2020a. Antibacterial and non-hemolytic cationic polyurethanes with N-carboxymethyl-N,N,N-triethylammonium groups for bacteremia-control in biomedical-using materials. Mater. Today Commun. 22, 100708. https://doi.org/10.1016/j.mtcomm.2019.100708.

Palencia, M., Mora, M.A., Lerma, T.A., Afanasjeva, N., Isaza, J.H., 2020b. Reinforcing agents based on cellulose fibers modified by insertion of end-alkyl groups obtained from pyrolytic bio-oil of sugarcane bagasse. Polym. Bull. 77, 5711–5724. https://doi.org/10.1007/s00289-019-03050-0.

Pawar, R.R., Ingole, P.G., Lee, S.M., 2020. Use of activated bentonite-alginate composite beads for efficient removal of toxic Cu^{2+} and Pb^{2+} ions from aquatic environment. Int. J. Biol. Macromol. 164, 3145–3154. https://doi.org/10.1016/j.ijbiomac.2020.08.130.

Pinho, E., Soares, G., 2018. Functionalization of cotton cellulose for improved wound healing. J. Mater. Chem. B 6, 1887–1898. https://doi.org/10.1039/C8TB00052B.

Popelka, A., Zavahir, S., Habib, S., 2020. Morphology analysis. In: Ali, M.A., Ponnamma, D., Carignano, M.A. (Eds.), Polymer Science and Innovative Applications, pp. 21–68. https://doi.org/10.1016/B978-0-12-816808-0.00002-0.

Putra, A.E.E., Renreng, I., Arsyad, H., Bakri, B., 2020. Investigating the effects of liquid-plasma treatment on tensile strength of coir fibers and interfacial fiber-matrix adhesion of composites. Compos. B Eng. 183, 107722. https://doi.org/10.1016/j.compositesb.2019.107722.

Rajeswari, A., Christy, E.J.S., Swathi, E., Pius, A., 2020. Fabrication of improved cellulose acetate-based biodegradable films for food packaging applications. Environ. Chem. & Ecotoxicol. 2, 107–114. https://doi.org/10.1016/j.enceco.2020.07.003.

Rivas, B.L., Palencia, M., 2011. Removal-concentration of pollutant metal-ions by water-soluble polymers in conjunction with double emulsion systems: a new hybrid method of membrane-based separation. Separ. Purif. Technol. 81, 435–443. https://doi.org/10.1016/j.seppur.2011.08.021.

Sabarinathan, P., Rajkumar, K., Annamalai, V.E., Vishal, K., 2020. Characterization on chemical and mechanical properties of silane treated fish tail palm fibres. Int. J. Biol. Macromol. 163, 2457–2464. https://doi.org/10.1016/j.ijbiomac.2020.09.159.

Sanchez, Y., García, A., Palencia, M., 2018. Determination and distribution of size of inorganic particles by spectral deconvolution of surface plasmon resonance. J. Sci. Technol. Appl. 5, 45–54. https://doi.org/10.34294/j.jsta.18.5.34.

Sanchez, Y., Mora, M., Palencia, M., 2019. A first approach to the development of new drugs with leishmanicidal activity via control of surface properties of silver nanoparticle. J. Sci. Technol. Appl. 6, 118–133. https://doi.org/10.34294/j.jsta.19.6.46.

Sanjay, M.R., Madhu, P., Jawaid, M., Senthamaraikannan, P., Senthil, S., Pradeep, S., 2018. Characterization and properties of natural fiber polymer composites: a comprehensive review. J. Clean. Prod. 172, 566–581. https://doi.org/10.1016/j.jclepro.2017.10.101.

Shihavudding, A., Basu, S., Rexhepaj, E., Delestro, F., et al., 2017. Smooth 2D manifold extraction from 3D image stack. Nat. Commun. 8, 15554. https://doi.org/10.1038/ncomms15554.

Thomason, J.L., 2019. Glass fibre sizing: a review. Compos. Appl. Sci. Manuf. 127, 105619. https://doi.org/10.1016/j.compositesa.2019.105619.

Wagner, T., Kroll, A., Haramagatti, C., Lipinski, L.G., Wiemann, M., 2017. Classification and segmentation of nanoparticle diffusion trajectories in cellular micro environments. Plos One 12, e017065. https://doi.org/10.1371/journal.pone.0170165.

Wang, H., 2017. Discovering and utilizing structure sensitivity: from chemical catalysis in the gas phase to electrocatalysis in the liquid phase. Stud. Surf. Sci. Catal. 177, 613–641. https://doi.org/10.1016/B978-0-12-805090-3.00016-4.

Wang, Q., Dommen, K., 2020. Particulate photocatalysts for light-driven water splitting: mechanisms, challenges, and design strategies. Chem. Rev. 120, 919–985. https://doi.org/10.1021/acs.chemrev.9b00201.

Yadav, D., Amini, F., Ehrmann, A., 2020. Recent advances in carbon nanofibers and their applications – a review. Eur. Polym. J. 138, 109963. https://doi.org/10.1016/j.eurpolymj.2020.109963.

Zielinska, A., Carreiro, F., Oliviera, A.M., Neves, A., Pires, B., et al., 2020. Polymeric nanoparticles: production, characterization, toxicology and ecotoxicology. Molecules 25, 3731. https://doi.org/10.3390/molecules25163731.

Zuluaga, R., Putaux, J.L., Cruz, J., Vélez, J., Mondragon, I., Gañán, P., 2009. Cellulose microfibrils from banana rachis: effect of alkaline treatments on structural and morphological features. Carbohydr. Polym. 76 (1), 51–59. https://doi.org/10.1016/j.carbpol.2008.09.024.

Part III

Applications of eco-friendly functional polymers

Chapter 12

Functional and eco-friendly polymers in agriculture

12.1 An overview of current problems in agriculture

The invention of agriculture was a key event for the development of human civilizations. Thus, 250,000 years ago agriculture was an innovation with which we have become accustomed, being today a regular activity with different degrees of technological evolution depending on the type of crop, production conditions, and the economic power of the farmers. Agriculture in its initial stage is usually described as an empirical process of domestication of plant organisms by relatively primitive humans, however, this is an oversimplification of the agricultural production process, which although it was a rudimentary activity compared with current agriculture, it is clear that it would not have been a simple or easy activity. This simplified view assumes that many of the current problems of agricultural production were previously nonexistent, which is clearly unrealistic. This is so not only because agricultural production encountered a knowledge gap, but also because many of the problems at that time remain the same currently, although possibly to greater or lesser extents. Pest control, the availability of water resources, and the conservation of natural resources are clear examples of the above. Therefore, the agricultural problems are not the same, they depend strongly on the crop, the location, and the resources available for the development of productive activities (Godfray et al., 2010; Reganold and Wachter, 2016; Bhargava and Srivastava, 2019).

Currently, agriculture faces problems associated with different components of the production process, which impact crop yields, commercial viability, and food security. In general terms, some of the most important challenges for agriculture are: the availability of suitable soils for agriculture, availability and efficient use of water resources, adequate control of plagues, efficient fertilization, adequate seed conservation, planting and propagation efficiency, crop conservation, adaptation to climate change, reduction of negative impact, ensuring the safety of the harvest, and reduction of production costs (Shokrzade and Saeedi, 2011; Amundson et al., 2015; Crowder and Reganold, 2015; Lee et al., 2015; Xu et al., 2020).

Eco-friendly Functional Polymers
https://doi.org/10.1016/B978-0-12-821842-6.00016-6
Copyright © 2021 Elsevier Inc. All rights reserved.

12.2 Functional polymers in agriculture

Functional polymers can be differentiated from structural polymers by the presence of heteroatoms in their structure, either at the level of their side chains or as part of the main chain. These functional groups are normally associated with carboxylic acids, alcohols, amines, amides, esters, sulfonates, phosphates, etc. The presence of these structural units gives the polymer a relatively high hydrophilicity, high adsorption capacity on charged surfaces, but also, high interaction capacity with ions and other species in solution. However, functional polymers are characterized as showing relatively poor mechanic properties compared to structural polymers. Typical examples of functional polymers are poly(acrylic acid), poly(sodium styrene sulfonate), poly(vinyl alcohol), poly(acrylamide), chitosan, starch, etc. (Rivas et al., 2011).

12.2.1 Availability of suitable soils for agriculture

Soils are substrates biogeochemically originated from meteorization processes of rocks, which act as support for plants, water reservoirs, and reservoirs of nutrients, and consequently, are necessary for agricultural production. In addition, soils are a nonrenewable resource. From a chemical point of view, soils are highly complex and dynamic systems, with high spatial and temporal variability (Bünemann et al., 2018; Abbas et al., 2020). In the agricultural context, the chemical, physical, and biological properties of soils, together with the climate and availability of water, determine their usefulness for a given crop. However, as a soil experiences degradation its capacity for the establishment of important agricultural systems diminishes. This loss of chemical, physical, and biological properties of soils in extreme situations can turn the soil into an inert system both for agricultural activities and for their ecosystem function (Ben-Hur, 2006; Bünemann et al., 2018; Abbas et al., 2020).

Functional polymers can be used both in the agricultural activity itself, as well as in soil remediation and conditioning activities. In the first case, the use of polymers is directed at the production system regardless of their environmental impact, while, when they are used in remediation activities, the polymers become eco-friendly materials even though the chemical nature can be exactly the same. Some examples are shown below.

Poly(acrylamide) have been used to avoid the erosion of soils, which can lead to the loss of arable land and cause desertification, but also, particulate-like polymeric resins of poly(acrylamide) have been used to improve the physical properties of soils such as structure or aggregation, aeration, hydraulic permeability, and water drainage (Seybold, 2008; Fijalkowska et al., 2020) and for remediation of petroleum waste from contaminated soils (Jung et al., 2016). Its addition to soils, it is considered innocuous due to its

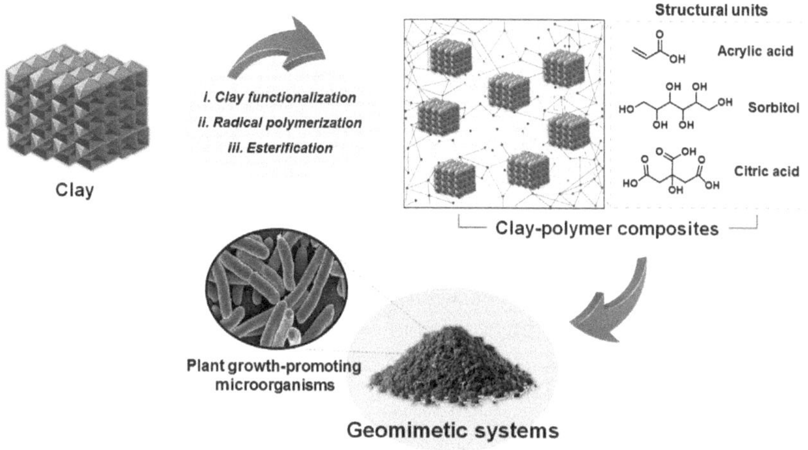

FIGURE 12.1 Preparation of biomimetic systems based on clay—polymer composites using poly(acrylic acid) and polyesters of citric acid and sorbitol as functional networks.

biodegradability and nontoxicity for plants and animals (Xiong et al., 2018; Nyyssölä and Ahlgren, 2019). Another interesting example is the making of geomimicry systems based on clays such as montmorillonite and functional polymers such as polyesters and poly(acrylic acid) (see Fig. 12.1); these systems have been designed to act functionally and structurally as soil particles, which includes permitting transport phenomena, growth of roots and microorganisms, release of nutrients, and other functional properties (Lerma et al., 2018a, 2018b, 2020). In particular, the polymeric phase is designed to be structurally analogous to humified organic matter of soils, including humic and fulvic acids, and humin, in order to achieve a similar functionality to the structural similarity (Garcia et al., 2018; Berrio et al., 2018; Mora et al., 2018; Restrepo et al., 2018). Other polymeric systems based on poly(acrylic acid), poly(styrene sulfonate), and the respective copolymers have been used for monitoring of the aqueous phase of soils (Garcés et al., 2018).

12.2.2 Availability and efficient use of water resources

Water is an essential resource for life, in the same way that it is for agricultural production. The amount of water and its availability are determining factors for the establishment of crops. The amount of water will depend on accessibility to water sources such as rivers, lakes, streams, rain, etc., and on the physical properties of the soils such as their texture (i.e., particle size distribution) and structure (i.e., aggregation). In this way, variations in water levels will determine the phreatic zone, and its fluctuation will depend on external factors such as climate. However, the chemical properties of the soils, such as the type of clay and the organic matter contents, will impact the water availability. In the first case, for example, in clay soils puddles and excessive accumulation of water are produced, but also, water is strongly absorbed by clays, restricting the amount available for the roots. In the second case, organic matter provides the soil with greater conservation of humidity, which is of great importance in higher altitudes and hot climates (Okorogbona et al., 2018; Velasco et al., 2018).

Hydrogels are one of the main types of polymer used in agriculture. Their main characteristic is that they can retain several times their mass in water, and therefore these polymers can be used as a reservoir of water and nutrients for plants (Mohamady, 2020). When hydrogels are biodegradable in nature, they are obtained from biopolymers or from biomolecules that are subsequently polymerized, in this way, after use, are not generally waste, however, they require periodic consumption and therefore become an additional cost (Neethu et al., 2018; Mohamady, 2020). On the other hand, if hydrogels are nonbiodegradable, they can fulfill their function as a water reserve, and at the same time contribute to the improvement of physical properties. In this case, these materials remain much longer in the soil, however, the degradation of their functional units by microbial action, oxidation processes, or chemical transformations little by little causes them to lose their properties and functionality. In addition, this type of hydrogel will generate a residuality and therefore produce a long-term negative impact due to its excessive use. However, in some production systems, they can also be used without their direct discharge to the ground, which would allow their reuse without necessarily becoming a contamination factor (Tanan et al., 2019). Finally, geomimetic polymeric materials have been developed and thought about for combining aspects of biopolymers such as biodegradability and at the same time having a nonbiodegradable synthetic component. In this sense, the nonbiodegradable component is visualized to imitate the humic and fulvic acids of the soils, so in principle, it is expected that the residuality generated by its use does not translate into a contamination process (Lerma et al., 2020). Biopolymer-based hydrogels have been made from carboxymethylcellulose and hydroxyethylcellulose, citric acid and sorbitol, whey protein concentrate, functionalized starches, etc. (Palencia et al., 2017; Durpekova et al., 2020). Due to the importance of hydrogels, these have been explored in other chapters of this book.

12.2.3 Adequate control of plagues

Plagues are one of the big problems in agriculture. This term does not refer to a specific biological entity, nor to its infectious nature or origin. In general, a plague is a biological entity capable of extensively affecting agricultural production. In this sense, plagues can be of animal origin where both higher animals and insects are included, of plant origin either due to nutrient competition, territoriality, or negative interactions between different species, and they can also be of microbiological origin (mainly viruses and fungi) (Sopeña et al., 2009; Perlatti et al., 2013; Nuruzzaman et al., 2016). In this context, pesticides, also called biocides, include a wide variety of compounds with a broad spectrum of chemical properties. A pesticide is defined as any substance used in order to prevent, destroy, attack, repel, or eliminate any plague, and in agriculture, this concept applies to any stage of agricultural production and/or storage, transportation, distribution, and elaboration of seeds or agricultural products. Thus, a pesticide can be a herbicide or any compound used as an insecticide, fungicide, defoliant, desiccant, or inhibitor of fruit thinning and germination (Sopeña et al., 2009; Milani et al., 2017). A main problem associated with the use of pesticides is its residual presence in soil, water, and foods, this being a problem identified in all geographic zones and in several products over large expanses of time from their applications (Sopeña et al., 2009; Martinez and Paez, 2016).

One of the main uses of polymers for plague control is for building barrier methods of insect control. In this sense, a polymeric cover is used to protect fruits. Typical examples are used in the production of bananas and other tropical fruits (Milani et al., 2017). By varying the composition of the polymer, different insecticides are added for greater effectiveness and control of the plague. Clearly, this effectiveness is accompanied by greater contamination, so the reuse of the material should be achieved as much as possible and, at the same time, the release of insecticide into the environment should be avoided as much as possible (Perlatti et al., 2013).

A main application of functional polymers for control of plagues is by the development of micro- and nanocapsules containing herbicides, insecticides, or fungicides. Starch, cellulose, chitin, chitosan, dextran, alginate, casein, albumin, gelatin, rubbers and waxes, poly(vinyl alcohol), polylatate, polyglycolate, polyamines, polyamides, poly(hydroxybutyrate), tannins, and other substances have been used to make capsules (Sopeña et al., 2009; Nuruzzaman et al., 2016).

12.2.4 Efficient fertilization

Fertilization is an agricultural operation that aims to deliver the nutrients required by a specific crop depending on the sowing conditions. Fertilizers, usually of inorganic nature, are supplied to crops depending on the type of plant, external planting factors, type of fertilizer, and nutritional capacity of the soils. The main nutrients are nitrogen, phosphorus, and potassium, because plants require these in large amounts (i.e., NPK-type fertilizers), however, other nutrients are named micronutrients (B, Fe, Cu, Mn, Zn, and Mn), and the amounts in which they must be added are relatively low compared with NPK-type fertilizers, in particular, and for these there is a small concentration range where the micronutrient is essential, in such a way that smaller quantities produce deficiency and larger quantities produce toxicity (Shaviv and Mikkelsen, 1993; Chen et al., 2018; Ashitha et al., 2021; Shahena et al., 2021). The use of fertilizers is currently recognized as an effective strategy for increasing crop production, however, in the fertilization process, fertilizer losses are common due to biodegradation, leaching, evaporation, adsorption, and other processes. These excesses and the subsequent losses have a negative impact on the environment due to the effect on neighboring lands and adjacent or close water sources to the crops (Zhang et al., 2015; Xu et al., 2020).

Functional polymers are used for the development of efficient fertilization, even "smart" to the extent that stimulus-sensitive polymers can be used for this purpose. In general, the use of polymers is directed toward two goals: (1) slow down and/or control the release of fertilizer and (2) reduce losses (Shaviv and Mikkelsen, 1993; Timilsena et al., 2015; Zhang et al., 2015; Chen et al., 2018; Al-Rawajfeh et al., 2021). Some examples are: cellulose acetate and gelatin in conjunction with CuO nanoparticles have been used for encapsulation of fertilizers and other agrochemicals such as pesticides and antimicrobial agents (Xu et al., 2020), functional polyurethanes have been proposed for smart release of boron (see Fig. 12.2) (Palencia et al., 2016a, 2016b), hydrogels based on citric acid and sorbitol as bacterial inoculum of *Azotobacter chroococcum* (Garcés et al., 2017; Lerma et al., 2020), multilayer films based on carboxymethylcellulose and zeolite have been proposed as smart materials for delivery of macro- and micronutrients (Pereira et al., 2020), and bacterial polyhydroxyalkanoates have been developed and proposed for the release of pesticides, phytohormones, and plant growth-promoting rhizobacteria (Dash et al., 2017). Other examples are starches, alginate, poly(vinyl alcohol), and poly(ethylene glycol) (Xu et al., 2020).

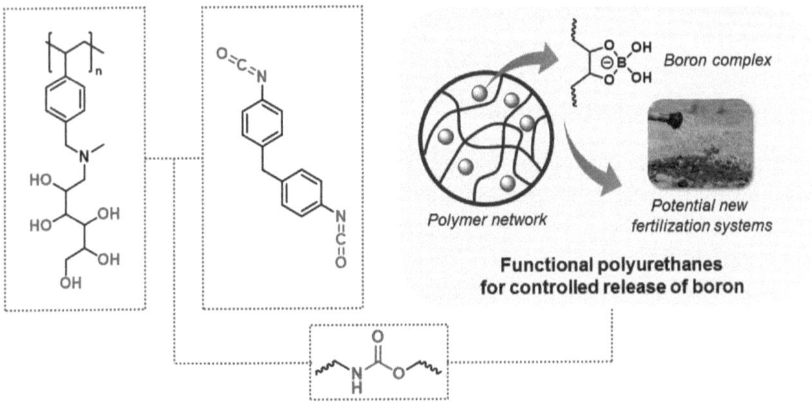

FIGURE 12.2 Preparation of boron-retaining polyurethanes for potential application as new fertilization systems.

12.2.5 Adequate seed and crop conservation

The storage and conservation of seeds and crops are aspects of great importance in agricultural production. Seed conservation consists of avoiding the loss of grains, cuttings, or germplasm used as seeds. In this operation, in some cases, drying conditions, humidity, and protection against insects, fungi and other possible biological, physical, or chemical entities that can affect the seed must be guaranteed, avoided, or controlled. Thus, seeds are essential to guarantee a future harvest and the type of seed varies according to the method of sowing and propagating each crop. Likewise, after harvesting, the produce must be stored, transported, and protected for its proper conservation (Pirzada et al., 2020).

The problem of seed conservation has different degrees of difficulty depending on the type of seed. The conservation process for orthodox seeds is directed toward reducing the moisture content ($<3-7\%$ fresh weight) and decreasing the temperature (Hay and Probert, 2013; Bayli 2004). For the conservation of this type of seeds and crops, structural polymers such as poly(ethylene) and poly(propylene) are usually used, due to their mechanical properties and ease of processing, as well as their hydrophobicity which prevents moisture and plagues, therefore these polymers are used as containers and barrier methods. However, functional polymers fulfill more specific functions within storage processes. For instance, dry hydrogels are used to capture moisture, and therefore, polymers act as a drying material. In addition, functional polymers are used for seed coating [e.g., starch, gelatin, poly(vinyl alcohol), cellulose diacetate] and encapsulation (chitosan, cyclodextrins, chitosan cinnamic acid, sodium alginate + calcium chloride, starch-alginate-clay-rice husk, alginate + lignin, lignin) (Ziani et al., 2010; Hu and Guan, 2018; Farias et al., 2019; Pathak and Ambrose, 2019; Vercelheze et al., 2019; Pirzada et al., 2020).

12.3 Conclusions and remarks

Agriculture is one of the most important productive activities for humanity; it supports food production and therefore the food security of the entire population. Although it is an activity with more than 4000 years experience, agriculture has had, and still has, to adapt to the changing dynamics of the environment, the climate, consumption trends, the availability of resources, and many other aspects. In this context, the potential of functional polymers for the sustainable and eco-friendly development of different agricultural processes and of agriculture itself is clear. The main uses of these materials, and others based on them, are directed toward the conditioning and remediation of soils, the efficient use of water, and improvement of the efficiency of fertilization programs. The specific application of functional polymers must be defined in terms of each crop, the corresponding sowing conditions, and the different factors associated either directly or indirectly. However, although at first it is possible that in a broad sense all the functional polymers used in agriculture are eco-friendly, in terms of their use, and in some cases in terms of their nature, as is the case of biopolymers, it is clear that their beneficial effects can be significantly diminished by their use in enormous quantities and by the lack of good production practices focused on correct waste management.

References

Al-Rawajfeh, A.E., Alrbaihat, M.R., AlShamaileh, E.M., 2021. Characteristics and types of slow- and controlled-release fertilizers. In: Lewu, F.B., Volova, T., Thomas, S., Rakhimol, K.R. (Eds.), Controlled Release Fertilizers for Sustainable Agriculture. Elsevier, pp. 57–78. https://doi.org/10.1016/B978-0-12-819555-0.00004-2.

Abbas, F., Mohkum, H., Ishaq, W., Ahsan, A., Faiq, H., Zia, Z., Fahad, S., Farhad, W., Cerda, A., 2020. A review of soil carbon dynamics resulting from agricultural practices. J. Environ. Manag. 268, 110319. https://doi.org/10.1016/j.jenvman.2020.110319.

Amundson, R., Berse, A.A., Hopmans, J.W., Olson, C., Sztein, A.E., Sparks, D.L., et al., 2015. Soil and human security in the 21st century. Science 348, 1261071. https://doi.org/10.1126/science.1261071.

Ashita, A., Rakhimol, K.R., Mathew, J., 2021. Fate of the conventional fertilizers in environment. In: Lewu, F.B., Volova, T., Thomas, S., Rakhimol, K.R. (Eds.), Controlled Release Fertilizers for Sustainable Agriculture. Elsevier, pp. 25–39. https://doi.org/10.1016/B978-0-12-819555-0.00002-9.

Bayli, C., 2004. Active oxygen species and antioxidants in seed biology. Seed Sci. Res. 14, 93–107. https://doi.org/10.1079/SSR2004159.

Bhargava, A., Srivastava, S., 2019. Human civilization and agriculture. In: Participatory Plant Breeding: Concept and Applications. Springer, Singapore, pp. 1–27. https://doi.org/10.1007/978-981-13-7119-6_1.

Ben-Hur, M., 2006. Using synthetic polymers as soil conditioners to control runoff and soil loss in arid and semi-arid regions - a review. Aust. J. Soil Res. 44, 191–204. https://doi.org/10.1071/SR05175.

Berrio, M.E., Palencia, S.L., Lerma, T.A., Mora, M.A., 2018. Bacterial colonization modelling on soil particles: effect of humic acids on the formation of nitrogen-fixing bacteria biofilms. J. Scie. Technol. Appl. 5, 33–44. https://doi.org/10.34294/j.jsta.18.5.33.

Bünemann, E.K., Bongiorno, G., Bai, Z., Creamer, R., De Deyn, G., et al., 2018. Soil quality – a critical review. Soil Biol. Biochem. 120, 105–125. https://doi.org/10.1016/j.soilbio.2018.01.030.

Chen, J., Lü, S., Zhang, Z., Zhao, X., Li, X., Ning, P., Liu, M., 2018. Environmentally friendly fertilizers: a review of materials used and their effects on the environment. Sci. Total Environ. 613–614, 829–839. https://doi.org/10.1016/j.scitotenv.2017.09.186.

Crowder, D.W., Reganold, J.P., 2015. Financial competitiveness of organic agriculture on a global scale. Proc. Natl. Acad. Sci. U.S.A 112, 7611–7616. https://doi.org/10.1073/pnas.1423674112.

Dash, A., Maity, S., Pati, S., Mohapatra, S., Samantaray, D.P., 2017. Biopolymer (polyhydroxyalkanoates)production by bacteria and its application in seed coating for sustainable agriculture. In: Conference: International Conference on Organic Farming for Sustainable Agriculture. https://doi.org/10.13140/RG.2.2.36671.82085.

Durpekova, S., Filatova, K., Cisar, J., Ronzova, A., Kutalkova, E., Sedlarik, V., 2020. A novel hydrogel based on renewable materials for agricultural application. Int. J. Poly. Sci. 8363418. https://doi.org/10.1155/2020/8363418.

Farias, B.V., Pirzada, T., Mathew, R., Sit, T.L., Opperman, C., Khan, S.A., 2019. Electrospun polymer nanofibers as seed coatings for crop protection. ACS Sustain. Chem. Eng. 7, 19848−19856. https://doi.org/10.1021/acssuschemeng.9b05200. 19848−56.

Fijalkowska, G., Szewczuk-Karpisz, K., Wiśniewska, M., 2020. Polyacrylamide soil conditioners: the impact on nanostructured clay minerals' aggregation and heavy metals' circulation in the soil environment. In: Fesenko, O., Yatsenko, L. (Eds.), Nanomaterials and Nanocomposites, Nanostructure Surfaces, and Their Applications. Springer Proceedings in Physics, pp. 111−127. https://doi.org/10.1007/978-3-030-51905-6_9.

Garcés, V., Palencia, M., Combatt, E., 2017. Development of bacterial inoculums based on biodegradable hydrogels for agricultural applications. J. Sci. Technol. Appl. 3, 5−14. https://doi.org/10.34294/j.jsta.17.2.11.

Garcés, V., Lerma, T.A., Palencia, M., Combatt, E., 2018. Building a probe-type passive sampler to study the contents of heavy metals in the aqueous phase in the soil. J. Sci. Technol. Appl. 4, 4−16. https://doi.org/10.34294/j.jsta.18.4.26.

Garcia, A., Combatt, E., Palencia, M., 2018. Structural study of humin and its interaction with humic acids by Fourier-transform mid-infrared spectroscopy. J. Sci. Technol. Appl. 4, 28−39. https://doi.org/10.34294/j.jsta.18.4.28.

Godfray, H.C.J., Beddington, J.R., Crute, I.R., Haddad, L., Lawrence, D., Muir, J.F., et al., 2010. Food security: the challenge of feeding 9 billion people. Science 327, 812−818. https://doi.org/10.1126/science.1185383.

Hay, F.R., Probert, R.J., 2013. Advances in seed conservation of wild plant species: a review of recent research. Conserv. Physiol. 1, 1−11. https://doi.org/10.1093/conphys/cot030.

Hu, J., Guan, Y., 2018. The application of high polymer materials in the aspects of seed technology. In: Thakur, V., Thakur, M.V.S. (Eds.), Polymeric Gels. Gels Horizons from Sci. To Smart Mater. Springer, Singapore, pp. 55−69.

Jung, J., Jang, J., Ahn, J., 2016. Characterization of a polyacrylamide solution used for remediation of petroleum contaminated soils. Materials 9, 16. https://doi.org/10.3390/ma9010016.

Lee, K.S., Choe, Y.C., Park, S.H., 2015. Measuring the environmental effects of organic farming: a meta-analysis of structural variables in empirical research. J. Environ. Manag. 162, 263−274. https://doi.org/10.1016/j.jenvman.2015.07.021.

Lerma, T., Combatt, E., Palencia, M., 2018a. Soil-mimicking hybrid composites based on clay, polymers and nitrogen-fixing bacteria for the development of remediation systems of degraded soil. J. Sci. Technol. Appl. 4, 17−27. https://doi.org/10.34294/j.jsta.18.4.27.

Lerma, T., Palencia, M., Combatt, E.M., 2018b. Soil polymer conditioner based on montmorillonite-poly(acrylic acid) composites. J. Appl. Polym. Sci. 135 https://doi.org/10.1002/app.46211.

Lerma, T., Garces, V., Palencia, M., 2020. Novel multi-and bio-functional hybrid polymer hydrogels based on bentonite-poly (acrylic acid) composites and sorbitol polyesters: structural and functional characterization. Eur. Polym. J. 128, 109627. https://doi.org/10.1016/j.eurpolymj.2020.109627.

Martínez, J.M., Paez, M.I., 2016. Evaluation of the QuEChERS method with GC-MS detection for the determination of pesticides into white grain corn. J. Sci. Technol. Appl. 1, 15−29. https://doi.org/10.34294/j.jsta.16.1.2.

Milani, P., Fanza, D., Gambaro, A., Faez, R., 2017. Polymers and its applications in agriculture. Polimeros 27. https://doi.org/10.1590/0104-1428.09316.

Mohamady, M., 2020. The application of natural polymer-based hydrogels for agriculture. In: Chen Y. Hydrogels Based on Natural Polymers, pp. 329–356. https://doi.org/10.1016/B978-0-12-816421-1.00013-6.

Mora, M., Palencia, M., Combatt, E., 2018. Effect of dissolved organic matter and humic substances on transport of ions and low molecular weight molecules by liquid membranes. J. Sci. Technol. Appl. 4, 40–48. https://doi.org/10.34294/j.jsta.18.4.29.

Neethu, T.M., Kumar, P., Kaswala, A., 2018. Prospects and applications of hydrogel technology in agriculture. Int. J. Curr. Microbiol. Appl. Sci. 7, 3155–3162. https://doi.org/10.20546/ijcmas.2018.705.369.

Nuruzzaman, M., Rahman, M.M., Liu, Y., Nanoencapsulation, N.R., 2016. nano-guard for pesticides: a new window for safe application. J. Agric. Food Chem. 64, 1447–1483. https://doi.org/10.1021/acs.jafc.5b05214.

Nyyssölä, A., Ahlgren, J., 2019. Microbial degradation of polyacrylamide and the deamination product polyacrylate. Int. Biodeterior. Biodegrad. 139, 24–33. https://doi.org/10.1016/j.ibiod.2019.02.005.

Okorogbona, A., Denner, F., Managa, L., Khosa, T., Maduwa, K., Adebola, P., Amoo, S., Ngobeni, H., Macevele, S., 2018. Water quality impacts on agricultural productivity and environment. In: Lichtfouse, E. (Ed.), Sustainable Agriculture Reviews 27. Springer. https://doi.org/10.1007/978-3-319-75190-0_1.

Palencia, M., Lerma, T., Córdoba, M., 2016a. Polyurethanes with boron retention properties for the development of agricultural fertilization smart systems. J. Sci. Technol. Appl. 1, 39–52. https://doi.org/10.34294/j.jsta.16.1.4.

Palencia, M., Mora, M., Palencia, S.L., 2017. Biodegradable polymer hydrogels based in sorbitol and citric acid for controlled release of bioactive substances from plants (polyphenols). Curr. Chem. Biol. 11, 36–43. https://doi.org/10.2174/2212796810666161028114432.

Palencia, M., Restrepo, D.F., Combatt, E., 2016b. Functional polymer from high molecular weight linear polyols and polyurethane-based crosslinking units: synthesis, characterization, and boron retention properties. J. Appl. Polym. Sci. 133, 43895–43905. https://doi.org/10.1002/app.43895.

Pathak, V., Ambrose, R.P.K., 2019. Starch-based biodegradable hydrogel as seed coating for corn to improve early growth under water shortage. J. Appl. Poly. Sci. 48523, 1–12. https://doi.org/10.1002/app.48523.

Pereira, T., Dias, V.D., Faez, R., 2020. Multilayer films of carboxymethylcellulose/zeolite as smart materials for macro and micronutrients delivery. Micropor. Mesopor. Mater. 302, 110195. https://doi.org/10.1016/j.micromeso.2020.110195.

Perlatti, B., de Souza, P.L., Fernandes da Silva, F., Batista, J., Rossi, M., 2013. Polymeric Nanoparticle-Based Insecticides: A Controlled Release Purpose for Agrochemicals, Insecticides - Development of Safer and More Effective Technologies. Stanislav Trdan, IntechOpen. https://doi.org/10.5772/53355.

Pirzada, T., de Francis, B.V., Mathew, R., Guenther, R.H., Byrd, M.V., et al., 2020. Recent advances in biodegradable matrices for active ingredient release in crop protection: towards attaining sustainability in agriculture. Curr. Opin. Colloid Interface Sci. 48, 121–136. https://doi.org/10.1016/j.cocis.2020.05.002.

Reganold, J., Wachter, J., 2016. Organic agriculture in the twenty-first century. Nat. Plants 2, 15221. https://doi.org/10.1038/nplants.2015.221.

Restrepo, D.F., Palencia, M., Palencia, V.J., 2018. Study by attenuated total reflectance spectroscopy of structural changes of humified organic matter by chemical perturbations via alkaline dissolution. J. Sci. Technol. Appl. 4, 49–59. https://doi.org/10.34294/j.jsta.18.4.30.

Rivas, B.L., Pereira, E.D., Palencia, M., Sánchez, J., 2011. Water-soluble functional polymers in conjunction with membranes to remove pollutant ions from aqueous solutions. Prog. Polym. Sci. 36, 294–322. https://doi.org/10.1016/j.progpolymsci.2010.11.001.

Seybold, C.A., 2008. Polyacrylamide review: soil conditioning and environmental fate. Commun. Soil Sci. Plant Anal. 25, 2171–2185. https://doi.org/10.1080/00103629409369180.

Shahena, S., Rajan, M., Chandran, V., Mathew, L., 2021. Conventional methods of fertilizer release. In: Lewu, F.B., Volova, T., Thomas, S., Rakhimol, K.R. (Eds.), Controlled Release Fertilizers for Sustainable Agriculture. Elsevier, pp. 1–24. https://doi.org/10.1016/B978-0-12-819555-0.00001-7.

Shaviv, A., Mikkelsen, R.L., 1993. Controlled-release fertilizers to increase efficiency of nutrient use and minimize environmental degradation - a review. Fert. Res. 35, 1–12. https://doi.org/10.1007/BF00750215.

Shokrzade, M., Saeedi, S.S., 2011. Pesticides in agricultural products: analysis, reduction, prevention. In: Pesticides- Formulations, Effects, Fate. IntechOpen. https://doi.org/10.5772/13419.

Sopeña, D., Maqueda, C., Morillo, E., 2009. Controlled release formulations of herbicides based on micro-encapsulation. Cienc. Investig. Agrar. 35, 27–42.

Tanan, W., Panichpakdee, J., Saengsuwan, S., 2019. Novel biodegradable hydrogel based on natural polymers: synthesis, characterization, swelling/reswelling and biodegradability. Eur. Polym. J. 112, 678–687. https://doi.org/10.1016/j.eurpolymj.2018.10.033.

Timilsena, Y.P., Adhikari, R., Casey, P., Muster, T., Gill, H., Adhikari, B., 2015. Enhanced efficiency fertilisers: a review of formulation and nutrient release patterns. J. Sci. Food Agric. 95, 1131–1142. https://doi.org/10.1002/jsfa.6812.

Velasco, J., Aznar-Sanchez, J.A., Belmonte, L.J., Roman-Sanchez, I.M., 2018. Sustainable water use in agriculture: a review of worldwide research. Sustainability 10. https://doi.org/10.3390/su10041084.

Vercelheze, A.E.S., Marim, B.M., Oliveira, A.L.M., Mali, S., 2019. Development of biodegradable coatings for maize seeds and their application for Azospirillum brasilense immobilization. Appl. Microbiol. Biotechnol. 103, 2193–2203. https://doi.org/10.1007/s00253-019-09646-w.

Xiong, B., Dettam, R., Shields, D., Pawlik, T., Hochreiter, R., Zydney, A., Kumar, M., 2018. Polyacrylamide degradation and its implications in environmental systems. NPJ Clean Water 1. https://doi.org/10.1038/s41545-018-0016-8.

Xu, T., Ma, C., Aytac, Z., Hu, X., Woei, K., White, J.C., Demokritou, P., 2020. Enhancing agrichemical delivery and seedling development with biodegradable, tunable, biopolymer-based nanofiber seed coatings. ACS Sustain. Chem. Eng. 8, 9537–9548. https://doi.org/10.1021/acssuschemeng.0c02696.

Zhang, X., Davidson, E.A., Mauzerall, D.L., Searchinger, T.D., Dumas, P., Shen, Y., 2015. Managing nitrogen for sustainable development. Nature 528, 51–59. https://doi.org/10.1038/nature15743.

Ziani, K., Ursúa, B., Maté, J.I., 2010. Application of bioactive coatings based on chitosan for artichoke seed protection. Crop Protect. 29, 853–859. https://doi.org/10.1016/j.cropro.2010.03.002.

Chapter 13

Functional and eco-friendly polymers for environmental applications

13.1 An overview of current problems in the environmental context

Environmental sciences are a set of disciplines focused on the study, diagnosis, conservation, and recovery of ecosystems. For the fulfillment of their objectives, the monitoring of pollutants or disturbances to natural systems and their removal or restoration is required. The main pollutant recipients are soil, water, and air. Pollution and soil deterioration can be of natural or anthropogenic origin; in the first case, it can be the result of events generating imbalances in the natural equilibrium of ecosystems both in short periods of time, for example, the explosion of a volcano or a forest fire, and over long periods of time, for example, wind and water erosion of soils, loss of fertility by compaction, or oxidation of pyritic minerals after exposure to air (Vallero, 2006; Place and Mitloehner, 2010; Vallero and Letcher, 2011; Kumar et al., 2018). In the case of water, contamination and deterioration of ecosystem quality are usually more evident because water is essential for life and because it is a fluid phase with high mobility. Characteristic examples include contamination by heavy metals, solid waste, and pesticides, eutrophication processes due to nutrients overload, and loss of hydric resources and the biotic component as a result of overexploitation, poor conservation practices, and excessive fishing (Vallero and Letcher, 2011; Kumar et al., 2018). On the other hand, air pollution is usually framed at two levels, the first level is of a local nature, which is more characteristic in urban and industrial areas due to the excessive emissions resulting from the burning of solid fuels (e.g., coal for the production of energy), liquid fuels (e.g., gasoline in internal combustion motors), and gaseous fuel (e.g., methane which is used as a source of heat energy for food cooking processes and heating systems in many homes around the world), in addition to the emissions generated by technological equipment either during its manufacture or use (e.g., refrigeration, paints, by-products of chemical transformations). On the other hand, from a global point of view,

Eco-friendly Functional Polymers
https://doi.org/10.1016/B978-0-12-821842-6.00026-9
Copyright © 2021 Elsevier Inc. All rights reserved.

modification of the average composition of the atmosphere by all the gaseous emissions generated puts on stage a larger scale problem associated with global warming, climate change, and the long-term continuity of our species on the planet (Vallero and Letcher, 2011; Kumar et al., 2018).

13.2 Functional polymers for environmental monitoring

An important point in the study of ecosystems is monitoring of the dynamics of their components, their disturbances, and changes to components. The complexity of the analysis depends on what you want to measure, and whether you want to establish a causal relationship or not, as it is not the same to measure the elemental composition of a body of water as the morphological variations of a population of birds due to some specific factor. Regardless of the above, the classic analytical measurement consists of a sampling stage, logistic operations like storage, conservation, and transport, and an analytical stage in the laboratory, including the analysis and interpretation of data. However, in some cases where sample processing is simple or null, measurements also can be made directly at the site of interest (Kalinowski et al., 2006; Kumar et al., 2013; Secerov et al., 2018).

Polymers can normally be used in any of the stages of the analysis since they can act as an active phase for the retention or absorption of contaminants, or the analyte of interest, as a container for storage, conservation, and transport, and as an extraction agent for the elimination of interferences or analyte concentration during analysis. Normally, functional polymers are used in stages of sampling, elimination of interferences, and concentration of analytes, while structural polymers are used in logistic stages. However, it is important to indicate that typical structural polymers such as polystyrene or polyethylene may have a functional application depending on the analyte, therefore, the structural differentiation of structural and functional polymers is not appropriate and, in many cases, the functionality of the polymeric material within the application is the main criterion to define a functional polymer (Rivas et al., 2003, 2011; Frechet, 2005). An example of this is the design of passive samplers for water, soils, and air with different configurations and different polymers as the active phase (Huang et al., 2013; Martínez et al., 2017; Garcés et al., 2018), which is a topic that will be discussed in more detail in Chapter 20.

In order to build a consistent development framework for the description of functional polymers in pollutant monitoring, the phase containing the analyte was selected for grouping polymers because, together with the characteristics of the analyte, it allows the main properties of the functional polymer to be delimited.

13.2.1 Monitoring of chemical species in soils

Soils are highly complex systems, made up of solid, liquid, and gaseous phases, all with high variability in terms of their composition and concentration of the chemical species that constitute them. Soils are also a system of high spatial and temporal variability, constituted by a metabolically active and heterogeneous biotic component (Amundson et al., 2015; Abbas et al., 2020).

For the monitoring of chemical species, both inherent components and pollutants, functional polymers are mainly used for the retention of one or more substances constituting the aqueous phase, or which has been transferred from the solid or gaseous phases to the liquid phase. Typical examples are chemical dissolution by acids, oxidizing agents, or extracting liquids. In the first case, the use of inorganic acids is common, sometimes accompanied previously or simultaneously with some thermal treatment and subsequent separation for the recovery of the analyte in the liquid phase (Kalinowski et al., 2006; Kumar et al., 2013; Secerov et al., 2018). The use of oxidizing agents promotes the transformation of organic matter, which acts as a retaining agent for chemical species, enabling its release in an aqueous medium (Abbas et al., 2020). Also, the extraction of the solid phase of the organic components is a common procedure when it is desired to decompose the matrix into an inorganic and an organic phase. A common example is the extraction of humic and fulvic acids by combining extraction solutions (García et al., 2018; Mora et al., 2018; Restrepo et al., 2018). On the other hand, components of the gas phase can be transferred to the aqueous phase through the physicochemical properties of the gas, for example, the greater solubility of CO_2 in alkaline solutions promotes its solubility, with this being used to establish the metabolic activity of soil microorganisms (Alef and Nannipieri, 1995; Chen et al., 2003; Bore et al., 2017; Guo et al., 2019). Further details are given in Section 13.2.2.

As can be inferred, due to the complex composition of soil, the presence of interferences is common and its removal is an important aspect for quantitative analysis, thus, these can be eliminated, when the interferer is cationic in nature, by using functional polymers such as poly(acrylic acid), poly(sodium styrene sulfonate), or poly(vinylpyrrolidone) among others, and, in the case of anions, polymers with quaternary ammonium groups are usually used; a typical example is poly(vinyl benzyl trimethyl ammonium) (Rivas et al., 2003; Palencia et al., 2009; Sánchez and Rivas, 2011; Sanchez et al., 2015). For the removal of neutral species such as boric acid, high-molecular-weight polyols such as poly(N-methyl-D-glucamine), poly(vinyl alcohol), and polymers derived from short-chain polyols like sorbitol are used (Palencia et al., 2016a, 2016b). After retention, separation must be carried out, for which centrifugation and subsequent decantation can be used. In some cases, more sophisticated methods are employed, such as liquid-phase polymer retention where, in addition to the use of linear functional polymers, polymeric membranes are used. Typical membranes are cellulose acetate, poly(sulfone), poly(ethersulfone), and poly(vinyl fluoride), among others (Rivas et al., 2011, 2016; Palencia et al., 2016). Details about the above are shown in Section 13.2.3. Furthermore, the monitoring of the aqueous phase of soils can be carried out directly through the use of sampling probes. In this case, the active polymeric

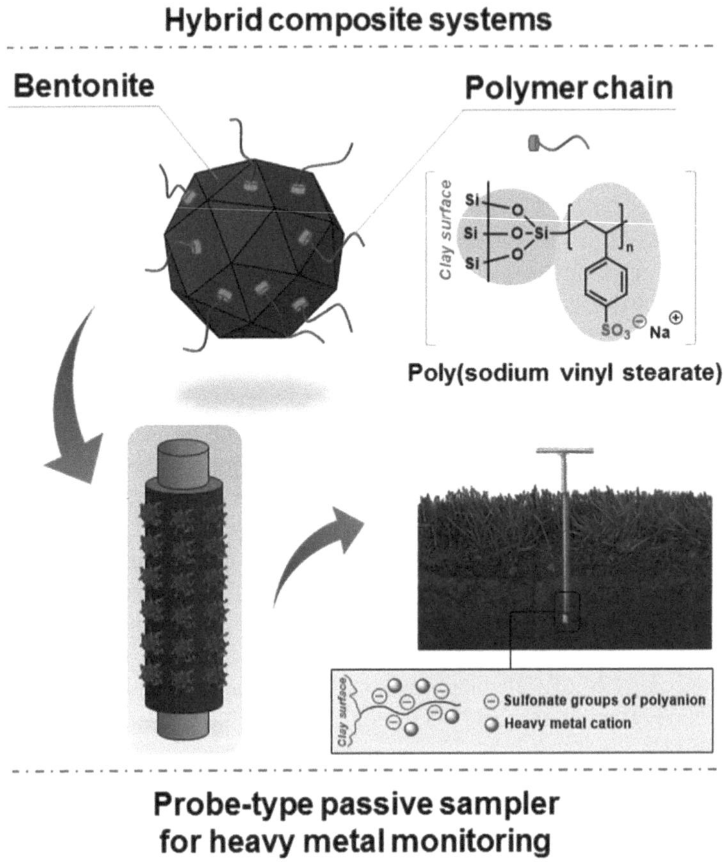

FIGURE 13.1 Design of a probe-type passive sampler for heavy metal monitoring in the aqueous phase of soils using hybrid composite systems based on bentonite and poly(sodium styrene sulfonate).

phase is usually the same: grafts of polycations and polyanions on polymeric support (see Fig. 13.1; Garcés et al., 2018).

13.2.2 Monitoring of air

Gas monitoring requires transfer of the gaseous molecules of interest, either by their direct capture, transfer to a liquid phase, or separation through transport phenomena using dense phases. In the first case, functional polymers can be used as adsorbents for some gases resulting from the high volatility of some compounds, which are usually hydrophobic in nature. After adsorption, they must be desorbed from the polymeric phase, usually using a solvent (Katulski et al., 2011; Wait et al., 2015). Details of this specific application are given in

Chapter 20. In the case of transfer to a liquid phase, the gas is first dissolved in the liquid and then retained by the polymer (Cheng et al., 2019). A very particular example is determination of the rate of respiration of soil microorganisms, which is detailed below.

The monitoring of the gaseous phase of the soils involves the transfer of the gases constituting the gaseous mixture to the liquid phase, and later, a functional polymer is used as active phase for removal of the target component (Wait et al., 2015). The gaseous phase of the soil is understood as the air contained in the interstices of the solid matrix, whether in pores, channels, or cracks present in it. The origin of this intricate set of cavities capable of containing gases is the result of physical properties such as texture, chemical composition such as the type of clays, the content of organic matter that influences the formation of aggregates, and the presence of micro- and macrobiota (insects, microorganisms, plant roots) (Vermoesen et al., 1991; Emerson and McGarry, 2002). On the other hand, the composition of the gaseous phase will be associated with the relative composition of the place, the chemical composition of the soil that through chemical transformations can increase, for example, the content of sulfides in highly reduced and sulfur-rich soils (Oertel et al., 2016; Tang et al., 2019), but also, the composition of soil gases is also influenced by the different metabolic processes of soil microorganisms and plants that can lead to production, and therefore to the enrichment of gases such as methane and carbon dioxide, and also to the reduction of gases such as oxygen or nitrogen (Oertel et al., 2016; Tang et al., 2019). However, since soil is an open system, the quantitative analysis of the gas phase is difficult and results usually show high variability. Therefore, monitoring of the gas phase is usually carried out using controlled conditions both for on-site and laboratory-level analyses (Wait et al., 2015).

On the other hand, the transport of gases through dense phases has been used as strategy for separation of gaseous mixtures. Basically, a polymer swollen with a liquid can be used as a dense membrane for the separation of gases. For that, a low polarity liquid should be used if the polymer is hydrophobic or has low hydrophilicity, whereas a high polarity liquid should be used if the polymer is hydrophilic, and obviously, intermediate situations are also possible. In relation to the polymer characteristics, the polymer is usually cross-linked to avoid its dissolution. The mechanism of separation is known as dissolution–diffusion and is characterized by two stages: the first is the dissolution of the gas molecules in the membrane phase as a function of their distribution constants, thus the amount dissolved is greater as the value of the constant is greater; but at the same time, the difference in chemical potentials on each side of the membrane promotes the diffusion of gas molecules, therefore, the diffusion of gas molecules is greater for the gas interacting less with the polymer (see Fig. 13.2) (Palencia et al., 2016). Other aspects that influence the diffusive process are temperature, molecular weight, and the geometry of the gas molecule (Jayakumar et al., 2015; Palencia et al., 2016).

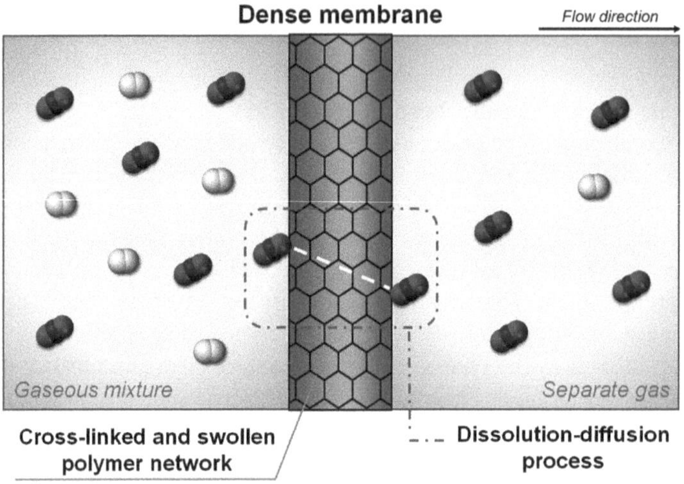

FIGURE 13.2 Gas separation mechanism using dense polymeric membranes. In this case, the greater solubility of one of the gases facilitates its diffusion and permeation through the swollen polymer network.

Another aspect of great relevance is the development of sensors for gases. These sensors are usually made of polymeric matrix nanocomposites or conductive polymers. In the both cases the active phase is made up of a polymer or some polymer nanocomposite containing usually an oxide or a nanostructured metal. In these sensors, the adsorption of gas at the polymeric interface produces a change in the electrical or optical response of the material, allowing its detection (Pandey 2016; Cheng et al., 2019). Multiple applications have been addressed under this strategy with the most common being the development of sensors for gases such as ammonia, nitrogen dioxide, carbon monoxide, and methanol using polymers such as poly(thiophene), poly(aniline), poly(pyrrole), and poly(3,4-ethylenedioxythiophene) (Persaud, 2005; Katulski et al., 2011; Cheng et al., 2019).

13.2.3 Monitoring of water

Aqueous phase monitoring is usually the most common and most widely used in environmental monitoring. This is not only because aqueous systems constitute one of the most important habitats and ecosystems in nature, but also because on many occasions the components present in solid or gaseous phases are transferred to aqueous phases. As previously indicated, one of the main applications of functional polymers is the elimination of interfering substances that may be present in solution, but also, they are used as the active phase of uptake agents. This is a direct result of the fact that they can interact with many substances using different mechanisms, with adsorption, ion exchange, complex formation, and electrostatic interaction being the most common (Rivas et al., 2003; Frechet, 2005).

Several techniques based on membranes, resins, and linear polymers using different mechanisms have been developed and can be easily revised from specialized papers (Rivas et al., 2011, 2016; Rivas and Palencia, 2011; Ali, 2012; Carpenter et al., 2015; Sanchez et al., 2015; Palencia 2015; Palencia et al., 2016; Gomez-Maldonado et al., 2019; Huang and Feng, 2019). A summary of characteristic functional polymers for removal of species from aqueous dissolution is shown in Table 13.1.

TABLE 13.1 Example of characteristic functional polymers used for the removal of inorganic and organic species from aqueous dissolution: poly(acrylic acid) (PAA), poly(styrene sulfonate) (PSS), poly(ethylenimine) (PEI), poly(vinylbenzyl-trimethylammonium) (PVBtMA), and poly(vinyl-benzyl-N-methyl-D-glucamine) (PVBNMDG) (Rivas et al., 2003; Palencia, 2015; Huang and Feng, 2019).

Functional polymer	Functional groups	Functional groups type	Target species	Interaction mechanisms
PAA	Carboxylic acid	Weak acid	Metal cations	Coordinating bonds, ionic exchange, electrostatic interaction depending on pH
			Organic cations	Adsorption, electrostatic interaction depending on pH
PSS	Sulfonate	Strong acid	Metal cations	Ionic exchange, electrostatic interaction
			Organic cations	Adsorption, electrostatic interaction
PEI	Amine	Weak base	Metal cations, organic anions	Adsorption, coordinating bonds, electrostatic interaction depending on pH
			Organic anions	Adsorption, electrostatic interaction depending on pH
PVBtMA	Quaternary ammonium	Strong base	Nonmetal anions	Ionic exchange, electrostatic interaction
			Organic anions	Adsorption, electrostatic interaction
PVBNMDG	Hydroxyl	Neutral group	Boric acid	Coordinating bonds and electrostatic interaction depending on pH

In the adsorption processes, dispersive forces, hydration phenomena, and diffusion processes intervene. In ion exchange, an electrostatic interaction is required between the charged analyte and the polymer surface with opposed surface charge. In this case, an ion on the surface of the polymer is exchanged for an ion in the solution, which is mediated for a parallel process such as the surface hydration and diffusion processes from bulk to surface of exchanger (Cyganowsky and Dzimitrowicz, 2020). For the formation of complexes, the nature of the ion and the functional groups or atoms on the polymer structure play a crucial role. Thus, for example, sodium and potassium ions are not expected to form complexes, and their interaction with the functional groups of the polymer surface will be mainly electrostatic. However, something different occurs with calcium, magnesium, mercury, copper, and cadmium ions to name a few. For these, the formation of coordination bonds with functional groups such as carboxylic acids and amines on the polymer is expected, but even for these ions only an electrostatic-type interaction is expected when the functional groups of the polymer are sulfonates or quaternary ammoniums (Chinachoti and Schmidt, 1991; Jacobasch and Simon, 1998; Slark and Hadgett, 1999; Cyganowsky and Dzimitrowicz, 2020). Thus, for interaction by the formation of coordination bonds, the formation of chemical bonds takes place through the creation of one covalent bond where two electrons are derived from the same atom, thus, the ion acts as an electron acceptor species, while the functional group on the polymer structure acts as an electron donor species, usually by atoms such as nitrogen, oxygen, phosphorous, etc. The coordinating bonds are reversible and usually unstable at strongly acidic or basic pH conditions, which is the reason why the regeneration and reuse of the polymer in many cases is a viable alternative (Rivas et al., 2003).

13.3 Functional polymers for environmental remediation

Environmental remediation is the restoration of the properties of ecosystems that have been affected by external factors or inherent to them. This intervention can be aimed at carrying out specific actions through the use of external agents (direct intervention), or through the isolation or cessation of external factors that led to its deterioration (indirect intervention). In the latter case, it is the equivalent of stopping the source of contamination and letting the ecosystem reestablish itself, consequently, it is clear that the use of functional polymers for remediation or mitigation activities of ecosystems is part of direct intervention strategies (Sharma et al., 2018; Hamadani et al., 2020). Remediation processes can also be described in terms of the ecosystem and its predominant phase (i.e., soil, water, or air for solid, liquid, or gas phases, respectively), therefore the same functional polymer can be used differently depending on whether the ecosystem is terrestrial, aquatic, or gaseous. The gas phase will not be discussed in this section because it is not an ecosystem in itself, but an inherent part of terrestrial ecosystems.

13.3.1 Remediation of terrestrial ecosystems

The remediation of terrestrial ecosystems using functional polymers is very versatile and is specifically circumscribed in each specific problem. Four key examples are: (1) immobilization of polluting species, for example heavy metals, (2) reduction of erosion and loss of topsoil, (3) reforestation activities, and (4) rehabilitation of degraded soils (Hamadani et al., 2020).

The immobilization of polluting species is a remediation strategy that starts from the premise that the bioactivity and toxicity of the pollutant is enhanced by its mobility and bioavailability within the ecosystem, either from a spatial point of view (mobility) or from the point of view of its absorption, translocation, and accumulation by living organisms (bioavailability). In the case of absorption, the main concern is its biomagnification within the food chain (Gomez-Maldonado et al., 2019). Consequently, innocuous additives such as some functional polymers are capable of immobilizing pollutants and reducing the risks associated with them by reducing their mobility and bioavailability. In principle, the strategy is simple and based on the visualization of a reverse process of liberation. However, in practice, soil dynamics can greatly hinder their effectiveness. For example, the release of arsenic can be a result of anthropogenic activities, but at the same time it can be caused by natural processes unconsciously triggered by human activities. Arsenic in nature is found as part of minerals such as arsenopyrite, which is a very common iron and nickel arsenide. The simultaneous presence of this mineral and pyrite, which is iron sulfide, is common, and consequently, due to the reduced state of sulfur, it is to be expected in soils rich in sulfur and in a state of high reduction, such as flooded soils. The exposure of these materials to oxygen, either by agricultural or mining activities, triggers the oxidation of pyrite, as products of this process, sulfuric acid is produced in large quantities, which is why the release of other minerals is promoted, that is, before the presence of arsenopyrite, arsenic will be released into the environment. Basically, arsenic increased its mobility and bioavailability due to the pyrite oxidation process, consequently, a high content of a very toxic species was indirectly released into the environment (Powell and Martens, 2005; Karimian et al., 2018; Ruprecht et al., 2018). The strategy of immobilizing the pollutant with functional polymers is based on reducing its mobility and availability, imitating the initial effect of the mineral, for which a polymer with high arsenic retention capacity and low solubility should be added to the soil. Some examples of polymers used for the retention of arsenic in laboratory tests are poly(4-vinyl-1-methyl pyridinium)bromide, poly[2-(acryloyloxy)ethyl]trimethylammonium chloride, and poly[3-(methacryloylamino)propyl]-dimethyl(3-sulfopropyl) ammonium (Sanchez and Rivas, 2011; Sanchez et al., 2015; Rivas et al., 2016). However, although in principle it may seem effective, in this specific example, the release of sulfuric acid will strongly acidify the environment, even overcoming the acid neutralization capacity of the soils (Powell and

Martens, 2005). In this scenario, the pH drops sharply and the ionic strength increases, causing a decrease in the arsenic retention capacity of the polymers. This effect of pH and ionic strength has been widely described in ion retention studies using soluble polymers and membrane systems. In conclusion, the immobilization strategy using soluble polymers is strongly dependent on current soil conditions, but also on future conditions generated by the dynamics of this type of ecosystem.

On the other hand, the reduction of erosion and the loss of topsoil is a problem that can also have both natural and anthropogenic origins. Normally, it has been approached by combining several technologies, including the use of polymeric soil conditioners, the most representative being poly(-acrylamide). Erosion can have different origins, the most common being wind erosion, water erosion, and the combined action of both, in addition to the loss of natural cementing agents such as organic matter humidified by forest fires. The use of soil conditioners seeks to reduce the impact of the aforementioned factors and improve the physical properties of the soils (Seybold, 2008; Lerma et al., 2018b; Fijalkowska et al., 2020). This strategy can be overlaid with reforestation activities, in which case, functional polymers can be used for restoration of the microbiological properties of soils through the development of bacterial inoculums (Garcés et al., 2017), or as containment materials during the growth of seedlings in greenhouses (Salachna, 2016). In the latter case, biodegradable polymers are used that allow the planting of seedlings without subjecting them to stress to improve their fixation and at the same time avoiding polymer residuality.

One of the recent applications with great potential to address soil remediation, reduce erosion and the loss of topsoil, and increase efficiency in reforestation programs is the use of geomimetic systems based on functional polymers. The concept of "geomimesis" was introduced by the Colombian researchers M. Palencia and T. Lerma, from the research group in sciences with technological applications GI-CAT of the department of chemistry of the Universidad del Valle in 2018, to describe polymer materials designed structurally and functionally to mimic soil particles (Lerma et al., 2018a). Geomimetic materials consist of a clay core, an interface analogous to humin, a supramolecular system based on functional polymers analogous to humic and fulvic acids, a matrix of biodegradable polymers that contain growth-promoting microorganisms for plants and give the particles properties of hydrogels which can be used for the storage of water and which can be released by the metabolic activity of the microorganisms inoculated in the material (Lerma et al., 2020). Detailed aspects of this technology were discussed in Chapter 8 and are expanded on in Chapter 22.

13.3.2 Remediation of aquatic ecosystems

The remediation of aquatic ecosystems using functional polymers has been widely studied at the laboratory level, and put into practice, till date, only in controlled aqueous systems as treatment prior to their discharge into the environment. In principle, from a theoretical approach, resins and linear polymers can be used for remediation of aquatic ecosystems through different mechanisms (i.e., adsorption, ion exchange, formation of a complexing bond, electrostatic interaction, and hydrophobic interactions) and in the absence of fouling (Slark and Hadgett, 1999; Rivas et al., 2003, 2011; Frechet, 2005; Cyganowsky and Dzimitrowicz, 2020). Among the most used systems are ion exchange resins which are mainly directed toward the removal of metal ions in solution, especially heavy metals. The disadvantage of these systems is the high implementation and operating costs. Among the most common polymers for this type of materials are polymers with carboxylic acid groups, sulfonates, and quaternary ammonium groups (Cyganowsky and Dzimitrowicz, 2020). On the other hand, linear functional polymers have been widely studied for the removal of heavy metals and other organic pollutants in aqueous solutions (Rivas et al., 2003, 2011, 2016; Huang and Feng, 2019). When these are used in flocculation operations, the main disadvantage is the formation of highly toxic sludge, due to its enrichment in the pollutants removed, which must be treated. In the 1980s, K. Geckeler et al. in Germany, proposed the combined use of water-soluble polymers and ultrafiltration membranes as an analytical strategy for the elimination of interferences associated with metals (Geckeler et al., 1986; Spivakov et al., 1985). Later, the same principle was extended to remediation processes for aqueous effluents. This application has been widely studied by the team of Professor B.L. Rivas from the polymer department of the Universidad de Concepción in Chile (Rivas et al., 2003, 2011; Palencia, 2015). In a generalized way this technique has been described by different authors using different names: liquid-phase polymer-based retention, polymer-enhanced ultrafiltration, ultrafiltration assisted by polymers, among others (Palencia, 2015).

Liquid-phase polymer-based retention is based on a two-stage mechanism: (1) the formation of a macromolecular species in the aqueous phase by contacting the solution to be treated with the polymer and (2) the of the polymer–metal ion species using a system ultrafiltration. Furthermore, it has the advantages that polymers can be synthesized to maximize their interaction with the contaminating species, the polymer can be reused, and the contaminant can be concentrated and purified for subsequent transformation or reuse. However, it has the disadvantage that the process is limited by the maximum retention capacity of the polymer, since the functional groups of the polymer must retain the metal ion and at the same time keep the polymer in solution to avoid fouling. Furthermore, the increase in the retention capacity cannot be effected at the expense of increasing the concentration of the polymer since this brings with it an increase in viscosity and a subsequent decrease in flow (Palencia 2015; Palencia et al., 2016). Complete reviews of the technique and its variants have been carried out (Rivas et al., 2011; Rivas and Palencia, 2011; Palencia 2015; Huang and Feng, 2019).

In contrast, another interesting application of functional polymers in the remediation of aquatic ecosystems is aimed at the removal of oil from fuel spills. Some of the polymers used for this purpose are: sodium salt of oleoyl carboxymethyl chitosan, chitosan microspheres modified with n-butyl acrylate, carboxymethyl chitosan modified with monochloroacetic acid, blends of poly(vinyl alcohol) nanoparticles with chitosan or starch, polyvinylpyrrolidone-coated magnetite nanoparticles, polystyrene-coated magnetic hollow Fe_3O_4 nanoparticles, etc. (Doshi et al., 2018).

13.4 Conclusions and remarks

The use of functional polymers for environmental applications is varied and promising. Clearly the spectrum of technologies that can be developed is very wide, and the approaches for their use are very diverse. From an environmental point of view, and in a general context, all functional polymers are in principle eco-friendly when they are used to contribute to the protection and conservation of natural environments. This is supported by the fact that these materials allow not only existing problems to be addressed from the point of view of diagnosis and the study phase, but also the fact that they can be used to remedy different pollution problems. However, one must be careful with the previous generalization, since each problem has its distinctions and specificities, and therefore, the use of a functional polymer must be weighed in terms of the advantages and disadvantages of its use in the specific context of the application. On the other hand, the susceptibility to degradation by physical, chemical, and biological agents is an aspect to be taken into account. Finally, the fact that natural environments can generate fouling and biofouling in the polymer surface, promoting its deterioration and ineffectiveness, should not be overlooked.

References

Abbas, F., Mohkum, H., Ishaq, W., Ahsan, A., Faiq, H., Zia, Z., Fahad, S., Farhad, W., Cerda, A., 2020. A review of soil carbon dynamics resulting from agricultural practices. J. Environ. Manag. 268 https://doi.org/10.1016/j.jenvman.2020.110319, 110319.

Alef, K., Nannipieri, P., 1995. Estimation of microbial activities. In: Alef, K., Nannipieri, P. (Eds.), Methods in Applied Soil Microbiology and Biochemistry. Academic Press, pp. 193–270. https://doi.org/10.1016/B978-012513840-6/50020-3.

Ali, I., 2012. New generation adsorbents for water treatment. Chem. Rev. 112, 5073–5091. https://doi.org/10.1021/cr300133d.

Amundson, R., Berse, A.A., Hopmans, J.W., Olson, C., Sztein, A.E., Sparks, D.L., et al., 2015. Soil and human security in the 21st century. Science 348. https://doi.org/10.1126/science.1261071, 1261071.

Bore, E.K., Apostel, C., Halicki, S., Kuzyakov, Y., Dippold, M.A., 2017. Soil microorganisms can overcome respiration inhibition by coupling intra- and extracellular metabolism: 13C metabolic tracing reveals the mechanisms. ISME J. 11, 1423–1433. https://doi.org/10.1038/ismej.2017.3.

Carpenter, A.W., De Lannoy, C.F., Wiesner, M.R., 2015. Cellulose nanomaterials in water treatment technologies. Environ. Sci. Technol. 49, 5277–5287. https://doi.org/10.1021/es506351r.

Cheng, Y., Chin, B., Haseeb, A.S.M.A., Aqilah, A., Hoong, Y., 2019. Review—conducting polymers as chemiresistive gas sensing materials: a review. J. Electrochem. Soc. 167, 037503.

Chen, G., Zhu, H., Zhang, Y., 2003. Soil microbial activities and carbon and nitrogen fixation. Res. Microbiol. 154, 393–398. https://doi.org/10.1016/S0923-2508(03)00082-2.

Chinachoti, P., Schmidt, S.J., 1991. Solute-polymer-water interactions and their manifestations. In: Levine, H., Slade, L. (Eds.), Water Relationships in Foods, Advances in Experimental Medicine and Biology, 302. Springer, pp. 561–583. https://doi.org/10.1007/978-1-4899-0664-9_30.

Cyganowsky, P., Dzimitrowicz, A., 2020. A mini-review on anion exchange and chelating polymers for applications in hydrometallurgy, environmental protection, and biomedicine. Polymers 12, 784. https://doi.org/10.3390/polym12040784.

Doshi, B., Sillanpaa, M., Kalliola, S., 2018. A review of bio-based materials for oil spill treatment. Water Res. 135, 262–277. https://doi.org/10.1016/j.watres.2018.02.034.

Emerson, W.W., McGarry, D., 2002. Organic carbon and soil porosity. Aust. J. Soil Res. 41, 107–118. https://doi.org/10.1071/SR01064.

Fijalkowska, G., Szewczuk-Karpisz, K., Wiśniewska, M., 2020. Polyacrylamide soil conditioners: the impact on nanostructured clay minerals' aggregation and heavy metals' circulation in the soil environment. In: Fesenko, O., Yatsenko, L. (Eds.), Nanomaterials and Nanocomposites, Nanostructure Surfaces, and Their Applications. Springer Proceedings in Physics, pp. 111–127. https://doi.org/10.1007/978-3-030-51905-6_9.

Frechet, J.M.J., 2005. Functional polymers: from plastic electronics to polymer-assisted therapeutics. Prog. Polym. Sci. 30 https://doi.org/10.1016/j.progpolymsci.2005.06.005, 884–857.

Garcés, V., Lerma, T.A., Palencia, M., Combatt, E., 2018. Building a probe-type passive sampler to study the contents of heavy metals in the aqueous phase in the soil. J. Sci. Technol. Appl. 4, 4–16. https://doi.org/10.34294/j.jsta.18.4.26.

Garcés, V., Palencia, M., Combatt, E., 2017. Development of bacterial inoculums based on biodegradable hydrogels for agricultural applications. J. Sci. Technol. Appl. 3, 5–14. https://doi.org/10.34294/j.jsta.17.2.11.

Garcia, A., Combatt, E., Palencia, M., 2018. Structural study of humin and its interaction with humic acids by Fourier-transform mid-infrared spectroscopy. J. Sci. Technol. Appl. 4, 28–39. https://doi.org/10.34294/j.jsta.18.4.28.

Geckeler, K.E., Bayer, E., Spivakov, B., Shkinev, V.M., Vorobeva, G.A., 1986. Liquid-phase polymer-based retention, a new method for separation and preconcentration of elements. Anal. Chim. Acta 189, 285–292. https://doi.org/10.1016/S0003-2670(00)83731-9.

Gomez-Maldonado, D., Vega Erramuspe, I.B., Peresin, M.S., 2019. Natural polymers as alternative adsorbents and treatment agents for water remediation. BioResources 14, 10093–10160. https://doi.org/10.15376/biores.14.4.Gomez-Maldonado.

Guo, H., Hu, Z., Zhang, H., Hou, Z., Min, W., 2019. Soil microbial metabolic activity and community structure in drip-irrigated calcareous soil as affected by irrigation water salinity. Water, Air and Soil Pollut. 230 https://doi.org/10.1007/s11270-019-4094-0.

Hamadani, H., Mudasir, S., Rehman, M., Ali, R., Rashid, M., et al., 2020. Global scenario of remediation techniques to combat environmental pollution. In: Hakeen, K.R., et al. (Eds.), Bioremediation and Biotechnology, Sustainable Approaches to Pollution Degradation. Springer. https://doi.org/10.1007/978-3-030-35691-0_5.

Huang, Y., Feng, X., 2019. Polymer-enhanced ultrafiltration: fundamentals, applications and recent developments. J. Membr. Sci. 586, 53–83. https://doi.org/10.1016/j.memsci.2019.05.037.

Huang, J., Lyman, S.N., Stamenkovic, J., Sexauer, M., 2013. A review of passive sampling systems for ambient air mercury measurements. Environ. Sci. Proces. Imp. 16, 374–392. https://doi.org/10.1039/c3em00501a.

Jacobasch, H.J., Simon, F., Weidenhammer, P., 1998. Adsorption of ions onto polymer surfaces and its influence on zeta potential and adhesion phenomena. Colloid Polym. Sci. 276, 434–442. https://doi.org/10.1007/s003960050263.

Jayakumar, A., Pethaiah, S., Ramos, M., RObertson, J., Al-Jumaily, A., 2015. A technical review on gas diffusion, mechanism and medium of PEM fuel cell. Ionics 21, 1–18. https://doi.org/10.1007/s11581-014-1322-x.

Kalinowski, M.B., Feichter, J., Nikkinen, M., Schlosser, C., 2006. Environmental sample analysis. In: Avenhaus, R., Kyriakopoulos, N., Richard, M., Stein, G. (Eds.), Verifying Treaty Compliance. Springer, pp. 367–387. https://doi.org/10.1007/3-540-33854-3_17.

Karimian, N., Johnston, S.G., Burton, E.D., 2018. Iron and sulfur cycling in acid sulfate soil wetlands under dynamic redox conditions: a review. Chemosphere 197, 803–816. https://doi.org/10.1016/j.chemosphere.2018.01.096.

Katulski, R.J., Namiesnik, J., Sadowski, J., Stefanski, J., Wardencki, W., 2011. Monitoring of gaseous air pollution, the impact of air pollution on health, economy, environment and agricultural sources. IntechOpen 143–158. https://doi.org/10.5772/20771. Mohamed K. Khallaf.

Kumar, N., Fatima, T., Mishra, I., Verma, M., Mishra, J., Mishra, V., 2018. Environmental sustainability: challenges and viable solutions. Environ. Sustain. 1, 309–340. https://doi.org/10.1007/s42398-018-00038-w.

Kumar, A., Kim, H., Hancke, G., 2013. Environmental monitoring systems: a review. IEEE Sensor. J. 13, 1329–1339. https://doi.org/10.1109/JSEN.2012.2233469.

Lerma, T., Combatt, E., Palencia, M., 2018a. Soil-mimicking hybrid composites based on clay, polymers and nitrogen-fixing bacteria for the development of remediation systems of degraded soil. J. Sci. Technol. Appl. 4, 17–27. https://doi.org/10.34294/j.jsta.18.4.27.

Lerma, T., Garcés, V., Palencia, M., 2020. Novel multi-and bio-functional hybrid polymer hydrogels based on bentonite-poly (acrylic acid) composites and sorbitol polyesters: structural and functional characterization. Europ. Polym. J. 128 https://doi.org/10.1016/j.eurpolymj.2020.109627, 109627.

Lerma, T., Palencia, M., Combatt, E.M., 2018b. Soil polymer conditioner based on montmorillonite-poly(acrylic acid) composites. J. Appl. Polym. Sci. 135 https://doi.org/10.1002/app.46211.

Martínez, J.M., Trochez, E., Paéz, M.I., 2017. Evaluation of poly(ethylene-co-vinylacetate) as polymeric adsorbent of polychlorinated biphenyls in water. J. Sci. Technol. Appl. 2, 94–107. https://doi.org/10.34294/j.jsta.17.2.18.

Mora, M., Palencia, M., Combatt, E., 2018. Effect of dissolved organic matter and humic substances on transport of ions and low molecular weight molecules by liquid membranes. J. Sci. Technol. Appl. 4, 40–48. https://doi.org/10.34294/j.jsta.18.4.29.

Oertel, C., Matschullat, J., Zurba, K., Zimmermann, F., Erasmi, S., 2016. Greenhouse gas emissions from soils—a review. Geochemistry 76, 327–352. https://doi.org/10.1016/j.chemer.2016.04.002.

Palencia, M., 2015. Liquid-phase polymer-based retention: theory, modeling, and application for the removal of pollutant inorganic ions. J. Chem. https://doi.org/10.1155/2015/965624, 965624.

Palencia, M., Córdoba, A., Vera, M., 2016. Membrane technology and chemistry. In: Visakh, P.M., Nazarenko, O. (Eds.), Nanostructured Polymer Membranes, vol. 1. Wiley, pp. 27–54.

Palencia, M., Lerma, T., Córdoba, M., 2016a. Polyurethanes with boron retention properties for the development of agricultural fertilization smart systems. J. Sci. Technol. Appl. 1, 39–52. https://doi.org/10.34294/j.jsta.16.1.4.

Palencia, M., Restrepo, D.F., Combatt, E., 2016b. Functional polymer from high molecular weight linear polyols and polyurethane-based crosslinking units: synthesis, characterization, and boron retention properties. J. Appl. Polym. Sci. 133, 43895–43905. https://doi.org/10.1002/app.43895.

Palencia, M., Rivas, B.L., Pereira, M., 2009. Metal ion recovery by polymer-enhanced ultrafiltration using poly(vinyl sulfonic acid): fouling description and membrane–metal ion interaction. J. Membr. Sci. 345, 191–200. https://doi.org/10.1016/j.memsci.2009.08.044.

Pandey, S., 2016. Highly sensitive and selective chemiresistor gas/vapor sensors based on polyaniline nanocomposite: a comprehensive review. Journal of Science: Adv. Mater. Dev. 1, 431–453. https://doi.org/10.1016/j.jsamd.2016.10.005.

Persaud, K.C., 2005. Polymers for chemical sensing. Mater. Today 8. https://doi.org/10.1016/S1369-7021(05)00793-5, 83–44.

Place, S.E., Mitloehner, F.M., 2010. Contemporary environmental issues: a review of the dairy industry's role in climate change and air quality and the potential of mitigation through improved production efficiency. J. Dairy Sci. 93, 3407–3416. https://doi.org/10.3168/jds.2009-2719.

Powell, B., Martens, M., 2005. A review of acid sulfate soil impacts, actions and policies that impact on water quality in Great Barrier Reef catchments, including a case study on remediation at East Trinity. Mar. Pollut. Bull. 51, 149–164. https://doi.org/10.1016/j.marpolbul.2004.10.047.

Restrepo, D.F., Palencia, M., Palencia, V.J., 2018. Study by attenuated total reflectance spectroscopy of structural changes of humified organic matter by chemical perturbations via alkaline dissolution. J. Sci. Technol. Appl. 4, 49–59. https://doi.org/10.34294/j.jsta.18.4.30.

Rivas, B.L., Palencia, M., 2011. Removal-concentration of pollutant metal-ions by water-soluble polymers in conjunction with double emulsion systems: a new hybrid method of membrane-based separation. Separ. Purif. Technol. 81, 435–443. https://doi.org/10.1016/j.seppur.2011.08.021.

Rivas, B.L., Pereira, E., Moreno-Villoslada, I., 2003. Water-soluble polymer–metal ion interactions. Prog. Polym. Sci. 28, 173–208. https://doi.org/10.1016/S0079-6700(02)00028-X.

Rivas, B.L., Pereira, E.D., Palencia, M., Sánchez, J., 2011. Water-soluble functional polymers in conjunction with membranes to remove pollutant ions from aqueous solutions. Prog. Polym. Sci. 36, 294–322. https://doi.org/10.1016/j.progpolymsci.2010.11.001.

Rivas, B.L., Sanchez, J., Palencia, M., 2016. Organic membranes and polymers for the removal of pollutants. In: Visakh, P.M., Nazarenko, O. (Eds.), Nanostructured Polymer Membranes, vol. 1. Wiley, pp. 203–236.

Ruprecht, J.E., Glamore, W.C., Rayner, D.S., 2018. Estuarine dynamics and acid sulfate soil discharge: quantifying a conceptual model. Ecol. Eng. 10, 172–184. https://doi.org/10.1016/j.ecoleng.2017.11.017.

Salachna, P., 2016. The use of biodegradable polymers to propagation of ornamental bulbous plants. Ecol. Eng. Environ. Technol. 46, 143–148. https://doi.org/10.12912/23920629/61477.

Sanchez, J., Butter, B., Rivas, B.L., Basaez, L., Santander, P., 2015. Electrochemical oxidation and removal of arsenic using water-soluble polymers. J. Appl. Electrochem. 45, 151–159. https://doi.org/10.1007/s10800-014-0785-9.

Sanchez, J., Rivas, B.L., 2011. Cationic hydrophilic polymers coupled to ultrafiltration membranes to remove chromium (VI) from aqueous solution. Desalination 279, 338–343. https://doi.org/10.1016/j.desal.2011.06.029.

Secerov, I., Dolinaj, D., Pavic, D., Milosevic, D., Savic, S., Popov, S., 2018. Zivanov Z. Environmental monitoring systems: review and future development. Wireless Eng. Technol. 10, 1–18. https://doi.org/10.4236/wet.2019.101001.

Seybold, C.A., 2008. Polyacrylamide review: soil conditioning and environmental fate. Commun. Soil Sci. Plant Anal. 25, 2171–2185. https://doi.org/10.1080/00103629409369180.

Sharma, S., Tiwari, S., Hasan, A., Saxena, V., Pandey, L.M., 2018. Recent advances in conventional and contemporary methods for remediation of heavy metal-contaminated soils. 3 Biotech 8, 216. https://doi.org/10.1007/s13205-018-1237-8.

Slark, A.T., Hadgett, P.M., 1999. The effect of polymer structure on specific interactions between dye solutes and polymers. Polymer 40, 1325–1332. https://doi.org/10.1016/S0032-3861(98)00346-2.

Spivakov, B., Geckeler, K., Bayer, E., 1985. Liquid-phase polymer-based retention — the separation of metals by ultrafiltration on polychelatogens. Nature 315, 313–315. https://doi.org/10.1038/315313a0.

Tang, J., Schurgers, G., Rinnan, R., 2019. Organic gases released and taken up by soil lack quantification. Eos 100. https://doi.org/10.1029/2019EO131663.

Vallero, D.A., 2006. Pollution revisited. In: Vallero, D.A. (Ed.), Paradigms Lost: Learning from Environmental Mistakes, Mishaps, and Misdeeds. ScienceDirect, pp. 55–94. https://doi.org/10.1016/B978-075067888-9/50003-X.

Vallero, D.A., Letcher, T.M., 2011. Regulation of wastes. In: Vallero, D.A., Letcher, T.M. (Eds.), Wastes: A Handbook for Management. ScienceDirect, pp. 23–59. https://doi.org/10.1016/B978-0-12-381475-3.10003-8.

Vermoesen, A., Ramon, H., Van Cleemput, O., 1991. Composition of the soil gas phase. Perm. Gases hydrocarb. Pedologie 119–132.

Wait, D., Ramsey, C., Maney, J., 2015. The measurement process. In: Murphy, B.L., Morrison, R.D. (Eds.), Introduction to Environmental Forensics, third ed., pp. 65–97. https://doi.org/10.1016/B978-0-12-404696-2.00004-7.

Chapter 14

Functional and eco-friendly polymers in food: edible polymers

14.1 Introduction: an overview of polymers in the food industry

At the food industry level, polymeric materials have been used extensively in the manufacture of packaging materials, and nonrenewable (petroleum-based) polymers are the most widely used materials due to their molding versatility, wide availability, cost, etc. However, the excessive use of this type of material leads to environmental problems due to its resistance to degradation. Poly(ethylene terephthalate), low- and high-density poly(ethylene), polypropylene, poly(vinyl chloride), and poly(styrene) are the most widely used polymers; it was estimated that, by 2020, around eight million metric tons of this type of plastic would have been produced and accumulated (Geyer et al., 2017; Mangaraj et al., 2019; Tajeddin and Arabkhedri, 2020). Many of these uses are directly or indirectly associated with the food industry, either in the early stages of production, due to its safety, resistance, adaptability to operations and functionality, etc., or in stages of processing, logistics, and conservation, etc.

Currently, there is growing demand for the use of materials that do not contribute to pollution by using sustainable and efficient processes. As a consequence, there are various investigations and processes aimed at the development of materials with these characteristics and that, in addition, equal or improve the functional capacities of conventional materials regarding food safety and quality.

Bio-based and biodegradable polymers are functional and eco-friendly alternatives that can be used to solve the problems inherent in traditional polymers, that is, excessive contamination by solid waste resulting from its residuality, excessive use, and nonbiodegradability. These types of compounds are usually classified into three groups according to the source of origin; the first category includes materials obtained from natural sources or biomass, secondly, materials produced by microorganisms, and the third category includes

Eco-friendly Functional Polymers
https://doi.org/10.1016/B978-0-12-821842-6.00011-7
Copyright © 2021 Elsevier Inc. All rights reserved.

materials produced from bio-based monomers (Petkoska et al., 2021). Generally speaking, all types previously mentioned are used in the food industry.

On the other hand, edible polymers are a subgroup of bio-based polymer materials characterized as being biodegradable, although they belong to the first category mentioned, as their name implies, they can be consumed or metabolized by different organisms including humans. These materials have good availability and profitability as they come from renewable raw materials of animal or vegetable origin that can be obtained from agro-industrial waste (Cazón et al., 2017; Kouhi et al., 2020). Additionally, they present different molecular characteristics and functional properties that can be used for the elaboration of materials tailored to the needs or with added capacities. Research in this field has focused mainly on obtaining packaging, packaging material, and coatings, where in addition to its use as a container or protective barrier to control the transfer of moisture or gases, it can, through additives, prolong the shelf-life of food by incorporating antimicrobial and antioxidant components (Kouhi et al., 2020; Zhong et al., 2020) or, alternately, fully edible packages can be prepared that are eaten together with the food, minimizing waste (Patel, 2019). The use of this type of polymer extends to the development of new food products of high quality and functionality with improved organoleptic and nutritional characteristics. They are also used in food processing where they improve stability, texture, physicochemical, and rheological properties; in biomedical applications as a means of controlled drug release; as safe materials for incorporation in the production of cosmetics and personal care products, etc. (Bais et al., 2005; Marto et al., 2015).

This chapter is focused on the chemical description of edible polymers, due to their inherent requirement of nontoxicity, digestibility, and processability needed for this application, their functional properties, and new trends for specific applications in the food industry.

14.2 Types of edible polymers according to composition

Edible polymers are totally biodegradable materials classified as safe for consumption by humans, animals, and microorganisms. Edible polymers are usually divided into three categories: (1) polysaccharides (hydrocolloids), (2) polypeptides (proteins), and (3) composite polymers (Shit and Shah, 2014; Shah, 2020). It should be mentioned that, in various articles and books, lipids appear as a category of edible polymers, however, despite the fact that these compounds have a relatively high molecular weight, they do not reach the molecular weights of a polymer, additionally, they are not shaped by monomers as they can contain different molecules linked to the hydrocarbon chain. However, lipids in conjunction with edible polymers can be used as a coating material or mixture that can improve the barrier properties of these materials (Shah, 2020).

14.2.1 Polysaccharides

Polysaccharides are obtained from several sources, including animals, plants, microorganisms, and by synthetic pathways. They structurally consist of hydrophilic macromolecules with abundant hydroxyl groups and polar groups in their structures. Chemically, polysaccharides are made up of structural subunits called monosaccharides, which are linked between them by glycosidic bonds (Kouhi et al., 2020; Shah, 2020). According to their composition, two types of polysaccharides are usually distinguished; homopolysaccharides, made up of the repetition of a monosaccharide (e.g., starch, glycogen, and cellulose) and, heteropolysaccharides made up of the ordered repetition of two different monosaccharides (e.g., pectin, agar—agar, gum arabic); therefore, heteropolysaccharides are copolymers. According to their charge, they can be classified as polyelectrolytes (i.e., positively or negatively charged by ionic or ionizable functional groups) and nonpolyelectrolytes (i.e., neutral polymers) (Shit and Shah, 2014; Guo et al., 2017).

The potential use of a given polysaccharide depends on both the physical and chemical properties of the material (e.g., molecular weight, composition, charge density, and flexibility); as well as external conditions of the preparation medium (e.g., temperature, pH, and ionic strength). For example, for the formation of films or packages, the formation mechanism involves a rupture and solubilization of the polymer segments with a subsequent reforming of the matrix in the form of a film or gel, promoted by intermolecular interactions that will depend on the composition of the polysaccharide. Therefore, multiple hydrogen bonds can be generated, as occurs with agar—agar, ionic interactions such as alginate (with carboxyl groups) or chitosan (with amino groups), or combinations of the above as occurs with carrageenan (hydroxyl and sulfonate groups). Additionally, electrolytic and ionic cross-linking is part of the preparation mechanisms of materials such as hydrogels (Le et al., 2017; Ansorena et al., 2018).

Polysaccharide-based materials have been proven to be effective barriers against lipids and gases such as CO_2 and O_2, however, given their chemical composition, they are ineffective barriers for water vapor that remains trapped between the polymer structure due to different factors in intermolecular interactions. This disadvantage has been taken advantage of in the coating area, using them as "sacrificial agents" that delay the loss of moisture from food products where the moistened polysaccharide is deposited on the outside of the food to reduce its loss of moisture and keep it fresh until the coating dehydrates. Another disadvantage of this type of material is its low mechanical resistance, it is brittle due to interactions between polymer chains (Cazón et al., 2017; Nieto, 2009); however, the functional properties of polysaccharides can be modified and improved after their combination with other biopolymers or compounds to obtain more resistant materials with better barrier characteristics.

14.2.2 Proteins

Proteins are macromolecules made up of monomers called amino acids, consequently, they are copolymers; amino acids are linked by covalent linking, usually referred to as peptide bonds. They can contain multiple different types of amino acids, therefore, proteins are complex copolymers compared to synthetic copolymers obtained by classical procedures of synthetic chemistry. In addition, the sequence and number of amino acids defines the identity of the protein. A protein is made up of a polypeptide skeleton containing side chains resulting in the portions of amino acids that do not participate in the formation of the peptide structure, these chains are called side groups or side chains (R). Since some of these side chains are polar, nonpolar, negatively or positively charged, or reactive in nature (with -SH, -COOH or -NH_2 groups), many of the properties of a protein are directly associated with the number and type of side chains; for instance, the solubility, net charge, reactivity, functionality, potential for hydrogen bonding, or ability of the protein to carry out dispersive or ionic interactions. A more specific illustration of the above is, for example, the water solubility of a protein, this property increases with the number of amino acids with polar or hydrophilic R groups, such as, amino acids with hydroxyl groups (e.g., serine, threonine), amide groups (e.g., asparagine, glutamine), or with negative or positive charges (e.g., aspartic acid, glutamic, lysine, and arginine); on the contrary, low solubility can be associated with a greater presence of nonpolar or hydrophobic neutral groups, as occurs with the R groups present in amino acids like alanine, valine, or leucine (Benbettaïeb et al., 2016; Garrido et al., 2018; Shah et al., 2020).

In terms of structure and mainly under the more complex and lower energy level of structuring that a protein can adopt (quaternary structure), each protein has a unique conformation that will depend on the chemical identity, number, and order of the amino acids that are part of the polypeptide skeleton. Its stabilization mainly comes from hydrophobic, electrostatic, and van der Waals interactions, hydrogen bonds, or disulfide bonds between the different available groups of the protein (Garrido et al., 2018).

Its structure and, therefore, the diversity of functional groups with interaction capacity, allow proteins to have good adhesion to hydrophilic surfaces and the ability to form films or coating materials. Similarly, structural differences affect the thermal, rheological, viscoelastic, and mechanical properties of materials prepared from proteins (Ansorena et al., 2018). The main mechanism for the formation of this type of material comes from denaturation of the protein with a subsequent structural reconstitution, promoting new protein—protein intermolecular interactions, increasing the cohesion between molecules. Typically, denaturation of proteins can be achieved by physical methods such as heating, shearing, hydrostatic pressure, or irradiation; chemical methods that involve alkylation, acylation, acetylation, pH alteration, or interaction with solvents; and biochemical methods through the use of enzymes. This type of procedure results in a structural or conformational change in the native structure, without altering the amino acid sequence (Ansorena et al., 2018).

In general, materials prepared based on proteins tend to have good sensory properties (e.g., tasteless and odorless), they can be heat-sealed and, like polysaccharides, they are good barriers against lipids and O_2 (less effective for CO_2); however, although they have better mechanical properties than polysaccharides, their mechanical resistance remains low–moderate. Additionally, the hydrophilic character granted by the R residues of amino acids considerably limits their use as a moisture barrier material (Shit and Shah, 2014; Hassan et al., 2018). The unfavorable characteristics mentioned can be overcome by working on protein–protein interactions; for example, to improve mechanical resistance, cross-linking procedures with natural cross-linking agents such as caffeic acid, genipin, enzymes such as transglutaminase (Garrido et al., 2018), and physical cross-linking by heat treatment and electromagnetic radiation (UV, γ, and β; Benbettaïeb et al., 2016). Another way to modify the interactions of proteins and, therefore, of the manufactured materials, is through the use of plasticizers such as mannitol, sorbitol, glycerol, and even water; the function of these compounds consists basically in embedding themselves in the polymer chains, destroying the crystalline domains, and creating spaces or increasing the free volume of the structure, which results in increasing the flexibility of the material, which also improves its processability (Cazón et al., 2017).

14.2.3 Synthetics and composites

Composite edible polymers are basically a mixture of polysaccharides, polypeptides, lipids, or synthetic polymers that seek to improve the characteristics of materials by taking advantage of the functional properties of each component of the system. The materials obtained can be used with different functionalities, among them, the formation of resistant packaging, formation of edible films, encapsulation of nutrients or flavors, formation of new food products, etc. (Le et al., 2017; Shah, 2020). For example, with regard to the manufacture of coatings or packaging, what is mainly sought is to improve the barrier and mechanical properties; in this case, the use of lipids that are hydrophobic materials can increase the permeability of water by improving the behavior of polysaccharides and polypeptides against humidity. Edible polymers have also been mixed with low-concentration synthetic polymers such as poly(ethylene), poly(vinyl alcohol), or poly(styrene) (Benbettaïeb et al., 2016); however, at this point it should be taken into account that for this type of materials only the biodegradable components degrade in the environment and it may be suitable or not for human or animal consumption depending on the synthetic polymer of the mixture (Iwata, 2015). In addition, composite materials can be prepared by extruding mixtures of polysaccharides, proteins, and/or lipids; forming emulsions or suspensions or through the formation of successive layers or laminates (e.g., polysaccharide–lipid bilayers) (Hassan et al., 2018).

Among the most studied composite materials are polysaccharide–protein complexes; the formation of these materials is mainly caused by electrostatic interactions that can occur between both macromolecules; although the formation of hydrogen bonds and hydrophobic interactions is also relevant. The interaction between both biopolymers depends mainly on factors such as temperature, pH, ionic strength, the protein–polysaccharide ratio, and the structural charge of the molecules (Le et al., 2017). For example, depending on the temperature, the protein in solution can be folded or unfolded and these configurations define the ability to interact with polysaccharides; an unfolded protein would have more reactive sites (R residues) exposed to the solvent phase, which would promote a greater interaction with polysaccharides, in turn dispersive interactions and hydrogen bonds may increase. Similarly, the net charge of the molecules at a given pH influences electrostatic interactions. Proteins, for example, have a positive charge if the $pH < pI$ or negative if the $pH > pI$, this charge will depend on the tendency to ionization of the R groups of the constituent amino acids. As for polysaccharides, these can also ionize at a certain pH, thus, those that have carboxylate groups can deprotonate and become negatively charged at $pH \sim pKa$. Therefore, if both molecules are at a pH in the solvent medium where they are given an opposite charge and the charges are totally neutralized, the complex will tend to form insoluble or soluble aggregates; in contrast, if repulsive interactions prevail, the system can form two-phase systems (each phase comprises a different polymer) or a single phase (where the polymers are uniformly distributed throughout the medium). Concentration is also an important factor since soluble complexes can be formed when one or another of the polymers involved is in excess, leaving positive or negative charges without neutralizing (Ghosh and Bandyopadhyay, 2012). In food applications, these complexes are used to make fibers or films with better mechanical characteristics; they are used to impart stability to food products in emulsion, to encapsulate active ingredients, and they can be used to create particles that replace fat in certain products, among other applications (Le et al., 2017; Ghosh and Bandyopadhyay, 2012).

Other alternatives to composite materials are based on the addition of nanosystems or nanomaterials to the polymeric matrices with the aim of improving their characteristics or granting new capabilities; these materials are known as bionanocomposites. Nanosystems involve nanoparticles (nanoparticles of silica, metal, and metal oxide), nanoemulsions, polymeric nanoparticles, inorganic/organic nanocomposites, nanotubes, nanocrystals, polymeric nanofibers, etc. (Zambrano-Zaragoza et al., 2018; Moustafa et al., 2019). In the food industry, in combination with edible polymers, they have been used as transport vehicles for active products (antioxidants, antimicrobials, nutraceutical compounds) that allow maintaining or improving the quality of food. In the elaboration of packaging, they are usually used to improve the mechanical and thermal properties, offer improved barrier characteristics against CO_2, O_2, and water vapor, UV absorption, and some are used to provide active and intelligent functions to the packaging (Rhim and Kim, 2014; Kuswandi, 2017). With regard to the use of inorganic nanoparticles, the incorporation of nanoclays such as montmorillonite has been used as a reinforcing agent as well as SiO_2 nanoparticles; it is used as a thickening food additive, and ZnO nanoparticles for their part, in addition to strengthening the polymeric matrix provide antimicrobial character (Klangmuang and Sothornvit, 2016; Xiong et al., 2015; Shi et al., 2008). In general, the performance of these nanomaterials is mainly due to their high surface–volume ratio that can modify significantly the properties of the polymeric matrix at very low doses.

An example of the above is described by Rhim (2011), who used agar/nanoclay bionanocomposites (natural montmorillonite-chloisite Na^+) for the production of films; evaluated the effect of the amount of added nanoclay (0, 2.5, 5, 10, 15, and 20 g of clay/100 g of agar) on the mechanical and barrier properties of the material. Agar is a fibrous carbohydrate extracted from the cell wall of marine algae of the class *Rhodophyceae* of the genera Galidium and Gracilaria. This material, although it has a high gelling power, is not suitable for preparing packaging films since its thermal instability, medium gas barrier properties, and low resistance to water do not allow it. The reason being, as the researcher describes it, the addition of nanoclay to the polymeric matrix allowed it to acquire a higher mechanical resistance compared to the pristine polymeric material (mainly when the percentage of nanoclay was kept at 10%); likewise, the permeability to water vapor and the solubility in water decreased with the nanoclay content (Rhim, 2011; Shit and Shah, 2014).

A possible explanation for the improvement of the mechanical characteristics of the agar is associated with the characteristics of nanoclays; this type of material has a scaly platelet structure that gives it a surface area of more than 750 m^2/g and thickness of $\sim$1 nm with lateral dimensions between 100 and 500 nm (Moustafa et al., 2019). This allows strong interactions (e.g., ionic or hydrogen bonds) to develop between the nanoclay layers and the polymer matrix. Additionally, since clays can be considered as layered, ordered, and dispersed silicates, they can cause gases and water molecules to travel through a "tortuous" path through the polymer matrix that surrounds the silicate particles; which results in an increase in the effective diffusion path length (Rhim, 2011; Rhim and Kim, 2014). The foregoing demonstrates how adding a nanocomposite to a biopolymeric material can modify its functional characteristics and make it more attractive for a specific use.

14.3 Applications

14.3.1 Functional edible films and coatings on food packaging

The preservation of a packaged food product, in addition to depending on external conditions of temperature, pressure, relative humidity, and asepsis, also depends on intrinsic mass transfer processes that take place between the food, the container (or coating), and the exterior. According to this dynamic, and depending on the type of food, it is feasible that, between the phases involved, mass transfer processes take place, which involve exchange of water, CO_2, O_2, volatile/nonvolatile organic compounds, microorganisms, etc. Therefore, the packaging material must have the chemical and structural characteristics and sufficient properties that allow extending of the useful life of the food, by controlling these processes in an adequate way. For example, when preserving fresh fruits or dry foods, barriers against water should be used, to avoid either hydration or dehydration. In the same way, the barrier material can regulate the permeability of gases (such as oxygen) or the loss of organic vapors, which then allows, through packaging or coating, to control the ripening of the fruits, the oxidation of components (fats, pigments), control browning reactions, retention of organoleptic properties, reduce the penetration of solvents or toxic components, control the entry or proliferation of microorganisms, among other functionalities (Cazón et al., 2017; Garrido et al., 2018; Petkoska et al., 2021). On the other hand, in terms of mechanical properties, the packaging material must have adequate strength to maintain its integrity during use. Additionally, the sensory compatibility between the food and the coating or packaging material is not something that should be ignored, since aesthetic appearance, taste, and smell are important characteristics in the perception of consumers (Kouhi et al., 2020).

Edible films and coatings are defined as thin layers that are arranged on the surface of food with the aim of constituting a container or primary barrier that, in addition to fulfilling specific functions of containment, conservation, and/or protection, are totally biodegradable and/or can be eaten with food (Hassan et al., 2018; Ansorena et al., 2018). Films are preformed, independent materials that are often used to make bags, capsules, wrappers, and casings; for their part, coatings are applied directly to the surface of the food and are an integral part of the food (Mohamed et al., 2020). The composition of the polymeric container material will depend on the food in which it is used, the processing conditions required, and the functions that the container must fulfill. According to the structural characteristics of the material, its use, and the way the food is stored, it is important to know the melting and glass transition temperatures, since the stability of the material will depend on these (Mangaraj et al., 2019).

For both films and coatings, it is necessary to prepare a film-forming solution or emulsion from the solubilization or dispersion of the raw materials in water, ethanol, or a mixture of both, the food additives of choice (e.g., plasticizers, antimicrobial agents, coloring, or flavoring) are introduced here, and pH or temperature conditions are adjusted with subsequent homogenization (Cazón et al., 2017; Kouhi et al., 2020; Mohamed et al., 2020). As mentioned previously, the formation of edible polymer composites is an alternative to improve the functional characteristics of coatings and films through the combination of different materials in the film-forming solution (e.g., polysaccharides, polypeptides, lipids, fibers, nanoparticles, nanosystems, etc.) (Ansorena et al., 2018).

In the case of coatings, the prepared solution is deposited on the surface of food in liquid form either by immersion or direct spraying followed by drying; in this case, it is important to take into account the characteristics of the edible material to be coated, since there must be compatibility between its surface properties (wettability—contact angle, surface tension) and those of the coating material (Petkoska et al., 2021). On food, the use of single or bilayers is traditional, with the latter done through layer-by-layer coating (LbL) where the food is alternately coated in different film-forming solutions, to provide improved functional characteristics to the coating (Kouhi et al., 2020; Kim et al., 2018). A bilayer coating can be created from materials of different natures, for example, to improve the barrier properties to water vapor, a first layer based on polysaccharides or protein can be used and on top of this, a second layer is formed by lipids (Mohamed et al., 2020). Another example of two-layer coatings is that provided by Bilbao-Sainz et al. (2018), who, to coat fruit bars enriched with ascorbic acid, formed an electrostatic LbL using two solutions prepared with oppositely charged polyelectrolytes, chitosan (polycation) obtained from mushroom (*Agaricus bisporus*) by-products, and alginate (polyanion) obtained from brown algae; the authors showed that compared to single-layer coatings, the LbL coating allowed the preservation of a higher content of ascorbic acid in the food, improving other aspects such as firmness and the prevention of fungal growth during storage (Bilbao-Sainz et al., 2018). Despite the advantages that have been evidenced, in the industry, the use of bilayers is less common due to the fact that two stages of coating and drying are necessary to obtain them, which implies more time and operational control of the process (Mohamed et al., 2020).

For films, wet or dry processes are usually used. The wet process or solvent casting requires solvents that allow adequate dispersion of the polymer and then spread the solution on a flat surface, over time, under controlled conditions, the solvent evaporates and the film is formed. Regarding dry methods, depending on the material, extrusion, injection, blowing, and hot pressing (thermoforming) techniques can be used (Kouhi et al., 2020; Garrido et al., 2018). In Fig. 14.1, the process for preparing the films and coatings is illustrated.

For the preparation of films, for packaging and coating, various sources of polysaccharides and proteins have been used. The most widely used polysaccharides are: starch, pectin, agar, carrageenan alginate, chitosan, xanthan gum, cellulose derivatives, guar gum, acacia, pullulan, and for proteins, gelatin, collagen, albumin, milk proteins (casein and whey protein), corn zein, soy protein, peanuts, peas, wheat gluten, etc. (Kouhi et al., 2020). At present, in addition to the improvement of “passive” attributes such as barrier properties, mechanical or thermal resistance of packaging materials, many researches are focused in giving new active and functional properties; for example, packaging materials able to release bioactive substances like

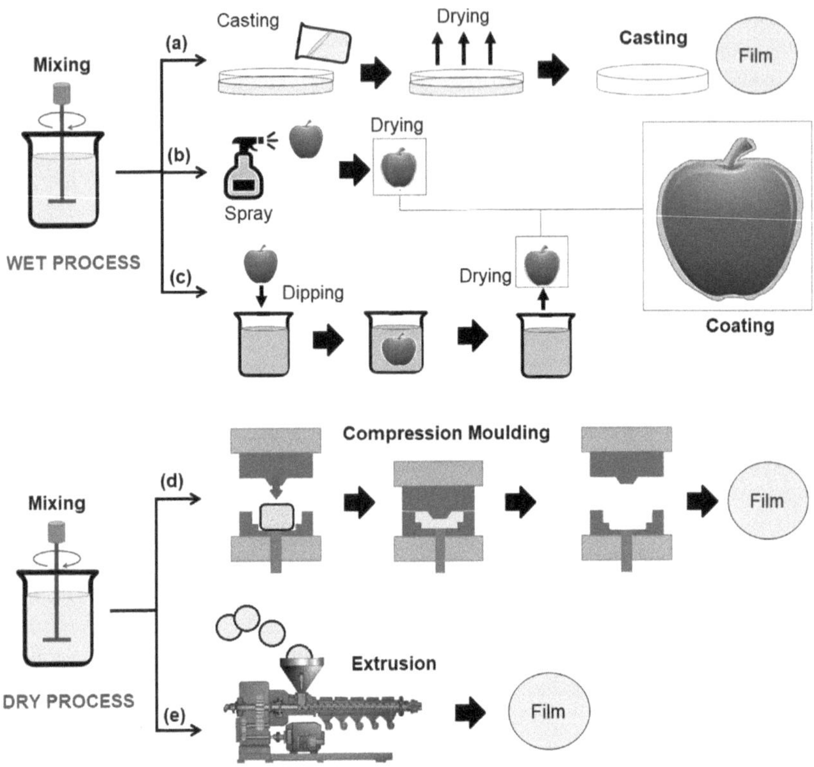

FIGURE 14.1 Wet and dry processes to obtain films and coatings.

vitamins, minerals, nutraceuticals, prebiotics, colorants, flavorings, antioxidants, or antimicrobial agents are highly desired. Usually this improvements aim of improving, preserving, and/or protecting the quality of food (Ansorena et al., 2018; Quirós-Sauceda et al., 2014).

The incorporation of these active compounds can be done on the external surface of the food after coating, between polymeric multilayers, at the food–edible polymer interface, or dispersed between the film (Quirós-Sauceda et al., 2014). However, some of these active and bioactive compounds are sensitive to light, heat, humidity, and the presence of oxygen, or are unstable to the conditions of preparation of the film-forming solution or the storage conditions; this results in a degradation or loss of its functionalities that can also influence the taste, smell, and color of the food that is in direct contact with it (Luo et al., 2020; Brandelli et al., 2016).

To overcome these difficulties, nanoencapsulation and microencapsulation are alternatives that seek to protect these components, since they can be considered to be reservoirs or physicochemical barriers against moisture, gases, or pro-oxidant agents such as free radicals, oxygen, and UV radiation (Azevedo et al., 2014; Brandelli et al., 2016; Petkoska et al., 2021). In addition to improving stability, the release of active and bioactive components to the food can be controlled (Sarkar et al., 2016; Brandelli et al., 2016). The encapsulation systems can be diverse, including capsules, spheres, and liposomes; also, they can be prepared from a wide range of wall materials (Brandelli et al., 2016). Among the techniques used for the preparation of micro- and nanoencapsulates are spray drying, coacervation, liposomes, co-crystallization, nanoemulsions, solid lipid nanoparticles (SLNs), nanoprecipitation, electrospinning, spray-drying, etc. (Hosseini et al., 2021; Quirós-Sauceda et al., 2014). Nano-emulsions have been used for the encapsulation of antimicrobial agents and antioxidants such as carvacol, cinnamaldehyde, α-tocopherol, or essential oils of garlic, clove, oregano, basil, mandarin, lemongrass, or resveratrol, and they have been used in coatings based on chitosan, pectin, sodium alginate, or methyl cellulose (Zambrano-Zaragoza et al., 2018).

In addition, intelligent packaging is another line of research that has gained importance recently, because it can have the ability to monitor in situ the production of gases, humidity, temperature, and growth of microorganisms in food. This is achieved through the use of materials capable of generating observable changes as a result of changes in the food environment (Imran et al., 2010; Ghaani et al., 2016). An example of the latter is shown by Zhai et al. (2018), who prepared an edible pH-sensitive film using as an indicator pigment an extract of anthocyanins obtained from red radish (*Raphanus sativus* L.) and a protein/polysaccharide mixture composed of gelatin/gellan gum as a film-forming agent; the film exhibited a color change from red to yellow in the pH range 2−12 as samples of milk and fish deteriorated over time. The release of acidic and basic gases from packaged foods led to a structural change in the active anthocyanins in the package, which allowed the observation of color changes that can be associated with the freshness and quality of the food (Zhai et al., 2018). Similarly, Pereira et al. (2015), designed an edible indicator packaging, with anthocyanins extracted from red cabbage (*Brassica oleraceae*) and poly(vinyl alcohol)/chitosan as polymeric material. The material was used to monitor the degradation of milk subjected to unfavorable temperature conditions that favor the accumulation of lactic acid product of the metabolism of the microbes, resulting in a change in the color of the packaging in the face of the new acidic conditions of the food (Pereira et al., 2015). Enzymes (Wu et al., 2015; Yan et al., 2008), bacteria (Kim et al., 2012), and silver nanocrystals (Zhang et al., 2013), among other materials, have been introduced to act as time−temperature indicators in packaging and even food labels; however, their selection will depend on the use and final destination that is given to the packaging, with natural, safe, and ecological materials always being the best alternative.

A brief description of examples of functional packaging, coatings, or films with food applications is given next.

Kurek et al. (2020) used extracts of olive leaf and sodium ascorbate as natural antioxidants added to individual film-forming solutions of carboxy-methylcellulose and gum arabic, which were used as a coating on freshly cut potatoes (*Solanum tuberosum* L.). The objective of this research, in addition to seeking to reduce the browning that arises after the mechanical stress of potato cutting (inactivation of polyphenol oxidase), or due to the increase in temperature (Maillard reaction), sought, through the application of the coating, create a compact barrier that slows the loss of water vapor from the surface of the food, which consequently results in a decrease in the absorption of oil in the fabric during frying and indirectly helps to reduce the amount of acrylamide formed and free trans fatty acids, without any effect on the organoleptic properties of the food. The authors reported a 45% reduction in fat compared to uncoated samples (Kurek et al., 2020). The coatings used did not significantly change the color, pH, or moisture content of the potatoes after 7 days of storage at 10°C. In general, both proteins and polysaccharides, given their thermogelling properties, many of them odorless and invisible, have been seen as interesting materials to reduce the excessive absorption of oil in frying; cross-linked starch from cassava root (tapioca), bamboo fiber, pectin-methylcellulose, corn zein, cornstarch, chitosan-carrageenan, okra, rice flour, xanthan gum, locust bean gum, gellan gum, white protein of dried egg, whey protein-pectin, sodium caseinate, among others, have been used in chicken strips, fish balls, potato chips, cereal products, cookies, donuts, chicken strips, nuggets, shrimp, carrot slices, sausages, chickpea flour, etc. (Liberty et al., 2019).

Li et al. (2018) prepared a hydrophobic interface film based on gum arabic, gelatin, and beeswax that can be used at the interface of packaging to reduce the loss of residual liquid foods such as milk, yogurt, and honey. The process carried out consisted in the deposition of the polymeric film-forming solution on the surface of a solid substrate with subsequent spraying of a microemulsion prepared from beeswax. The superhydrophobic interface exhibited high contact angles ($\geq$150 degrees) for the various liquids evaluated; additionally, it showed reasonable mechanical durability since it was able to maintain its structure and functionality after repeated folding and washing (Li et al., 2018).

Cui et al. (2020) prepared active zein films which were doped with chitosan polymeric nanoparticles inside which pomegranate peel extracts were encapsulated through the ionic gelling method (Cui et al., 2020). In this case, in addition to the antimicrobial effect of the chitosan nanoparticles, the extracts of pomegranate, rich in polyphenols, have been shown to have the same capacity because their hydroxyl groups can interact with the cell membrane of bacteria through hydrogen bonds or other mechanisms, and can cause alteration of the structure and permeability of the membrane, and induce the loss of cellular content and subsequent death of the microorganism (Bouarab et al., 2019). The authors reported a uniform distribution of the nanoparticles in the film, with improved thermal stability, and the controlled release of polyphenols to meat samples was corroborated, showing strong antimicrobial activity against *L. monocytogenes*. Clove essential oil and zinc oxide rods and gelatin in shrimp (Ejaz et al., 2018), clove essential oil-soy protein/montmorillonite in muscle fillets of bluefin tuna (*Thunnus thynnus*) (Echeverría et al., 2018), lysozyme nanofibers-pullulan (Silva et al., 2018), lactoferrin-cellulose in fresh sausages (Padrão et al., 2016), have been used as antimicrobial films, additionally, most have antioxidant properties.

Similarly, Otoni et al. (2014) prepared methylcellulose films doped with nanoemulsions of essential oils of clove and oregano to increase the shelf-life of sliced bread. The presence of compounds such as eugenol and carvacol (phenols) in essential oils made it possible to improve the antimicrobial properties (against yeasts and molds) of the films in comparison with antifungal agents for commercial use (Otoni et al., 2014). Additionally, reducing the size of the nanoemulsion droplets, in addition to increasing the antimicrobial activity, made it possible to improve the plasticity of the polymeric material. Nanoemulsions of carvacol in chitosan, of cinnamaldehyde in pectin matrix, cinnamaldehyde-α-tocopherol (essential oil of garlic) in gelatin/gelatin-chitosan/gelatin-sodium caseinate films, essential oil of lemongrass or mandarin in sodium alginate matrix or chitosan, among others, have been used as functional polymeric materials with antimicrobial or antioxidant activity, extending the shelf-life of foods such as cucumber, sliced bread, freshly cut apples, grape berries, and green beans (Zambrano-Zaragoza et al., 2018).

On the other hand, due to the fact that its action as an absorbent or scrubber of gases or undesirable substances in food may be one of the active functions of a container, some research has focused on the preparation of bio-based polymeric materials for the capture of ethylene and oxygen. The importance of these investigations lies in the fact that both compounds play an important role in the dynamics of food preservation. Ethylene, for example, can act as a phytohormone responsible for the ripening and aging of fruits and vegetables, and oxygen can generate sensory and nutritional changes in the product, as it promotes oxidation and the growth of aerobic microorganisms (Vilela et al., 2018). In this context, Byun et al. (2012) used hot water fish gelatin as a film-forming matrix activated with an oxygen-uptake system based on nanoparticles loaded with α-tocopherol and iron (II) chloride, the material was exposed to high relative humidity (>50%) to increase oxygen permeability. Although the material presented undesirable characteristics in conventional packaging films (high roughness, permeability to water vapor and oxygen, and low tensile strength), it can be considered as a complement material, since it means a significant reduction in the content of oxygen (20.9% at the start, 4.56% after 50 days), which suggests its possible applications (Byun et al., 2012). Similarly, Johansson et al. (2014) used organosoluble lignin and lignosulfonates as packaging coating materials, adding the laccase enzyme from the white rot fungus *Trametes versicolor* to the film-forming solution, to promote the cross-linking reaction of the lignin derivative after consuming the oxygen in the head space, the modified films, in addition to increasing the storage modulus and mechanical resistance after cross-linking, a decrease in oxygen content of up to 60% was observed in 4 days (Johansson et al., 2014).

Regarding the uptake of ethylene, Kaewlin et al. (2018) synthesized a nanocomposite film of chitosan–TiO_2 with potential applications in post-harvest handling during the fruit ripening process or climacteric vegetables. Although the antimicrobial effect has been proven for films of this composition, the authors sought to give it a different functionality, taking advantage of the fact that TiO_2, when exposed to UV radiation, is capable of generating hydroxyl radicals (•OH) and active oxygen species on its surface that react with organic molecules, therefore, when food is packaged in this type of nanocomposite film, UV activation and subsequent reaction with the surrounding atmosphere, allows the gradual elimination of the ethylene. The prepared film presented better tensile strength and barrier properties, as well as a photodegradation capacity of ethylene superior to that of pristine chitosan when applied to tomato samples (Kaewlin et al., 2018).

14.3.2 Functional food

The International Life Sciences Institute of North America (ILSI) has defined functional foods as, "foods that by virtue of physiologically active food components provide health benefits beyond basic nutrition" (Henry, 2010). A functional food can be: (1) a natural food, (2) a food to which an active component is added, (3) a food to which a component has been removed, (4) a food of which one or more components are modified, (5) a food of which the bioavailability has been modified, or (6) any combination of these. It does not classify pills, capsules, or dietary supplements within this category (Peso-Echarri et al., 2015). This type of food has become important due to the growing awareness and concern about health.

Regarding the addition of active components in food, adding vitamins, minerals, prebiotics, or anticoagulants, antiinflammatories, antivirals, antitumorals, antioxidants, appetite regulators, gastrointestinal regulators, pressure regulators, cholesterol leveling agents, among others, can be aggregates. In this context, edible polymers, depending on their characteristics, can themselves be active components that can be added directly to food; thus, for example, whey protein concentrates or isolates (β-lactoglobulins, α-lactoalbumin) added to a food product, would provide a source of essential amino acids such as leucine, isoleucine, or valine, which can act as antioxidants, antiinflammatories, antitumorals, or blood pressure regulators (Patel, 2015; Bessada et al., 2019). The direct use of polysaccharides with functional benefits has also been documented, for example, sulfated polysaccharides such as carrageenans, fucoidans, and ulvans extracted from algae, have been shown to have important biological activities due to structural similarities with sulfated polysaccharides present in animals and humans (such as free glycosaminoglycans and proteoglycans), which are involved in a wide range of biological processes (Kidgell et al., 2019). Additionally, nondigestible polysaccharides such as cellulose, hemicellulose, β-glucan, pectin, or lignin, can be used as dietary fiber, which when degraded by intestinal bacteria are converted into short-chain fatty acids (propionate, butyrate) that provide energy for the colon, promote epithelial proliferation, regulate immune responses, and protect against cancers, among other functions (Zhang et al., 2018; Slavin, 2013).

On the other hand, as mentioned in the previous section, edible polymers can be used as a material for the encapsulation of active components; however unlike their use in coatings or packaging, the encapsulated active component is introduced directly to the food, therefore, the encapsulated compounds for food application must not be toxic, biodegradable, or cause any adverse effect when passing through the human body; they must also use safe excipients. A toxicological safety evaluation is required before the material can be used in food (Gutiérrez et al., 2013). Micro- and nano-level delivery systems have been used; liposomes, emulsions stabilized by biopolymers, micelles, inclusion complexes, and nanoparticles have aroused great interest as transport vehicles for a diversity of active components compatible with many food formulations without presenting negative sensory effects (Đorđević et al., 2015; Sarkar et al., 2016; Luo et al., 2020; Delshadi et al., 2020).

Chitosan/sodium alginate nanoliposomes and nanocapsules, nanoparticles of whey protein isolates, nanohydrogels based on bovine soil albumin and pectin or pullulan, nanoemulsions prepared based on whey protein isolates/acacia and mustard oil, nanoaggregates of soy protein isolates and canola oil have been used for the encapsulation of lipophilic and hydrophilic vitamins (vitamins C, D3, B1, B2, B6) (Assadpour and Jafari, 2019). In these cases, the encapsulates were aimed at guaranteeing the protection and bioavailability of the active components, since it is known that most fat-soluble vitamins and essential fatty acids are very sensitive to oxidation and inappropriate environmental conditions.

Similarly, Mahalakshmi et al., 2020, encapsulated β-carotene (a lipophilic bioactive precursor of vitamin A and antioxidant) in zein matrices with glycerol as stabilizer, preparing micro- and nanosystems by means of spray drying and electrospray, respectively. The results showed better performance of the nanoencapsulates in terms of better dissolution and release (permeability) of the active component under gastrointestinal conditions in vitro; which demonstrates the potential of these systems to improve the bioavailability of active components such as β-carotene (Mahalakshmi et al., 2020). Nanoencapsulation by supercritical fluids, solid lipid nanoparticles, and casein micelles by self-assembly are some methods for the nanoencapsulation of this compound (Gutiérrez et al., 2013).

The incorporation of vegetable oils through nanoencapsulation has also attracted attention due to its compositional and functional diversity, since, for example, some vegetable oils are rich in omega-3 and omega-6 fatty acids, vitamins, polyphenols, and other compounds that can act as antioxidants, antiinflammatories, antivirals, or, that can reduce cholesterol and improve heart health, among other benefits (Bai et al., 2019; Fioramonti et al., 2019). Different types of oils have been microencapsulated, for example, almond oil in spherical microencapsulated of Taro starch (Hoyos-Leyva et al., 2019); soybean oil in Karifim microcapsules and sodium caseinate by spray drying (Bai et al., 2019), or linseed oil in double-layer emulsions of whey protein and alginate (Fioramonti et al., 2019). Functional emulsion gels based on soy protein/inulin and soy protein/inulin/chia flour isolate have also been used to encapsulate soybean oil (i.e., rich in healthy fatty acids), with the aim of being a substitute for pork loin fat used in the preparation of sausages (Paglarini et al., 2019).

Fish and shellfish oils, which are rich sources of eicosapentaenoic acid (EPA) and docosahexaenoic acid (DHA), have also been encapsulated to prevent oxidation and subsequent formation of volatile molecules that generate unpleasant tastes and odors, as well as loss of functionality fatty acid. Ilyasoglu and El (2014), and previously, Zhong et al. (2009), nanoencapsulated fish oil by means of coacervation and precipitation methods, using sodium caseinate/gum arabic and zein, respectively; likewise, Zimet et al. (2011) used casein nanomicelles as encapsulating material for isolated DHA fatty acid (Zimet et al., 2011).

For the encapsulation of minerals and subsequent use in food, El-Sayed et al. (2015) used chitosan–protein complexes through ionic gelation to produce nanoparticles loaded with $FeSO_4$ to improve Fe bioavailability under conditions of enzymatic and acid degradation, and the results showed an increase in stability in gastrointestinal tests in vitro (El-Sayed et al., 2015). Sharifi et al. (2013) synthesized alginate nanoparticles to which $FeCl_3$ and $ZnCl_2$ were incorporated to enrich ice cream (Sharifi et al., 2013).

On the other hand, encapsulation can also help to preserve the functional capacities of active components that are used in food products subjected to heating. For example, Davidov-Pardo et al. (2012), used microencapsulated grape seed extracts in two types of polymeric matrices, mesquite gum/zein and maltodextrin/zein, and the material was used in cookies subjected to baking (180 °C). According to the authors' results, encapsulation allowed less thermal degradation of the extract, since the antioxidant activity of cookies with microcapsules was 10 times greater than that of a cookie with the free active component, in addition, consumer perception was satisfactory (Davidov-Pardo et al., 2012). A similar application was reported by Rocha et al. (2012), who microencapsulated lycopene in modified food starch (Capsul) for its application in cakes; although the encapsulation efficiency was low (21% and 29%), after the baking process, greater protection of the lycopene and a homogeneous release of the pigment were evidenced, allowing the food to be colored in a homogeneous way (Rocha et al., 2012). The γ- cyclodextrins, considered as oligomers produced by the enzymatic conversion of starch, have also been used to create inclusion complexes for the thermal protection of vegetable and essential oils (Yoshikiyo et al., 2019).

Additionally, through encapsulation, it is possible to increase the shelf-life of the food or the elimination of harmful microbial agents, through the encapsulation of natural preservatives (Zhu et al., 2021); however, the organoleptic characteristics of the food can be improved by introducing natural colorants such as chlorophylls, flavonoids, betalains, or compounds such as citral, menthol, vanillin, geraniol, or essential oils that can improve the taste and smell of food (Assadpour and Jafari, 2019). Although these functionalities do not provide health benefits for those who consume the food, they can help to improve the sensory or safety perception of a food, and thereby encourage its consumption.

14.3.3 Edible polymer-like emulsifying food additives

As described in the previous sections, both in packaging and in functional foods, additives can be used to fulfill physiological, biological, nutritional, and/or organoleptic functions. However, in this context, edible polymers as such, or under specific configurations, can be used as additives that act as stabilizing and microstructuring agents in the preparation of emulsions, foams, gels, or as thickeners in natural or processed foods (Rayner et al., 2016); in this sense, they have been used to modify the rheological properties, texture, or stability of soups, sauces, salad dressings, ice cream, cakes, desserts, bread, dairy products, juices, soft drinks, candies ,and dairy products, among many other applications, while also contributing to the improvement of organoleptic properties (Lam and Nickerson, 2013).

In the case of emulsions and foams (dispersions), edible polymers act as emulsifiers to reduce the interfacial energy of these types of structures, achieving their stabilization. In this type of system, the interaction of two immiscible phases is involved in which one of the phases (dispersed phase) is distributed in the form of small drops or bubbles in another phase (continuous phase) (Lam and Nickerson, 2013; Kim et al., 2020). In the food industry, emulsions are used to disperse small drops of oil in water (o/w) as in the case of milk, creams, or mayonnaise; or to disperse small drops of water in oil (w/o) as in the case of butter or margarine. More complex or advanced systems such as w/o/w, o/w/o systems or nanoemulsions, also can be used. The foams, for their part, are dispersions of air in water (a/w), examples of these systems are whipped cream, meringues, marshmallows, ice cream, and beer foam, and they are even used in bakeries to improve the texture and appearance of the bread (Lam and Nickerson, 2013; Patel, 2018).

Therefore, emulsifiers must have a dual capacity that allows, depending on the case, to interact with both the dispersed phase and the continuous phase to stabilize the interface between both and avoid grouping, flocculation, and subsequent total separation of the phases. In the case of proteins, this condition is achieved due to their amphiphilic nature and film-forming capacity; they have been widely used under their molecular distribution or in the form of aggregates, micelles, colloidal polysaccharide–protein complexes or nanoparticles (Lam and Nickerson, 2013; Patel, 2018). In the case of polysaccharides, their polar or nonpolar character depends on whether they are found to be charged or neutral, and a higher affinity for lipophilic phases can be achieved by the inclusion of hydrophobic side groups, as reported by Simsek et al. (2015), who through esterification with octenyl succinic anhydride incorporated alkenyl groups into the hydrophilic starch molecule (Simsek et al., 2015). In the case of polysaccharides, these are usually used more as viscosifying agents to slow down the gravitational instability of emulsions and foams, or to improve the stabilization properties of proteins after the formation of surface complexes. Some examples are gum arabic, modified starches, modified cellulose, pectins, and galactomannans. In addition, regarding proteins, caseins and whey proteins (β-lactoglobulin, α-lactalbumin) are the most used; in the same way, proteins of vegetable origin such as those found in peas, wheat, and soybeans, are composed predominantly of albumins and globulins (Lam and Nickerson, 2013; Dickinson, 2009; Kim et al., 2020).

14.3.4 Synthetic meat

In the field of food science and technology, in vitro meat is a concept that has been making its way as an alternative to meat from slaughtered animals as a source of nutritional muscle tissue. Basically, in vitro meat is a selectively obtained and molded muscle tissue, using biotechnological tools (Datar and Betti, 2010; Orellana et al., 2020). There are currently two methods of in vitro meat production, the first method is called "self-organizing techniques" and consists of the use of an explant of muscle tissue of animal origin that grows in a specific culture medium (Sharma et al., 2015; Datar and Betti, 2010). The second method, which is more accepted and with high potential for use on a large scale, is known as "scaffold-based," and is based on the use of myoblasts or muscle stem cells from live animals (cells from bovines, pigs, turkeys, chickens, ducks, fish) or pluripotent cells of an embryo, which proliferate on scaffolds or supports, in the presence of a suitable culture medium within a bioreactor, under strict control of environmental factors (see Fig. 14.2; Tuomisto, 2019; Datar and Betti, 2010). Regarding the second method, in the early stages of the process, the muscle cells multiply, and cell differentiation and subsequent fusion occur, giving rise to myotubes or strands of small cells that line up in the scaffolds; later, after electrical stimulation, the production of proteins and muscle tissue is boosted in vitro (Tuomisto, 2019). Current research revolves around the three main components of this process: satellite cells, scaffolds, and improvement of the culture conditions in the bioreactor.

As for the scaffolds, they are basically substrates that allow the union, proliferation, and cell differentiation of myoblasts (Sharma et al., 2015), generally polymeric materials are used, preferably edible polymers, such as collagen, fibronectin, cellulose, alginate, or chitosan (Riboldi et al., 2005); however, the use of nonanimal materials would be the best option, to be consistent with the overall objective of in vitro meat. Materials with rigidity

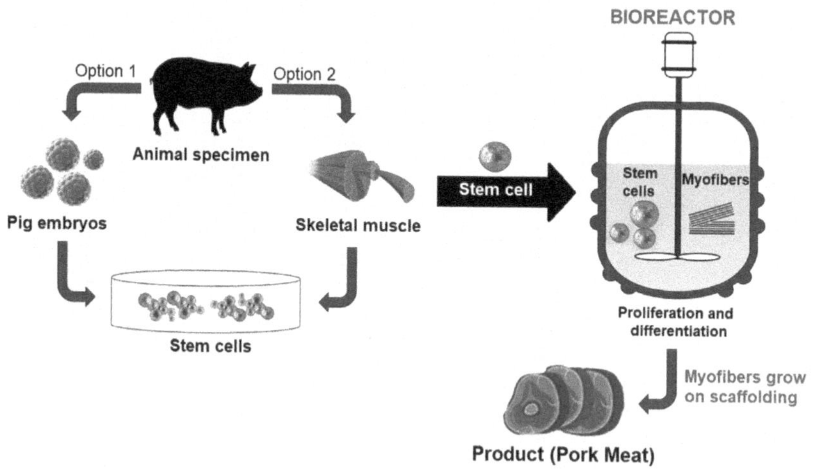

FIGURE 14.2 In vitro meat production process, under the "scaffold-based" method.

similar to that of a tissue, flexible and with a large surface area, are desirable characteristics, since they allow better cell differentiation and contraction, and at the same time their diffusion, fusion, proliferation, and nutrition would be maximized (Datar and Betti, 2010). Different textures and microstructures have been used, among them substrates with wavy surfaces with micropatterns or microchannels, electrospun microfiber mesh membranes, or artificial networks of sheet channels (Orellana et al., 2020; Riboldi et al., 2005; Lam et al., 2006). An example of the manufacture of scaffolds for potential uses for the production of meat in vitro is shown by Orellana et al. (2020), who for the development of a micromolded scaffold, initially manufactured a microstructured mold with parallel microchannels that were 300 μm wide and height made of acrylic material and engraved using a laser cutter; subsequently, they used a filmogenic solution based on salmon gelatin (promotes cell proliferation), calcium alginate, agarose, and glycerol, which was poured into the mold completely covering the structure to obtain the microstructured polymeric template after evaporation of the solvent (Orellana et al., 2020). The authors reported that the fabricated scaffold promoted the alignment and proliferation of the seeded cells, which is essential for the formation of muscle fibers.

References

Ansorena, M.R., Pereda, M., Marcovich, N.E., 2018. Edible films. In: Gutiérrez, T. (Ed.), Polymers for Food Applications. Springer, pp. 5–24. https://doi.org/10.1007/978-3-319-94625-2_2.

Assadpour, E., Jafari, S.M., 2019. A systematic review on nanoencapsulation of food bioactive ingredients and nutraceuticals by various nanocarriers. Crit. Rev. Food Sci. Nutr. 59, 3129–3151. https://doi.org/10.1080/10408398.2018.1484687.

Azevedo, M.A., Bourbon, A.I., Vicente, A.A., Cerqueira, M.A., 2014. Alginate/chitosan nanoparticles for encapsulation and controlled release of vitamin B2. Int. J. Biol. Macromol. 71, 141–146. https://doi.org/10.1016/j.ijbiomac.2014.05.036.

Bai, X., Li, C., Yu, L., Jiang, Y., Wang, M., et al., 2019. Development and characterization of soybean oil microcapsules employing kafirin and sodium caseinate as wall materials. Lebensm. Wiss. Technol. 111, 235–241. https://doi.org/10.1016/j.lwt.2019.05.032.

Bais, D., Trevisan, A., Lapasin, R., Partal, P., Gallegos, C., 2005. Rheological characterization of polysaccharide–surfactant matrices for cosmetic O/W emulsions. J. Colloid Interface Sci. 290, 546–556. https://doi.org/10.1016/j.jcis.2005.04.044.

Benbettaïeb, N., Gay, J.P., Karbowiak, T., Debeaufort, F., 2016. Tuning the functional properties of polysaccharide–protein bio-based edible films by chemical, enzymatic, and physical cross-linking. Compr. Rev. Food Sci. Food Saf. 15 https://doi.org/10.1111/1541-4337.12210.

Bessada, S.M., Barreira, J.C., Oliveira, M.B., 2019. Pulses and food security: dietary protein, digestibility, bioactive and functional properties. Trends Food Sci. Technol. 93, 53–68. https://doi.org/10.1016/j.tifs.2019.08.022.

Bilbao-Sainz, C., Chiou, B.S., Punotai, K., Olson, D., Williams, T., et al., 2018. Layer-by-Layer alginate and fungal chitosan based edible coatings applied to fruit bars. J. Food Sci. 83, 1880–1887. https://doi.org/10.1111/1750-3841.14186.

Bouarab, C.L., Degraeve, P., Ferhout, H., Bouajila, J., Oulahal, N., 2019. Plant antimicrobial polyphenols as potential natural food preservatives. J. Sci. Food Agric. 15, 1457–1474. https://doi.org/10.1002/jsfa.9357.

Brandelli, A., Brum, L.F., dos Santos, J.H., 2016. Nanobiotechnology methods to incorporate bioactive compounds in food packaging. In: Ranjan, S., Dasgupta, N., Lichtfouse, E. (Eds.), Nanoscience in Food and Agriculture 2. Sustainable Agriculture Reviews. Springer, pp. 27–58. https://doi.org/10.1007/978-3-319-39306-3_2.

Byun, Y., Bae, H.J., Whiteside, S., 2012. Active warm-water fish gelatin film containing oxygen scavenging system. Food Hydrocoll. 27, 250–255. https://doi.org/10.1016/j.foodhyd.2011.06.010.

Cazón, P., Velazquez, G., Ramirez, J.A., Vázquez, M., 2017. Polysaccharide-based films and coatings for food packaging: a review. Food Hydrocoll. 68, 136–148. https://doi.org/10.1016/j.foodhyd.2016.09.009.

Cui, H., Surendhiran, D., Li, C., Lin, L., 2020. Biodegradable zein active film containing chitosan nanoparticle encapsulated with pomegranate peel extract for food packaging. Food Packag. Shelf Life 24, 100511. https://doi.org/10.1016/j.fpsl.2020.100511.

Datar, I., Betti, M., 2010. Possibilities for an in vitro meat production system. Innovat. Food Sci. Emerg. Technol. 11, 13–22. https://doi.org/10.1016/j.ifset.2009.10.007.

Davidov-Pardo, G., Moreno, M., Arozarena, I., Marín-Arroyo, M.R., Bleibaum, R.N., et al., 2012. Sensory and consumer perception of the addition of grape seed extracts in cookies. J. Food Sci. 77, S430–S438. https://doi.org/10.1111/j.1750-3841.2012.02991.x.

Delshadi, R., Bahrami, A., Tafti, A.G., Barba, F.J., Williams, L.L., 2020. Micro and nano-encapsulation of vegetable and essential oils to develop functional food products with improved nutritional profiles. Trends Food Sci. Technol. 104, 72–83. https://doi.org/10.1016/j.tifs.2020.07.004.

Dickinson, E., 2009. Hydrocolloids as emulsifiers and emulsion stabilizers. Food Hydrocoll. 23, 1473–1482. https://doi.org/10.1016/j.foodhyd.2008.08.005.

Đorđević, V., Balanč, B., Belščak-Cvitanović, Lević, S., Trifković, K., et al., 2015. Trends in encapsulation technologies for delivery of food bioactive compounds. Food Eng. Rev. 7, 452–490. https://doi.org/10.1007/s12393-014-9106-7.

Echeverria, I., López-Caballero, M.E., Gómez-Guillén, M.C., Mauri, A.N., Monterio, M.P., 2018. Active nanocomposite films based on soy proteins-montmorillonite- clove essential oil for the preservation of refrigerated bluefin tuna (*Thunnus thynnus*) fillets. Int. J. Food Microbiol. 266, 142–149. https://doi.org/10.1016/j.ijfoodmicro.2017.10.003.

Ejaz, M., Arfat, Y.A., Mulla, M., Ahmed, J., 2018. Zinc oxide nanorods/clove essential oil incorporated Type B gelatin composite films and its applicability for shrimp packaging. Food Packag. Shelf Life 15, 113–121. https://doi.org/10.1016/j.fpsl.2017.12.004.

El-Sayed, M.M., Hassan, Z.M., Awad, M.I., Salama, H.H., 2015. Chitosan-whey protein complex (CS-WP) as delivery systems to improve bioavailability of iron. Int. J. Appl. Pure Sci. Agricul. 1, 34–46.

Fioramonti, S.A., Stepanic, E.M., Tibaldo, A.M., Pavón, Y.L., Santiago, L.G., 2019. Spray dried flaxseed oil powdered microcapsules obtained using milk whey proteins-alginate double layer emulsions. Food Res. Int. 119, 931–940. https://doi.org/10.1016/j.foodres.2018.10.079.

Garrido, T., Uranga, J., Guerrero, P., de la Caba, K., 2018. The potential of vegetal and animal proteins to develop more sustainable food packaging. In: Gutiérrez, T. (Ed.), Polymers for Food Applications. Springer, pp. 25–59. https://doi.org/10.1007/978-3-319-94625-2_3.

Geyer, R., Jambeck, J.R., Law, K.L., 2017. Production, use, and fate of all plastics ever made. Sci. Adv. 3, 1–6. https://doi.org/10.1126/sciadv.1700782.

Ghaani, M., Cozzolino, C.A., Castelli, G., Farris, S., 2016. An overview of the intelligent packaging technologies in the food sector. Trends Food Sci. Technol. 51, 1–11. https://doi.org/10.1016/j.tifs.2016.02.008.

Ghosh, A.K., Bandyopadhyay, P., 2012. Polysaccharide-protein interactions and their relevance in food colloids. In: Karunaratne, D.N. (Ed.), The Complex World of Polysaccharides. Desiree Nedra Karunaratne. IntechOpen., pp. 395–408. https://doi.org/10.5772/50561

Guiérrez, F.J., Albillos, S.M., Casas-Sanz, E., Cruz, Z., García-Estrada, C., et al., 2013. Methods for the nanoencapsulation of β-carotene in the food sector. Trends Food Sci. Technol. 32, 73–83. https://doi.org/10.1016/j.tifs.2013.05.007.

Guo, M.Q., Hu, X., Wang, C., Ai, L., 2017. Polysaccharides: structure and solubility. In: Xu, Z. (Ed.), Solubility of Polysaccharides. IntechOpen, pp. 7–20. https://doi.org/10.5772/intechopen.71570.

Hassan, B., Chatha, S.A., Hussain, A.I., Mahmood, K., Zia, K.M., et al., 2018. Recent advances on polysaccharides, lipids and protein based edible films and coatings: a review. Int. J. Biol. Macromol. 109, 1095–1107. https://doi.org/10.1016/j.ijbiomac.2017.11.097.

Henry, C., 2010. Functional foods. Europ. J. Clin. Nutrit. Nat. 64, 657–659. https://doi.org/10.1038/ejcn.2010.101.

Hosseini, S.F., Ramezanzade, L., McClements, D.J., 2021. Recent advances in nanoencapsulation of hydrophobic marine bioactives: bioavailability, safety, and sensory attributes of nano-fortified functional foods. Trends Food Sci. Technol. 109, 322–339. https://doi.org/10.1016/j.tifs.2021.01.045.

Hoyos-Leyva, J.D., Bello-Perez, L.A., Agama-Acevedo, J.E., Alvarez-Ramirez, J., Jaramillo-Echeverry, L.M., 2019. Characterization of spray drying microencapsulation of almond oil into taro starch spherical aggregates. Lebensm. Wiss. Technol. 101, 526–533. https://doi.org/10.1016/j.lwt.2018.11.079.

Ilyasoglu, H., El, S.N., 2014. Nanoencapsulation of EPA/DHA with sodium caseinate–gum Arabic complex and its usage in the enrichment of fruit juice. LWT-Food Sci. Technol. 56, 461–468. https://doi.org/10.1016/j.lwt.2013.12.002.

Imran, M., Revol-Junelles, A.M., Martyn, A., Tehrany, E.A., Jacquot, M., et al., 2010. Active food packaging evolution: transformation from micro- to nanotechnology. Crit. Rev. Food Sci. Nutr. 50, 799–821. https://doi.org/10.1080/10408398.2010.503694.

Iwata, T., 2015. Biodegradable and bio-based polymers: future prospects of eco-friendly plastics. Angew. Chem. Int. 54 (11), 3210–3215. https://doi.org/10.1002/anie.201410770.

Johansson, K., Gillgren, T., Winestrand, S., Järnström, L., Jönsson, L.J., 2014. Comparison of lignin derivatives as substrates for laccase-catalyzed scavenging of oxygen in coatings and films. J. Biol. Eng. 8, 1–10. https://doi.org/10.1186/1754-1611-8-1.

Kaewklin, P., Siripatrawan, U., Suwanagul, A., Lee, Y.S., 2018. Active packaging from chitosan-titanium dioxide nanocomposite film for prolonging storage life of tomato fruit. Int. J. Biol. Macromol. 112, 523–529. https://doi.org/10.1016/j.ijbiomac.2018.01.124.

Kidgell, J.T., Magnusson, M., de Nys, R., Ulvan, G.C.R., 2019. A systematic review of extraction, composition and function. Algal Res. 39 https://doi.org/10.1016/j.algal.2019.101422.

Kim, M.J., Jung, S.W., Park, H.R., Lee, S.J., 2012. Selection of an optimum pH-indicator for developing lactic acid bacteria-based time–temperature integrators (TTI). J. Food Eng. 113, 471–478. https://doi.org/10.1016/j.jfoodeng.2012.06.018.

Kim, J.H., Hong, W.S., Oh, S.W., 2018. Effect of layer-by-layer antimicrobial edible coating of alginate and chitosan with grapefruit seed extract for shelf-life extension of shrimp (Litopenaeus vannamei) stored at 4°C. Int. J. Biol. Macromol. 120, 1468–1473. https://doi.org/10.1016/j.ijbiomac.2018.09.160.

Kim, W., Wang, Y., Selomulya, C., 2020. Dairy and plant proteins as natural food emulsifiers. Trends Food Sci. Technol. 105, 261–272. https://doi.org/10.1016/j.tifs.2020.09.012.

Klangmuang, P., Sothornvit, R., 2016. Combination of beeswax and nanoclay on barriers, sorption isotherm and mechanical properties of hydroxypropyl methylcellulose-based composite films. LWT - Food Sci. Technol. 65, 222–227. https://doi.org/10.1016/j.lwt.2015.08.003.

Kouhi, M., Prabhakaran, M.P., Ramakrishna, S., 2020. Edible polymers: an insight into its application in food, biomedicine and cosmetics. Trends Food Sci. Technol. 103, 248–263. https://doi.org/10.1016/j.tifs.2020.05.025.

Kurek, M., Repajić, M., Marić, M., et al., 2020. The influence of edible coatings and natural antioxidants on fresh-cut potato quality, stability and oil uptake after deep fat frying. J. Food Sci. Technol. 1–13. https://doi.org/10.1007/s13197-020-04811-w.

Kuswandi, B., 2017. Environmental friendly food nano-packaging. Environ. Chem. Lett. 15, 205–221. https://doi.org/10.1007/s10311-017-0613-7 (2017).

Lam, R.S., Nickerson, M.T., 2013. Food proteins: a review on their emulsifying properties using a structure–function approach. Food Chem. 141, 975–984. https://doi.org/10.1016/j.foodchem.2013.04.038.

Lam, M., Sim, S., Zhu, X., Takayama, S., 2006. The effect of continuous wavy micropatterns on silicone substrates on the alignment of skeletal muscle myoblasts and myotubes. Biomaterials 27, 4340–4347. https://doi.org/10.1016/j.biomaterials.2006.04.012.

Le, X.T., Rioux, L., Turgeon, S.L., 2017. Formation and functional properties of protein–polysaccharide electrostatic hydrogels in comparison to protein or polysaccharide hydrogels. Adv. Colloid Interface Sci. 239, 127–135. https://doi.org/10.1016/j.cis.2016.04.006.

Li, Y., Bi, J., Wang, S., Zhang, T., Xu, X., et al., 2018. Bio-inspired edible superhydrophobic interface for reducing residual liquid food. J. Agric. Food Chem. 66, 2143–2150. https://doi.org/10.1021/acs.jafc.7b05915.

Liberty, J.T., Dehghannya, J., Ngadi, M.O., 2019. Effective strategies for reduction of oil content in deep-fat fried foods: a review. Trends Food Sci. Technol. 92, 172–183. https://doi.org/10.1016/j.tifs.2019.07.050.

Luo, Y., Wang, Q., Zhang, Y., 2020. Biopolymer-based nanotechnology approaches to deliver bioactive compounds for food applications: a perspective on the past, present, and future. J. Agric. Food Chem. 68, 12993–13000. https://doi.org/10.1021/acs.jafc.0c00277.

Mahalakshmi, L., Leena, M.M., Mosesa, J.A., Anandharamakrishnan, C., 2020. Micro- and nano-encapsulation of β-carotene in zein protein: size-dependent release and absorption behavior. Food and Function 11, 1647–1660. https://doi.org/10.1039/C9FO02088H.

Mangaraj, S., Yadav, A., Bal, L.M., Dash, S.K., Mahanti, N., 2019. Application of biodegradable polymers in food packaging industry: a comprehensive review. J. Packag. Technol. Res. 3, 77–96. https://doi.org/10.1007/s41783-018-0049-y.

Marto, J., Gouveia, L., Jorge, I.M., Duarte, A., Gonçalves, L.M., et al., 2015. Starch-based pickering emulsions for topical drug delivery: a QbD approach. Colloids Surf. B Biointerfaces 135, 183–192. https://doi.org/10.1016/j.colsurfb.2015.07.024.

Mohamed, S.A., El-Sakhawy, M., El-Sakhawy, M.A., 2020. Polysaccharides, protein and lipid -based natural edible films in food packaging: a review. Carbohydr. Polym. 238 https://doi.org/10.1016/j.carbpol.2020.116178.

Moustafa, H., Youssef, A.M., Darwish, N.A., Abou-Kandil, A.I., 2019. Eco-friendly polymer composites for green packaging: future vision and challenges. Compos. B Eng. 172, 16–25. https://doi.org/10.1016/j.compositesb.2019.05.048.

Nieto, M.B., 2009. Structure and function of polysaccharide gum-based edible films and coatings. In: Huber, K., Embuscado, M. (Eds.), Edible Films and Coatings for Food Applications. Springer, pp. 57–112. https://doi.org/10.1007/978-0-387-92824-1_3.

Orellana, N., Sanchez, E., Benavente, D., Prieto, P., Enrione, J., 2020. A new edible film to produce. In: vitro meat. Foods 9, 1–14. https://doi.org/10.3390/foods9020185.

Otoni, C.G., Pontes, S.F., Medeiros, E.A., Soares, N.F., 2014. Edible films from methylcellulose and nanoemulsions of clove bud (syzygium aromaticum) and oregano (origanum vulgare) essential oils as shelf life extenders for sliced bread. J. Agric. Food Chem. 62, 5214–5219. https://doi.org/10.1021/jf501055f.

Padrão, J., Gonçalves, S., Silva, J.P., Sencadas, V., Lanceros-Méndez, S., et al., 2016. Bacterial cellulose-lactoferrin as an antimicrobial edible packaging. Food Hydrocoll. 58, 126–140. https://doi.org/10.1016/j.foodhyd.2016.02.019.

Paglarini, C.S., Furtado, G.F., Honório, A.R., Mokarzel, L., Vidal, V.A., et al., 2019. Functional emulsion gels as pork back fat replacers in Bologna sausage. Food Struct. 20, 100105. https://doi.org/10.1016/j.foostr.2019.100105.

Patel, S., 2015. Functional food relevance of whey protein: a review of recent findings and scopes ahead. J. Funct. Foods 19, 308–319. https://doi.org/10.1016/j.jff.2015.09.040.

Patel, A.R., 2018. Edible foams stabilized by food-grade polymers. In: Gutiérrez, T. (Ed.), Polymers for Food Applications. Springer, pp. 251–269. https://doi.org/10.1007/978-3-319-94625-2_10.

Patel, P., 2019. Edible Packaging. Can you really have your cake and eat the wrapper, too? ACS Cent. Sci. 5, 1907–1920. https://doi.org/10.1021/acscentsci.9b01251.

Pereira, V.A., Queiroz, I.N., Stefani, R., 2015. Active chitosan/PVA films with anthocyanins from Brassica oleraceae (Red Cabbage) as Time–Temperature Indicators for application in intelligent food packaging. Food Hydrocoll. 43, 180–188. https://doi.org/10.1016/j.foodhyd.2014.05.014.

Peso-Echarri, P., González-Bermúdez, A., Ros-Berruezo, G., Martínez-Graciá, C., Frontela-Saseta, C., 2015. Polysaccharides as bioactive components of functional food. In: Cirillo, G., Spizzirri, U.G., Iemma, F. (Eds.), Functional Polymers in Food Science: From Technology to Biology. Wiley. https://doi.org/10.1002/9781119108580.ch7.

Petkoska, A.T., Daniloski, D., D'Cunha, N.M., Naumovski, N., Broach, A.T., 2021. Edible packaging: sustainable solutions and novel trends in food packaging. Food Res. Int. 140, 1–15. https://doi.org/10.1016/j.foodres.2020.109981.

Quirós-Sauceda, A.E., Ayala-Zavala, J.F., Olivas, G.I., González-Aguilar, G., 2014. Edible coatings as encapsulating matrices for bioactive compounds: a review. J. Food Sci. Technol. 51, 1674–1685 https://dx.doi.org/10.1007%2Fs13197-013-1246-x.

Rayner, M., Östbring, K., Purhagen, J., 2016. Application of natural polymers in food. In: Olatunji, O. (Ed.), Natural Polymers. Springer, pp. 115–161 https://doi.org/10.1007/978-3-319-26414-1_5.

Rhim, J.W., 2011. Effect of clay contents on mechanical and water vapor barrier properties of agar-based nanocomposite films. Carbohydr. Polym. 86, 691–699. https://doi.org/10.1016/j.carbpol.2011.05.010.

Rhim, J.W., Kim, Y.T., 2014. Biopolymer-based composite packaging materials with nanoparticles. In: Han, J.H. (Ed.), Innovations in Food Packaging. Academic Press, pp. 413–442. https://doi.org/10.1016/B978-0-12-394601-0.00017-5.

Riboldi, S.A., Sampaolesi, M., Neuenschwander, P., Cossu, G., Mentero, S., 2005. Electrospun degradable polyesterurethane membranes: potential scaffolds for skeletal muscle tissue engineering. Biomaterials 26, 4606–4615. https://doi.org/10.1016/j.biomaterials.2004.11.035.

Rocha, G.A., Fávaro-Trindade, C.S., Grosso, C.R., 2012. Microencapsulation of lycopene by spray drying: characterization, stability and application of microcapsules. Food Bioprod. Proc. 90, 37–42. https://doi.org/10.1016/j.fbp.2011.01.001.

Sarkar, P., Irshaan, S., Sivapratha, S., Choudhary, R., 2016. Nanotechnology in food processing and packaging. In: Ranjan, S., Dasgupta, N., Lichtfouse, E. (Eds.), Nanoscience in Food and Agriculture 1. Sustainable Agriculture Reviews. Springer, pp. 185–227. https://doi.org/10.1007/978-3-319-39303-2_7.

Shah, P., 2020. Polymers in food. In: AlMaadeed, M.A., Ponnamma, D., Carignano, M.A. (Eds.), Polymer Science and Innovative Applications. Elsevier, pp. 567–592. https://doi.org/10.1016/B978-0-12-816808-0.00018-4.

Sharifi, A., Golestan, L., Baei, M.S., 2013. Studying the enrichment of ice cream with alginate nanoparticles including Fe and Zn salts. J. Nanopart. 2013, 1–5. https://doi.org/10.1155/2013/754385.

Sharma, S., Thind, S.S., Kaur, A., 2015. In vitro meat production system: why and how? J. Food Sci. Technol. 52, 7599–7607. https://doi.org/10.1007/s13197-015-1972-3.

Shi, L., Zhou, J., Gunasekaran, S., 2008. Low temperature fabrication of ZnO–whey protein isolate nanocomposite. Mater. Lett. 62, 4383–4385. https://doi.org/10.1016/j.matlet.2008.07.038.

Shit, S.C., Shah, P.M., 2014. Edible polymers: challenges and opportunities. J. Polym. 2014, 1–13. https://doi.org/10.1155/2014/427259.

Silva, N.H., Vilela, C., Almeida, A., Marrucho, I.M., Freire, C.R., 2018. Pullulan-based nanocomposite films for functional food packaging: exploiting lysozyme nanofibers as antibacterial and antioxidant reinforcing additives. Food Hydrocoll. 77, 921–930. https://doi.org/10.1016/j.foodhyd.2017.11.039.

Simsek, S., Ovando-Martinez, M., Marefati, A., Sjöö, M., Rayner, M., 2015. Chemical composition, digestibility and emulsification properties of octenyl succinic esters of various starches. Food Res. Int. 75, 41–49. https://doi.org/10.1016/j.foodres.2015.05.034.

Slavin, J., 2013. Fiber and prebiotics: mechanisms and health benefits. Nutrients 5, 1417–1435 https://dx.doi.org/10.3390%2Fnu5041417.

Tajeddin, B., Arabkhedri, M., 2020. Polymers and food packaging. In: AlMaadeed, M.A., Ponnamma, D., Carignano, M.A. (Eds.), Polymer Science and Innovative Applications. Elsevier, pp. 525–543. https://doi.org/10.1016/B978-0-12-816808-0.00016-0.

Tuomisto, H.L., 2019. The eco-friendly burger. EMBO Rep. 20, 1–6. https://doi.org/10.15252/embr.201847395.

Vilela, C., Kurek, M., Hayouka, Z., Röcker, B., Yildirim, S., et al., 2018. A concise guide to active agents for active food packaging. Trends Food Sci. Technol. 80, 212–222. https://doi.org/10.1016/j.tifs.2018.08.006.

Wu, D., Hou, S., Chen, J., Sun, Y., Ye, X., et al., 2015. Development and characterization of an enzymatic time-temperature indicator (TTI) based on Aspergillus Niger lipase. LWT - Food Sci. Technol. 60, 1100–1104. https://doi.org/10.1016/j.lwt.2014.10.011.

Xiong, W., Yang, D., Zhong, R., Li, Y., Zhou, H., Qiu, X., 2015. Preparation of lignin-based silica composite submicron particles from alkali lignin and sodium silicate in aqueous solution using a direct precipitation method. Ind. Crop. Prod. 74, 285–292. https://doi.org/10.1016/j.indcrop.2015.05.021.

Yan, S., Huawei, C., Limin, Z., Fazheng, R., Zhang, L., et al., 2008. Development and characterization of a new amylase type time–temperature indicator. Food Contr. 19, 315–319. https://doi.org/10.1016/j.foodcont.2007.04.012.

Yoshikiyo, K., Yoshioka, Y., Narumiya, Y., Oe, S., Kawahara, H., et al., 2019. Thermal stability and bioavailability of inclusion complexes of perilla oil with γ-cyclodextrin. Food Chem. 294, 56–59. https://doi.org/10.1016/j.foodchem.2019.04.093.

Zambrano-Zaragoza, M.L., González-Reza, R., Mendoza-Muñoz, N., Miranda-Linares, V., Bernal-Couoh, T.F., et al., 2018. Nanosystems in edible coatings: a novel strategy for food preservation. Int. J. Mol. Sci. 19, 1–24. https://doi.org/10.3390/ijms19030705.

Zhai, X., Li, Z., Zhang, J., Shi, J., Zou, X., et al., 2018. Combined with electrochemical writing for intelligent food packaging. J. Agric. Food Chem. 66, 12836–12846. https://doi.org/10.1021/acs.jafc.8b04932.

Zhang, C., Yin, A.X., Jiang, R., Rong, J., Dong, L., et al., 2013. Time–temperature indicator for perishable products based on kinetically programmable Ag overgrowth on Au nanorods. ACS Nano 7, 4561–4568. https://doi.org/10.1021/nn401266u (2013).

Zhang, T., Yang, Y., Liang, Y., Jiao, X., Zhao, C., 2018. Beneficial effect of intestinal fermentation of natural polysaccharides. Nutrients 10, 1055 https://dx.doi.org/10.3390%2Fnu10081055.

Zhong, Q., Tian, H., Zivanovic, S., 2009. Encapsulation of fish oil in solid zein particles by liquid-liquid dispersion. J. Food Process. Preserv. 33, 255–270. https://doi.org/10.1111/j.1745-4549.2009.00390.x.

Zhong, Y., Godwin, P., Jin, Y., Xiao, H., 2020. Biodegradable polymers and green-based antimicrobial packaging materials: a mini-review. Adv. Indus. Eng. Polym. Res. 3, 27–35. https://doi.org/10.1016/j.aiepr.2019.11.002.

Zhu, Y., Li, C., Cui, H., Lin, L., 2021. Encapsulation strategies to enhance the antibacterial properties of essential oils in food system. Food Contr. 123, 107856. https://doi.org/10.1016/j.foodcont.2020.107856.

Zimet, P., Rosenber, D., Livney, Y.D., 2011. Re-assembled casein micelles and casein nanoparticles as nano-vehicles for ω-3 polyunsaturated fatty acids. Food Hydrocoll. 25, 1270–1276. https://doi.org/10.1016/j.foodhyd.2010.11.025.

Chapter 15

Functional and eco-friendly polymers in structural applications

15.1 Functional polymers versus structural polymers: a conceptual approach

In previous chapters, the differentiation between functional and structural polymers has been made, evidencing that the distinction between them is very subjective and greatly dependent on the application. To facilitate the development of this chapter, we delve a little more into the conceptual formalism and differentiation between these two groups of polymers. In the first place, a functional polymer could be understood to be a polymer containing in its structure a significant number of heteroatoms, usually atoms of oxygen, nitrogen, sulfur, and phosphorus, from which the polymer molecules acquire physical properties (i.e., mechanical and thermal) and chemical properties (i.e., properties associated with molecular interactions with its surroundings). That is, the chemical functionality is taken as a molecular criterion for the definition of a functional polymer (Rivas et al., 2003). Thus, the resulting properties influence the behavior of the entire material, and significantly differentiate it from its polyolefin-type counterpart with greater structural similarity. On the other hand, typically, a structural polymer is intuitively defined by its mechanical properties, and its classification as thermoplastic and thermosets, properties such as its glass transition temperature, features of solid states like its crystallinity, elasticity, or its intrinsic anisotropy (Plummer, 2016; Li, 2018).

The previous definition is a structural definition since it uses the structural characteristics to differentiate the different polymers. For example, poly(ethylene) (PE) corresponds to ethylene units covalently inserted to form a larger molecule. The absence of heteroatoms on the PE molecule leads to the conclusion that it is not a functional polymer according to the previous definition. Furthermore, PE can be linear or cross-linked and, although it is still PE in its chemical nature, the physical and chemical properties can differ significantly between them. However, in both cases, the linear PE and the cross-linked PE follow being structural

Eco-friendly Functional Polymers
https://doi.org/10.1016/B978-0-12-821842-6.00030-0
Copyright © 2021 Elsevier Inc. All rights reserved.

polymers. To achieve the cross-linking of the PE chains, a cross-linker is used, which can be formed only by atoms of C and H (e.g., divinylbenzene) or be formed by C, H, or one or more heteroatoms (e.g., O, N, C, S, or P). For convenience we will call this last one the functional cross-linker. Assuming that in both cases the cross-linking of PE molecules is possible, the first question to answer is when can the cross-linked PE be defined as a functional polymer? The answer to this question is found in the definition initially given, and consistently with it, it is to say, only when the amounts of cross-linker added to the PE to carry out the cross-linking significantly impact the physical and chemical properties of the cross-linked PE using a functional cross-linker compared with a structural cross-linker. Therefore, the impact of the functional cross-linker on PE properties is expected to be small as the amount of functional cross-linker is small; and in that direction, as the amounts of functional cross-linker in the polymer structure increase, it is possible that the polymer becomes a functional polymer. In fact, as cross-linker concentrations increase, it must be evaluated whether the resulting polymer is a homopolymer or a copolymer, a fact usually implying a certain degree of ambiguity. Therefore, it can be concluded that the previous definition of a functional polymer is consistent with the expected behavior in terms of its properties and composition. The counterpart to this conclusion is the definition of the structural polymer, which could be defined as: a structural polymer is a polymer that has a significant amount of carbon and hydrogen atoms in its structure. Because the difference between functional and structural polymers was defined in terms of atomic composition, it is clear that any quantitative approximation must be made considering the number of atoms and not their mass, since the relatively small molar mass of carbon and hydrogen atoms could influence the analysis. A first quantitative analysis is presented in Table 15.1.

From Table 15.1 it can be concluded that the atomic composition is not a sufficient criterion to make a quantitative classification between structural and functional polymers, as it is necessary for the insertion of other factors such as the composition in order to try to include the impact of the nature of the comonomer in the case of copolymers. Also, the effect of heteroatoms is not the same, for example, the electronegativity of oxygen is greater than the electronegativity of sulfur and silicon, and consequently, it should be introduced in the analysis as some factor capable of describing the contribution of the nature of heteroatoms.

On the other hand, all typical structural polymers have excellent thermal and mechanical properties under normal operation conditions, and therefore they are used for applications where materials are subjected to mechanical or thermal stress, but also, for applications of containment, storage, or when a specific shape is required (Plummer, 2016; Li, 2018). However, it is well-known that polymers show different behaviors depending on the use conditions; for instance, poly(vinyl chloride) (PVC) can be used to store water but cannot be used to store cyclohexanone. Thus, the other approach to provide the definition of functional and structural polymers is based on their specific

TABLE 15.1 Quantitative analysis for the differentiation of functional and structural polymers.

Polymer	Linear/resin	Cross-linker	Polymer type	(C + H) (%)	Polymer class
PE	Linear	–	Addition	100	Structural polymer
	Cross-linked	DVB	Addition	100	Structural polymer
	Cross-linked	MBAm	Addition	86	Functional polymer
PET	Linear	–	Condensation	82	Structural polymer
	Cross-linked	MSiOx	Condensation	83	Structural polymer
PAA	Linear	–	Addition	78	Functional polymer
	Cross-linked	DVB	Addition	94	Structural polymer
	Cross-linked	MBAm	Addition	78	Functional polymer
PSS	Linear	–	Addition	79	Functional polymer
	Cross-linked	DVB	Addition	90	Structural polymer
	Cross-linked	MBAm	Addition	78	Functional polymer

Note 1: *DVB*, divinylbenzene, *MBAm*, N,N′-methylenebisacrylamide, *MSiOx*, methylsiloxane, *PAA*, poly(acrylic acid), *PE*, Poly(ethylene), *PET*, poly(ethylene terephthalate), *PSS*, poly(styrene sulfonate). Note 2: (C + H) (%) denotes the molar percentage of carbon and oxygen covalently linked to the polymer. For copolymers and cross-linked polymers the molar relation assumed was 1:1. Note 3: Combination of monomer and cross-linker is illustrative and does not reflecting the real synthetic aspects. In addition, composition of chain ends was not considered.

application; that is to say, a polymer can be structural in one application, but at the same time, it can be a functional polymer in another application. For example, PVC is a structural polymer when it is used as a recipient for the storage or transport of polar liquids such as water or ethanol (Makris et al., 2019), as a material for making membranes (Kazemi et al., 2020), or as a material for making a blood storage bag (Palencia et al., 2016a,b), however, when PVC is used as an adsorbent for the sorption of lead from an aqueous solution it is undoubtedly a functional polymer (Wankasi and Dixon, 2014).

Something similar happens with elastomers depending on the approach in which their properties are visualized. In a first approach, the resistance of the elastomers to a maximum longitudinal stress may be of interest, in this way, its deformation capacity is understood in terms of its resistance to elongation. However, under another approach, the same capacity for deformation and recovery of its initial shape can be understood as a functional property: shape memory (Xia et al., 2020). In conclusion, a functional polymer can be defined under two criteria, the first is based on its chemical structure, while the second is based on its specific application.

In order to achieve a single definition, the following definitions are proposed. A functional polymer is a polymer or polymeric material performing a function within a specific application, which is not directly related to its mechanical properties, but with changes or disturbances at the molecular level as a result of a stimulus or its interactions with the environment. Consequently, a structural polymer is a polymer or polymeric material performing a function directly related to its mechanical properties within a specific application. It is important to clarify that polymers can be multifunctional materials, and in this way, within the same use they can have more than one application, and therefore, there is the possibility that a polymer can be simultaneously defined as a structural and functional polymer. Typical examples of this are described in detail in Chapters 14, 16, and 19.

15.2 An overview of multifunctional structural polymers for technological applications

As indicated in the previous section, the functionality of a structural polymer can be achieved through multifunctionality, thus, a first function is evident and related to its use as a structural or support material. Some examples of usual structural polymers are: poly(propylene) (PP) (Maddah, 2016), poly(ethylene terephthalate) (PET) (Jankauskaite et al., 2008; McKeen, 2017), poly(urethane) (PU) (Akindoyo et al., 2016), poly(styrene) (PS) (Lynwood, 2014), PVC (Abdul et al., 2013; Makris et al., 2019), and poly(ethylene-vinyl-acetate) (EVA) (Alothman, 2012; Schneider et al., 2017). However, these materials have also been used as functional polymers in very specific applications. From the point of view of structural applications, the previously exemplified materials are not eco-friendly for different reasons, among which the following stand out: they are synthetic polymers obtained from petroleum by-products, in addition, their use is frequently directed toward the making of one-use products, for example, PS and PP are frequently used as disposable containers, PU is used as material for cushioning, PET is used in the making of beverage or medicine bottles, and PVC is used for making bags for biomedical use. These polymers are not biodegradable materials, with the exception of PET, and as a consequence, they are a generalized source of highly persistent solid waste, associated with the contamination of terrestrial and aquatic ecosystems. This is related to their low cost and excessive production; however, these polymers also are frequently used with a series of additives such as plasticizers, dyes, antioxidants, flame retardants, among others, and as a consequence, their residues can also act as pollutant-releasing vehicles (Hermabessiere et al., 2017; Hahladakis et al., 2018). However, it should not be overlooked that their excellent mechanical and thermal properties, low density, high resistance, and ease of adapting their shape to multiple applications make them materials that are frequently used for multiple applications within industry and daily life, and also, for the advancement of science and the development of new technologies in all areas of current society. Therefore, the discussion about their use or not is irrelevant, since the focus of attention should be directed on consumption habits, their proper use, and the inclusion of appropriate waste management in all facets of daily life.

On the other hand, structural polymers have advanced toward new applications due to the better understanding their properties, their use combined with other polymers, or their use as a polymeric phase in the manufacture of nanocomposites. It is important to remember that each polymer has properties intrinsic to its nature, and therefore, the knowledge of these, the inclusion of new approaches, and the development of processing techniques have made possible the development of new applications and multifunctionality of these materials. In the particular case of PP, it is a low-density polymer (0.90 g/cm^3) with many industrial applications, in addition, it has high chemical resistance, long life for electrical and mechanical applications, and excellent thermal and optical properties, and, compared to PVC, PP is expected to be more environmentally friendly because it does not release toxic gases when burned (Maddah, 2016). Several studies have proposed the use of PP, its copolymers, and modified PP as an adsorbent for various pollutants including oil spills, organic matter, and Hg by poly(acrylamide)-PP hollow fibers (Wei et al., 2003; Liu et al., 2018; Abd-Aziz et al., 2019).

PET is a biodegradable thermoplastic polyester with excellent mechanical and thermal properties, chemical resistance including solvents as alcohols, aliphatic hydrocarbons, oils, greases, and diluted acids, and water and moisture barrier properties. Its use is very widespread in different industries including the food, the automobile, and electronic device industries due to its insulating properties (Jankauskaite et al., 2008; McKeen, 2017). In contrast to the above, thin-film PET in conjunction with gold electrodes has been developed as glucose sensors showing good sensitivity, reproducibility, and with no interference from biomolecules such as uric acid and dopamine (Wang et al., 2019); in addition, PET has been used as a substrate in the making of nanostructured flexible chemiresistors (Wang et al., 2010), and it has been suggested that its functionalization with phenolic compounds such as phenol, p-chlorophenol, and hydroxyquinone permits its use as an adsorbent for removal of Cu^{2+} ions from aqueous media (see Fig. 15.1; Ungureanu et al., 2020).

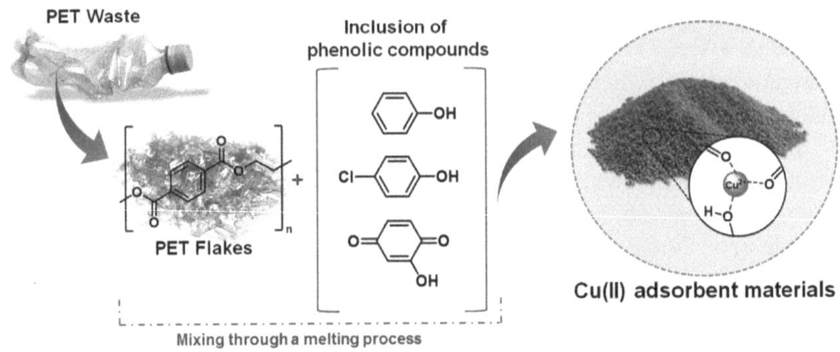

FIGURE 15.1 Preparation of adsorbent materials of Cu (II) ions from aqueous media employing PET waste and phenolic compounds such as phenol, p-chlorophenol, and hydroxyquinone.

PUs are a material family characterized to contain urethane units on their structure. Usually PUs are synthesized as block polymers in order to control their properties that depend on the length and nature of segments. Thus, elastic properties, hydrophilicity, hydrophobicity, degradation rates, adsorption capacity, porosity, and other properties can be included or controlled in the material. The use of PUs is widespread, they can be used as barriers for adhesion processes, scaffolds for the repair and regeneration of various tissues, and liquid-injectable materials which solidify after injection, among other applications (Akindoyo et al., 2016). Functional PUs have been developed for agricultural applications, in particular, for the retention and release of boron in agricultural fertilization smart systems (Palencia et al., 2016a,b).

Currently, PS is a homopolymer with many uses. It shows a wide spectrum of functional properties on small structural differences associated with the orientation of the side groups, interchain space, molecular weight, and cross-linking degree. Usually PS is characterized as a transparent thermoplastic with high tensile strength and poor barrier properties to moisture, vapor, and gases, however, it can be a thermoset material also, that is hard and fairly brittle, characterized as being nonpermeable, hydrophobic, and with a closed-cell structure. It is commonly used in the packing of food for applications such as protective packaging, cartons and containers, and disposable plastic silverware and food trays, among others (Lynwood, 2014). In addition, expanded PS has been identified as a suitable material for the construction industry (Ramli at al., 2019). As for other structural polymers, PS can be used for functional applications in conjunction with other materials either forming blends, multilayer systems, composites, nanocomposites, copolymers, or by surface functionalization (Lynwood, 2014). For example, nanofibers of PS coated with poly(aniline)-graphene nanocomposites have been developed for use as gas sensors (Bhadra et al., 2019). In the same context, PS has been used in making optical sensors from PVC-graft-PS copolymers and pyrene as a fluorescent indicator, which exhibited high sensitivity and selectivity toward

Fe^{3+} over Cu^{2+}, Mg^{2+}, Co^{2+}, Zn^{2+}, Ni^{2+}, and Ag^{+} (Panawong et al., 2015). In the same way, PS crystals have been proposed as a sensor for volatile organic compounds by reflectance peak changes (Burratti et al., 2018).

PVC is a very popular material that is present in multiple daily-life applications. An important characteristic of PVC is that its properties change significantly in the presence of plasticizers. Thus, for instance, plasticized PVC has lower chemical resistance to organic solvents like chlorinated hydrocarbons, ketones, and cyclic ethers compared with unplasticized PVC. In addition, it is well known that the chemical resistance of plasticized PVC is dependent on the type and amount of plasticizer used (Abdul et al., 2013; Makris et al., 2019). Some functional applications of PVC include the making of ion-selective electrodes for pharmaceutical analysis (Lenik, 2015). In addition, PVC is an innocuous material with high biocompatibility, therefore, it has been used for biomedical applications in blood storage (Blass 1992), however, adverse reactions have been described when it is in contact with body tissue and fluids (Yiani, 1995). In order to solve this problem, PVC has been modified using poly(ethylene glycol) (Balakrishnan et al., 2005), natural polymers like collagen to obtain blends (Lungu et al., 2010), and by functionalization of PVC/layered double-hydroxide nanocomposites with thiosulfate, thiourea, and sulfite (Singh et al., 2018), among others. In addition, bioactive properties have been included in PVC films with the use of nanoparticles (Palencia et al., 2016a,b) and surface functionalization with ammonium quaternary groups (see Fig. 15.2) (Palencia et al., 2019).

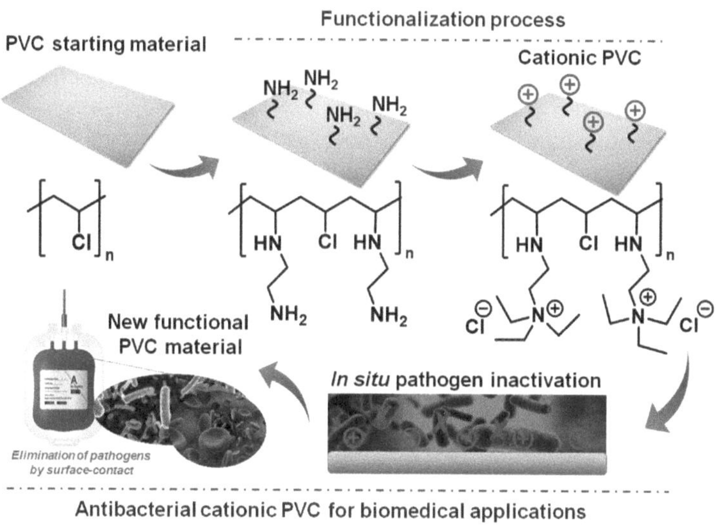

FIGURE 15.2 Development of a new functional PVC material through the superficial inclusion of cationic groups. In this case, the new material presented good antibacterial properties and low hemolysis, such that it was postulated for its implementation on biomedical surfaces, for example, blood storage bags.

EVA is a copolymer resulting from low-density polyethylene polymerized with vinyl acetate, and is characterized to show high mechanical strength, flexibility at low temperatures, and high permeability to water vapor and gases. In general, structural properties of EVA depend on the vinyl acetate content, molecular weight and distribution, and molecular branching (Alothman, 2012; Schneider et al., 2017). EVA has been used widely for biomedical applications as a component of drug-in-adhesive matrices and transdermal delivery systems like quinupramine, furosemide, ethinylestradiol, and loratadine, among others (Schneider et al., 2017). In environmental applications it has been used as a passive sampler for adsorption of polychlorinated biphenyls (Martinez et al., 2017). More details related to the use of EVA as a passive sampler are given in Chapter 20.

15.3 Conclusions and remarks

The difference between functional and structural polymers must be analyzed very carefully because the adjective "functional" applied to a polymer is strongly linked to its application and context of use. Obviously, many applications emerge from structural characteristics of polymers, and consequently, many applications are associated with mechanical and thermal properties since these are commonly circumscribed in a set of structural characteristics, such as high hydrophobicity by the absence of highly polar groups, which is an appropriate quality for outdoor use of some polymers, or a high mechanical resistance and low density resulting from the intermolecular interactions of their chains and the molecular ordering, which gives a polymer the ability to be used in applications that involve constant stresses on the material. As is shown throughout this chapter, by changing the context, it is possible that the same properties needed for structural uses can be used for functional applications, such as the adsorption of hydrophobic organic compounds, or their inclusion in biomedical components such as prosthetics, when the biocompatibility criteria are satisfied. An important aspect to highlight is the existence of strategies allowing expansion of the functional use of polymeric materials regardless of their initial nature; these strategies mainly include the making of blends, multilayer systems, composites, nanocomposites, copolymers, or functionalization of the polymeric surface.

References

Abd-Aziz, N.H., Alias, S., Bashar, N.A.M., Amir, A., Abdul-Talib, S., Tay, C.C., 2019. A short review: potential use of plastic waste as adsorbent for various pollutant. AIP Conf. Proc. 2124, 020034. https://doi.org/10.1063/1.5117094.

Abdul, H.P.S., Tehrani, M.A., Davoudpour, Y., Bhat, A.H., Jawaid, M., Hassan, A., 2013. Natural fiber reinforced poly(vinyl chloride) composites: a review. J. Reinforc. Plast. Compos. 32. https://doi.org/10.1177/0731684412458553.

Akindoyo, J.O., Beg, M.D.H., Ghazali, S., Islam, M.R., Jeyaratnam, N., Yuvaraj, A.R., 2016. Polyurethane types, synthesis and applications – a review. RSC Adv. 115, 114453–114482. https://doi.org/10.1039/C6RA14525F.

Alothman, O.Y., 2012. Processing and characterization of high density polyethylene/ethylene vinyl acetate blends with different VA contents. Adv. Mater. Sci. Eng. 635693. https://doi.org/10.1155/2012/635693.

Balakrishnan, B., Kumar, D.S., Yoshida, Y., Jayakrishnan, A., 2005. Chemical modification of poly(vinyl chloride) resin using poly(ethylene glycol) to improve blood compatibility. Biomaterials 26, 3495–3502. https://doi.org/10.1016/j.biomaterials.2004.09.032.

Bhadra, J., Popelka, A., Abdulkareem, A., Ahmad, Z., Touati, F., Al-thari, N., 2019. Fabrication of polyaniline–graphene/polystyrene nanocomposites for flexible gas sensors. RSC Adv. 9, 12496–12506. https://doi.org/10.1039/C9RA00936A.

Blass, C.R., 1992. PVC as a biomedical polymer–plasticizer and stabilizer toxicity. Med. Dev. Technol. 3, 32–40.

Burratti, L., De Matteis, F., Casalboni, M., Francini, R., Pizzoferrato, R., Prosposito, P., 2018. Polystyrene photonic crystals as optical sensors for volatile organic compounds. Mater. Chem. Phys. 212, 274–281. https://doi.org/10.1016/j.matchemphys.2018.03.039.

Hahladakis, J.N., Velis, C.A., Weber, R., Iacovidou, E., Purnell, P., 2018. An overview of chemical additives present in plastics: migration, release, fate and environmental impact during their use, disposal and recycling. J. Hazard Mater. 344, 179–199. https://doi.org/10.1016/j.jhazmat.2017.10.014.

Hermabessiere, L., Dehaut, A., Paul-Pont, I., Lacroix, C., Jezequel, R., Soudant, P., Duflos, G., 2017. Occurrence and effects of plastic additives on marine environments and organisms: a review. Chemosphere 182, 781–793. https://doi.org/10.1016/j.chemosphere.2017.05.096.

Jankauskaite, V., Macijauskas, G., Lygaitis, R., 2008. Polyethylene terephthalate waste recycling and application possibilities: a review. Mater. Sci. 14, 119–127.

Kazemi, F., Jafarzadeh, Y., Masoumi, S., Rostamizadeh, M., 2020. Oil-in-water emulsion separation by PVC membranes embedded with GO-ZnO nanoparticles. J. Environ. Chem. Eng. 104992. https://doi.org/10.1016/j.jece.2020.104992.

Lenik, J., 2015. Application of PVC in construction of ion-selective electrodes for pharmaceutical analysis: a review of polymer electrodes for nonsteroidal, anti-inflammatory drugs. In: Kumar, V., Kumari, M. (Eds.), Handbook of Polymers for Pharmaceutical Technologies: Processing and Applications, vol. 2. https://doi.org/10.1002/9781119041412.ch8.

Li, Y., 2018. Challenges and issues of using polymers as structural materials in MEMS: a review. J. Microelectromech. Syst. 27, 581–598. https://doi.org/10.1109/JMEMS.2018.2837684.

Liu, C., Jia, J., Liu, J., Liang, X., 2018. Hg selective adsorption on polypropylene-based hollow fiber grafted with polyacrylamide. Adsorpt. Sci. Technol. 36, 287–299. https://doi.org/10.1177/0263617416689480.

Lungu, M., Moldovan, L., Craciunescu, O., Doicin, C., 2010. Biocompatible blends based on polyvinyl chloride and natural polymers for medical device fabrication. Mater. Plast. 47, 278–281.

Lynwood, C., 2014. Polystyrene: Synthesis, Characteristics and Applications. Nova Science Publishers, pp. 1–320.

Maddah, H.A., 2016. Polypropylene as a promising plastic: a review. Am. J. Polym. Sci. 6, 1–11. https://doi.org/10.5923/j.ajps.20160601.01.

Makris, K.F., Langeveld, J., Clemens, F., 2019. A review on the durability of PVC sewer pipes: research vs. practice. Struct. Infrastruct. Eng. 16, 880–897. https://doi.org/10.1080/15732479.2019.1673442.

Martínez, J.M., Troches, E., Paez, M.I., 2017. Evaluation of poly(ethylene-co-vinylacetate) as polymeric adsorbent of polychlorinated biphenyls. J. Sci. Technol. Appl. 2, 94–107. https://doi.org/10.34294/j.jsta.17.2.18.

McKeen, L., 2017. Polyester plastics. In: Permeability Properties of Plastics and Elastomers Permeability Properties of Plastics and Elastomers, fourth ed. Elsevier, pp. 95–114. https://doi.org/10.1016/B978-0-323-50859-9.00006-3.

Palencia, M., Berrio, M.E., Melendres, M.F., 2016a. Nanostructured polymer composites with potential applications into the storage of blood and hemoderivates. J. Sci. Technol. Appl. 1, 4–14. https://doi.org/10.34294/j.jsta.16.1.1.

Palencia, M., Lerma, T.A., Córdoba, A., 2016b. Polyurethanes with boron retention properties for the development of agricultural fertilization smart systems. J. Sci. Technol. Appl. 1, 39–52. https://doi.org/10.34294/j.jsta.16.1.4.

Palencia, M., Lerma, T., Afanasjeva, N., 2019. Antibacterial cationic poly(vinyl chloride) as an approach for in situ pathogen-inactivation by surface contact with biomedical materials. Eur. Polym. J. 115, 212–220. https://doi.org/10.1016/j.eurpolymj.2019.03.041.

Panawong, C., Pandhumans, T., Youngme, S., Martwiset, S., 2015. Enhancing performance of optical sensor through the introduction of polystyrene and porous structures. J. Appl. Polym. Sci. 132, 41759. https://doi.org/10.1002/app.41759.

Plummer, C.J.G., 2016. Fracture assessment of polymers. In: Plummer, C.J.G. (Ed.), Reference Module in Materials Science and Materials Engineering, pp. 588–606. https://doi.org/10.1016/B978-0-12-803581-8.00943-7.

Ramli, N.H., Syaerah, S.A., Abdul, M.K., 2019. Application of expanded polystyrene (EPS) in buildings and constructions: a review. J. Appl. Polym. Sci. 136, 47529. https://doi.org/10.1002/app.47529.

Rivas, B.L., Pereira, E., Moreno-Villoslada, I., 2003. Water-soluble polymer–metal ion interactions. Prog. Polym. Sci. 28, 173–208. https://doi.org/10.1016/S0079-6700(02)00028-X.

Schneider, C., Langer, R., Loveday, D., Hair, D., 2017. Applications of ethylene vinyl acetate copolymers (EVA) in drug delivery systems. J. Contr. Release 262, 284–295. https://doi.org/10.1016/j.jconrel.2017.08.004.

Sigh, M., Kumar, R., Kumar, S., Kumar, S., Misra, N., 2018. In vitro biocompatibility analysis of functionalized poly(vinyl chloride)/layered double hydroxide nanocomposites. RSC Adv. 8, 40611–40620. https://doi.org/10.1016/10.1039/C8RA06175K.

Ungureanu, O.I., Bulguriu, D., Mocanu, A.M., Bulgariou, L., 2020. Functionalized PET waste based low-cost adsorbents for adsorptive removal of Cu(II) ions from aqueous media. Water 12, 2624. https://doi.org/10.3390/w12092624.

Wang, L., Luo, J., Yin, J., Zhang, H., Wu, J., Shi, X., et al., 2010. Flexible chemiresistor sensors: thin film assemblies of nanoparticles on a polyethylene terephthalate substrate. J. Mater. Chem. 20, 907–915. https://doi.org/10.1039/B920957C.

Wang, Y., Wang, X., Lu, W., Yuan, Q., Zheng, Y., Yao, B., 2019. A thin film polyethylene terephthalate (PET) electrochemical sensor for detection of glucose in sweat. Talanta 198, 86–92. https://doi.org/10.1016/j.talanta.2019.01.104.

Wankasi, D., Dixon, E., 2014. Polyvinulchloride Waste as an Adsorbent for the Sirption of Pb2+ from Aqueous Solution. https://doi.org/10.1155/2014/817527, 817527.

Wei, Q., Mather, R., Fotheringham, A.F., Yang, R.D., 2003. Evaluation of nonwoven polypropylene oil sorbents in marine oil-spill recovery. Mar. Pollut. Bull. 46, 780–783. https://doi.org/10.1016/S0025-326X(03)00042-0.

Xia, Y., He, Y., Zhang, F., Liu, Y., Leng, J., 2020. A review of shape memory polymers and composites: mechanisms, materials, and applications. Adv. Mater. 33 (6), 1–33. https://doi.org/10.1002/adma.202000713.

Yiani, J.P., 1995. Making PVC more biocompatible. Med. Dev. Technol. 6, 20–26.

Chapter 16

Functional and eco-friendly polymers in coatings and surfaces

16.1 General aspects about surface, coatings, and adhesion

A surface is a solid–gas or liquid–gas interface. However, in everyday language a solid–liquid interface is sometimes wrongly called a surface. This aspect is irrelevant when the solid phase is the center of study, and the liquid, being a fluid phase, is part of the environment directly in contact with the solid. The inner surface of a tubular membrane in contact with the liquid that moves through it is familiar and in consequence usually the conceptual formalism is neglected. Conversely, in volume liquid membranes where two liquid phases are in contact, it would be confusing to refer to the interface between the liquids as surfaces, an analogous example to the above is a multiphasic solid. The double interpretation of the concept surface, that is, the interface between a condensed phase and a gas phase, and the upper part or border delimiting a three-dimensional system, contributes to the confusion behind the use of the term, and as a consequence, it is always recommended to analyze the context of its use for the correct interpretation (Lüth, 2001).

Surfaces are important from a technological overview because they are part of the contact line of objects, or if it is preferred, they are the part of system mediating in the contact and associated interactions of a system with its surrounding. Surface interactions are linked to coatings, adhesives, biocompatibility, adsorption and catalytic processes, hydration and fouling phenomena, and optic phenomena such as light reflection and refraction, among others (Krishnan et al., 2008; Awaja et al., 2009; Chen et al., 2010; Pattanaik et al., 2012; Pocin-Epaillard et al., 2012; Hosseini et al., 2014; Smith et al., 2014; Hasan and Pandey, 2015; Sun et al., 2015; Bag et al., 2017; Li et al., 2018; Chakraborty et al., 2019; Gong et al., 2019). A surface can be described from a macroscopic approach from thermodynamics principles including the surface free energy in the case of solids or surface tension for liquids, its hydrophobicity or hydrophilicity, surface charge density, hardness, roughness, thermal

Eco-friendly Functional Polymers
https://doi.org/10.1016/B978-0-12-821842-6.00003-8
Copyright © 2021 Elsevier Inc. All rights reserved.

properties, refractive index, permeability, etc. On the other hand, from a microscopic approach, surfaces are characterized by molecular structure of materials, but also, by nanoscopic properties such as nano-roughness and nano-hardness, differing from macroscopic properties in the scale and resolution achieved during the analysis (Lüth, 2001).

Advances in the study and control of surfaces have been recently achieved due to the contributions of nanotechnology and the development of increasingly advanced analytical techniques. Typical methods for the study and evaluation of surface properties of polymers are based on microscopy techniques (e.g., scanning electron microscopy or SEM, transmission electron microscopy or TEM, atomic force microscopy or AFM, scanning tunneling microscopy) (St John et al., 2003; Xu et al., 2009), spectroscopic techniques (e.g., solid ultraviolet spectroscopy, diffuse reflectance infrared spectroscopy with Fourier transform or DRIFT, attenuated total reflectance or ATR, Rutherford backscattering spectrometry, X-ray photoelectron spectroscopy, secondary ion mass spectrometry) (Composto et al., 2002; St John et al., 2003; Xu et al., 2009; Palencia, 2018), electrochemical techniques (e.g., cyclic voltammetry and electrochemical impedance spectroscopy) (Palencia et al., 2015), and others such as contact angle in conjunction with the van Oss–Chaudhury–Good (vOCG) model (Xu et al., 2009; Palencia, 2017), digital image analysis using chemical indicators for color production (Palencia et al., 2016), analysis based on adsorption isotherms (Ayawei et al., 2017), and derivatization of surface chemical groups (St John et al., 2003).

On the other hand, a coating is a surface made or deposited on a surface or substrate. Therefore, if a coating is formed on a polymer surface then two new interfaces are produced: (1) the polymer–coating interface and (2) the coating–gas surface. Coatings can be metallic, polymeric, ceramic, vitreous, etc. In addition, their function can be decorative (e.g., paintings), functional (e.g., to make a surface biocompatible, to provide catalytic, optical, magnetic, or conductive properties, and to protect it from radiation, moisture, or fire), or both (De Rosa et al., 2012; Kausar, 2018). In this book, polymer coatings are the main focus, and so, coating by metals, nanoparticles, ceramics, and others are not considered.

For a satisfactory coating, adequate adhesion is required. Adhesion is one of the most important factors in the construction of coatings. This can be mediated by physical or chemical interactions, or both. In its initial stage, the interfacial phenomena are decisive for adequate contact between the substrate and coating. Good contact promotes the homogeneous distribution of the coating, both in cases where the coating agent is deposited on the surface, and in those situations where the coating is built by transference of mass from the phase adjacent to the target surface. To promote the formation of the coating, an "adhesion promoter" or "adhesive" usually is used in order to increase the affinity between the substrate surfaces and the coating agent. Therefore, adhesion promoters, also called "coupling agents" act analogously to surfactants in the sense that they promote the compatibility of the immiscible liquids by decreasing the surface energy, however, in this case, coupling agents improve the compatibility of phases by creating favorable intermolecular interactions, such as hydrogen bonds, dipole–dipole interactions, covalent bonds, electrostatic interactions, etc. (Wypych, 2018). It should be noted that, while in emulsion technology surfactants act through noncovalent molecular interactions, in coatings technology coupling agents can act through both covalent and noncovalent interactions.

Adhesion mechanisms are varied and cover a wide spectrum of possibilities. This variability is explained by the large number of materials, surfaces, and applications in which it is necessary to implement the use of a coating, and consequently, adhesion processes. The main mechanisms of adhesion are: (1) mechanical interlocking (this is based on the exploitation of voids, pores, deep valleys, holes, craters, and other imperfections of substrate, and consequently, requires adequate rheological properties of the deposited material), (2) surface condition and shape (this is based on the elimination of adhesion inhibitors on the surface, therefore, cleaning processes should be performed in order to increase the traction of the adhesive, a typical example is the removal of iron oxides on metallic surfaces), (3) diffusion and entanglement (this mechanism is based on two stages, diffusion and entanglement formation, thus, initially penetration of phases occurs through diffusion, and after, the entanglement acts by physical interactions; this mechanism is usual in polymers), (4) the adsorption mechanism includes acid–base and electrostatic interactions and hydrogen bonding, (5) surface free energy and wetting (this is based on a decrease in the interfacial tension and includes a combination of dispersive forces, i.e., van der Waals forces, and polar forces such as hydrogen binding), (6) crystalline properties (these are based on the formation of so-called "crystallographic bonding" which must be understood as a physical linking resulting from strong crystallographic relations among two different phases in contact), (7) interphase formation (this is based on the addition of coupling agents which produce an interphase promoting the adhesion), (8) chemical bonding (this is based on the formation of chemical bonds between the substrate and the coating), (9) hydrogen bonding (this is based on the formation of hydrogen interaction and therefore the substrate and coating require a surface molecular structure compatible with the formation of these links, i.e., one phase requires hydrogen atoms which should be linked to electronegative atoms such as N or O, and the counterpart should be a surface with electronegative atoms to make possible a strong interaction with hydrogen atoms characterized by a strongly positive partial charge), (10) microbiological and biological adhesion (this is based on the secretion of biomolecules from microbiological metabolism in order to achieve colonization of the surface, as a consequence, it is directly related to the formation of biofilms), and (11) cellular adhesion (this is based on the cell–extracellular matrix interactions analogous to the formation of biofilms, thus it is possible promote the anchoring, traction, and release of bioadhesives) (Wypych, 2018; Espinosa et al., 2020).

16.2 Eco-friendly polymer surfaces and coatings

As has been emphasized throughout this book, the eco-friendly concept is a broad concept, which refers not only to manufacturing methods, reusability, durability, residuality, biodegradability, and the origin of materials and raw materials, but also includes the uses given to them in the field of conservation and protection of the environment, in aspects such as monitoring, conservation, and remediation. Therefore, polymeric surfaces and coatings must be analyzed for each of these aspects.

16.2.1 Eco-friendly polymer surfaces

A polymeric surface should be understood as the solid–gas interface delimiting a polymeric phase, which for a specific application can be used in processes associated with solid–gas, solid–liquid, or solid–solid interfaces. Therefore, although polymer solutions are not included, polymer dispersions do meet the above definition. The same happens with polymeric materials in the molten state, since their practical existence occurs at very high temperatures, and so their conditions of use do not conform to normal environmental temperatures (i.e., approximately −20 to 60°C).

In relation to the origin of a surface, it is evident that its chemical composition is directly related to the nature of the material. In general, polymer surfaces with low functionalization should be used for applications that do not require chemical modifications. The reason for this is simple, a modification involves the use of solvents, reagents, purification steps, energy consumption, etc,; consequently, if these steps can be avoided or are not required, the impact of the formation of a functional surface will be eco-friendlier. However, it is important to clarify that the individual impacts of each modification stage must be weighed against the impact generated by the modified material. Many polymeric surfaces are used for the removal of ions and other polluting species dissolved in water such as acrylic, cationic and anionic resins, neutral adsorbents, etc. (see Fig. 16.1; Rivas et al., 2003; Bajdur et al., 2016; Sajid et al., 2018). Normally, in many cases the surface functional groups result from the nature of the polymer, which can be nonbiodegradable like poly(acrylic acid), poly(styrene sulfonate), and poly(acrylamide). These are obtained from the polymerization of acrylic acid, styrene sulfonate, and acrylamide, respectively (Rivas et al., 2003). Also, surface functional groups can come from biodegradable polymers, such as poly(lactic acid) (Garlotta, 2001), alginate (Szekalska et al., 2016), or chitosan (Kou et al., 2021).

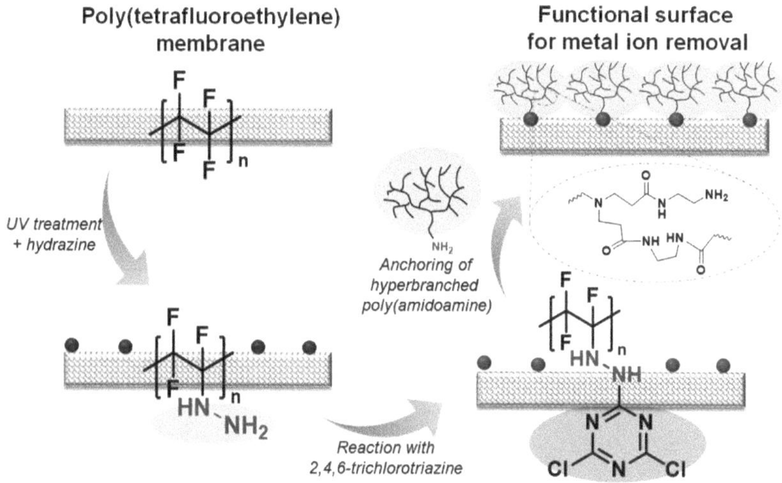

FIGURE 16.1 Modification of polymeric membrane for obtaining new functional polymeric surfaces based on dendrimers, more specifically, a hyperbranched poly(amidoamine). These new surfaces were employed for cation removal in an aqueous environment.

On the other hand, there are also cases in which the surfaces must be modified to achieve the desired objective, because the initial polymeric raw material does not offer the required functionality. In this way, the functional characteristic of the surface can be obtained by: (1) modifying the polymer and subsequent construction of the surface, (2) modifying the monomer and subsequent polymerization to finally carry out the construction of the surface, or (3) making the surface carry out its subsequent modification later. In the latter case, it can be carried out by chemical modification of the functional groups or by applying a coating (see Section 16.2.2.). Some examples of the making of an active surface include allyl isocyanate reacts with alcohols to produce alkyl-N-carbamates, resulting in the monomer being polymerized in the presence the benzoyl peroxide at 80°C; if this reaction is performed inside porous polypropylene it is possible to obtain interpenetrating polymer networks. As a consequence of this reaction, a porous hydrophobic surface with short length of alkyl chains ($-CH_3$) can be modified to obtain a new surface with a longer length of alkyl chains ($-CH_3$, $-C_2H_5$, $-C_6H_{13}$, $-C_8H_{17}$, and $CH_2C_6H_5$) (see Fig. 16.2; Lerma et al., 2016). Vinyl-benzyl-triethylammonium chloride can be obtained from 4-vinylbenzyl chloride and N,N,N-triethylamine in the same way, resulting in the monomer being polymerized inside the glass fiber membrane in order to obtain a surface with positively charged functional groups (Benitez et al., 2017). Analogously, by interpenetrating the polymer network, the insertion of different functional groups has been described such as short-chain hydroxyl groups (Arbelaez et al., 2017), thiolated groups (Lerma et al., 2017), and units of N-methyl-D-glucamine (Palencia et al., 2017).

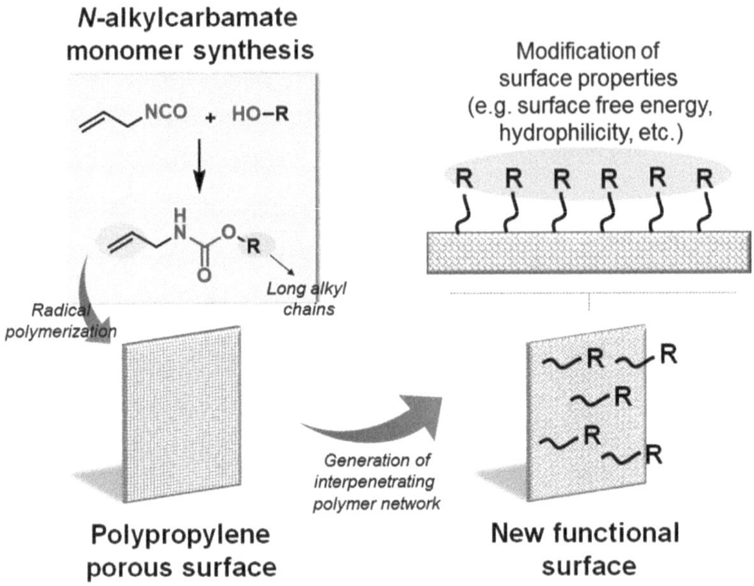

FIGURE 16.2 Generation of a new surface employing interpenetrating polymer networks based on polypropylene and allyl N-alkylcarbamates. See text for further details.

16.2.2 Eco-friendly polymer coatings

Coatings can be eco-friendly depending on their application. They are used to increase durability by protection surfaces, and in this way, they decrease the production of waste. In addition, coatings can be used to introduce new properties of adsorption or retention of pollutants, and as a consequence, their use includes activation of the surface for monitoring and remediation activities. There are a large number of functional polymer-based coating alternatives, such as acrylic, cellulosic, urethane, epoxy, and polyester resins. In some cases, coating using polymers involves polymerization reactions [i.e., polycondensation for poly(urethanes), epoxides, polyamide, and polyesters]. In other cases, it is necessary to apply coupling agents which can be based on water or nonaqueous solvents. In general, coupling agents containing functional groups react, bind, or strongly interact with the substrate surface (Gilliam, 2015). Some functional coating applications include nonwettable coatings, conductive coatings, biodegradable coatings, antifouling coatings, sensory coatings, and adsorbent coatings (Krishnan et al., 2008; Kim et al., 2020; Peralta-Ruiz et al., 2020; Predoi et al., 2020); but also, growing interest has emerged in coatings based on polymer nanocomposites, inorganic nanoparticles, montmorillonite, bentonite, titanium oxide, magnetite, carbon nanotubes, and graphene oxide, which have been used for reinforcing polymers employed in coatings, such as poly(ethylene glycol), polyurethanes, poly(3-hydroxyalkanoate), acrylic resins, poly(lactide-co-glycolide), poly(-glycolic acid), etc. (Kausar, 2018).

Two main points are related to the origin and residuality of coatings. In some sectors, for example electronics, thermoplastic polymer coatings based on conductive paints are used; however, this process is associated with the emission of volatile organic compounds (VOCs) (Randall, 1992; Kim, 2011; Kozicki et al., 2018). Keeping with the example of the paints, some common operations can be identified in all paint-making processes: premixing of pigment or dye, resin, and solvents, dispersion only for pigments, thinning-down and adjustments, filtration and packaging. As is expected, multiple residues are produced in these operations, however, equipment cleaning is the main source of waste. In addition, if the process has been designed in an eco-friendly fashion then wastes can be recycled and treated on-site using methods such as distillation. Some methods used in order to reduce the wastes in paint manufacturing are from government, for example the implementation and enforcement of regulations, and from industry, for example the implementation of good manufacturing practices, green process (e.g., high-pressure wash systems, reuse of equipment and raw materials, countercurrent rinse methods, green cleaning agents, improving of pigment dispersion, installing of baghouse systems, etc.), and recycling on-site and off-site. However, during paint application wastes are also produced resulting from raw materials, removal of organic substances, scale, dirt, cleaning solutions used prior to application, and overspray from paint applications (this is the result of transfer efficiency not being complete for all methods of application for conventional methods, with air being atomized at around 30%–60% and 85%–95% for electrostatic and centrifugally atomized methods, respectively) (Randall, 1992). In conclusion, when eco-friendly features of polymeric coatings are analyzed, this must be carried out carefully due to the characteristics of the different processes for each application sector.

16.3 Conclusions and remarks

The current use and potential uses of surfaces in polymer science and technology are very broad and promising. It is evident that many functional properties of polymers are mediated by their surface interactions with their environment. Regardless of their origin, other aspects such as their residuality, ease of recovery and reuse during the manufacturing stage, toxicity, and use are determining factors of their environmental impact, and therefore, these are important factors to visualize them as eco-friendly systems. In terms of their manufacture, eco-friendly features are strongly dependent on good environmental practices that are implemented throughout their life process, that is, from synthesis or obtaining to their final use and discharge. Nevertheless, not all surfaces are adapted to the desired functions, and it is at this point that coatings enter the picture. It is evident that a coating operationally speaking is the change from one surface for another, and in this sense, everything previously stated for surfaces also extends to coatings. However, in this context, adhesives and surface adhesion techniques are relevant and must be considered.

One limitation that usually goes unnoticed is the negative impact of coatings when their durability is low. After they are applied, their reuse is null, with the most obvious example being paints. As a consequence, durability is an important eco-friendly factor, since, if the coating loses its function and must be replaced, the removal process from the substrate involves the release of a large amount of residues, including solvents, pulverized material and particulates of small size, surfactants, and other cleaning agents. Therefore, it can be generalized that it is easier to recycle and reuse a functional polymer with a uniform composition, and even its surface, than a functional polymer coating. Consequently, it is very important to define the post-use destination of a coating in terms of residuality, dangerousness, bio-assimilation, biodegradability, and/or ease of chemical transformation into harmless components for the correct management of this type of waste.

References

Arbelaez, N., Lerma, T.A., Córdoba, A., 2017. Modification of membranes by insertion of short-chain alcohols on reactive porous substrates: effect of chain length on the surface free energy. J. Sci. Technol. Appl. 2, 75–83. https://doi.org/10.34294/j.jsta.17.2.16.

Ayawei, N., Newton, A., Wankasi, D., 2017. Modelling and interpretation of adsorption isotherms. J. Chem. 3039817. https://doi.org/10.1155/2017/3039817.

Awaja, F., Gilber, M., Kelly, G., Fox, B.L., Pigram, P., 2009. Adhesion of polymers. Prog. Polym. Sci. 34, 948–968. https://doi.org/10.1016/j.progpolymsci.2009.04.007.

Bag, M.A., Valenzuela, L.M., 2017. Impact of the hydration states of polymers on their hemocompatibility for medical applications: a review. Int. J. Mol. Sci. 18, 1422. https://doi.org/10.3390/ijms18081422.

Bajdur, W.M., Wlodarczyk-Makula, M., Idzikowski, A., 2016. A new synthetic polymers used in removal of pollutants from industrial effluents. Desalin. Water Treat. 57, 1038–1049. https://doi.org/10.1080/19443994.2015.1043495.

Benitez, E., Lerma, T.A., Córdoba, A., 2017. Making of porous ionic surfaces by sequential polymerization: polyurethanes + grafting of polyelectrolytes. J. Sci. Technol. Appl. 2, 44–53. https://doi.org/10.34294/j.jsta.17.2.13.

Chakraborty, J., Nath, I., Song, S., Mohamed, S., Kan, A., Heyderickx, P.M., Verpoort, F., 2019. Porous organic polymer composites as surging catalysts for visible-light-driven chemical transformations and pollutant degradation. J. Photochem. Photobiol. C Photochem. Rev. 41, 100319. https://doi.org/10.1016/j.jphotochemrev.2019.100319.

Chen, S., Li, L., Zhao, C., Zheng, J., 2010. Surface hydration: principles and applications toward low-fouling/nonfouling biomaterials. Polymer 51, 5283–5293. https://doi.org/10.1016/j.polymer.2010.08.022.

Composto, R.J., Walters, R.M., Genzar, J., 2002. Application of ion scattering techniques to characterize polymer surfaces and interfaces. Mater. Sci. Eng. R Rep. 38, 107–180. https://doi.org/10.1016/S0927-796X(02)00009-8.

De Rosa, E., Fernandez-Moure, J., Tasciotti, E., 2012. Polymer coatings. In: Bhushan, B. (Ed.), Encyclopedia of Nanotechnology. Springer. https://doi.org/10.1007/978-90-481-9751-4_105.

Espinosa, A., Palencia, S.L., García, A., Palencia, M., 2020. Polymicrobial Biofilms: fundamentals, diseases, and treatments - a review. J. Sci. Technol. Appl. 8, 4–54. https://doi.org/10.34294/j.jsta.20.8.54.

Garlotta, D., 2001. A literature review of poly(lactic acid). J. Polym. Environ. 9, 63–84. https://doi.org/10.1023/A:1020200822435.

Gilliam, M., 2015. Polymer surface treatment and coating technologies. In: Nee, A. (Ed.), Handbook of Manufacturing Engineering and Technology. Springer, pp. 99–124. https://doi.org/10.1007/978-1-4471-4670-4_20.

Gong, L., Xiang, L., Zhang, J., Chen, J., Zeng, H., 2019. Fundamentals and advances in the adhesion of polymer surfaces and thin films. Langmuir 35, 15914–15936. https://doi.org/10.1021/acs.langmuir.9b02123.

Hasan, A., Pandey, L.M., 2015. Review: polymers, surface-modified polymers, and self assembled monolayers as surface-modifying agents for biomaterials. Polym. Plast. Technol. Eng. 54, 1358–1378. https://doi.org/10.1080/03602559.2015.1021488.

Hosseini, S., Ibrahim, F., Djordjevic, I., Koole, L.H., 2014. Recent advances in surface functionalization techniques on polymethacrylate materials for optical biosensor applications. Analyst 139, 2933–2943. https://doi.org/10.1039/C3AN01789C.

Kausar, A., 2018. Polymer coating technology for high performance applications: fundamentals and advances. J. Macromol. Sci., Part A. 55, 440–448. https://doi.org/10.1080/10601325.2018.1453266.

Kozicki, M., Piasecki, M., Goljan, A., Deptula, H., Nieslochowski, A., 2018. Emission of volatile organic Compounds (VOCs) from dispersion and cementitious waterproofing products. Sustainability 10, 2178. https://doi.org/10.3390/su10072178.

Kim, B.R., 2011. VOC emissions from automotive painting and their control: a review. Environ. Eng. Res. 16, 1–9. https://doi.org/10.4491/eer.2011.16.1.001.

Kim, H., Nam, K., Lee, D.Y., 2020. Fabrication of robust superhydrophobic surfaces with dual-curing siloxane resin and controlled dispersion of nanoparticles. Polymers 12, 1687. https://doi.org/10.3390/polym12061420.

Kou, S., Peters, L.M., Mucalo, M.R., 2021. Chitosan: a review of sources and preparation methods. Int. J. Biol. Macromol. 169, 85–94. https://doi.org/10.1016/j.ijbiomac.2020.12.005.

Krishnan, S., Weinman, C.J., Ober, C.K., 2008. Advances in polymers for anti-biofouling surfaces. J. Mater. Chem. 18, 3405–3413. https://doi.org/10.1039/b801491d.

Lerma, T.A., Collazos, S., Córdoba, A., 2016. Effect of side chain length of carbamates on the surface properties of porous interpenetrating networks. J. Sci. Technol. Appl. 1, 30–38. https://doi.org/10.34294/j.jsta.16.1.3.

Lerma, T.A., Martínez, G., Palencia, M., 2017. Eneration of thiolated porous surfaces by interpenetrating polymeric networks: study of their surface properties. J. Sci. Technol. Appl. 3, 56–65. https://doi.org/10.34294/j.jsta.17.3.24.

Li, L.Y., Cui, L.Y., Zeng, R.C., Li, S.Q., Chen, X.B., Zheng, Y., Kannan, M.B., 2018. Advances in functionalized polymer coatings on biodegradable magnesium alloys – a review. Acta Biomater. 79, 23–36. https://doi.org/10.1016/j.actbio.2018.08.030.

Lüth, H., 2001. Surface and interface physics: its definition and importance. In: Solid Surfaces, Interfaces and Thin Films. Advanced Texts in Physics. Springer, pp. 1–31. https://doi.org/10.1007/978-3-662-04352-3_1.

Palencia, M., 2017. Surface free energy of solids by contact angle measurements. J. Sci. Technol. Appl. 2, 84–93. https://doi.org/10.34294/j.jsta.17.2.17.

Palencia, M., 2018. Functional transformation of Fourier-transform mid-infrared spectrum for improving spectral specificity by simple algorithm based on wavelet-like functions. J. Adv. Res. 14, 53–62. https://doi.org/10.1016/j.jare.2018.05.009.

Palencia, M., Arrieta, A.A., Melendrez, M., 2015. Study by impedance spectroscopy of polymeric membranes modified by adsorption of anthraquinone-type redox mediators and conductive polymers. Colloid Polym. Sci. 293, 3025–3032. https://doi.org/10.1007/s00396-015-3705-6.

Palencia, M., Lerma, T.A., Berrio, M.E., 2017. Thin-film composite by in situ polymerization of 4-chloromethyl styrene functionalized with N-methyl-D-glucamine in pore-type microreactors. J. Sci. Technol. Appl. 3, 66–76. https://doi.org/10.34294/j.jsta.17.3.25.

Palencia, M., Lerma, T., Palencia, V., 2016. Description of fouling, surface changes and heterogeneity of membranes by color-based digital image analysis. J. Membr. Sci. 510, 229–237. https://doi.org/10.1016/j.memsci.2016.02.057.

Pattanaik, B., Pawar, S., Pattanaik, S., 2012. Biocompatible implant surface treatments. Indian J. Dent. Res. 23, 398–406. https://doi.org/10.4103/0970-9290.102240.

Peralta-Ruiz, Y., Grande, C.D., Sinning-Mangonez, A., Coronell, E.A., Marino, M.F., Chaves-Lopez, C., 2020. Reduction of postharvest quality loss and microbiological decay of tomato "chonto" (*Solanum lycopersicum* L.) using chitosan-E essential oil-based edible coatings under low-temperature storage. Polymers 12, 1822. https://doi.org/10.3390/polym12081822.

Predoi, D., Iconaru, S.L., Predoi, M.V., Motelica-Heino, M., 2020. Removal and oxidation of as(III) from water using iron oxide coated CTAB as adsorbent. Polymers 12, 1420. https://doi.org/10.3390/polym12081687.

Pocin-Epaillard, F., Vrlinic, T., Debarnot, D., Mozetic, M., Coudreuse, A., Legeay, G., Moualij, B., Zorzi, W., 2012. Surface treatment of polymeric materials controlling the adhesion of biomolecules. J. Funct. Biomater. 3, 528–543. https://doi.org/10.3390/jfb3030528.

Randall, P.M., 1992. Pollution prevention methods in the surface coating industry. J. Hazard Mater. 29, 275–295. https://doi.org/10.1016/0304-3894(92)85073-a.

Rivas, B.L., Pereira, E., Moreno-Villoslada, I., 2003. Water-soluble polymer–metal ion interactions. Prog. Polym. Sci. 28, 173–208. https://doi.org/10.1016/S0079-6700(02)00028-X.

Sajid, M., Khaed, M., Ihsanullah, Baig, N., Osman, A.M., 2018. Removal of heavy metals and organic pollutants from water using dendritic polymers based adsorbents: a critical review. Separ. Purif. Technol. 191, 400–423. https://doi.org/10.1016/j.seppur.2017.09.011.

Smith, J., Lamprou, D.A., 2014. Polymer coatings for biomedical applications: a review. Trans. IMF 92, 9–19. https://doi.org/10.1179/0020296713Z.000000000157.

St John, H.A.W., Gengenbach, T.R., Hartley, P.G., Griesser, H.J., 2013. Surface Analysis of Polymers. In: O'Connor, D.J., Sexton, B.A., Smart, R.S.C. (Eds.), Surface Analysis Methods in Materials Science, Springer Series in Surface Sciences, vol. 23. Springer, pp. 519–551. https://doi.org/10.1007/978-3-662-05227-3_23.

Sun, Q., Dai, Z., Meng, X., Xiao, F.S., 2015. Porous polymer catalysts with hierarchical structures. Chem. Soc. Rev. 44, 6018–6034. https://doi.org/10.1039/C5CS00198F.

Szekalska, M., Pucilowska, A., Szymanska, E., Ciosek, P., Winnicka, K., 2016. Alginate: current use and future perspectives in pharmaceutical and biomedical applications. Int. J. Polymer Sci. 1–17. https://doi.org/10.1155/2016/7697031.

Wypych, G., 2018. Mechanisms of adhesion. In: Wypych G. Handbook of Adhesion Promoters. ScienceDirect, pp. 5–44. https://doi.org/10.1016/B978-1-927885-29-1.50004-3.

Xu, A., Wan, L., Huang, X., 2009. Techniques for membrane surface characterization. In: Surface Engineering of Polymer Membranes. Advanced Topics in Science and Technology in China. Springer. https://doi.org/10.1007/978-3-540-88413-2_2.

Chapter 17

Functional and eco-friendly polymers in medical and biomedical applications

17.1 Polymer materials in medicine and biomedicine: a general overview

In order to accomplish its objective, medicine requires not only new knowledge, in its day-to-day practice, it requires instruments, medicines, data, materials, etc. In particular, both structural and functional medicine polymers, with their versatility of properties, have become one of the most important types of materials in medicine and biomedicine. They offer mechanical and thermal resistance, shape adaptability, and processability, and in some cases biocompatibility, innocuousness, and biodegradability, among other important characteristics (Ulery et al., 2011). Curiously, many natural biopolymers are inherently biodegradable, biocompatible, and nontoxic, and sometimes their limitations are in terms of processability, hydration, and susceptibility to microbial colonization. At a more specialized level, some natural biopolymers act as specific protein-binding sites and biochemical signals and, consequently, they can assist in tissue healing, regeneration, and integration (Davidenko et al., 2018).

Due to the nature of the applications in medicine and biomedicine, the synthetic and natural polymeric materials used often cannot be disposed of through normal procedures, in order to avoid biological contamination, the proliferation of diseases, and to minimize risks to public health. For these same reasons, in some cases materials for biomedical use are single-use materials. Therefore, it is important to know what happens after they are used. For example, if these materials are subjected to thermal degradation, then their composition must be such that it avoids the generation of toxic gases, greenhouse gases, or other substances that may put the health of people and the environment at risk. Poly(vinylidene fluoride) (PVDF) is a synthetic polymer which has been described as being biocompatible and nonbiodegradable, and it has been used in the making of scaffolds as tissue reinforcement agents (Houis et al., 2010). Nevertheless, PVDF by thermal scission produces HF,

Eco-friendly Functional Polymers
https://doi.org/10.1016/B978-0-12-821842-6.00007-5
Copyright © 2021 Elsevier Inc. All rights reserved.

halogenated, and oxygenated organic compounds, and other alkyl radicals (Silva et al., 2020a,b). Poly(vinyl chloride) (PVC) is a widely used synthetic polymer used for biomedical applications, including blood-storage bags and other devices in contact with tissues and biological fluids such as catheters and probes (Palencia et al., 2016). By thermal degradation, PVC undergoes dehydrochlorination of the polymer producing HCl, and small amounts of benzene, toluene, and other Cl-compounds (Yu et al., 2016).

However, not all polymers used in biomedicine are for single-use materials, in fact, the spectrum of applications is so great that a wide number of polymers and properties are required. For example, for the controlled release of drugs using polymers, the structural characteristics depend on the type of drug, the route of release, the residence time, etc. More specifically, an innocuous and nonbiodegradable polymer is not a recommended material for use in controlled drug release by the oral route, regardless of the excellent affinity that the material may have for the drug. A polymer susceptible to enzymatic biodegradation, or capable of undergoing acid hydrolysis at stomach pH, would be a more suitable material for such an application. Another example is tissue engineering, in particular, for the case of bone tissue, where not only good safety and biocompatibility properties are required, but other properties such as adequate mechanical resistance are essential (Ulery et al., 2011; Velema and Kaplan, 2006; Silva et al., 2020a,b). Synthetic polymers frequently used in medicine and biomedicine include polyacrylates, polysiloxanes, synthetic polyamides, polycarbonates, synthetic polyesters, polyurethanes, polystyrenes, synthetic polypeptides, polyolefins, polyketones, etc. (Palencia et al., 2014; Lerma et al., 2017; Otálora et al., 2019). In contrast, biopolymers are mainly proteins and carbohydrates such as collagen, keratin, fibroin, natural polyesters [e.g., polyhydroxyalkanoates (PHAs)], and polysaccharides (e.g., chitin/chitosan, alginate, cellulose, and hyaluronate) (Davidenko et al., 2018; Garcés and Chate-Galvis, 2020; Silva et al., 2020a,b; see Fig. 17.1).

The healthcare business is a market worth hundreds of billions of dollars per year. Therefore, the technological potential of biopolymers in the medicine and biomedicine field, as well as in replacing commonly used plastics, has enormous commercial potential. Thus, with regard to bioplastics, it has been estimated that the size of the market in 2017 was US$ 6000 billion, with a growth estimation of US$ 14.92 billion by 2023 (Silva et al., 2020a,b).

17.2 Biopolymer materials in biomedical applications

As previously indicated, biopolymers are promising biomaterials for the development of biomedical applications due to their functional properties. Their main areas of application include implantable devices, wound management products, controlled drug-delivery systems, and tissue engineering scaffolds (Velema and Kaplan, 2006; Davidenko et al., 2018). A number of applications are described below.

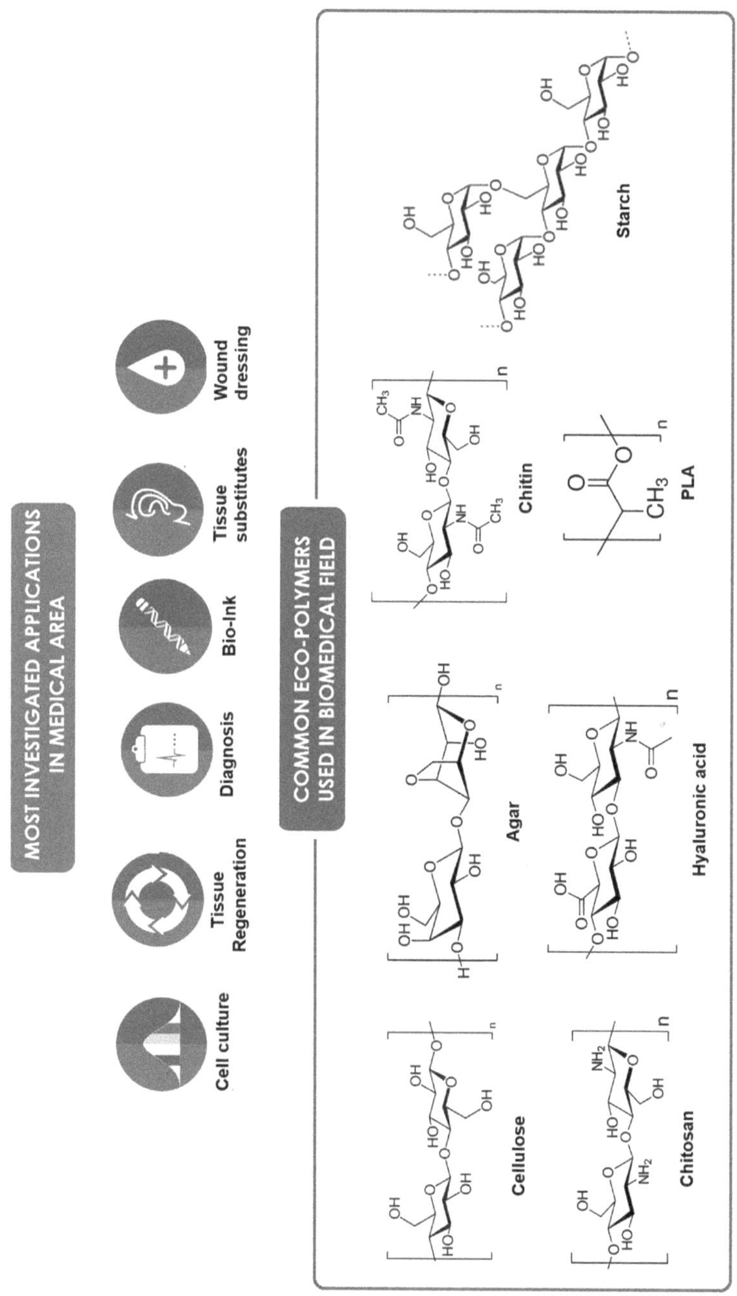

FIGURE 17.1 Some of the common eco-polymers and their uses in medical and biomedical applications.

The use of biopolymers can be carried out directly, that is, without modifications, or through physical and chemical modifications in order to impart new properties to the material. These modifications include the addition of nanomaterials, functionalization of side chains by chemical reactions, mixing of polymers, incorporation of additives, copolymerization through grafting, etc., and they aim to produce a synergistic effect in order to enhance the biological performance of the biopolymer (Davidenko et al., 2018). For example, active molecules can be electrostatically anchored to fibroin at physiological pH in order to promote the requirements of certain applications. The cellulose surface can be modified by multiple methodologies to obtain cellulose grafted with chains or units of poly(ethylene glycol), xylol, acetyl, poly(lactic acid), urethanes, poly(*N*-isopropyl-acrylamide), etc. (Kunduru et al., 2016; Garcés et al., 2017; Palencia et al., 2017a,b, 2018; Silva et al., 2020a,b). Grafting copolymers based on cellulose have been developed as micelles and drug carriers by combining the properties of cellulose matrix and functional properties of the grafting. Moreover, "smart" properties can be added to cellulose by grafting with stimuli-responsive side polymer chains, which can include surface properties stimulating cell division, adhesion, and motility (Velema and Kaplan, 2006; Matricardi et al., 2013; Davidenko et al., 2018; Silva et al., 2020a,b).

In the field of nanostructured biomaterials, great interest has arisen due to multiple and new properties which can be incorporated into the polymer phase by the physical or chemical addition of nanoparticles. Polymer nanocomposites are characterized as being constituted by one polymer phase and one nanostructured phase, such as metals, ceramics, oxides (e.g., CuO, ZnO, and TiO_2), or any other nanomaterial (e.g., graphene, carbon nanotubes, and nanocellulose). Thus, the inclusion of nanoparticles is motivated by the improvement of mechanical, thermal, spectral, and dielectric properties, renewability, catalytic and surface bioactivity, durability, and biodegradability with lower cost, and lower density (Al-Ghraibah et al., 2020). In particular, silver nanoparticles have been widely studied and used to impart antibacterial properties to biopolymers and synthetic polymers (Córdoba and Palencia, 2017; Muñoz-Bonilla et al., 2019).

17.2.1 Biopolymers in tissue engineering

In recent decades, tissue engineering has established itself as a line of research where multiple disciplines converge, including mechanical engineering, polymer chemistry, and medicine, among others (Al-Ghraibah et al., 2020). Such is its importance that it is possible to state that most uses of biopolymers in biomedical applications are related to tissue engineering. The objective of this area is the manufacture of materials through the individual and combined use of biopolymers, cells, growth factors, and even synthetic materials such as plastics, nanomaterials, etc. In addition, under this approach, these materials are used in the construction, reshaping, or treatment of damaged tissues and organs (Jahangirian et al., 2018; Curvello et al., 2019). The above implies the development of structures that functionally mimic tissues and the extracellular matrix, in such a way that these artificial systems can replace biological tissues without causing adverse effects or rejection by neighboring tissues and adjacent cells. In this context, the biocompatibility of the material at the surface level plays an important role as it is the interface in contact with the surroundings (Ulery et al., 2011; Jahangirian et al., 2018; Al-Ghraibah et al., 2020).

The ideal polymer biomaterial for this type of application should be characterized by adequate mechanical properties and a degradation time consistent with the desired function, not producing an inflammatory response, their degradation products must be nontoxic, and they should be easily reabsorbed or excreted, also, these materials should have adequate permeability and processability properties (Ulery et al., 2011). Some examples are described below.

The inclusion of aloe vera to biopolymer scaffolds increased their thermal stability as well as their hydrophilicity. In addition, from cell culture studies it was observed that scaffolds showed enhanced growth and proliferation of fibroblasts without evidence of toxicity (Jithendra et al., 2013). Chondroitin sulfate, which is a glycosaminoglycan structurally similar to hyaluronic acid with a sulfate group in at least one of its side chains of biopolymer, has been identified as a biomaterial able to help in healing processes. In addition, for this polymer, antiinflammatory properties able to stimulate the cartilage's metabolic response and the linkings among extracellular matrix components have been described. As a consequence, it has been studied extensively, in conjunction with collagen, hyaluronic acid, and chitosan, in the manufacture of biopolymer hydrogels for wound dressings and cartilage tissue engineering (Velema and Kaplan, 2006; Ulery et al., 2011).

The control of surface properties is based on a combination of materials in order to produce combined properties for the bulk and for the interface. Thus, for example, hydrogels based on alginate have been studied for the delivery of specialized-tissue cells such as chondrocytes, osteoblasts, myoblasts, fibroblasts, keratinocytes, and adipose-derived stem cells, with the aim of increasing the compatibility, neural differentiation, and biological activity of biomaterials. As a consequence, with the combination of cells and scaffolds coated with bioactive hydrogels it has been possible to advance the development of different strategies of regenerative medicine and cell therapy that are useful in tissue engineering (Ulery et al., 2011; Chen et al., 2015; Hassan et al., 2019; Bharadwaz and Jayasuriya, 2019). Three-dimensional (3D) tissue matrices are an important technology for the in vitro culture of some cell lines such as stem and tumor cells, and for the regeneration or replacement of organs or tissues, and tissue repair. In addition, these matrices are a robust model to investigate some kinetic aspects related to cancer progression and drug screening (Curvello et al., 2019). Hydrogels based on cellulose and nanocellulose have been synthesized and evaluated as scaffolds for 3D matrices in cultures of mesenchymal stem cells, human pluripotent stem cells, and hepatic tumor cells (Andersen et al., 2015; Rijal and Li, 2017; Curvello et al., 2019). In Table 17.1 some applications of biopolymers in the tissue engineering field are shown.

TABLE 17.1 Examples of eco-polymers in tissue engineering and their most common uses in this field.

Biopolymer	Applications in tissue engineering	References
Cellulose	Bioscaffolds for bone, cartilage, and soft tissues, blood vessel replacement, cartilage replacements, skin tissue repair, regeneration of soft-tissue Cell culture	Palencia et al. (2018), Hassan et al. (2019)
Chitosan	Regeneration of tissues such as bone, ligament, tendon, neural, skin, and cartilage Heart valves and vascular tissue Antibacterial platform Wound dressing	Rebelo et al. (2017), Jahangirian et al. (2018)
Collagen	Regeneration of tissues like bone, tendon, and ligament Wound dressing. Scaffolds for bone and cartilage regeneration	Ulery et al. (2011), Ramanathan et al. (2018)
Alginate	Bioink for 3D printing Regeneration of tissues by scaffolds and delivery of different cellular lines such as chondrocytes, osteoblasts, and fibroblasts, etc.	Hassan et al. (2019), Ulery et al. (2011)
Silk	Bioink for 3D printing Basis for extracellular matrix-like materials	Aljohani et al. (2017)
Hyaluronic acid	Scaffolds for cartilage regeneration Component of bioink hydrogels for 3D printing	Velema and Kaplan (2006)
PHAs	Scaffolds for tissue engineering Diagnosis Medical sutures	Ward et al. (2020), Ulery et al. (2011)
Poly(lactic acid)	Tissue regeneration scaffolds Cell culture Porous scaffolds for bone tissue	Bharadwaz and Jayasuriya (2019), Hassan et al. (2019)
Albumin	Bioadhesive for surgery Component of copolymeric scaffolds	Ulery et al. (2011)

In the field of tissue engineering, one of the most researched applications is the development of platforms for bone tissue and cartilage tissue. Commonly, the treatment or replacement of bones affected by some type of injury or health disorder has been an aspect of great importance in the medical sciences. Many biomaterials have been proposed as substitutes for bone tissue, including not only polymers, but also metals and ceramics, and a combination of these (Jahangirian et al., 2018; Al-Ghraibah et al., 2020; Bharadwaz and Jayasuriya, 2019). However, their efficiency is evaluated by characterization of properties such as cell anchoring, infiltration, morphogenesis, proliferation, and differentiation of cells, besides adequate biomechanical properties (Jahangirian et al., 2018). Regardless of the nature of the bulk material, the biological compatibility of many of these materials has been addressed through the construction of biointerfaces analogous to the real system. Thus, due to collagen being the main component of extracellular matrix in conjunction with hydroxyapatite, that stimulates cellular adhesion and provides mechanical strength to tissues, it is defined as an excellent candidate to achieve the objective of biocompatibilization (Bharadwaz and Jayasuriya, 2019). Different sources of collagen have been evaluated for obtaining this biomaterial, for example, type I collagen isolated from tilapia scale has shown similar results to those obtained from porcine skin collagen, in terms of biocompatibility (Tang and Saito, 2015). Other biomaterials have been studied for this application, for example, biopolymer materials based on chitosan, hydroxyapatite, and magnetite, pectin modified with Arg−Gly−Asp (RGD), and polysaccharide-based components, poly(lactic-co-glycolic acid) modified with phospholipids, and fucoidan, which is sulfated polysaccharides obtained from algae (Jahangirian et al., 2018; Bharadwaz and Jayasuriya, 2019).

On the other hand, cartilage has been another tissue receiving a great deal of attention. The facts behind this are mainly the lack of blood vessels, low metabolic requirements, low capacity for self-generation, and the preponderance of degenerative health effects such as arthritis (osteoarthritis and rheumatoid arthritis) (Velema and Kaplan, 2006). The most popular biomaterials for these applications have been collagen, hyaluronic acid, and alginate. The first two are natural components of cartilage tissue and are able to interact with the endogenous cartilage material, whereas alginate is an anionic polysaccharide obtained from the cell wall of algae; however, although this biomaterial is able to contribute to chondrocyte viability, it does not possess the optimal mechanical properties compared with collagen and hyaluronic acid (Velema and Kaplan, 2006; Park et al., 2017).

Three-dimensional (3D) printing is a relatively new technology that has revolutionized the manufacturing process in tissue engineering. This technique allows, through a computerized and high-precision extrusion system, the construction of 3D tissues (Axpe and Oyen, 2016; Bharadwaz and Jayasuriya, 2019). Biomaterials such as gelatin, hyaluronic acid, nanocellulose, fibrin, collagen, fibronectin, alginate, chitosan, starch, etc., have been used in the implementation of this technology, and are generally described as bioinks (Aljohani et al., 2017). Cartilage tissue based on nanocellulose-alginate bioink was developed using 3D bioprinting with live cells, that is, human chondrocytes, which showed a cell viability of 73% (after 1 day) and 86% (after 7 days) (Markstedt et al., 2015). The main limitations of biopolymers for their use in 3D printing are low thermal stability, inadequate mechanical properties, and poor barrier properties (Hassan et al., 2019). However, the inclusion of an exogenous component to the biopolymer matrix has been proposed to improve the properties of scaffolds, including nanoparticles of silica, hydroxyapatite, zirconia, alumina, etc. (Rebelo et al., 2017; Davidenko et al., 2018; Hassan et al., 2019).

17.2.2 Wound dressing

The skin is an organ that, due to its function, is exposed to multiple biological and nonbiological impacts (e.g., microbial infections, burns, trauma, and lacerations). In particular, a wound is a physical or chemical injury to the skin with different levels of severity, from superficial wounds such as scrapes, to deep or extensive wounds, which can also be accompanied by coinfections and inflammation. A typical example is the formation of polymicrobial biofilms in a wound produced by mechanical rupture of the tissues and with little or protection from infectious vectors (Abdo et al., 2020; Espinosa et al., 2020).

Taking into account the above, a wound dressing is defined as a tissue, coating, or artificial material that is applied to a wound for its restoration, either promoting the continuity of living tissue, generating a response that stimulates cellular colonization, or performing a remodeling of the skin matrix. For any of the above mechanisms, inflammation is required as a mechanism of innate immunity and neovascularization of the affected tissue for its subsequent healing. The biomaterial for wound dressing is usually designed with a multiple-part configuration, generally its internal part consists of a gel, or a scaffold layer in direct contact with the wound, and then an external elastic covering whose function is to fix the device to the tissue. At present, hyaluronic acid, chitosan, collagen, and a combination of these or other biomaterials are the main biopolymers for application directed toward wound healing (Sen et al., 2009; Zomer and Treting, 2018). Currently, more than 3000 wound treatment products are available in the market, of which more than 900 products, including gels, pads, particles, pastes, powders, sheets, or solutions, are based on biopolymers such as collagen, hyaluronic acid, chitosan, alginate, and elastin (e.g., Xcell, Bioprocess, Biofill, Gengiflex, XCell, Dermafill, Alloderm, Promogran, Biobrane, OrCel, HemCon, AlgiDERM, AlgiSite M, Tegaderm, etc.) (Park et al., 2017; Davidenko et al., 2018; Silva et al., 2020a,b).

Collagen matrices containing gelatin microspheres impregnated with organic–metallic complexes of Fe^{3+} with low molecular weight produced by bacteria and fungi (i.e., siderophores) have been proposed as delivery systems to control infections and protease levels (Ramanathan et al., 2018). Wound dressings based on bacterial cellulose and chitosan have shown good mechanical, antibacterial, and barrier properties, and high moisture retention for treating various kinds of wounds, burns, and ulcers (Silva et al., 2020a,b). Pectin is a biopolymer with low toxicity, adequate barrier properties against bacteria or fungi, and adequate hydrophilicity for use as a biomaterial in wound dressings; however, chitosan and collagen are oxygen permeable and have biocompatibility. In particular, chitosan has greater bioactivity then many biopolymers. The particular advantages of chitosan are induction of interleukin-8 production which is a neutrophil chemotactic factor regulating the production of adhesion molecules, and bioactive lipids, and that it stimulates angiogenesis (Kean and Thanou, 2011; Ulery et al., 2011).

Nanocomposites based on hydroxyethyl cellulose and silver nanoparticles have been identified as a promissory candidate for skin regeneration or improvement due to morphology, antibacterial properties, degradability, water absorption capacity, nontoxicity, and ease of in situ preparation of silver nanoparticles (Jahangirian et al., 2018). Regarding this last aspect, it is important to point out that, according to multiple investigations, some toxicity can be produced by silver nanoparticles. Thus, although a low hemolytic activity was reported in a wide range of concentrations, for a mean particle size of 10–15 and 20–25 nm (Golubeva et al., 2010), more recent studies have concluded otherwise. Huang et al. (2016) concluded that silver nanoparticles can cause hemolysis and seriously affect the proliferation and viability of lymphocytes at all concentrations investigated by them (10, 20, 40 μg/mL). However, they also reported that these nanoparticles did not show any effect on platelet aggregation and the coagulation process for concentrations lower than 40 μg/mL (Huang et al., 2016). In the same way, size-dependent nanotoxicity of silver nanoparticles has been identified to be greater, the smaller the nanoparticles are. Thus, using particles with sizes of 15, 50, and 100 nm, greater toxicity was identified for nanoparticles of 15 nm, and their action mechanism was associated with the induction of oxidative stress, membrane injury, and subsequently hemolysis (Chen et al., 2015). In addition, other authors have described the nonspecific interaction of silver nanoparticles with biomolecular systems (Palencia et al., 2015; Sanchez et al., 2018). However, studies modifying the stabilizing agent have made it possible to show the influence of the surface composition of nanoparticles on their hemolytic activity (Palencia et al., 2016).

17.2.3 Diagnosis and detection

One of the key components in modern medicine is analytical instrumentation. This includes, in many cases, devices made of inorganic and organic materials for different types of interventions, but also includes instruments from which the collection of information is performed. Thus, diagnostic devices are analytical tools capable of quantifying one or more physiological properties, monitoring vital signs, and detecting the presence of viruses or other microbial agents. Among the diagnostic and detection tools are tests for blood typing, sugar, metabolites, HIV, dengue, and many others (Curvello et al., 2019).

In the field of biomaterials, some examples include: the immobilization of specific proteins with enzymatic activity on cellulose that has allowed their use as a biosensor (Curvello et al., 2019); PHAs of medium chain length combined with graphene used in electrochemical sensing for the detection of pathogens such as *Pseudomonas aeruginosa* (Ward et al., 2020); superparamagnetic nanoparticles coated with carboxymethyl starch that have been designed for cancer diagnosis (Saboktakin et al., 2009); air-filled human albumin microspheres, commercially known as Albunex, used as contrast agents during echocardiography by ultrasound analysis, since they have the ability to enhance the dispersion of the ultrasonic signal of the blood through variations in the acoustic impedance of the gas inside the microspheres compared to that of the blood of the surroundings (Patil, 2003); cellulose surfaces modified by urethanization that have been used as surfaces for the detection of spectroscopic properties of biofilms of pathogenic microorganisms such as *Helicobacter pylori*, which is a bacterium capable of colonizing the stomach and the first carcinogenic biological agent (Palencia et al., 2021) and *Candida albicans*, which is a yeast colonizing different parts of body, and which is considered on many occasions to be an opportunistic pathogen microorganism (Palencia et al., 2020a,b,c).

17.3 Antimicrobial surfaces and platforms

Microbial infections are an important aspect related to the biosecurity of foods, healthcare products, medical devices, and many other systems. As a result of resistance to antimicrobial drugs, it is necessary for the development of ever more efficient alternatives to avoid or minimize the effects of microorganisms. In the case of polymers, different polyelectrolytes have been identified as showing antimicrobial activity, in particular, polymers containing amino groups and quaternary ammonium groups on their structure (Khaira et al., 2013). Chitosan is a typical example as it has on its structure units of N-acetyl glucosamine (Silva et al., 2020a,b). However, they can be chemically modified for the incorporation of changes to increase or decrease one or more properties, including antibacterial activity (Muñoz-Bonilla et al., 2019; Palacio et al., 2019). Other biopolymers investigated for antimicrobial applications are poly(L-lysine), poly γ-glutamic acid, cationic polyurethanes based on sorbitol, N-methyl-D-glucamine, and cationic aminoglycosides (Ulery et al., 2011; Khaira et al., 2013; Bharadwaz and Jayasuriya, 2019; Palencia et al., 2020a,b,c). In addition, biopolymers can enhance their antimicrobial activity by acting as a support or substrate for the inclusion of bioactive substances, for example, plant extracts, drugs, secondary metabolites, nanomaterial, essential oils, antioxidants, probiotics, etc. (Ninan et al., 2016; Córdoba and Palencia, 2017; Muñoz-Bonilla et al., 2019; Silva et al., 2020a,b).

17.4 Conclusions and remarks

Many polymers are used in medicine and biomedicine. These include structural and functional polymers and their applications cover a wide spectrum of systems and contexts. It has been concluded that biopolymers are suitable materials for many areas in medicine and biomedicine, with the most significant being tissue engineering, wound dressing, diagnosis and detection, the development of antimicrobial surfaces, and drug delivery. However, it has been shown also that on many occasions that alone they do not satisfy the necessary requirements of many of the problems in the biomedical field. Therefore, their combined use and the inclusion of other components have been established as promising strategies for new and innovative alternatives. Collagen and chitosan have undoubtedly demonstrated their versatility and aptitude for applications where physiological processes of cell growth are necessary, and new advances based on them are expected to emerge in the future. Finally, it is important to say that the growth of possible and potential research, applications, and technologies will undoubtedly stimulate the biopolymers sector, and more so in a field where they emerge naturally as the main option.

References

Abdo, J.M., Sopko, N.A., Milner, S.M., 2020. The applied anatomy of human skin: a model for regeneration. Wound Med. 28, 100179. https://doi.org/10.1016/j.wndm.2020.1001.

Al-Ghraibah, A.M., Al-Qudah, M., AL-Oqla, F.M., 2020. Medical implementations of biopolymers. In: Al-Oqla, F.M., Sapuan, S.M. (Eds.), Advanced Processing, Properties, and Applications of Starch and Other Bio-Based Polymers, pp. 157–171. https://doi.org/10.1016/b978-0-12-819661-8.00010-x.

Aljohani, W., Ullah, M.W., Zhang, X., Yang, G., 2017. Bioprinting and its applications in tissue engineering and regenerative medicine. Int. J. Biol. Macromol. 107, 261–275. https://doi.org/10.1016/j.ijbiomac.2017.08.171.

Andersen, T., Auk-Emblem, P., Dornish, M., 2015. 3D cell culture in alginate hydrogels. Microarrays 4 (2), 133–161. https://doi.org/10.3390/microarrays4020133.

Axpe, E., Oyen, M., 2016. Applications of alginate-based bioinks in 3D bioprinting. Int. J. Mol. Sci. 17 (12), 1976. https://doi.org/10.3390/ijms17121976.

Bharadwaz, A., Jayasuriya, A.C., 2019. Recent trends in the application of widely used natural and synthetic polymer nanocomposites in bone tissue regeneration. Mater. Sci. Eng. C 110698. https://doi.org/10.1016/j.msec.2020.110698.

Chen, L.Q., Fang, L., Ling, J., Ding, C., Kang, B., Huang, C.Z., 2015. Nanotoxicity of silver nanoparticles to red blood cells: size dependent adsorption, uptake, and hemolytic activity. Chem. Res. Toxicol. 28, 501–509. https://doi.org/10.1021/tx500479m.

Córdoba, A., Palencia, M., 2017. Development of nanostructured crosslinkers with antimicrobial properties for free-radical polymerization. J. Sci. Technol. Appl. 2, 54–64. https://doi.org/10.34294/j.jsta.17.2.14.

Curvello, R., Raghuwanshi, V.S., Garnier, G., 2019. Engineering nanocellulose hydrogels for biomedical applications. Adv. Colloid Interface Sci. 267, 47–61. https://doi.org/10.1016/j.cis.2019.03.002.

Davidenko, N., Cameron, R., Best, S., 2018. Natural biopolymers for biomedical applications. Ref. Modul. Biomed. Sci. 1–14. https://doi.org/10.1016/b978-0-12-801238-3.11026-8.

Espinosa, A., Palencia, S.L., García, A., Palencia, M., 2020. Polymicrobial biofilms: fundamentals, diseases, and treatments – a review. J. Sci. Technol. Appl. 8, 4–33. https://doi.org/10.34294/j.jsta.20.8.54.

Garcés, V., Chate-Galvis, N., 2020. Biosynthesis and biotransformations of biopolymers: starch and poly(hydroxyalkanoates). J. Sci. Technol. Appl. 9, 4–9. https://doi.org/10.34294/j.jsta.20.9.60.

Garcés, V., Palencia, M., Combatt, E., 2017. Development of bacterial inoculums based on biodegradable hydrogels for agricultural applications. J. Sci. Technol. Appl. 2, 13–23. https://doi.org/10.34294/j.jsta.17.2.10.

Golubeva, O.Y., Shamova, O.V., Orlov, D.S., Pazina, T.Y., Boldina, A.S., Kokryakov, V.N., 2010. Study of Antimicrobial and Hemolytic Activities of Silver Nanoparticles Prepared by Chemical Reduction, vol. 36, pp. 628–634. https://doi.org/10.1134/S1087659610050135.

Hassan, M., Dave, K., Chandrawati, R., Dehghani, F., Gomes, V.G., 2019. 3D printing of biopolymer nanocomposites for tissue engineering: nanomaterials, processing and structure-function relation. Eur. Polym. J. 109340. https://doi.org/10.1016/j.eurpolymj.2019.109340.

Houis, S., Engelhardt, E.M., Wurm, F., Gries, T., 2010. Application of polyvinylidene fluoride (PVDF) as a biomaterial in medical textiles. Med. & Healthc. Text. 342–352. https://doi.org/10.1533/9780857090348.342.

Huang, H., Lai, W., Cui, M., Liang, L., Lin, Y., et al., 2016. An evaluation of blood compatibility of silver nanoparticles. Sci. Rep. 6, 25518. https://doi.org/10.1038/srep25518.

Jahangirian, H., Ghasemian Lemraski, E., Rafiee-Moghaddam, R., Webster, T., 2018. A review of using green chemistry methods for biomaterials in tissue engineering. Int. J. Nanomed. 13, 5953–5969. https://doi.org/10.2147/ijn.s163399.

Jithendra, P., Rajam, A.M., Kalaivani, T., Mandal, A.B., Rose, C., 2013. Preparation and characterization of aloe vera blended collagen-chitosan composite scaffold for tissue engineering applications. ACS Appl. Mater. Interfaces 5 (15), 7291–7298. https://doi.org/10.1021/am401637c.

Kean, T., Thanou, M., 2011. Chitin and chitosan: sources, production and medical applications. In: Williams, P.A. (Ed.), Renewable Resources for Functional Polymers and Biomaterials, pp. 292–318. https://doi.org/10.1039/9781849733519-00292.

Khaira, G.K., Ganguli, A., Ghosh, M., 2013. Synthesis and evaluation of antibacterial activity of quaternized biopolymer from *Klebsiella terrigena*. J. Appl. Microbiol. 116, 511–518. https://doi.org/10.1111/jam.12400.

Kunduru, K.R., Basu, A., Domb, A.J., 2016. Biodegradable polymers: medical applications. In: Mank, H.F. (Ed.), Encyclopedia of Polymer Science and Technology, pp. 1–22. https://doi.org/10.1002/0471440264.pst027.pub2.

Lerma, T., Martínez, G., Palencia, M., 2017. Generation of thiolated porous surfaces by interpenetrating polymeric networks: study of their surface properties. J. Sci. Technol. Appl. 3, 56–65. https://doi.org/10.34294/j.jsta.17.3.24.

Markstedt, K., Mantas, A., Tournier, I., Martínez Ávila, H., Hägg, D., Gatenholm, P., 2015. 3D bioprinting human chondrocytes with nanocellulose–alginate bioink for cartilage tissue engineering applications. Biomacromolecules 16 (5), 1489–1496. https://doi.org/10.1021/acs.biomac.5b00188.

Matricardi, P., Di Meo, C., Coviello, T., Hennink, W.E., Alhaique, F., 2013. Interpenetrating polymer networks polysaccharide hydrogels for drug delivery and tissue engineering. Adv. Drug Deliv. Rev. 65 (9), 1172–1187. https://doi.org/10.1016/j.addr.2013.04.002.

Muñoz-Bonilla, A., Echeverria, C., Sonseca, Á., Arrieta, M.P., Fernández-García, M., 2019. Bio-based polymers with antimicrobial properties towards sustainable development. Materials 12 (4), 641. https://doi.org/10.3390/ma12040641.

Ninan, N., Forget, A., Shastri, V.P., Voelcker, N.H., Blencowe, A., 2016. Antibacterial and anti-inflammatory pH-responsive tannic acid-carboxylated agarose composite hydrogels for wound healing. ACS Appl. Mater. Interfaces 8 (42), 28511–28521. https://doi.org/10.1021/acsami.6b10491.

Otálora, A., Lerma, T.A., Palencia, M., 2019. Synthesis and characterization of polyurea-based hydrogels by multicomponent polycondensation of 1,6-hexametilene-diisocyanate, sorbitol and cysteine. J. Sci. Technol. Appl. 7, 5–16. https://doi.org/10.34294/j.jsta.19.7.47.

Palacio, D.A., Urbano, B.F., Palencia, M., Rivas, B.L., 2019. Preparation of alkylated chitosan-based polyelectrolyte hydrogels: the effect of monomer charge on polymerization. Eur. Polym. J. 118, 551–560. https://doi.org/10.1016/j.eurpolymj.2019.06.024.

Palencia, M., Vera, M., Rivas, B.L., 2014. Modification of ultrafiltration membranes via interpenetrating polymer networks for removal of boron from aqueous solution. J. Membr. Sci. 466, 192–199. https://doi.org/10.1016/j.memsci.2014.05.003.

Palencia, S.L., Buelvas, A.M., Palencia, M., 2015. Interaction mechanisms of inorganic nanoparticles and biomolecular systems of microorganisms. Curr. Chem. Biol. 9, 11–23. https://doi.org/10.2174/2212796809666151022201811.

Palencia, M., Berrio, M.E., Meledrez, M.F., 2016. Nanostructured polymer composites with potential applications into the storage of blood and hemoderivates. J. Sci. Technol. Appl. 1, 4–14. https://doi.org/10.34294/j.jsta.16.1.1.

Palencia, M., Córdoba, A., Melendrez, M.F., 2017a. Nanocomposites based on cationic polyelectrolytes and silver nanoparticles: synthesis, characterization, molybdate retention and antimicrobial activity. Arab. J. Chem. 12 (6), 825–834. https://doi.org/10.1016/j.arabjc.2017.11.020.

Palencia, M., Mora, M.A., Palencia, S.L., 2017b. Biodegradable polymer hydrogels based in sorbitol and citric acid for controlled release of bioactive substances from plants (polyphenols). Curr. Chem. Biol. 11 (1) https://doi.org/10.2174/2212796810666161028114432.

Palencia, M., Berrío, M.E., García-Quintero, A., Lerma, T.A., 2018. Cellulose-based nanocomposites for pharmaceutical and biomedical applications. In: Kumar, T., Ghosh, B. (Eds.), Polysaccharide Based Nano-Biocarrier in Drug Delivery, pp. 39–61. https://doi.org/10.1201/9780429449437-3.

Palencia, M., Lerma, T.A., Arrieta, Á.A., 2020a. Antibacterial and non-hemolytic cationic polyurethanes with N-carboxymethyl-N,N,N-triethylammonium groups for bacteremia-control in biomedical-using materials. Mater. Today Commun. 100708. https://doi.org/10.1016/j.mtcomm.2019.100708.

Palencia, M., Mora, M.A., Lerma, T.A., Afanasjeva, N., Hisaza, J.H., 2020b. Reinforcing agents based on cellulose fibers modified by insertion of end-alkyl groups obtained from pyrolytic bio-oil of sugarcane bagasse. Polym. Bull. 17, 5711–5724. https://doi.org/10.1007/s00289-019-03050-0.

Palencia, S.L., García, A., Palencia, M., 2020c. Vibrational spectrum characterization of outer surface of *Helicobacter pylori* biofilms by functionally-enhanced derivative spectroscopy (FEDS). J. Chil. Chem. Soc. 65 https://doi.org/10.4067/S0717-97072020000405015.

Palencia, S.L., García, A., Palencia, M., 2021. Mid-infrared vibrational spectrum characterization of outer surface of Candida Albicans by functionally-enhanced derivative spectroscopy (FEDS). J. Appl. Spectrosc. 81 (1), 166–180. https://doi.org/10.1007/s10812-021-01155-x.

Park, S.B., Lih, E., Park, K.S., Joung, Y.K., Han, D.K., 2017. Biopolymer-based functional composites for medical applications. Prog. Polym. Sci. 68, 77–105. https://doi.org/10.1016/j.progpolymsci.2016.12.003.

Patil, G.V., 2003. Biopolymer albumin for diagnosis and in drug delivery. Drug Dev. Res. 58 (3), 219–247. https://doi.org/10.1002/ddr.10157.

Ramanathan, G., Thyagarajan, S., Sivagnanam, U.T., 2018. Accelerated wound healing and its promoting effects of biomimetic collagen matrices with siderophore loaded gelatin microspheres in tissue engineering. Mater. Sci. Eng. C 93, 455–464. https://doi.org/10.1016/j.msec.2018.08.026.

Rebelo, R., Fernandes, M., Fangueiro, R., 2017. Biopolymers in medical implants: a brief review. Procedia Eng. 200, 236–243. https://doi.org/10.1016/j.proeng.2017.07.034.

Rijal, G., Li, W., 2017. A versatile 3D tissue matrix scaffold system for tumor modeling and drug screening. Sci. Adv. 3, e1700764. https://doi.org/10.1126/sciadv.1700764.

Saboktakin, M.R., Maharramov, A., Ramazanov, M.A., 2009. Synthesis and characterization of superparamagnetic nanoparticles coated with carboxymethyl starch (CMS) for magnetic resonance imaging technique. Carbohydr. Polym. 78 (2), 292–295. https://doi.org/10.1016/j.carbpol.2009.03.042.

Sanchez, Y., García-Quintero, A., Palencia, M., 2018. Determination and distribution of size of inorganic particles by spectral deconvolution of surface plasmon resonance. J. Sci. Technol. Appl. 5, 45–54. https://doi.org/10.34294/j.jsta.18.5.34.

Sen, C.K., Gordillo, G.M., Roy, S., Kisner, R., Lambert, L., et al., 2009. Human skin wounds: a major and snowballing threat to public health and the economy. Wound Repair Regen. 17, 763–771. https://doi.org/10.1111/j.1524-475X.2009.00543.x.

Silva, A., Contreras, M.M., Nascimiento, C.R., Fereira da Costa, M., 2020a. Kinetics of thermal degradation and lifetime study of poly(vinylidene fluoride) (PVDF) subjected to bioethanol fuel accelerated aging. Helyon 6, e04573. https://doi.org/10.1016/j.heliyon.2020.e04573.

Silva, S.S., Rodrigues, L.C., Fernandes, E.M., Reis, R.L., 2020b. Fundamentals on biopolymers and global demand. In: Agostini De Moraes, M., Ferreira Da Silva, C., Silveria Vieira, R. (Eds.), Biopolymer Membranes and Films: Health. Food, Environment, and Energy Applications, pp. 3–34. https://doi.org/10.1016/b978-0-12-818134-8.00001-8.

Tang, J., Saito, T., 2015. Biocompatibility of novel type I collagen purified from tilapia fish scale: an in vitro comparative study. BioMed Res. Int. 2015, 1–8. https://doi.org/10.1155/2015/139476.

Ulery, B.D., Nair, L.S., Laurencin, C.T., 2011. Biomedical applications of biodegradable polymers. J. Polym. Sci. B Polym. Phys. 49 (12), 832–864. https://doi.org/10.1002/polb.22259.

Velema, J., Kaplan, D., 2006. Biopolymer-based biomaterials as scaffolds for tissue engineering. In: Lee, K., Kaplan, D. (Eds.), Tissue Engineering I. Advances in Biochemical Engineering/Biotechnology, 102. Springer, pp. 187–238.

Ward, A.C., Dubey, P., Basnett, P., Lika, G., Newman, G., Corrigan, D.K., Russell, C., Kim, J., Chakrabarty, S., Connolly, P., Roy, I., 2020. Toward a closed loop, integrated biocompatible biopolymer wound dressing patch for detection and prevention of chronic wound infections. Front. Bioeng. & Biotechnol. 8, 1039. https://doi.org/10.3389/fbioe.2020.01039.

Yu, J., Sun, L., Ma, C., Qiao, Y., Yao, H., 2016. Thermal degradation of PVC: a review. Waste Manag. 48, 300–314. https://doi.org/10.1016/j.wasman.2015.11.041.

Zomer, H.D., Treting, A.G., 2018. Skin wound healing in humans and mice: challenges in translational research. J. Dermatol. Sci. 90, 3–12. https://doi.org/10.1016/j.jdermsci.2017.12.009.

Chapter 18

Functional and eco-friendly polymers in pharmaceutical applications

18.1 Pharmaceutical industry: an environmental look

The pharmaceutical industry has become one of the most developed sectors in recent times. In environmental terms, this industry is generally considered an eco-friendly sector due to its small environmental impact, the implementation of good manufacturing practices, and the relatively low amount of waste and emissions (Taylor, 2010). In a broader sense, that is, beyond the manufacturing process, the products of this industry, their use, and the consequent release of their metabolic by-products, as well as those excess amounts nonassimilated by the body, have triggered what is currently known as emerging pollutants. This "emerging" denomination is related to the growing and reasonable interest in guaranteeing safety at all levels, which is directly connected with a greater awareness that, for decades, a constant release of pharmaceutical waste has been carried out, which by their nature has an inherent bioactivity, and therefore, doubts about its effect are an imminent risk that could have enormous negative repercussions in worst case scenarios (more details are given in Chapter 21) (Clark et al., 2010; Sauve and Desrosiers, 2014). In the same way, it is evident that the ecological vision of the pharmaceutical industry is based on a traditional denomination instead of an updated vision under the concepts of green chemistry and integral sustainability (that is, social, environmental, and economic). Approximately 90 percent of the organic chemicals used in the industry are derived from the oil industry, and so it is desirable that these inputs can be obtained from sustainable sources, for example, biomass, while not forgetting that renewable does not necessarily imply sustainable (Clark et al., 2010). Sustainable implies a vision of the complete life cycle during the evaluation of a supply chain, from the origin of the raw materials, including the process, its use and the end of the useful life of the product, but also including a complete material balance and energy efficiency. For example, often as researchers we wonder about the effects of a particular drug, or the by-products derived from it during its manufacture or use. However, we usually

Eco-friendly Functional Polymers
https://doi.org/10.1016/B978-0-12-821842-6.00020-8
Copyright © 2021 Elsevier Inc. All rights reserved.

forget to take into account the final disposition of the packages of that same medicine, which are one-use materials and can be made of multiple materials, including paper, cardboard, glass, plastic, metal, rubber, etc. (Sumpter, 2010; Neffe et al., 2013; Toma and Crişan, 2018).

For several decades, the concept "green pharmacy" has been introduced into the pharmaceutical industry. It is not strictly a new concept as it is nothing more than the conceptual extrapolation of green chemistry applied to the specific field of the pharmaceutical industry. Consequently, it can be defined as a classification of those products and processes that reduce or eliminate the use and generation of dangerous substances, under the premises of prevention, conservation, and protection of the environment (Kümmerer, 2010; Neffe et al., 2013; Toma and Crişan, 2018). In this sense, one of the most promising alternatives to advance the generation of a more sustainable pharmaceutical industry is the use of biopolymers in the manufacturing processes, because these are associated with properties such as biocompatibility, biodegradability, and low residuality, and they can be obtained from renewable sources. However, it must be borne in mind that the use of biopolymers for the administration of drugs, in most cases, if not always, requires adaptation at a structural level, through synthetic strategies, which allow the correct and precise adaptation of the material properties (Neffe et al., 2013). Among these adaptations are, for example, the improvement of cell adhesion, biocompatibility, or the generation of specific responses to physiological stimuli, which is possible through the inclusion or modification of the polymer side chains, cross-linking reactions, insertion or elimination of functional groups, manufacture of mixtures, composites, or nanocomposites, copolymerizations (i.e., block formation, chain end reactions, graft formation), etc. (Pal et al., 2013; Jacob et al., 2018).

18.2 Biopolymers in the pharmaceutical industry

As previously indicated, biopolymers are ideal materials for medical and biomedical applications, including their use as materials for diagnostics, production of devices, and regenerative medicine, but also, they have an important role in the development of new drugs. Biopolymers can perform passive functions, and therefore they act as biopolymer excipients or as transport vehicles of the active substances, however, they can also perform active functions such as controlling the release of drugs or locating the drug at the site of interest. At present, one of the applications of biopolymers in the pharmaceutical industry is directed toward improving the efficiency of commercial drugs as it is more economically beneficial to improve the efficiency of a drug with a proven effect than advances in the development of new drugs. For example, polymeric coatings based on polysaccharides, acting as modulators of the release of the active ingredient, minimizing its losses and controlling the dosage, have been developed (Pal et al., 2013; Das and Pal, 2015; de La Torre et al., 2017; Chen et al., 2019a,b,c,d).

Among the characteristics that a biopolymer must have for its application in the controlled release of drugs, in addition to those mentioned above, are that it must interact favorably with the active agent, releasing only in the conditions appropriate under which it must fulfill its function, in addition, it is ideal for the biopolymer to act specifically with the tissues of interest, so that the active agent can be efficiently released in the vicinity of the target site (Pal et al., 2013; Das and Pal, 2015; de La Torre et al., 2017).

18.2.1 Biopolymer-based drug-delivery systems and delivery pathways

Drug-delivery systems aim to transport the active agent to the target site, in the required concentration for the required time, and consequently, improve the pharmacokinetics of a drug. For this, an adequate interaction between the drug and the delivery system is necessary, which should reduce toxicity during treatment, increase the therapeutic index of the drug (i.e., the ratio between the amount necessary to produce the treatment and the amount at which toxicity occurs), and decrease the frequency in the implementation of dosing procedures, which in some cases can be expensive, invasive, and unpleasant. In addition, they must be biocompatible, nontoxic, biodegradable, and assimilable or easily excretable (Anirudhan and Rejeena, 2015).

Biopolymer-based drug-delivery systems can be microspheres, liposomes, micelles, emulsions, hydrogels, nanoparticles, films, etc. (Mahdavinia et al., 2017; Palencia et al., 2017; Jacob et al., 2018). In addition, the advantage of using polymers is that their functional groups can be used as synthetic modification platforms for the design and improvement of interactions with organs, tissues, cells, biomolecules, biological fluids, etc. (Park et al., 2017). Biopolymer properties can be modified not only through chemical reactions but also by associations with other species. For example, biopolymer-based drug-delivery systems have been developed using dextran, carboxymethyl cellulose, chitosan-carboxymethyl cellulose, and peptin in conjunction with silver nanoparticles and plant extracts (e.g., *Anacardium occidentale*, *Acacia nilotica*, and *Punica granatum*) (Jacob et al., 2018; Papagiannopoulos and Pipas, 2019). In addition, different synthetic strategies have been proposed to carry out these modifications with nanostructured systems, from physical inclusion in the polymer phase to using the nanoparticles like a cross-linking agent (Pal et al., 2013; Palencia et al., 2016; Córdoba and Palencia, 2017). In particular, the use of cellulose is not associated with its original state but with derivatives like nanocellulose (Palencia et al., 2018; Gopinath et al., 2018). For example, alginate-nanocellulose nanocomposites have been identified as an excellent delivery system for ampicillin (Yin et al., 2018; Palencia et al., 2018). Hydrogels based on chitosan and hydroxyl-ethyl cellulose have been combined with vegetable polyhydroxylated oils acting as a filler and cross-linking agent (Vashist et al., 2012). Another strategy for the making of a bioactive surface or drug-delivery surface is the building of interpenetrating polymer networks (Palencia et al., 2014; Lerma et al., 2017).

In particular, the biopolymeric nature offers several advantages compared to synthetic biopolymers, such as biocompatibility, biodegradability, and in many cases low or nonexistent immunogenicity. Some examples of these biopolymers include fibrins, chitosan, gelatin, PHAs, albumin, starch, cellulose, and nanocellulose, peptides, among others (see Fig. 18.1) (Kean and Thanou, 2011; Jacob et al., 2018). Some biopolymers and bioactive compounds are listed in Table 18.1.

On the other hand, a relevant aspect in the design of biopolymer-based drug-delivery systems is the release route to be used, which can be ocular, transdermal, oral, parenteral, etc. (see Fig. 18.1). An important factor for determination of the efficiency of drug carriers is to establish how the interaction between the carrier and physiological components in the different administration pathways takes place (Pal et al., 2013; Chen et al., 2019a,b,c,d).

The oral pathway is the usual and most convenient route in terms of ease, with the action time required for its activity being between 30 min and 2 h. However, this route implies passage through several varied physiological environments. In this way, the drug must enter the gastrointestinal tract and experience the strongly acidic environment characteristic of the stomach, where it is exposed to chemical degradation, followed by subsequent passage to the intestine where it must possess adequate intestinal motility to reach the mucosal surface and the specific sites for absorption, before permeating through the mucosa and reaching the liver. As a consequence, biopolymers should have adequate mucoadhesive properties like chitosan and xanthan gum. If the drug is used to act in the stomach, the gastric residence time should be considered. For example, when in contact with gastric fluid, collagen sponges can increase their size (Pal et al., 2013; Park et al., 2017). In addition, alginate, cellulose nanofibers, and κ-carrageenan oligosaccharide nanoparticles, and gelatin microparticles, are other biopolymers that frequently use this administration route (Bobde and Ghosh, 2018; Johnson et al., 2020).

The sublingual route is an attractive method compared with intravenous administration. This alternative is based on the high permeability existing in the oral mucosa and high accessibility to the blood due to high blood irrigation (Kean and Thanou, 2011; Das and Pal, 2015; López-Cebral et al., 2018). However, the nasal epithelium is a highly permeable monolayer and also is a tissue with high blood irrigation, and as a consequence, nasal administration is another interesting alternative (Chaturvedi et al., 2011). Several effects are directly associated with this administration pathway, such as nasal congestion, infections, acute respiratory distress, asthma, allergic rhinitis, etc. Soluble cellulose derivatives have been used for this route, including hydroxypropyl methyl-, hydroxypropyl-, methyl-, and carboxymethyl cellulose (Chaturvedi et al., 2011; Kean and Thanou, 2011; Nagpal et al., 2019; Nayak et al., 2019a,b).

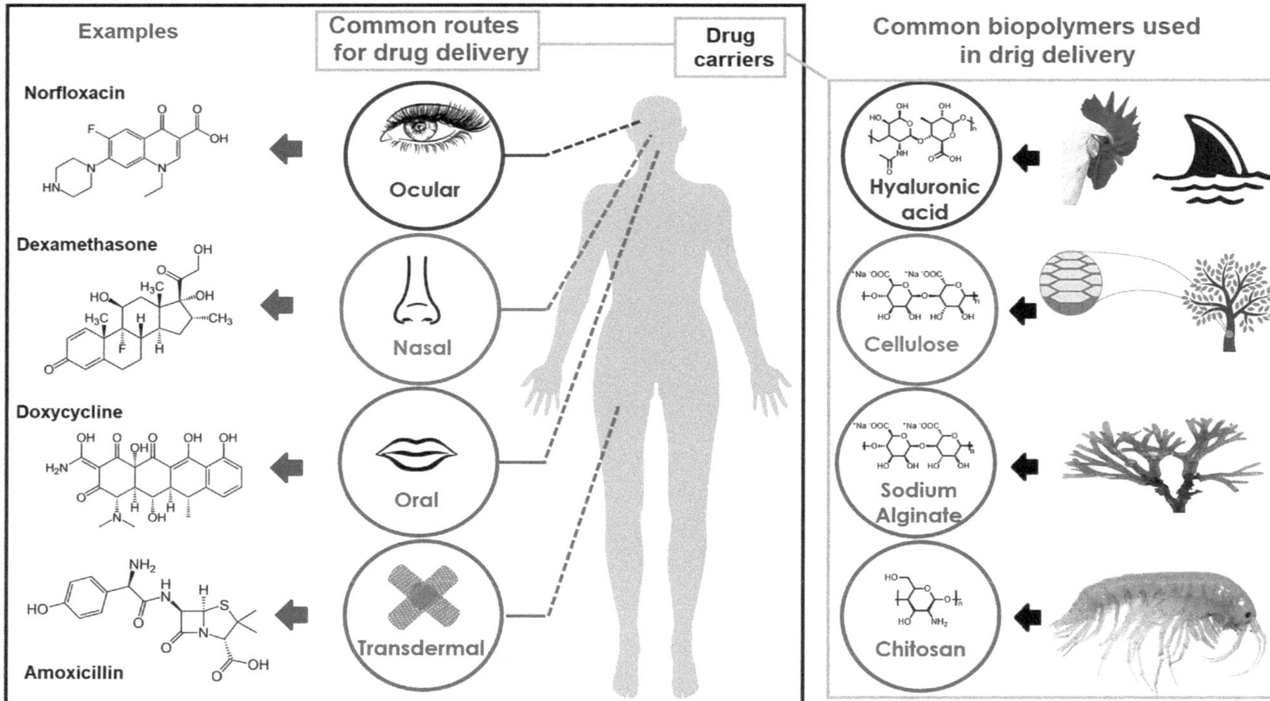

FIGURE 18.1 Some of the common drug-delivery routes and the most representative biopolymers used for this purpose and exemplification of the biological origin.

TABLE 18.1 Examples of the eco-polymers in drug delivery and some of the drugs released.

Type of basic biopolymer	Released bioactive compounds	References
Cellulose	Insulin, ciprofloxacin, lidocaine, ibuprofen	Gao et al. (2013), YerriSwamy et al. (2010), Trovatti et al. (2012), Palencia et al. (2018)
Chitosan	Carbamazepine, levothyroxine, atorvastatin	Gavini et al. (2006), Rostami et al. (2013), Anwar et al. (2011), Kean and Thanou (2011)
Collagen	Glucocorticoids, gentamicin, chondrocyte	Jacob et al. (2018)
Alginate	Metformin, carvedilol, metoprolol, omeprazole	Nayak et al. (2011), Patil et al. (2011), Rajinikanth et al. (2003), Del Gaudio et al. (2015), Bobde and Ghosh (2018)
Silk	Bovine serum albumin, curcumin, rhodamine B, heparin	Jacob et al. (2018)
Hyaluronic acid	Sorafenib, cisplatin, dexamethasone	Paşcalău et al. (2020), Parkkola et al. (2014), Na et al. (2007)

Ocular drug delivery is a very different administration route to the oral, sublingual, and nasal routes because the cornea is a non-vascularized tissue acting as a barrier, and the eyes are hydrated, cleaned, and lubricated continuously by the lachrymal film. Usually, the release systems are aqueous systems containing drugs that are dissolved or dispersed. However, the improvement of the bioavailability of drugs have been significantly enhanced using drug-loaded collagen (Pal et al., 2013; Maiti and Jana, 2017). Biopolymer-based colloids and hydrogels have been used also for drug administration. The risk of tissue damage is very high when solid carriers are used. In particular, for use in ophthalmic delivery of drugs, biopolymers should transport a large amount of active agent in the smallest possible volume, not induce changes in pH or enzymatic degradation, and in addition, be optically transparent within the vitreous humor to avoid any interference with vision (Jemni-Damer et al., 2020). The most used for ocular applications is hyaluronic acid as it is a natural component of the eye (Maiti and Jana, 2017; Nayak et al., 2019a,b). Other biopolymers used include chitosan-based gels, alginate with residues of hyaluronic acid, albumin, and gellan gum (i.e., water-soluble polysaccharide obtained from microbiological fermentation of glucose) (Kean and Thanou, 2011; Pal et al., 2013; Nagpal et al., 2019;

Nayak et al., 2019a,b; Chowhan and Giri, 2020). Biodegradable copolymers like poly(lactic-co-glycolic acid) have been used as solid implants for controlled release of drugs, for example, dexamethasone, however, its use is very limited as it can increase the intraocular pressure and result in cataract formation (Jemni-Damer et al., 2020).

The transdermal route is used with relatively high frequency for systemic drug delivery since it is a noninvasive method and allows a prolonged therapeutic window. However, skin is formed by epithelial tissue acting as an impermeable barrier which can limit drug administration (Badwaik, 2018). Three stages are required for adequate drug transdermal administration, including drug diffusion, drug permeation, and exhibition of the desired pharmacological response. Chitosan, polysaccharide nanoparticles, and nanofiber resins based on chitosan and poly(ethylene glycol) have been widely employed and studied for drug administration using the transdermal pathway due to its biodegradability, bioadhesion, and permeability across the skin (Galipoğlu et al., 2014; Atta et al., 2015; Badwaik, 2018; Seeli and Prabaharan, 2019).

18.2.2 Stimuli-responsive biopolymers

One of the most interesting and promising applications of biopolymers is the controlled release of drugs. A biopolymer can be used for this purpose when it is able to retain the drug under certain conditions and release it under others. Whereas the retention process is mediated by physical and chemical intermolecular interactions under controlled conditions (e.g., hydrogen bonds, electrostatic interactions, coordination bonds, occlusion within the polymer phase, etc.), the release process is carried out at physiological conditions and, therefore, with few exceptions, it is not possible to carry out a control. Among the release mechanisms are dissolution, diffusion, degradation, etc. (Lee and Yeo, 2015; Choiri and Ainurofiq, 2018). However, these mechanisms are general mechanisms and can even be referred to as "classical" On the other hand, other mechanisms are directly connected to specific stimuli that produce a change in the interaction and trigger the release (Pal et al., 2013). Since the release is controlled by external stimuli, biopolymers that operate under this type of mechanism are called "smart materials" The stimuli usually used are temperature, pH, ionic strength, radiation, a magnetic field, redox potential, ultrasonic, enzymatic activity, and molecular recognition (Pal et al., 2013; de La Torre et al., 2017; Palencia et al., 2021). In particular, polymers sensitive to stimuli acquire great relevance when the stimuli are exclusive to the disease, allowing a specific response to the pathological factors associated with the effect of interest (Chen et al., 2019a,b,c,d).

A physiological approach allows three types of stimulus to be defined: triggered at the organ level, triggered by pathological conditions, and triggered by factors specific to the cell compartment. However, from a chemical approach, a better understanding of the response to the stimulus with respect to the polymer structure is possible. For example, the normal pH of tissue cells is 7.4, whereas the extracellular pH of tumor tissue is about 5.0−6.0. As a consequence, there is a small difference in the pHs of these systems. Thus, if a biopolymer contains functional groups able to form acid−base chemical equilibrium in an aqueous environment, such as to −COOH or−NH of hyaluronic acid, the charge density of polymer changes as a function of pH, and consequently, its intra- and intermolecular interaction, but also, any interaction with any charged substance, is affected by this pH variation. An advantage of pH as a stimulus is that this is a reversible interaction which can be exploited to modulate the release. Similar changes to pH are observed for lysosomes, Golgi apparatus, and cytosol. Stronger variations are observed in the epithelial tissue of the stomach where it is possible to carry out acid hydrolysis of polyesters as poly(alkanoates) (PHAs) and other biopolymers. Other examples are gelatin, anionic and cationic poly(amino acids), chitosan, and alginate, which can be used in the making of biocompatible drug-delivery vesicles with different surface charge density sensitivities to pH (de La Torre et al., 2017; Efthimiadou et al., 2018).

Redox reactions are another usual activation mechanism for drug delivery. This stimulus is possible due to the extracellular and intracellular microenvironments having a large redox potential difference. Thus, redox potential determines the tendency of chemical species to be reduced and their stability in a redox environment. One example is the thiol/disulfide exchange reactions mediated by glutathione, which is a tripeptide formed by glutamate-cysteine-glycine, which makes possible the cleavage of disulfide bonds in biopolymeric matrix containing this type of link (de La Torre et al., 2017; Bobde and Ghosh, 2018). Thus, biopolymers like starch can be modified using N,N-bisacryloylcystamine to incorporate redox-sensitive properties (Chen et al., 2019a,b,c,d). Biological stimuli can be induced by the integration of biomolecules with nanoparticles, which, by an affinity mechanism, guide the drug to the target cells. As a part of the biological designs of living organisms, some molecules show special affinity for some active sites of enzymes or other biomolecular systems. Thus, glucose, DNA, ATP, or any biomolecule, in conjunction with biopolymer, due to low toxicity and biocompatibility, can be used for the transportation of drugs by taking advantage of this affinity (de La Torre et al., 2017). For example, chitosan nanoparticles containing glucose oxidase, which is an oxidoreductase that catalyzes the oxidation of glucose to H_2O_2 and D-glucono-δ-lactone is used to control the glucose levels in diabetics (de La Torre et al., 2017). Also, nanoparticles can be directed by external stimuli, for example, the use of a magnetic field. However, for this, nanoparticles are required to possess magnetic properties (Bobde and Ghosh, 2018; Chen et al., 2019a,b,c,d). In this way, among the many examples available in the specialized literature, the most widely used magnetic component is iron oxide. For example, core−shell particles of iron oxide coated with chitosan and carboxymethyl cellulose have been synthesized (Li et al., 2015). Other external stimuli include ultrasound, light, temperature, moisture, microbiological activity, etc. (Pal et al., 2013; Das and Pal, 2015; Garcés et al., 2017; Maiti, 2017; Chen et al., 2019a,b,c,d; Otálora et al., 2019). In addition, systems for drug delivery based on dual or multidirectional response have been developed, such as pH–temperature, pH–redox, pH–magnetic field, temperature–reduction, temperature–magnetic field, temperature–enzyme, temperature–pH–redox, temperature–pH–magnetic, and pH–redox–magnetic, among others (Cheng et al., 2013). For example, the release of flurbiprofen, which is a nonsteroidal antiinflammatory drug, from mesoporous silica nanoparticles shows a dual response to chiral micro-environments and pH (Wu et al., 2021).

18.3 Conclusions and remarks

Biopolymers are promissory materials for pharmaceutical applications. They allow the design and modulation of properties to achieve adequate drug transportation and release. These modification and adaptation abilities allow their use in several drug administration routes. In addition, nanostructured and smart systems greatly broaden the spectrum of possibilities for substantial efficiency improvements in key areas such as delivery, transport, therapeutic window, and drug directionality. In the coming years, research will be directed toward the development of new biopolymers, new structural configurations, and new combinations of systems with special properties. Therefore, it is expected that advances in this field will reach high levels of specialization and interdisciplinary work, in conjunction with the exploitation of nano-technological characteristics and the multifunctionality of treatments. In conclusion, biopolymers, nanotechnology, and multifunctionality are the key points for the building of a new outlook for the development and creation of new and novel pharmacological treatments.

References

Anirudhan, T.S., Rejeena, S.R., 2015. Biopolymer-based stimuli-sensitive functionalized graft copolymers as controlled drug delivery systems. In: Thakur, V.K., Singh Singha, A. (Eds.), Surface Modification of Biopolymers, pp. 291–334. https://doi.org/10.1002/9781119044901.ch12.

Anwar, M., Warsi, M.H., Mallick, N., Akhter, S., Gahoi, S., Jain, G.K., Talegaonkar, S., Ahmad, F.J., Khar, R.K., 2011. Enhanced bioavailability of nano-sized chitosan–atorvastatin conjugate after oral administration to rats. Eur. J. Pharmaceut. Sci. 44 (3), 241–249. https://doi.org/10.1016/j.ejps.2011.08.001.

Atta, S., Khaliq, S., Islam, A., Javeria, I., Jamil, T., Athar, M.M., Shafiq, M.I., Ghaffar, A., 2015. Injectable biopolymer based hydrogels for drug delivery applications. Int. J. Biol. Macromol. 80, 240–245. https://doi.org/10.1016/j.ijbiomac.2015.06.044.

Badwaik, H.R., 2018. Polysaccharide-based nanocarriers for drug delivery into and through the skin. In: Kumar, T., Ghosh, B. (Eds.), Polysaccharide Based Nano-Biocarrier in Drug Delivery, pp. 227–250. https://doi.org/10.1201/9780429449437-12.

Bobde, Y.S., Ghosh, B., 2018. Environmental responsive polysaccharide nanocarriers for controlled drug delivery. In: Kumar, T., Ghosh, B. (Eds.), Polysaccharide Based Nano-Biocarrier in Drug Delivery, pp. 169–190. https://doi.org/10.1201/9780429449437-9.

Chaturvedi, M., Kumar, M., Pathak, K., 2011. A review on mucoadhesive polymer used in nasal drug delivery system. J. Adv. Pharm. Technol. Res. (JAPTR) 2 (4), 215. https://doi.org/10.4103/2231-4040.90876.

Chen, J., Chen, L., Xie, F., Li, X., 2019a. Material nature and physicochemical properties for high performance of carriers. In: Chen, J., Chen, L., Xie, F., Li, X. (Eds.), Drug Delivery Applications of Starch Biopolymer Derivatives, pp. 19–27. https://doi.org/10.1007/978-981-13-3657-7_2.

Chen, J., Chen, L., Xie, F., Li, X., 2019b. Physiological and pathological bases for designing high performance drug delivery carriers. In: Chen, J., Chen, L., Xie, F., Li, X. (Eds.), Drug Delivery Applications of Starch Biopolymer Derivatives, pp. 1–17. https://doi.org/10.1007/978-981-13-3657-7_1.

Chen, J., Chen, L., Xie, F., Li, X., 2019c. Starch. In: Chen, J., Chen, L., Xie, F., Li, X. (Eds.), Drug Delivery Applications of Starch Biopolymer Derivatives, pp. 29–40. https://doi.org/10.1007/978-981-13-3657-7_3.

Chen, J., Chen, L., Xie, F., Li, X., 2019d. Starch-based DDSs with stimulus responsiveness. In: Chen, J., Chen, L., Xie, F., Li, X. (Eds.), Drug Delivery Applications of Starch Biopolymer Derivatives, pp. 41–99. https://doi.org/10.1007/978-981-13-3657-7_4.

Choiri, S., Ainurofiq, A., 2018. Understanding the drug release mechanism from a montmorillonite matrix and its binary mixture with a hydrophilic polymer using a compartmental modelling approach. IOP Conf. Ser. Mater. Sci. Eng. 333, 012065. https://doi.org/10.1088/1757-899X/333/1/012065.

Cheng, R., Meng, E., Deng, C., Klok, H.A., Zhong, Z., 2013. Dual and multi-stimuli responsive polymeric nanoparticles for programmed site-specific drug delivery. Biomaterials 34, 3647–3657. https://doi.org/10.1016/j.biomaterials.2013.01.084.

Chowhan, A., Giri, T.K., 2020. Polysaccharide as renewable responsive biopolymer for in situ gel in the delivery of drug through ocular route. Int. J. Biol. Macromol. 150, 559–572. https://doi.org/10.1016/j.ijbiomac.2020.02.097.

Clark, J.H., Breeden, S.W., Summerton, L., 2010. Green(er) pharmacy. In: Kümmerer, K., Hempel, M. (Eds.), Green and Sustainable Pharmacy, pp. 37–59. https://doi.org/10.1007/978-3-642-05199-9_4.

Córdoba, A., Palencia, M., 2017. Development of nanostructured crosslinkers with antimicrobial properties for free-radical polymerization. J. Sci. Technol. Appl. 2, 54–64. https://doi.org/10.34294/j.jsta.17.2.14.

Das, D., Pal, S., 2015. Modified biopolymer-dextrin based crosslinked hydrogels: application in controlled drug delivery. RSC Adv. 5 (32), 25014–25050. https://doi.org/10.1039/c4ra16103c.

De La Torre, L.G., Sipoli, C.C., Oliveira, A.F., Eş, I., Pessoa, A.C.S.N., Vitor, M.T., Vit, F.F., Naves, T.F., 2017. Biopolymers for gene delivery applications. In: Jana, S., Maiti, S. (Eds.), Biopolymer-Based Composites, pp. 289–323. https://doi.org/10.1016/b978-0-08-101914-6.00010-7.

Del Gaudio, P., De Cicco, F., Sansone, F., Aquino, R.P., Adami, R., Ricci, M., Giovagnoli, S., 2015. Alginate beads as a carrier for omeprazole/SBA-15 inclusion compound: a step towards the development of personalized paediatric dosage forms. Carbohydr. Polym. 133, 464–472. https://doi.org/10.1016/j.carbpol.2015.07.064.

Efthimiadou, E.K., Theodosiou, M., Toniolo, G., Abu-Thabit, N.Y., 2018. Stimuli-responsive biopolymer nanocarriers for drug delivery applications. In: Abu-Thabit, N.Y., Hamdy Makhlouf, A.S. (Eds.), Stimuli Responsive Polymeric Nanocarriers for Drug Delivery Applications, vol. 1, pp. 405–432. https://doi.org/10.1016/b978-0-08-101997-9.00019-9.

Galipoğlu, M., Erdal, M.S., Güngör, S., 2014. Biopolymer-based transdermal films of donepezil as an alternative delivery approach in Alzheimer's disease treatment. AAPS Pharm. Sci. Tech. 16 (2), 284–292. https://doi.org/10.1208/s12249-014-0224-6.

Gao, X., Cao, Y., Song, X., Zhang, Z., Zhuang, X., He, C., Chen, X., 2013. Biodegradable, pH-responsive carboxymethyl cellulose/poly(acrylic acid) hydrogels for oral insulin delivery. Macromol. Biosci. 14 (4), 565–575. https://doi.org/10.1002/mabi.201300384.

Garcés, V., Palencia, M., Combatt, E., 2017. Development of bacterial inoculums based on biodegradable hydrogels for agricultural applications. J. Sci. Technol. Appl. 2, 13–23. https://doi.org/10.34294/j.jsta.17.2.10.

Gavini, E., Hegge, A.B., Rassu, G., Sanna, V., Testa, C., Pirisino, G., Karlsen, J., Giunchedi, P., 2006. Nasal administration of carbamazepine using chitosan microspheres: in vitro/in vivo studies. Int. J. Pharm. 307 (1), 9–15. https://doi.org/10.1016/j.ijpharm.2005.09.013.

Gopinath, V., Saravanan, S., Al-Maleki, A.R., Ramesh, M., Vadivelu, J., 2018. A review of natural polysaccharides for drug delivery applications: special focus on cellulose, starch and glycogen. Biomed. Pharmacother. 107, 96–108. https://doi.org/10.1016/j.biopha.2018.07.136.

Jacob, J., Haponiuk, J.T., Thomas, S., Gopi, S., 2018. Biopolymer based nanomaterials in drug delivery systems: a review. Mater. Today Chem. 9, 43–55. https://doi.org/10.1016/j.mtchem.2018.05.002.

Jemni-Damer, N., Guedan-Duran, A., Fuentes-Andion, M., Serrano-Bengoechea, N., Alfageme-Lopez, N., Armada-Maresca, F., Guinea, G.V., Pérez-Rigueiro, J., Rojo, F., Gonzalez-Nieto, D., Kaplan, D.L., Panetsos, F., 2020. Biotechnology and biomaterial-based therapeutic strategies for age-related macular degeneration. Part I: biomaterials-based drug delivery devices. Front. Bioeng. & Biotechnol. 3 (8), 549089. https://doi.org/10.3389/fbioe.2020.549089.

Johnson, A., Kong, F., Miao, S., Lin, H.V., Thomas, S., Huang, Y.C., Kong, Z.L., 2020. Therapeutic effects of antibiotics loaded cellulose nanofiber and κ-carrageenan oligosaccharide composite hydrogels for periodontitis treatment. Sci. Rep. 10, 18037. https://doi.org/10.1038/s41598-020-74845-9.

Kean, T., Thanou, M., 2011. Chitin and chitosan: sources, production and medical applications. In: Williams, P.A. (Ed.), Renewable Resources for Functional Polymers and Biomaterials, pp. 292–318. https://doi.org/10.1039/9781849733519-00292.

Kümmerer, K., 2010. Why green and sustainable pharmacy? In: Kümmerer, K., Hempel, M. (Eds.), Green and Sustainable Pharmacy, pp. 3–10. https://doi.org/10.1007/978-3-642-05199-9_1.

Lee, J.H., Yeo, Y., 2015. Controlled drug release from pharmaceutical nanocarriers. Chem. Eng. Sci. 125, 75–84. https://doi.org/10.1016/j.ces.2014.08.046.

Lerma, T., Martínez, G., Palencia, M., 2017. Generation of thiolated porous surfaces by interpenetrating polymeric networks: study of their surface properties. J. Sci. Technol. Appl. 3, 56–65. https://doi.org/10.34294/j.jsta.17.3.24.

Li, J., Wang, F., Shi, D., Zhang, Y., Shao, Z., 2015. Multifunctional biopolymer nanoparticles for drug delivery and protein immobilization. Ferroelectrics 486 (1), 156–167. https://doi.org/10.1080/00150193.2015.1100907.

López-Cebral, R., Peng, G., Reys, L.L., Silva, S.S., Oliveira, J.M., Chen, J., Silva, T.H., Reis, R.L., 2018. Dual delivery of hydrophilic and hydrophobic drugs from chitosan/diatomaceous earth composite membranes. J. Mater. Sci. Mater. Med. 29 (2). https://doi.org/10.1007/s10856-018-6025-9.

Mahdavinia, G.R., Mosallanezhad, A., Soleymani, M., Sabzi, M., 2017. Magnetic- and pH-responsive κ-carrageenan/chitosan complexes for controlled release of methotrexate anticancer drug. Int. J. Biol. Macromol. 97, 209–217. https://doi.org/10.1016/j.ijbiomac.2017.01.012.

Maiti, S., 2017. Nanometric biopolymer devices for oral delivery of macromolecules with clinical significance. In: Grumezescu, A.M. (Ed.), Multifunctional Systems for Combined Delivery Biosensing and Diagnostics, pp. 109–138. https://doi.org/10.1016/b978-0-323-52725-5.00006-x.

Maiti, S., Jana, S., 2017. Biocomposites in ocular drug delivery. In: Maiti, S., Jana, S. (Eds.), Biopolymer-Based Composites, pp. 139–168. https://doi.org/10.1016/b978-0-08-101914-6.00006-5.

Na, K., Kim, S., Woo, D.G., Sun, B.K., Yang, H.N., Chung, H.M., Park, K.H., 2007. Combination material delivery of dexamethasone and growth factor in hydrogel blended with hyaluronic acid constructs for neocartilage formation. J. Biomed. Mater. Res. 83A (3), 779–786. https://doi.org/10.1002/jbm.a.31374.

Nagpal, S., Dubey, S.K., Rapalli, V.M., Singhvi, G., 2019. Pharmaceutical applications of gellan gum. In: Nayak, A.K., Pal, D., Hasnain, M.S. (Eds.), Natural Polymers for Pharmaceutical Applications, pp. 87–110. https://doi.org/10.1201/9780429328299-4.

Nayak, A.K., Ansari, M.T., Pal, D., Hasnain, M.S., 2019a. Hyaluronic acid (hyaluronan): pharmaceutical applications. In: Nayak, A.K., Pal, D., Hasnain, M.S. (Eds.), Natural Polymers for Pharmaceutical Applications, pp. 1–32. https://doi.org/10.1201/9780429328350-1.

Nayak, A., Jain, S.K., Pandey, R.S., 2011. Controlling Release of metformin HCl through incorporation into stomach specific floating alginate beads. Mol. Pharm. 8 (6), 2273–2281. https://doi.org/10.1021/mp2001395.

Nayak, A.K., Nanda, S.K., Yi, D.K., Hasnain, M.S., Pal, D., 2019b. Pharmaceutical applications of tamarind gum. In: Nayak, A.K., Pal, D., Hasnain, M.S. (Eds.), Natural Polymers for Pharmaceutical Applications, pp. 1–20. https://doi.org/10.1201/9780429328251-1.

Neffe, A.T., Wischke, C., Racheva, M., Lendlein, A., 2013. Progress in biopolymer-based biomaterials and their application in controlled drug delivery. Expet Rev. Med. Dev. 10 (6), 813–833. https://doi.org/10.1586/17434440.2013.839209.

Otálora, A., Lerma, T.A., Palencia, M., 2019. Synthesis and characterization of polyurea-based hydrogels by multicomponent polycondensation of 1,6-hexametilene-diisocyanate, sorbitol and cysteine. J. Sci. Technol. Appl. 7, 5–16. https://doi.org/10.34294/j.jsta.19.7.47.

Pal, K., Paulson, A.T., Rousseau, D., 2013. Biopolymers in controlled-release delivery systems. In: Ebnesajjad, S. (Ed.), Handbook of Biopolymers and Biodegradable Plastics, pp. 329–363. https://doi.org/10.1016/b978-1-4557-2834-3.00014-8.

Palencia, M., Berrío, M.E., García-Quintero, A., Lerma, T.A., 2018. Cellulose-based nanocomposites for pharmaceutical and biomedical applications. In: Kumar, T., Ghosh, B. (Eds.), Polysaccharide Based Nano-Biocarrier in Drug Delivery, pp. 39–61. https://doi.org/10.1201/9780429449437-3.

Palencia, M., Berrío, M.E., Melendrez, M.F., 2016. Nanostructured polymer composites with potential applications into the storage of blood and hemoderivates. J. Sci. Technol. Appl. 1, 4–14. https://doi.org/10.34294/j.jsta.16.1.1.

Palencia, M., Mora, M.A., Lerma, T.A., 2021. Environment-friendly stimulus sensitive polyurethanes based on cationic aminoglycosides for the controlled release of phytohormones. Smart & Sustain. Built Environ 10, 1–17. https://doi.org/10.1108/SASBE-09-2019-0126.

Palencia, M., Mora, M.A., Palencia, S.L., 2017. Biodegradable polymer hydrogels based in sorbitol and citric acid for controlled release of bioactive substances from plants (polyphenols). Curr. Chem. Biol. 11 (1), 36–43. https://doi.org/10.2174/2212796810666161028114432.

Palencia, M., Vera, M., Rivas, B.L., 2014. Modification of ultrafiltration membranes via interpenetrating polymer networks for removal of boron from aqueous solution. J. Membr. Sci. 466, 192–199. https://doi.org/10.1016/j.memsci.2014.05.003.

Papagiannopoulos, A., Pipas, S., 2019. Pharmaceutical applications of carrageenan. In: Nayak, A.K., Pal, D., Hasnain, M.S. (Eds.), Natural Polymers for Pharmaceutical Applications, pp. 111–140. https://doi.org/10.1201/9780429328299-5.

Park, S.B., Lih, E., Park, K.S., Joung, Y.K., Han, D.K., 2017. Biopolymer-based functional composites for medical applications. Prog. Polym. Sci. 68, 77–105. https://doi.org/10.1016/j.progpolymsci.2016.12.003.

Parkkola, H., Amieva, L.S., Vivero, L., Miñana, R., Sendra, J., 2014. Controlled release of cisplatin using hyaluronic oligosaccharides-coated gold nanoparticles as an efficient delivery system applied to the treatment of pancreatic tumors. Eur. J. Canc. 50, 87–88. https://doi.org/10.1016/s0959-8049(14)70389-2.

Paşcalău, V., Tertis, M., Pall, E., Suciu, M., Marinca, T., Pustan, M., Merie, V., Rus, I., Moldovan, C., Topala, T., Pavel, C., Popa, C., 2020. Bovine serum albumin gel/polyelectrolyte complex of hyaluronic acid and chitosan based microcarriers for Sorafenib targeted delivery. J. Appl. Polym. Sci. 137 (34), 49002. https://doi.org/10.1002/app.49002.

Patil, S.B., Kaul, A., Babbar, A., Mathur, R., Mishra, A., Sawant, K.K., 2011. In vivo evaluation of alginate microspheres of carvedilol for nasal delivery. J. Biomed. Mater. Res. B Appl. Biomater. 100B (1), 249–255. https://doi.org/10.1002/jbm.b.31947.

Rajinikanth, P.S., Sankar, C., Mishra, B., 2003. Sodium alginate microspheres of metoprolol tartrate for intranasal systemic delivery: development and evaluation. Drug Deliv. 10 (1), 21–28. https://doi.org/10.1080/713840323.

Rostami, E., Kashanian, S., Askari, M., 2013. The effect of ultrasound wave on levothyroxine release from chitosan nanoparticles. Adv. Mater. Res. 829, 284–288. https://doi.org/10.4028/www.scientific.net/amr.829.284.

Sauve, S., Desrosiers, M., 2014. A review of what is an emerging contaminant. Chem. Cent. J. 8–15. https://doi.org/10.1186/1752-153X-8-15.

Seeli, D.S., Prabaharan, M., 2019. Guar gum and its derivatives: pharmaceutical applications. In: Nayak, A.K., Pal, D., Hasnain, M.S. (Eds.), Natural Polymers for Pharmaceutical Applications, pp. 115–138. https://doi.org/10.1201/9780429328251-5.

Sumpter, J.P., 2010. Pharmaceuticals in the environment: moving from a problem to a solution. In: Kümmerer, K., Hempel, M. (Eds.), Green and Sustainable Pharmacy, pp. 11–22. https://doi.org/10.1007/978-3-642-05199-9_2.

Taylor, D.E., 2010. Ecopharmacostewardship - a pharmaceutical industry perspective. In: Kümmerer, K., Hempel, M. (Eds.), Green and Sustainable Pharmacy, pp. 105–126. https://doi.org/10.1007/978-3-642-05199-9_7.

Toma, A., Crişan, O., 2018. Green pharmacy - a narrative review. Clujul Med. 91 (4), 391–398. https://doi.org/10.15386/cjmed-1129.

Trovatti, E., Freire, C.S.R., Pinto, P.C., Almeida, I.F., Costa, P., Silvestre, A.J.D., Neto, C.P., Rosado, C., 2012. Bacterial cellulose membranes applied in topical and transdermal delivery of lidocaine hydrochloride and ibuprofen: in vitro diffusion studies. Int. J. Pharm. 435 (1), 83–87. https://doi.org/10.1016/j.ijpharm.2012.01.002.

Vashist, A., Gupta, Y.K., Ahmad, S., 2012. Interpenetrating biopolymer network based hydrogels for an effective drug delivery system. Carbohydr. Polym. 87 (2), 1433–1439. https://doi.org/10.1016/j.carbpol.2011.09.030.

Wu, L., Gou, K., Guo, X., Guo, Y., Chen, M., Hou, J., Li, S., Li, H., 2021. Dual response to pH and chiral microenvironments for the release of a flurbiprofen-loaded chiral self-assembled mesoporous silica drug delivery system. Colloids Surf. B Biointerfaces 199, 111501. https://doi.org/10.1016/j.colsurfb.2020.111501.

YerriSwamy, B., Venkata Prasad, C., Reedy, C.L.N., Mallikarjuna, B., Rao, K.C., Subha, M.C.S., 2010. Interpenetrating polymer network microspheres of hydroxy propyl methyl cellulose/ poly (vinyl alcohol) for control release of ciprofloxacin hydrochloride. Cellulose 18 (2), 349–357. https://doi.org/10.1007/s10570-010-9475-x.

Yin, Z.C., Wang, Y.L., Wang, K., 2018. A pH-responsive composite hydrogel beads based on agar and alginate for oral drug delivery. J. Drug Deliv. Sci. Technol. 43, 12–18. https://doi.org/10.1016/j.jddst.2017.09.009.

Chapter 19

Functional and eco-friendly polymers in textile applications

19.1 Background to the textile industry and advanced fiber applications

After the food industry, the textile industry is one of the most important to human civilization. This industry is associated with an inherent need that differentiates us from animals and that identifies us as humans. However, in addition to providing clothing, the textile industry impacts industrial and technical applications for specialized sectors such as medicine, military, agriculture, etc., and also domestic applications. In general, as human activities become more specialized, the characteristics and specifications expected of textile materials become increasingly demanding, and consequently, their products also become more specialized (Gruyter, 2018; Madhav et al., 2018; Islam et al., 2020; Luján-Ornelas et al., 2020). In the textile industry, multiple technologies converge such as fibers, materials, the transformation of raw materials into threads, the production and application of dyes, the manufacture of chemical agents for the transformation and conservation of materials, etc. (Madhav et al., 2018; Uddin, 2019). At the present, other technologies as nanotechnology, biotechnology, and electronic technology must be included (Luján-Ornelas et al., 2020; Provin et al., 2020).

Textiles are clearly very diverse materials that, depending on their characteristics, are obtained through processes that can be very varied. However, their generalization is useful when attempting to describe this industry, its environmental impact, and its economic importance. In that way, the first objective is obtaining of the fibers, which can be natural or synthetic, however, this stage includes substages depending on their origin, in the case of natural fibers, substages include the large-scale production of the raw material, the extraction of fiber, its purification, storage, conservation, and transport, etc. In the case of synthetic fibers, chemical precursors must be transformed into the material that later becomes the fiber. After obtaining the fiber, the formation of threads takes place, with the production of the textile, wet processing, and multiple other stages such as sizing, scouring, bleaching, dying, etc. (Madhav et al., 2018; Uddin, 2019). In this chapter the main focus is on the obtaining of

Eco-friendly Functional Polymers
https://doi.org/10.1016/B978-0-12-821842-6.00024-5
Copyright © 2021 Elsevier Inc. All rights reserved.

fibers and their modification, since this is the most important aspect from the point of view of material science and technology.

An example of the evolution of fiber technology in the application sector is sportswear. Thus, cotton and wool are widely used for the manufacture of sportswear, which can even be described as "high-performance clothing" since it is subject to special conditions of use. However, innovations in fibers for this type of clothing have gradually entered the general market. For example, ultra-high-molecular-weight polyethylene is used for the production of the so-called "spectra fiber" which is characterized by being very light, having high mechanical strength, resistance to abrasion, and high durability, exceeding fibers such as nylon, polyester, cellulose, silk, and fiberglass in these characteristics. Some metamorphic rocks, like nephrite and jadeite, known as jade, which is characterized as having a hardness greater than that of steel and a larger heat transfer rate compared with polyester fibers, are used in the production of fibers of jade-polyesters by the insertion of nanometric particles of rocks in the fibers (Yip and Chang, 2020). More examples are shown in Table 19.1.

Some authors have defined several categories for a textile classification based on their applications and functionality: Agrotex (i.e., materials based on fibers used for agricultural applications including greenhouse covers, crop protection, bags for plants and fruits, fertilization, geotextiles, etc.); Buildtex (i.e., materials based on fibers used for building activities including visual protection systems, textiles for sound and thermal insulation, etc.); Clothtex (i.e., materials based on fibers for clothing); Geotex (i.e., materials based on fibers which are used in construction activities like filters, drainage systems,

TABLE 19.1 Examples of natural and synthetic fibers.

Fiber	Description	References
Cotton fiber	Natural fiber conventionally used	Rajasundaram et al. (2014); Shad et al. (2019)
Wool fiber	Natural fiber conventionally used	Fiore et al. (2019); Allafi et al. (2020)
Spectra fiber	Fiber based on ultra-high-molecular-weight polyethylene	Murali and Sankar (2017)
Jade fiber	Fiber based on polyester nanostructured with rocks (jade)	Yip and Chang (2020)
Synthetic insulation fiber	Fibers for thermal insulation for long-term use	Bianco et al. (2017); Yip and Chang (2020)
Hygroscopic synthetic fibers	Polyester fibers designed with channels along the fibers to absorb moisture and ease transpiration	Célino et al. (2012)

soil sealing that are usually resistant, durable, and hydrophobic); Meditex (i.e., materials based on fibers for medical and biomedical uses), Ecotex (i.e., materials based on fibers for their use in environmental protection in applications including prevention of erosion, pollution, cleaning of air, etc.); Protex (i.e., materials used for protection of external factors in the industry or dangerous activities, the use of which is directed to minimize the risk of negative effects of radiation, chemicals, fires, etc.), etc. (Caicedo et al., 2019).

On the other hand, smart clothing has been an interesting and recent innovation. As a consequence, to achieve this goal, the textiles used obviously differ markedly from conventional textiles in terms of the materials and applications (e.g., application in electronics), and therefore, the fiber that constitutes them also can be denominated to be nonconventional fibers or, if preferred, smart fibers. Smart clothing has been directed toward the production of textiles able to collect (Chen et al., 2017; Gong et al., 2019) and store energy (Zhang et al., 2016), and change color and shape (Shi et al., 2019), among other nonconventional properties. A short illustration of functional smart fibers is shown in Table 19.2.

TABLE 19.2 Examples of functional fibers for the making of smart devices in advanced applications.

Fiber type	Application	Description	References
Functional fibers for energy capture	Photoelectric fibers for solar cells	Fibers able to convert solar energy into electric energy	Pu et al. (2017); Shi et al. (2019); Chen et al. (2020)
	Fibers based on triboelectric nanogenerators	Fibers able to convert mechanical energy into electric energy	Pu et al. (2017); Shi et al. (2019); Chen et al. (2020)
	Fibers for thermo-electric devices	Fibers able to recover residual heat and energy transfer systems	Ryan et al. (2017); Chen et al. (2020)
Functional fibers for energy storage	Batteries based on fibers	Fiber used for the making lithium ion batteries with fiber shape	Yang et al. (2018); Li et al. (2019)
	Supercapacitors based on fibers	Fibers used for making wearable electronic devices based on graphene using sandwich-like fibers	Shao et al. (2015); Li et al. (2018a)

Continued

TABLE 19.2 Examples of functional fibers for the making of smart devices in advanced applications.—cont'd

Fiber type	Application	Description	References
Functional fibers for chromatic devices	Electrochromic fibers	Fibers used for making materials that are able to change their visible absorption spectra in response to electrical change	Li et al. (2018b)
	Functional fibers for thermochromic devices	Fibers used for making materials able to change their visible absorption spectra in response to thermal change	Zhang et al. (2019)
Functional fibers for devices with changing shape	Electrically controlled deformable fibers	Fibers based on dielectric elastomers, e.g., nanocomposite based on elastic styrene-(ethylene-butylene)-styrene rubber and carbon nanotubes, which are able to change their shape by application of a high voltage	Liu et al. (2015); Shi et al. (2019)
	Solvent-responsive deformable fibers	Fibers able to change their shape or length in response to surface adsorption or desorption of solvent	Chen et al.,(2015); Fang et al. (2019)
	Light-induced deformable fibers	Fibers able to show rotation and bending in response to electromagnetic stimulus based on nanocomposites of sodium polyacrylate and graphene oxide	Shi et al. (2017); Liu et al. (2018)

19.2 Eco-friendly polymer-based fibers

Polymer-based fibers acquire their connotation of eco-friendliness as a result of their functionality, however, when fibers are based on bio-based polymers or biological polymers this quality is increased due to their sustainable origin. In general, it is possible to define four types of polymer-based fibers: (1) natural polymer-based fibers, (2) synthetic polymer-based fibers, (3) polymer-based fibers obtained from the combination of natural and synthetic polymers (i.e., "hybrid" polymer-based fibers), and (4) composite or nanocomposite fibers (Chad and Fajim, 2008; John and Anandjiwala, 2009; Vigneswaran et al., 2014; Shireesha and Nandipati, 2019).

Natural polymer-based fibers have a biological origin, usually animal or vegetable, however, it is important to know that these fibers also can be obtained from biotechnological techniques, and therefore, it is possible to state that natural polymer-based fibers can have a microbiological origin (e.g., bacterial cellulose). The textile fibers of vegetable origin are: (1) those that originate in the hair of seeds, such as cotton, (2) in foliage such as pineapple sisal, stone pine, and abaca, (3) in stem such as jute and flax, and (4) fibers originated from the shell, such as coconut; whereas, natural fibers from animal origin include wool, hair, and secretions, such as silk (Chad and Fajim, 2008; Shireesha and Nandipati, 2019). On the other hand, synthetic fibers are very common, with products such as nylon, polyesters, poly(ethylene), poly(propylene), and polyacrylates being well-known. The use of these fibers not only is a result of their low costs at the manufacturing level, but also a consequence of their mechanical and chemical resistances, and versatility for their application in different fields (John and Anandjiwala, 2009; Vigneswaran et al., 2014; see Fig. 19.1).

"Hybrid" fibers based on polymers are strictly synthetic fibers since these, in their final state, are not found in nature nor are they produced by a biological entity. They are based on the adequate compatibility of polymeric phases of different origin, but in turn, they keep similarities in their properties in such a way that phenomena such as phase segregation do not occur. Note that, although these mixtures can also be made from two or more biopolymers, in these cases, the fibers should be classified as natural polymer-based fibers. A characteristic example of "hybrid" fibers is those based on collagen and chitosan, collagen is a biopolymer that can have various sources, from animal connective tissue to fish scales; for example, collagen is mixed with poly(vinylpyrrolidone) to produce hybrid fibers for medical uses. In the second case, chitosan is also a biopolymer produced by insects and crustaceans, which has been mixed with different synthetic polymers to make fibers, such as poly(vinyl alcohol), poly(vinylpyrrolidone), poly(ethylene glycol), etc. (Sionkowska, 2011).

Composite or nanocomposite fibers are materials with two phases, with one polymeric phase containing one nanostructured phase dispersed in the polymer matrix. Thus, the addition of nanomaterials to polymers produces an increase in mechanical, thermal, optical, dielectric, and magnetic properties, among other interesting aspects, with the improvement depending on nature of nanoparticles and nature of the polymer, and also depending on efficiency of dispersion, nanoparticle size and distribution, nanoparticle shape, etc. Carbon nanotubes, graphene, metal and metal oxides, nanocellulose, etc. have been used in making these fibers (Girdhar et al., 2015). Fundamental details about fibers are given in Chapter 11.

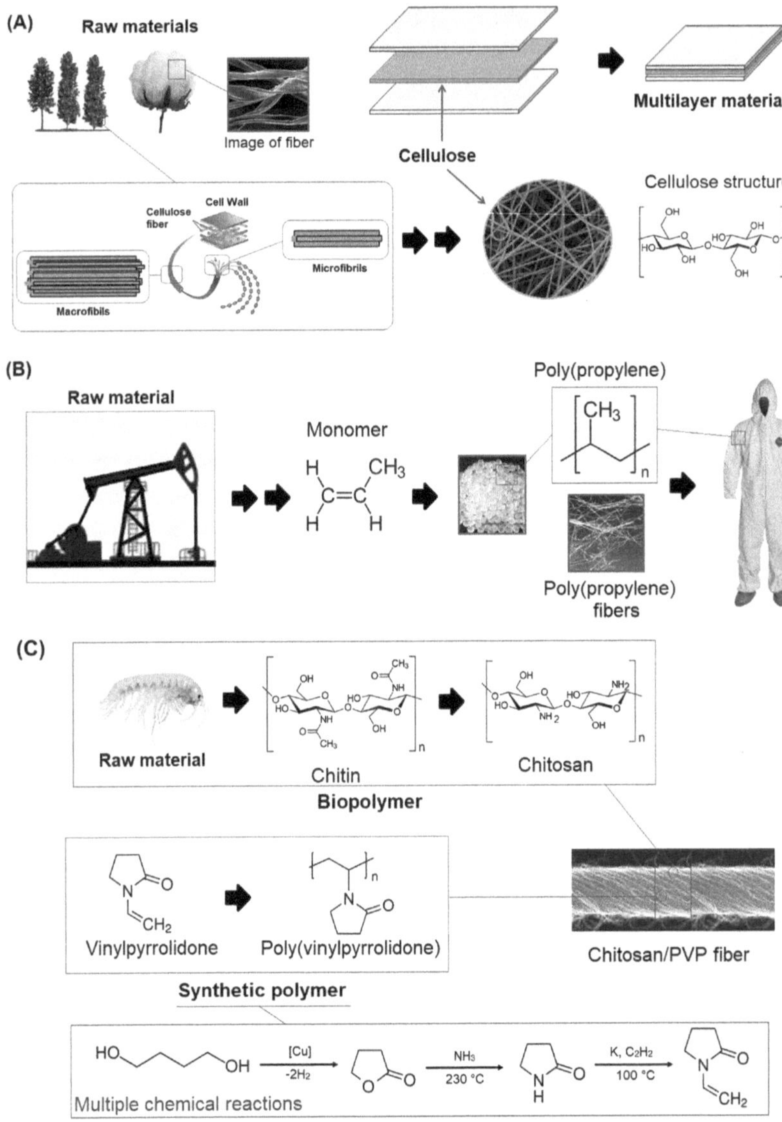

FIGURE 19.1 (A) Examples of natural fibers: raw materials, illustration of fibers, chemical structure, and applications of cellulose fibers in multilayer systems; (B) examples of synthetic fibers: raw materials, chemical structure, shape, illustration of fiber and applications of poly(propylene) fibers; and (C) examples of fibers based on polymer blends: raw material for extraction of chitin, chemical transformation of chitin in chitosan, and application in the fiber; and sequential chemical reaction for the obtaining of vinylpyrrolidone which is the raw material for the synthesis of poly(vinylpyrrolidone), for the formation of chitosan/poly(vinylpyrrolidone) blends.

19.3 Conclusions

The application potential of polymer-based fibers is enormous. Their combined use with nanomaterials has directed their applications toward the development of technologies in fields that were alien previously. However, at present, the combination of different disciplines has enhanced the technological scope of these materials. Also, technological advances must be analyzed correctly, since from the perspective of sustainability criteria, not only the functionality or origin is important, but also the useful life cycle, the management and processing of waste, and the impact of the industry that manufactures these types of materials on the environment and human populations. One observation, which may seem recurring, is that the trend of technological developments in polymer-based fibers is toward the use of biopolymers and nanotechnology for the development of multifunctional materials.

References

Allafi, F., Hossai, M.D., Lalung, J., Shaad, M., Salehabadi, A., et al., 2020. Advancements in applications of natural wool fiber: review. J. Nat. Fibers 1–16. https://doi.org/10.1080/15440478.2020.1745128.

Bianco, L., Serra, V., Pollo, R., 2017. Wood fiber vs synthetic thermal insulation for roofs energy retrofit: a case study in turin, Italy. Energy Proc. 111, 347–356. https://doi.org/10.1016/j.egypro.2017.03.196.

Caicedo, C., Melo-López, L., Cabello-Alvarado, C., Cruz-Delgado, V., Avila-Orta, C., 2019. Biodegradable polymer nanocomposites applied to technical textiles: a review. DYNA 86. https://doi.org/10.15446/dyna.v86n211.80230.

Célino, A., Freour, S., Jacquemin, F., Casari, P., 2012. The hygroscopic behavior of plant fibers: a review. Front. Chem. 1, 43. https://doi.org/10.3389/fchem.2013.00043.

Chand, N., Fahim, M., 2008. Tribology of natural fibers polymer composites. In: Chand, N., Fahim, M. (Eds.), Natural Fibers and Their Composites. Woodhead Publishing Limited, pp. 1–58. https://doi.org/10.1533/9781845695057.1.

Chen, B., Chen, S., Dong, B., Gao, X., Xiao, X., et al., 2017. Electrical heating-assisted multiple coating method for fabrication of high-performance perovskite fiber solar cells by thickness control. Adv. Mater. Interf. 4, 1700833. https://doi.org/10.1002/admi.201700833.

Chen, G., Li, Y., Bick, M., Chen, J., 2020. Smart textiles for electricity generation. Chem. Rev. 120, 3668–3720. https://doi.org/10.1021/acs.chemrev.9b00821.

Chen, P., Xu, Y., He, S., Sun, X., Pan, S., et al., 2015. Hierarchically arranged helical fibre actuators driven by solvents and vapours. Nat. Nanotechnol. 10, 1077. https://doi.org/10.1038/nnano.2015.198.

Fang, B., Xiao, Y., Xu, Z., Chang, D., Wang, B., et al., 2019. Handedness-controlled and solvent-driven actuators with twisted fibers. Mater. Horizons (6), 1207–1214. https://doi.org/10.1039/c8mh01647j.

Fiore, V., Di Bella, G., Valenza, A., 2019. Effect of sheep wool fibers on thermal insulation and mechanical properties of cement-based composites. J. Nat. Fibers 1532–1543. https://doi.org/10.1080/15440478.2019.1584075.

Girdhar, M., Mohan, A., Sharma, A., 2015. Blending strategies of natural polymers: a review. Trends Biomater. Artif. Organs 30, 61–76.

Gruyter, D., 2018. A Review on the Sustainability of Textile Industries Wastewater with and without Treatment Methodologies, vol. 33. https://doi.org/10.1515/reveh-2018-0013.

Gong, W., Hou, C., Zhou, J., Guo, Y., Zhang, W., et al., 2019. Continuous and scalable manufacture of amphibious energy yarns and textiles. Nat. Commun. 10, 868. https://doi.org/10.1038/s41467-019-08846-2.

Islam, M.M., Perry, P., Gill, S., 2020. Mapping environmentally sustainable practices in textiles, apparel and fashion industries: a systematic literature review. J. Fash. Mark. Manag.: Int. J. 1−23. https://doi.org/10.1108/JFMM-07-2020-0130.

John, M.J., Anandjiwala, R.D., 2009. Surface modification and preparation techniques for textile materials. In: Wei, Q. (Ed.), Surface Modification of Textiles. Woodhead Publishing Limited, pp. 1−25. https://doi.org/10.1533/9781845696689.1.

Li, R., Li, K., Wang, G., Li, L., Zhang, Q., Yan, J., et al., 2018b. Ion-transport design for high-performance Na+-based electrochromics. ACS Nano 12, 3759. https://doi.org/10.1021/acsnano.8b00974.

Li, L., Lou, Z., Chen, D., Jiang, K., Han, W., Shen, G., 2018a. Recent advances in flexible/stretchable supercapacitors for wearable electronics. Small 14, 1702829. https://doi.org/10.1002/smll.201702829.

Li, H., Tang, Z., Liu, Z., Zhi, C., 2019. Evaluating flexibility and wearability of flexible energy storage devices. Joule 3, 613. https://doi.org/10.1016/j.joule.2019.01.013.

Liu, Z.F., Fang, S., Moura, F.A., Ding, J.N., Jiang, N., et al., 2015. Stretchy electronics. Hierarchically buckled sheath-core fibers for superelastic electronics, sensors, and muscles. Science 349, 400. https://doi.org/10.1126/science.aaa7952.

Liu, L., Onck, P.R., 2018. Topographical changes in photo-responsive liquid crystal films: a computational analysis. Soft Matter 14, 2411. https://doi.org/10.1039/C7SM02474F.

Luján-Ornelas, C., Güereca, L.P., Franco-García, M.L., Heldeweg, M., 2020. A life cycle thinking approach to analyse sustainability in the textile industry: a literature review. Sustainability 12, 10193. https://doi.org/10.3390/su122310193.

Madhav, S., Ahamad, A., Singh, P., Kumar, P., 2018. A review of textile industry: wet processing, environmental impacts, and effluent treatment methods. Environ. Qual. Manag. 27, 31−41. https://doi.org/10.1002/tqem.21538.

Murali, R., Sankar, S., 2017. Spectra fibre reinforced composite -A review. Int. J. Sci. Res. 6, 689−692.

Pu, X., Hu, W., Lin-Wang, Z., 2017. Toward wearable self-charging power systems: the integration of energy-harvesting and storage devices. Small 14, 1702817. https://doi.org/10.1002/smll.201702817.

Provin, A.P., Dutra, A.R., Machado, M.M., Vieira, A.L., 2020. New materials for clothing: rethinking possibilities through a sustainability approach - a review. J. Clean. Prod. 282, 124444. https://doi.org/10.1016/j.jclepro.2020.124444.

Rajasundaram, D., Runavot, J.L., Guo, X., Willats, W.G., Meulewaeter, F., Selbig, J., 2014. Understanding the relationship between cotton fiber properties and non-cellulosic cell wall polysaccharides. PloS One 9, e112168. https://doi.org/10.1371/journal.pone.0112168.

Ryan, J.D., Alemu, D., Gabrielsson, R., Lund, A., Müller, C., 2017. Machine-washable PEDOT:PSS dyed silk yarns for electronic textiles. ACS Appl. Mater. Interf. 9, 9045−9050. https://doi.org/10.1021/acsami.7b00530.

Shad, M., Yaqoob, A., Usmaan, M., Shakoor, S., Salisu, I.B., et al., 2019. Mini-Review cotton fibre quality management for sustainable textile industry. The ICAC Record. 22−27.

Shao, Y., El-Kady, M.F., Wang, L.J., Zhang, Q., Li, Y., Wang, H., Mousavi, M.F., Kaner, R.B., 2015. Graphene-based materials for flexible supercapacitors. Chem. Soc. Rev. 44, 3639. https://doi.org/10.1039/C4CS00316K.

Shireesha, Y., Nandipati, G., 2019. State of art review on natural fibers. Mater. Today Proc. 18, 15–24. https://doi.org/10.1016/j.matpr.2019.06.272.

Shi, Q., Li, J., Hou, C., Shao, Y., Zhang, Q., Li, Y., Wang, H., 2017. A remote controllable fiber-type near-infrared light-responsive actuator. Chem. Commun. 253, 11118. https://doi.org/10.1039/C7CC03408C.

Shi, Q., Sun, J., Hou, C., Li, Y., Zhang, Q., Wang, H., 2019. Advanced functional fiber and smart textile. Adv. Fiber Mater. 1, 3–31. https://doi.org/10.1007/s42765-019-0002-z.

Sionkowska, A., 2011. Current research on the blends of natural and synthetic polymers as new biomaterials: Review. Prog. Polym. Sci. 36, 1254–1276. https://doi.org/10.1016/j.progpolymsci.2011.05.003.

Uddin, F., 2019. Introductory chapter: textile manufacturing processes. In: Faheem, U. (Ed.), Textile Manufacturing Processes. IntechOpen. https://doi.org/10.5772/intechopen.87968.

Vigneswaran, C., Ananthasubramanian, M., Kandhavadivu, P., 2014. Bioprocessing of synthetic fibres. In: Vigneswaran, C., Ananthasubramanian, M., Kandhavadivu, P. (Eds.), Bioprocessing of Textiles: Fundamentals for Applications and Research Perspective. Woodhead Publishing India, pp. 189–250. https://doi.org/10.1016/B978-93-80308-42-5.50004-4.

Yang, Y., Zhong, Y., Shi, Q., Wang, Z., Sun, K., Wang, H., 2018. Electrocatalysis in lithium sulfur batteries under lean electrolyte conditions. Angew. Chem. Int. Ed. 130, 15775–15778. https://doi.org/10.1002/anie.201808311.

Yip, J., Chan, W.Y., 2020. Textile fibers and fabrics. Developments for Activewear. In: Yip, J. (Ed.), Latest and Technological Development for Activewear. Elsevier, pp. 47–72. https://doi.org/10.1016/B978-0-12-819492-8.00003-X.

Zhang, Y., Zhao, Y., Ren, J., Weng, W., Peng, H., 2016. Advances in wearable fiber-shaped lithium-ion batteries. Adv. Mater. 28, 4524. https://doi.org/10.1002/adma.201503891.

Zhang, Y., Hu, Z., Xiang, H., Zhai, G., Zhu, M., 2019. Fabrication of visual textile temperature indicators based on reversible thermochromic fibers. Dyes Pigm. 162, 705. https://doi.org/10.1016/j.dyepig.2018.11.007.

Part IV

Selected applications of eco-friendly functional polymers

Chapter 20

Passive sampler of organochloride compounds in water and air

20.1 Sampling of pollution in water and air: an overview

Air and water are essentially two abiotic components of supreme importance for the survival of human beings and ecosystems in general. However, due to anthropogenic activities mainly related to technological and industrial processes, many chemical compounds are released into the environment directly, indirectly, or as a result of secondary reactions with the environment (Kjellstrom et al., 2006; Inyinbor et al., 2018). This leads to a contamination problem that has serious environmental consequences.

Among the pollutant organic compounds that can be found in water and air, organochlorine compounds (OCs) are molecules with a recurring presence, since they are produced for industrial, agricultural, pharmaceutical, and domestic uses, among others (Huang et al., 2014; Nuro, 2018; Kanan and Samara, 2018). Organochlorine pesticides, polychlorinated biphenyls (PCBs), dioxins, chlorobenzenes, chlorophenols, and chlorinated alkanes are some of the types of OCs with the greatest presence in the environment. Many of these compounds have particular characteristics such as high toxicity, stability, high bioaccumulation factor, transport capacity, and high persistence (Adu-Kumi et al., 2010; Martínez et al., 2017a,b). A large number of these compounds have been related to adverse processes in humans and wildlife, mainly affecting reproduction processes, acting as endocrine, neurological, and immunological disruptors or causing different diseases and abnormalities (Kleanthi et al., 2008; Sruthi et al., 2017; Kanan and Samara, 2018). This is why they are included within current environmental concerns due to their ability to generate great damage if their emissions are not controlled or if successful elimination is not achieved after their use.

The adverse effects and physicochemical characteristics of compounds such as PCBs, dioxins, furans, and chlorinated pesticides, have resulted in them being classified as persistent organic compounds (POPs). The evaluation of the effectiveness of the measures imposed to comply with this treaty

Eco-friendly Functional Polymers
https://doi.org/10.1016/B978-0-12-821842-6.00022-1
Copyright © 2021 Elsevier Inc. All rights reserved.

includes both obtaining monitoring data on the presence of polluting chemicals, as well as the record of information on their regional and long-range transport (Stockholm Convention, 2001; Fiedler, 2007).

Consequently, with regard to air monitoring (one of the priority matrices in the agreement), monitoring plans have been developed, such as the Global Monitoring Plan (GMP) that seeks to identify trends, assess risks, and establish related priorities with POPs, and has received support from the international scientific community. However, this network feeds on regional monitoring programs that cannot provide representative information as they use active samplers to determine concentration levels (Klánová and Harner, 2013). These samplers use pumps for their operation, therefore reliable and available energy sources are required, and additionally, the cost of the equipment is considerable, which makes their use limited.

Given the limitations of active sampling and the need to obtain reliable information, with greater deployment in the field and at a lower cost, researchers developed the passive air-sampler (PAS) in the 1990s. This type of sampler is based on processes of passive diffusion of pollutants to an active or receptor phase in charge of accumulating the compounds without the need for the use of pumps or additional energy (Forbes, 2020; Francisco et al., 2017; Vrana et al., 2005; St George et al., 2010). Currently there are monitoring networks that use passive samplers, one of the most important networks created since 1994 is the Global Atmospheric Passive Sampling (GAPS) Network which has about 60 locations worldwide that use polyurethane foam (PUF) disks and nonpolar resins as receiving phases of the passive samplers (e.g., Amberlite XAD resin TM); similarly, there are other programs for Europe, Australia, Japan, the United Kingdom, and Latin America (Klánová and Harner, 2013). The use of passive samplers for monitoring contaminants in water is more recent, however, given its advantages, it is gaining importance at an international scientific level since it enables, in conjunction with conventional punctual sampling, to establish contamination trends or to verify compliance with environmental quality criteria (Mills et al., 2011).

The receptor phase is the heart of a passive sampler; among the materials most used for this function, polymers have been of great importance due to their relatively low cost, diversity of presentation (solids, liquids, thin films, pellets, etc.), adsorption capacity, ease of preparation, escalation, functionalization, and the possibility of reuse (Forbes, 2020; Taylor et al., 2019). In this context, polymers have the function of capturing or receiving pollutant molecules (Sanders et al., 2018); and based on this, environmental concentration levels can be quantitatively determined; such information can be used to establish levels of exposure, routes, and sources of contamination, and also can serve as scientific support for the establishment of environmental regulations in order to help improve risk assessments. The above allows the polymers used for this purpose to be classified as functional polymers and in turn as eco-friendly materials, because their function is focused on contributing to resolving an environmental problem.

This chapter deals with some generalities of the most used sampling devices for the sampling of chlorinated organic pollutants, with an emphasis on the functional polymers that constitute them; it addresses the pollutant–receptor phase chemical interactions involved in the capture process, and finally some important considerations are presented.

20.2 General context of organochloride contaminants

Organochlorine compounds are, in essence, hydrocarbons with more than one chlorine atom in their structure. Although these compounds are mainly of synthetic origin or derived from secondary reactions, some have been isolated from natural sources (Kleanthi et al., 2008; Nuro, 2018). Such is their variety that there is a wide range of compounds with different applications; however, some of these chlorinated derivatives are of great environmental concern due to the adverse effects they cause in the environment. Among the compounds with the greatest environmental repercussions are organochlorine pesticides, PCBs, dioxins, furans, chlorobenzenes, chlorophenols, alkanes, and chlorinated alkenes (Huang et al., 2014; Singh et al., 2016; Jayaraj et al., 2016; Zubair and Adrees, 2019); some representative chlorinated pesticides and general structures of other types of chlorinated derivatives are shown in Fig. 20.1.

These compounds can be considered as ubiquitous in the environment, since they can be found in soils, air, and any fluvial medium (groundwater, rivers, lakes); they have moderate polarity, typically hydrophobic, which gives them a high lipophilic capacity that allows them to enter the fatty tissues of animals and humans (Chopra et al., 2011; Nuro, 2018; Yamazaki et al., 2020). They have high chemical stability that makes them resistant to degradation, which is why they persist in the environment for a long time after being released; additionally, some present a high volatility that, together with their persistence, makes it possible to transport them over long distances in different

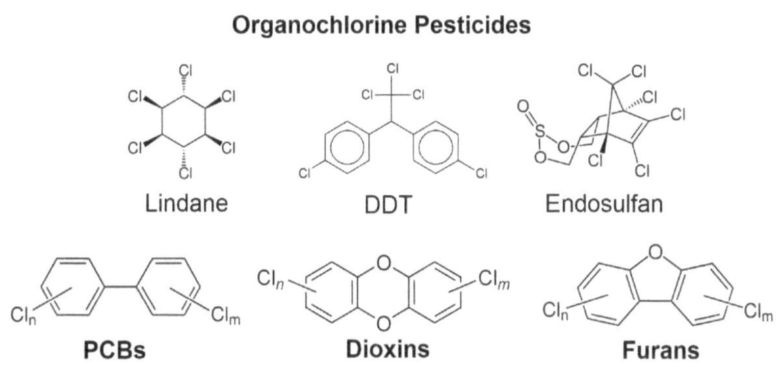

FIGURE 20.1 Representative organochlorine pesticides and general structures of PCBs, dioxins, and furans.

environmental media at different altitudes and geographic latitudes (Van Drooge et al., 2004; Zubair and Adrees, 2019; Krasnobaev et al., 2020). For example, the carbon tetrachloride used as a solvent and in the manufacture of extinguishers, refrigerants, and aerosols, can persist for up to 100 years, which makes it possible for it to spread great distances (Huang et al., 2014). For their part, some studies have reported organochlorines such as dichlorodiphenyltrichloroethane isomers (DDT) and their metabolites (DDE and DDD isomers), hexachlorobenzene (HCB) isomers, hexachlorocyclohexane (HCH) isomers, dieldrin, endrin, and heptachlor in ice cores, air, water, soil, and biota of the North Pole and Antarctic; while these places can be considered in principle "pristine" the presence of these pollutants is attributed to cold condensation and global fractionation during long-range atmospheric transport, which turns these places into pollutant deposit areas that, eventually, and due to climate change, may become re-emission zones for chlorinated compounds due to thawing (Taiwo, 2019; Potapowicz et al., 2020).

20.2.1 Organochlorine pesticides (OCPs)

OCPs are synthetic compounds of different types, including chlorinated cyclodienes, chlorodiphenylethanes, chlorinated benzenes, and cyclohexanes (Taiwo, 2019). They are used throughout the world in domestic applications and the public health sector to fight vector-borne diseases such as typhus and malaria (Snow et al., 2012; Li et al., 2017), in turn, they also have been used to combat destruction from insects, weeds, fungi, bacteria, etc. that attack agricultural crops to promote greater production (Jayaraj et al., 2016). The most popular and most widely used OCP was DDT, registered for the first time in 1946, and which was used in agriculture and livestock to combat pests or used in public health problems to control the vectors of malaria, bubonic plague, etc. The commercial technical mixture consisted of a mixture of its isomers (p,p′-DDT, p,p′-DDE, and p,p′-DDD) that could also contain isomers of its metabolites, dichlorodiphenyldichloroethane (DDD), dichlorodiphenyldichloroethylene (DDE), compounds also classified as dangerous (Berry-Cabán, 2011; Snow et al., 2012). It is considered dangerous, with high persistence and a half-life of 2–15 years, with a bioaccumulative capacity in adipose tissues so high that it is possible that there is no living organism that does not contain it (Jayaraj et al., 2016). Besides DDT, other chlorinated pesticides such as aldrin, toxaphene, DDT, chlordane, dieldrin, endrin, heptachlor, endosulfan, lindane, pentachlorobencene, metoxichlore, mirex, HCH, and HCB, through the Stockholm Convention, have been classified as POP whose production, import, and use are prohibited in more than 50 countries around the world, since these compounds threaten the long-term survival of ecosystems and human health, due to their high toxicity, persistence, and ability to bioaccumulate (Stockholm Convention, 2001). Despite the restrictions, in many developing countries they remain in use (Jayaraj et al., 2016).

20.2.2 Polychlorinated biphenyls

Another group of organochlorine compounds of environmental concern is PCBs, these compounds are made up of two biphenyl rings with different number and position of chlorine atoms; these compounds have been classified as POPs and as carcinogenic contaminants according to the International Agency for Cancer Research and the US Environmental Protection Agency (Eddy, 2020). Between 1930 and 1977, 209 congeners were produced, and for identification purposes, they have been numbered from PCB 1 to PCB 209 according to the number and position of the chlorine atoms around the biphenyl rings (Dhakal et al., 2018).

Due to their high boiling points, nonflammability, chemical stability, and insulating properties, they were used mainly as additives for insulating fluids, lubricants, condensers, plastics, among others; however, they were banned in 1980 due to their persistence, bioaccumulation, and toxicity (Martínez et al., 20017a; Dumanoglu et al., 2018; Nuro, 2018; Reddy et al., 2019). Coplanar and mono-ortho-substituted PCBs have been shown to have toxicological effects similar to dioxins, which are endocrine-disrupting compounds; on the other hand, PCBs with fewer chlorine atoms tend to be less persistent and can be metabolized in animals and plants, resulting in compounds with a multiplicity of biological effects (interference with hormonal signaling, inhibition of enzymes, electrophilic metabolites that form adducts with DNA, macromolecules, among others) (Nuro, 2018; Dhakal et al., 2018). The release into the environment of this type of compound is due to activities related to their use, mainly due to leakage of electrical transformers and capacitors; waste incineration also releases PCBs into the environment (Zeng et al., 2014; Reddy et al., 2019).

20.2.3 Dioxins and furans

As for dioxins and furans, they are polychlorinated compounds, which are generated indirectly due to volcanic eruptions or forest fires, incineration processes, or incomplete combustion of chlorinated aromatic hydrocarbons present in waste, or they can be emitted through chemical processes during the production of pesticides, fertilizers, bleaching of paper and wood pulp, among others (Kanan and Samara, 2018; Nuro, 2018). Additionally, it has been shown that they can be formed unintentionally during the implementation of remediation technologies in wastewater due to the presence of precursor compounds such as 2-chlorophenol (Schröder et al., 2021). Dioxins are known as polychlorodibenzene-p-dioxins (PCDD) and furans as polychlorodibenzofurans (PCDF); 210 possible congeners have been identified, of which 2,3,7,8-tetrachlorodibenzo-p-dioxin (TCDD) was identified as the most toxic compound (Kanan and Samara, 2018; Zubair and Adrees, 2019). PCDD/F have been identified as a toxic, carcinogenic, immunosuppressive, endocrine disruptor, which can affect neurodevelopment and reproduction (Rathna et al., 2018; Fernandes and Falandysz, 2021).

20.3 Brief description of accumulation of chemicals by passive samplers

Passive samplers are basically devices that extract solutes (organochlorine pollutants in this case) from the sampled matrix (water, air, soil) and concentrate them in the active phase of the sampler, all without additional energy requirement (Garcés et al., 2018; Lévy et al., 2020; Wania and Shunthirasingham, 2020; Valenzuela et al., 2020). This process is due to a difference in fugacity or chemical potentials between the phases involved, which facilitates the diffusion of the solute toward the uptake material (Rusina et al., 2010a,b; Taylor et al., 2019). However, the description of the free diffusion process that is mentioned is not that simple, and it is common that the transport of solutes to the receptor includes overcoming different "barriers."

To illustrate this process in a simple way, consider that a sampler is located in a contaminated water stream (see Fig. 20.2), when the sampler is exposed to water, according to film theory, a laminar layer or stagnant film of a certain thickness, also called the "water boundary layer" (WBL), diffusive layer, or diffusive limit, where the flow is clearly laminar. As the contaminant molecules approach the surface of the sampler, subsequent mass transfer will take place primarily by molecular diffusion; it is noteworthy that as the

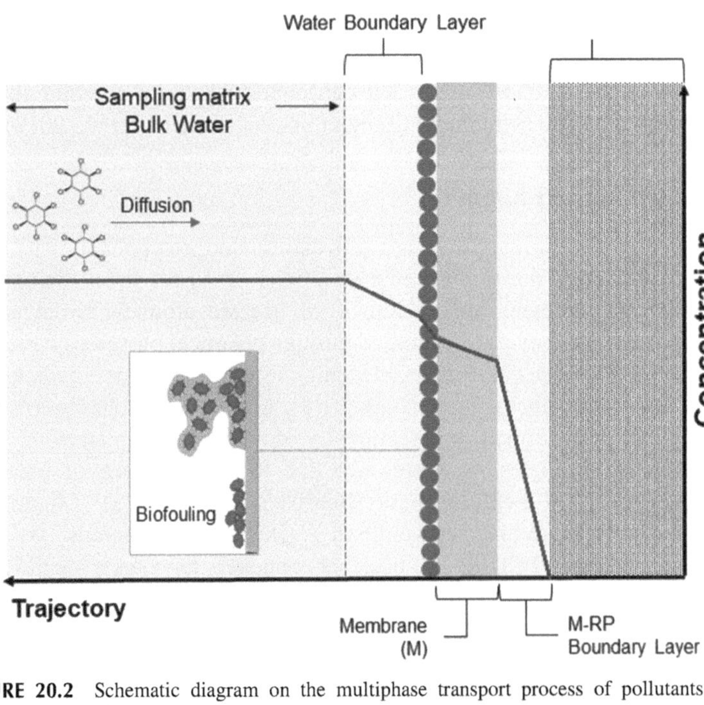

FIGURE 20.2 Schematic diagram on the multiphase transport process of pollutants during uptake by a sampler.

distance from the surface of the sampler to the bulk of the water increases, the turbulence becomes greater and therefore, the convection and swirl diffusion govern the mass transfer. Accordingly, the solute molecules must first be transported or diffused through the WBL (or gas film if air is considered) for later, and depending on the quality of the exposure water, they may encounter a complex layer of fouling, made up of organic material (biofilm if it corresponds to bacteria, algae, microfauna) and/or inorganic material (for example, calcareous precipitates). The thickness of this new layer and its composition vary (in fact, it may be absent) and it is considered that mass transport through it also occurs mainly by molecular diffusion (Valenzuela et al., 2020). After crossing this phase, as will be seen later, depending on the configuration of the sampler, the solute may meet directly with the receptor active phase or it must first pass through a protective membrane and/or a diffusion gel; for example, in the case of semipermeable membrane device (SPMD) samplers, the solute after passing through the WBL and the fouling layer, first encounters a low-density polyethylene surface to finally dissolve and concentrate in a pure lipid of high molecular weight, typically triolein (Esteve-Turrillas et al., 2008).

There are several models that try to quantitatively describe this type of process, some more complex than others. The Chemical Reaction Kinetics (CRK) Model and the Mass Transfer Coefficient (MTC) Model are usually used; in the literature, mathematical descriptions associated with the capture process can be found that allow determining the environmental concentrations of pollutants (Wania and Shunthirasingham, 2020; Valenzuela et al., 2020). Sampling rates and receptor phase-water (or air) partition coefficients are important parameters involved in the catchment process that are modeled or determined empirically at the laboratory or field level; however, the assumptions that each model establishes when deducing the equations that describe the concentrations in air and water must be taken into account.

20.4 Functional polymers using passive samplers

Due to the fact that one of the advantages of the use of passive samplers lies in the possibility of massive and extended deployment, this type of device, in principle, in addition to presenting adequate chemical affinity for the compounds of interest, must conceive a simple, economical structure, and be easy to prepare; for this reason, the vast majority of samplers include polymeric materials in some part of their components, due to their low cost, flexibility, and structural variability (Rusiana et al., 2007; Palencia et al., 2016; Taylor et al., 2019; Godlewska et al., 2021). Polymers are usually the preferred receptor phases since they can be used under different configurations (thin films, granules, or gel) and, depending on their structure and chemical nature, materials with different chemical affinity can be found (hydrophobic, hydrophilic, amphiphilic, anionic, or cationic); with regard to the latter, it is essential that between the receptor phase used and the monitored compounds, and sufficient chemical interactions are present for their extraction from the sampled matrix (see Section 20.5).

In general, for the sampling of water and air, in terms of the structure of the device used, the existing configurations are diverse, however, in this chapter a categorization is made according to the number of polymeric phases involved within the sampler. A first category corresponds to devices made up of a single polymeric phase (monophasic) that acts as a receptor material; a second category corresponds to two- or three-phase samplers where, in addition to the use of a polymeric receptor phase, a second or third polymeric phase is used with functions of mechanical protection, slowing down of the uptake or elimination of dependence on flow rates during the stage of sampling and/or improvement of the catchment (Vrana et al., 2016; Taylor et al., 2019; Wania and Shunthirasingham, 2020). Additionally, it should be noted that the use of polymers in the devices not only involves the aforementioned; in general, for devices that require protection or support structures, polymers such as Teflon or poly(vinyl chloride) (PVC) are also selected for their resistance to abrasion, durability, and reuse, mainly in water sampling.

20.4.1 Monophasic samplers

For the sampling of chlorinated organic pollutants of a hydrophobic nature, different polymers such as low-density poly(ethylene) (LDPE), poly(oxymethylene) (POM), silicone rubber (SR), poly(dimethyl siloxane) (PDMS) and its derivatives, poly(urethane) (PU), copolymers of styrene divinylbenzene (resins called XAD), poly(ether sulfone) (PES), poly(2,6-diphenyl-p-phenylene oxide) (PPP) marketed as Tenax, poly(ethylene-co-vinyl acetate) (EVA), poly(methyl methacrylate) (PMMA), nylon, carbon-based foams, and monophasic mixed polymeric materials (e.g., PDMS/Oasis HLB) have been used as simple or monophasic receptor phases (Rusiana et al., 2007; Ter Laak et al., 2008; Schummer et al., 2012; Lao et al., 2016; Can-Güven et al., 2019; Taylor et al., 2019; Lévy et al., 2020; Wania and Shunthirasingham, 2020). An advantage of this type of sampler is the possibility of designing the device's absorption capacities "custom-made," since the modification of the polymer surface, its area, thickness, or volume can affect the uptake kinetics (Gong et al., 2018). On the other hand, they are relatively inexpensive due to their structural simplicity, low cost, and possibility of reuse; and, depending on the exposure medium, the conformation (or presentation) of the polymer can be exposed naked or supported in some particular structure (Zhang et al., 2012; Esteve-Turrillas et al., 2012). For example, single-phase air samplers based on XAD-type resins, as they are powdered materials, are usually deposited in stainless steel mesh cylinders that do not obstruct the diffusion of analytes; these cylinders in turn are installed inside a second container that has an opening in the lower part that allows controlling the wind speed and protecting the capture material from solar radiation, the deposition of particulate material, or precipitation, factors that can affect sampling rates. In the case of samplers in water, polymers are also usually protected with canisters or suitable exposure chambers to minimize the physical damage produced by water currents or floating material (pieces of wood, small rocks, or gross sand) (Martínez et al., 2017b; Wania and Shunthirasingham, 2020).

Of the materials mentioned above, LDPE and SR samplers have been the most widely used. LDPE is a linear polyolefin with repeating ethylene units $(-CH_2-CH_2-)_n$, with long and short structural branches that give it a lower density and crystallinity; it is usually deployed in the field in the form of sheets or tubes in direct contact with the analysis matrix (Khairy and Lohmann, 2013; Wania and Shunthirasingham, 2020). Regarding the uptake of hydrophobic compounds in water, SR polymers, mainly derivatives with phenyl, methyl, and vinyl substituents, have been shown to be materials with high diffusion coefficients and low values of resistance to transfer mass (Rusina et al., 2010a,b; Pintado-Herrera et al., 2016); both in air and water, they are usually deployed in the field in the form of sheets, rods, or tubes or deposited as thick films in Stir Bar Sorptive Extraction (SBSE) systems, which are stirring bars arranged in glass tubes coated with a polymer layer (0.5−1 mm thick) of PDMS (Brockmeyer et al., 2015; Vrana et al., 2016). SR are popular as "personal" air pollutant samplers consisting of small, convenient devices adapted for use on humans or pets, in the form of handcuffs and collars or similar portable structures (Okeme et al., 2018; Aerts et al., 2018; Koelmel et al., 2021).

XAD resins, as mentioned above, are also used for both air and water sampling; they have a high capacity for absorbing volatile organic compounds and are used for long-term integrative sampling (months to years) (Schummer et al., 2012; Silva-Barni et al., 2019a). These are highly porous spherical polymers based on highly cross-linked, macroreticular polystyrene, aliphatic, or phenol-formaldehyde condensate polymers. They are usually divided according to the polymeric matrix of constitution into two groups, XAD resins based on polystyrene-divinyl benzene (including XAD-1, XAD-2, XAD-4, XAD-16, XAD-1180, XAD-2000, and XAD -2010) and polyacrylic acid ester-based resins (including XAD-7, XAD-8, and XAD-11) (Esteve-Turrillas et al., 2012; Amad et al., 2015); There are many other references to resins of this type where the polymeric matrix can be functionalized or modified according to the capture requirements (sampling medium, type of analyte). These materials can be directly exposed to the sampling medium deposited inside an enclosure that does not allow the loss of the material (stainless steel mesh tubes) or they can be used in conjunction with other adsorbent polymers or, as will be shown later, they are a fundamental part of biphasic samplers (Gong et al., 2018; Wania and Shunthirasingham, 2020).

Another sampler in recurring use today is the one known as solid-phase microextraction (SPME), which, more than the name given to a device, is the name of a versatile and economical technique initially used since 1990 for the preparation of analytical samples at the laboratory level (Namieśnik et al., 2005; Taylor et al., 2019; Valenzuela et al., 2020). One of the main advantages of SPME is related to its ease of analysis, since after its deployment in the field it can be analyzed directly by solvent desorption/liquid chromatography (LC) or thermal desorption/gas chromatography (GC); in the latter, the use of extraction solvents is completely eliminated. As a sampler, it is used for the capture of a wide range of analytes in gaseous, liquid, and solid media (Reyes-Garcés et al., 2018; Taylor et al., 2019).

Structurally, in its simplest form, it consists of a kind of syringe (with a plunger) that holds a needle or sheath inside which houses a narrow and elongated substrate (solid fused silica fiber, nichrome capillaries or stainless steel) coated with a thin layer of polymer (liquid or solid). The traditional coating for hydrophobic contaminants consists of PDMS; octadecylsilane (C18) and mixed materials such as poly(dimethyl siloxane)/divinylbenzene (PDMS/DVB), carboxen/poly(dimethyl siloxane) (Car/PDMS), and DVB/Car/PDMS (Esteve-Turrillas et al., 2012) have also been used. Additionally, polymeric sorbents with particular morphologies and physicochemical properties have been used to give the receptor phase improvements in wettability, porosity, surface area, enrichment, and selectivity; metal organic frameworks (MOFs), carbon nanotubes/polymers, and molecularly printed polymers (MIP) are examples of these materials (Reyes-Garcés et al., 2018; Lompart et al., 2019). Both the thickness of the coating and its affinity for the analytes largely determine the sensitivity of the method and the uptake kinetics (Taylor et al., 2019). Although there are a variety of designs, in conventional SPME sampling the coated substrate extends completely outside the needle to come into contact with the contaminant in the analysis matrix; under this setting equilibrium is reached quickly (within seconds to hours). Another sampling modality is to leave the coated substrate at the same level as the tip of the outer needle, which provides a very small diffusion path diameter, increasing the time in which equilibrium is achieved and turning the SPME into a kinetic sampler. Additionally, this methodology provides greater protection of the receptor phase and depending on the diameter of the needle opening, dependence on flow can be decreased (Esteve-Turrillas et al., 2012). It should be noted that there are SPME devices that use polymeric membranes arranged on the surface of the receiving phase to eliminate the interference of macromolecules or materials that could harm the capture process; therefore, this type of samplers would fall into the category of biphasic samplers.

20.4.2 Biphasic and triphasic samplers

As for the samplers with more than one structural phase, the devices that can be found are diverse; however, among the most used are: SPMD and SPME used in water and air; Chemcatcher, the polar organic chemical integrative sampler (POCIS) and organic-diffusive gradients in thin-film (o-DGT) used in water; and Radiello used in air (Namieśnik et al., 2005; Valenzuela et al., 2020; Godlewska et al., 2021). The classic SPMD is most used for the uptake in the linear or kinetic phase of organic compounds of medium and high hydrophobicity; and uses a first phase made up of an LDPE tube inside which is housed a second liquid phase made up of a thin layer of high-purity triolein [1,2,3-tri (*cis*-9-octadecenoyl) glycerol], which acts as a receptor phase and whose permeability through LDPE is very low, as is the triolein–LDPE interfacial tension (Esteve-Turrillas et al., 2008; Taylor et al., 2019; Valenzuela et al., 2020). In this case, the polymer corresponding to a material with reduced pores (5–10 Å) functions as a semipermeable membrane that allows the passage of contaminants (available fraction) toward the receptor phase through a diffusion dissolution mechanism. Initially, this device was introduced by Huckins et al. (1990) as a biomimetic material in the bioconcentration process of apolar organic pollutants in aquatic organisms, which is why triolein was selected as a filler material since it is the main storage lipid found in most organisms (Huckins et al., 1990); however, with respect to this initial conception, it must be taken into account that (1) the bioaccumulation processes between organisms is different, (2) in the sampler, unlike what happens in nature, equilibrium is not achieved for most organic compounds, and (3) in this case the semipermeable LDPE membrane has the ability to absorb and retain contaminants in its structure (Esteve-Turrillas et al., 2008). The changes in the triolein receptor phase by adsorbents such as XAD resins, activated carbon/Florisil, ionic liquids, solvents such as 2,2,4-trimethylpentane (isooctane), or magnetic bars impregnated with PMDS have been tested (Esteve-Turrillas et al., 2008; Wania and Shunthirasingham, 2020).

The Chemcatcher, POCIS, and o-DGT devices are three types of samplers used for sampling in water, whose structural conformation is similar, since they have reusable housings made of PTF or polycarbonate (PC) that serve as support and protection for the core of the device. All three use external polymeric membranes and are highly versatile, since a variation of the receptor material allows the monitoring of compounds of different polarity and nature (organic, inorganic, hydrophobic, hydrophilic). They generally operate in the kinetic/integrative mode for hours or days, or if required for longer periods (months) they enter the equilibrium phase (Gong et al., 2018; Valenzuela et al., 2020).

For its part, in the inside of plastic casing, the Chemcatcher has a solid-phase extraction disk (SPE) acting as a receptor phase, and also has a diffusion membrane with one of its sides exposed to water. For the sampling of chlorinated organic compounds, disks based on C18 (nonpolymeric) are used, and copolymers such as poly(styrene-co-divinylbenzene) (SBD-XC), poly(styrene-co-divinylbenzene) sulfonated in reverse phase (SBD-RPS), and the balanced hydrophilic–lipophilic copolymer poly(divinylbenzene-co-N-vinylpyrrolidone) known as HLB can be obtained in different commercial houses (Empore/AttractSPE/Atlantic/Oasis) (Gong et al., 2018; Valenzuela et al., 2020).

SDB-RPS and SBD-XC are 100% copolymeric, porous, cross-linked particles used for polar and low-polarity compounds (useful for the capture of chlorinated pesticides) (Charriau et al., 2016). On the other hand, for compounds of moderate and low polarity, polysulfone (PS), polyethersulfone (PES), and LDPE membranes are usually used; however, membrane selection is crucial, since in addition to providing protection to the disk, it confers selectivity and control of analyte absorption.

A different receptor/membrane phase combination can change the Log Kow domain of uptake; for example, C18 SPE disks with LDPE membranes may have higher uptake performance for compounds with a Log Kow of 4–8, while the same receptor phase with PES membranes manages to uptake compounds with a Log Kow between approximately 2–5 (Greenwood et al., 2007). On the other hand, if the analyte has a high affinity for the membrane, a diffusion delay will occur, which is why, for the analysis of highly hydrophobic compounds such as PCBs and chlorinated pesticides (e.g., p,p′-DDT, endosulfan, or chlorobenzenes) the C18/PS combination is not suitable due to the affinity of the analytes to the membrane (Charriau et al., 2016). It is usual to use PS or PES with disks based on PDB-RPS, SDB-XC, and PES or LDPE with C18 disks; additionally, it has been shown that for hydrophobic compounds such as DDT, PCBs, hexachlorobenzene, pp'-DDE, dieldrin, and lindane, the addition of n-octanol to the interstitial space between the receptor phase and a membrane can reduce the internal resistance to the transfer of mass which improves uptake (Vrana et al., 2005; Charriau et al., 2016). Chemcatchers can also use the "naked" receptor phase (monophasic samplers), however, it must be taken into account that the uptake kinetics becomes faster, the selectivity decreases, and the degree of biofouling and deterioration of the disk increases (Stephens et al., 2009).

The POCIS samplers were initially used to monitoring polar organic compounds (Log Kow<4), however, at present, they are being used like sorbents for the capture compounds of moderate hydrophobicity (e.g., chlorinated pesticides) (Gong et al., 2018). The device consists of a granular receptor phase located between two polymeric membranes (membrane–sorbent–membrane) and two stainless steel rings that hold and secure the system by means of screws or grips (Valenzuela et al., 2020; Godlewska et al., 2021). Regarding the receptor phases, granular Oasis HLB is one of the most used since its balanced hydrophilic–lipophilic character allows the uptake of polar and apolar organic compounds. Similarly, the adsorbent marketed as Strata-X, which corresponds to a surface modified co-polymer of styrene-divinylbenzene, also has the ability to retain analytes of different polarity (Kaserzon et al., 2014; Gong et al., 2018; Godlewska et al., 2021). The polymers used in the Chemcatcher can be used in this sampler in granular form, as well as a mixture of them; additionally, sorbents based on MIPs and combinations with carbonaceous material have been studied to improve the performance of the sampler (Berho et al., 2017).

The o-DGTs, for their part, have recently been used to capture organic compounds and consist of a receptor phase (or binding gel), a diffusing layer (diffusive hydrogel), and a protective membrane or filter (optional) (Valenzuela et al., 2020). The presence of the diffuser layer limits the mass transfer of the compounds during sampling, therefore, unlike POCIS and Chemcatcher, the influence of environmental conditions on sampling rates is lower. It has been used to capture chlorinated pesticides (2,4-D, Chlorpyrifos, metallochlor, 4-chlorophenol, acetachlor) or chlorinated drugs such as diclofenac; among the receptor phases used are Oasis HLB and XAD resins (XAD-18, XAD-2, XAD-4) or modified styrene divinylbenzene resins (Sepra-ZT). Mainly agarose and polyacrylamide have been used in the diffusing phase, and polyvinylidene fluoride (PVDF), PES, glass fiber, cellulose nitrate (CN) membranes, and cellulose acetate membranes as external filters. It should be noted that MIPs prepared from methacrylic acid as monomer and ethylene glycol dimethacrylate as cross-linker have been used as alternative receptor phases for the uptake of 4-chlorophenol (Dong et al., 2014; Guibal et al., 2019) and, recently, given the ability of cyclodextrins to promote host–host interactions, they have been used to functionalize polymeric membranes for the uptake of triclosan, triclocarban, and methyltriclosan (Wei et al., 2019).

For the capture of chlorinated volatile organic compounds (e.g., polychloromethanes, polychloroethanes, and polychloroethylenes, used as solvents and degreasing agents, among others), the Radiello sampler has been used. This sampler is light and small, and can be used as a personal, indoor, or outdoor sampler with short sampling times (minutes or hours); structurally it presents a radial symmetry with a 360 degrees capture window, externally composed of a cylindrical diffuser body consisting of a microporous PE or silicone polymeric membrane and inside it houses a cartridge with the receptor phase traditionally made of carbonaceous material (Namieśnik et al., 2005); although Tenax and Florisil coated with 2,4-dinitrophenylhydrazine have been used also (Esteve-Turrillas et al., 2012). Fig. 20.3 shows some passive sampling devices that have been used to capture OC in air and water, in addition, Table 20.1 shows some applications of these samplers.

20.5 Factors involved in the capture process during passive sampling

As mentioned in a general way in Section 20.3, passive sampling involves the diffusion and transport of the pollutant through different barriers, which also includes diffusion through the polymeric receptor phase. In this sense, the chemical potential or factors such as biofouling or external flow (winds or water currents) can modify the mass transfer process; however, the transport of the contaminant and the subsequent absorption into the polymeric receptor phase also depend, to a great extent, on the nature of the contaminant, the

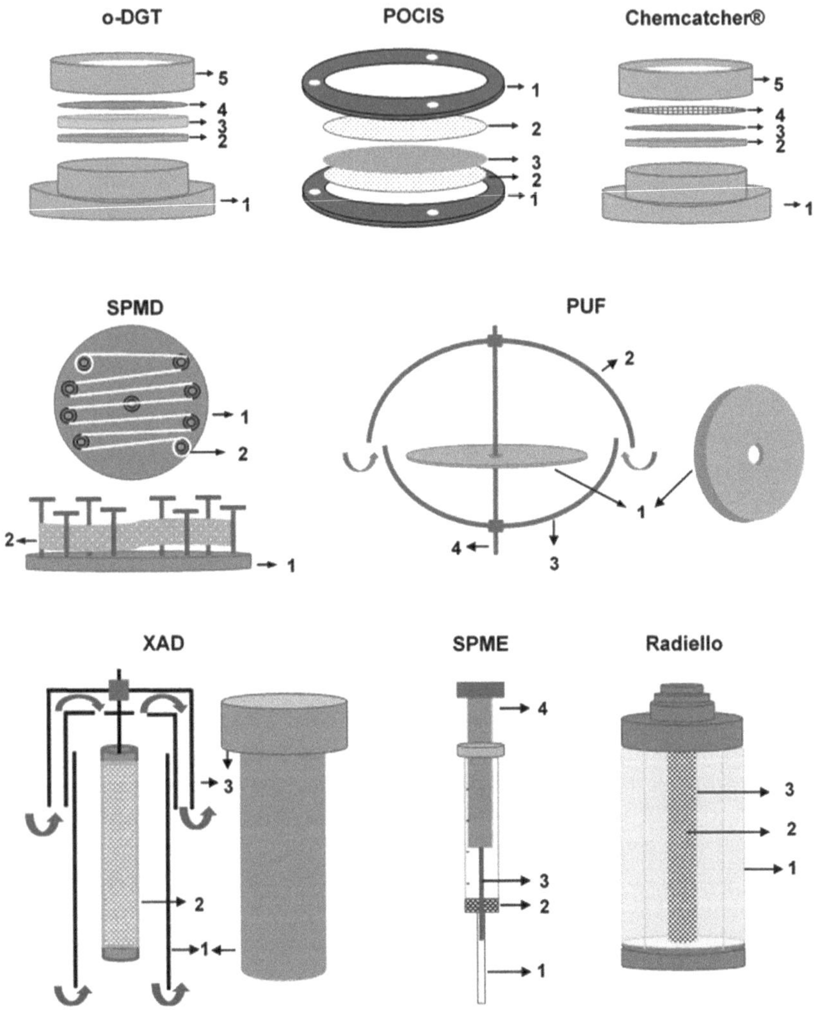

FIGURE 20.3 O-DGT (1: base; 2: receiving phase; 3: diffusive gel; 4: membrane); POCIS (1: stainless steel ring; 2: membrane; 3: receiving phase); Chemcatcher (1: base; 2: receiving phase; 3: membrane; 4: stainless mesh); SPMD (1: deployment rack; 2: LDPE flat tube with triolein as receptor phase); PUF (1: PUF disk; 2: lid; 3: bottom shelter); XAD (1: bottom shelter; 2: XAD resin in mesh cylinder; 3: lid); SPME (1: holder; 2: stainless steel nut; 3: receiving phase; 4: gastight syringe); Radiello (1: diffusive body; 2: receiving phase; 3: mesh cylinder).

properties of the material, and the contaminant–receptor phase interaction. Higher sampling rates and diffusion coefficients that improve the uptake process are obtained when the resistance to transport of the contaminant in the polymeric material is low (Rusina et al., 2010a,b).

In principle, linear, flexible, symmetrical, and low-molecular-weight molecules show high mobility and diffusion, which allows them to quickly

TABLE 20.1 Selected passive sampler applications for sampling chlorinated organic compounds.

Sampler/ polymer	Matrix	Compounds	References
LDPE	River water	PCB congeners	Estoppey et al. (2015, 2016)
	River water	PCB congeners	Apell and Gschwend (2017)
	River water and air	OCP	Meire et al. (2016)
	North Atlantic and tropical ocean	PCB congeners, and OCP	Sun et al. (2016)
	Coastal waters	PCDD, PCDF	Friedman et al. (2012), Friedman and lohmann (2014)
	Coastal waters	PCB congeners	Oziolor et al. (2018)
	Aire	PCB congeners	Donald and Anderson (2017)
	Fluvial channels	PCB congeners	Martínez et al. (2019)
SPMD	Aire	PCDD, PCDF, PCB congeners	Mari et al. (2008)
	Coastal waters	PCB congeners and OCP	Zhao et al. (2018)
	River water	PCDD, PCDF	Charlestra et al. (2008)
	Lake water	PCB congeners and OCP	O'Toole et al. (2006)
	Air (high mountain)	Isomers of HCB and PCB congeners	Van Drooge et al. (2005)
	Air	PCDD, PCDF congeners	Lohmann et al. (2001)
	Air and water	OCP	Lisouza et al. (2020)
	Water	PCDD, PCDF and dioxin-like PCBs	Roach et al. (2009)

Continued

TABLE 20.1 Selected passive sampler applications for sampling chlorinated organic compounds.—cont'd

Sampler/ polymer	Matrix	Compounds	References
Silicone rubber	Coastal waters	PCDD, PCDF congeners	Cornelissen et al. (2010)
	River water	OCP	Abbasi and Mannaerts (2018)
	River water	OCP and PCB congeners	Chepchirchir et al. (2017)
	River water	OCP and PCB congeners	Silva-Barni et al. (2019b)
	Air	OCP, PCB congeners, phthalates, organophosphate esters, among others.	Tromp et al. (2019)
	Air	PCB congeners	Vorkamp et al. (2016)
Chemcatcher®	Water	γ-HCH, dieldrin, PCB-52, PCB-153 (C18 as reception phase).	El-Shenawy et al. (2010)
	River water	Dimetachlor, metolachlor, dichlorvos, among other types of compounds (Atlantic HLB-L disk as reception phase)	Grodtke et al. (2020)
	River water	HCB, lindane, endosulfan I, dieldrin, among other types of compounds	Vrana et al. (2006)
XAD-2	Air	PCB congeners	Guéguen et al. (2013)
	Air in polar regions	PCB congeners and OCP	Choi et al. (2008)
	Air	OCP	Shen et al. (2005)
	Air	Endosulfan HCB, metolachlor, alachlor, among other types of compounds	Hayward et al. (2010)
	River water	PCB congeners, OCP	Silva-Barni et al. (2019b)
	Air	OCP (endosulfan isomers, γ-HCH, DDT, DDD, and DDE isomers, γ-chlordane, heptachlor epoxide, among others)	Silva-Barni et al. (2019a)

TABLE 20.1 Selected passive sampler applications for sampling chlorinated organic compounds.—cont'd

Sampler/ polymer	Matrix	Compounds	References
PUF	Air	PCDD, PCDF, dioxin-like PCBs, OCP (chlordane, α-HCH, γ-HCH, heptachlor, heptachlor epoxide, hexachlorobenzene, and DDT)	Toinaga et al. (2016)
	Air	PCDD, PCDF	Ren et al. (2011)
	Air	OCP, PCBs, Polybrominated Diphenyl Ethers (PBDE)	De la Torre et al. (2020)
	Air	17 PCDDs/PCDFs and 12 dL-PCBs	Yoonki et al. (2014)
	Air	OCP, PCBs, PCDD, and PCDF congeners among other type of compounds	White et al. (2020)
	Air	PCDD, PCDF and dioxin-like PCBs	Cortés et al. (2016)
	Air	PCB congeners, OCP (HCH isomers, chlordane compounds, dichlorodiphenyltrichloroethane DDT isomers, endosulfan isomers, among others)	Can-Güven et al. (2019)

penetrate through the polymer; likewise, the rate of diffusion of the pollutant through the polymer varies with the free volume (interchain space), the mobility of its molecular chains, and the presence of crystalline or amorphous domains in the polymer, a characteristic that is defined by the glass transition temperature (T_g). Thus, polymers with low T_g have greater diffusivity; for example, silicone rubbers (with T_g between −125 and −127) have shown higher penetration of organic compounds compared to other polymers such as polyethylene ($-125 > T_g < 100°C$), POM (T_g, −90 at −70°C), ethylene/octene copolymer (T_g, −75°C) (Rusina et al., 2007, 2010a,b).

On the other hand, depending on the chemical structure of the polymeric phase (sorbent) and the OC (sorbate), specific and nonspecific molecular interactions (dipole−dipole, induced dipole−dipole, hydrogen bond) may occur; nonspecific interactions occur in all systems and are mainly caused by the uneven distribution of electron density in the molecule, which generates temporary dipoles. On the other hand, specific interactions are fostered when both in the sorbent and in the sorbate, there are functional groups with negative charge density (lone electron pairs, π bonds) or positive charge (protonic hydrogen atoms of acidic hydroxyl groups, cations of small radius exchange) in which a donor–acceptor interaction predominates (Schaeffer, 2008; Martínez et al., 2017b). The functional groups of the OC are varied, since they can present different numbers of chlorines, or they can have simple structures formed by short carbon chains (e.g., polychloromethanes), or slightly more complex structures with partially or totally unsaturated cyclic carbon skeletons, with aromatic rings or oxygenated, nitrogenous functional groups, organophosphates, or functional groups containing sulfur (e.g., Ensosulfan, with a sulfinic acid ester) (Nuro, 2018; Taiwo, 2019).

With sorbents without functional groups, permanent dipoles, or charges at the structural level, as occurs with samplers based on LDPE or PDMS, the interactions between the OC and the sorbent are mediated by dispersive nonspecific interactions, induced dipole–dipole interactions, and linkages of weak hydrogen (for example, –CH ⋯ O– or –CH ⋯ Cl–C–) (Schaeffer, 2008). On the other hand, sorbents such as XAD-4 or SBD-XC based on poly(styrene-co-divinylbenzene) or poly2,6-diphenyl-p-phenylene oxide, in addition to their ability to interact in a dispersive/nonspecific way with OC, can through benzene rings (group rich in electrons) interact specifically with polar groups, aromatic rings (π–π stacking), cations, or hydrogen bond donor groups ($N^{+}H_3$, OH, NH, CH, SH) found in the pollutant structure (Schaeffer, 2008; Amad et al., 2015). Samplers based on EVA, POM, or OasisHLB can also present different types of dispersive interactions and nondispersive given the presence of apolar and polar groups in the structure (Hofman-Caris et al., 2015; Jeong et al., 2017; Martinez et al., 2017a; Martínez et al., 2017b), which gives it the capacity to capture compounds of different polarity. This is also achieved with mixed materials (PUF-XAD, PDMS/DVB).

Additionally, for solid receptor phases, the surface tension is a physicochemical parameter associated with the chemical structure and free energy of the surface of the polymeric material; therefore, knowledge of this parameter for each polymer can help to describe the adsorption of pollutants and even the formation of biofouling on sampler surfaces thermodynamically. In this context, materials with high surface energy are preferred that tend to reduce energy through the adsorption of pollutants present in the sampling phase (Palencia, 2017).

20.6 Chemical analysis of passive samplers

After sampling, extraction of the contaminants from the receiving phase and subsequent chemical analyses are essential for qualitative or quantitative eterminations of the OC in the sampling matrix. Regarding chemical analysis, chromatographic techniques (i.e., gas chromatography and high-efficiency liquid chromatography) coupled with mass spectrometry (MS) are the most commonly used instrumental techniques because they allow the separation and determination of different compounds in a single chromatographic run; in addition to coupling MS confers greater sensitivity and precision in determinations (Taylor et al., 2019). On the other hand, the extraction of pollutants is one of the crucial stages; the selection of the most appropriate extractive procedure will depend on the nature of the OC (volatility and polarity), the nature of the sampler (polymeric foam, polymeric membrane, solid support), or factors such as time availability, cost of analysis, or availability of an adequate instrumental equipment (Esteve-Turrillas et al., 2012).

Volatile OCs can be extracted from the receiving phase in a simple way if there are thermal desorption, purge and trap injection ports, or headspace (HS) that are traditionally coupled to GC equipment with flame ionization detectors, electron capture, or MS; the advantage of these processes lies in the possibility of miniaturization of the process and in extraction without the need for toxic solvents. On the other hand, for less volatile compounds, such as PCBs or some pesticides, it is necessary to sequentially extract the contaminants with appropriate organic solvents (hexane, methanol, acetonitrile, ethyl acetate, cyclohexane, acetone/hexane, ether of petroleum, toluene, dichloromethane, etc.).The selection of the extracting solvent will depend on the affinity for the sampled OC, however the said solvent must not have the ability to dissolve the polymer or cause considerable swelling in it, as this could affect the physical manipulation of the receiving phase, in some cases the possibility of its reuse would be eliminated, and the quality of the analysis extracts would be compromised. This last aspect would lead to the addition of exhaustive cleaning steps without guarantees of complete elimination of the polymer or analytical interferences (plasticizers, impurities, or residual uncured oligomers) released, damaging the quality of the results and compromising the functioning of the instrumental equipment (Rusina et al., 2007; Mills et al., 2014).

Dialysis, solvent extraction, pressurized solvent extraction (PSE), microwave-assisted extraction (MAE), accelerated solvent extraction (ASE), and Soxhlet are different extraction techniques that can be implemented depending on the receptor phase. Similarly, since different matrix components are also adsorbed by the polymers in the samplers (mainly in water), extracts are usually cleaned to reduce or eliminate interferences in the chromatographic measurement; solid-phase extraction and dispersion (SPE and d-SPE), liquid–liquid extraction, solid–liquid column extraction, and gel permeation chromatography are methods used (Esteve-Turrillas et al., 2012; Tucca et al., 2014; Brockmeyer et al., 2015; Martínez et al., 2017b).

For selection of the best extractive method, in addition to taking into account the selection of methods with the best analytical quality parameters (sensitivity, reproducibility, precision, accuracy, selectivity), it is important to evaluate other factors such as the safety of the analyst or operator and the environmental impact of the analytical procedure; since it is controversial that the entire sampling, extraction, and analysis process has meritorious environmental objectives, but that during this process waste is used or produced equal to or more toxic than the chemical pollutants that are intended to be determined. For example, the extraction of an SPMD or a PUF through dialysis and Soxhlet, respectively, uses around 100–900 mL of organic solvents, some of which are very toxic, such as petroleum ether. If it is taken into account that the determinations are usually carried out in duplicate or triplicate with large spatial distributions, the number of samplers to analyze and therefore the amount of solvent used is considerable, and most of it is not reused. This is the reason why, in that sense, one must opt for more environmentally friendly methodologies, which allow a reduction in waste and the use of toxic solvents while preserving analytical quality. Direct injection techniques such as TD, HS, or automated extraction and cleaning methods, or the possibility of miniaturization (MAE, PSE, ASE, d-SPE, QuEChERS), would allow, in addition to reducing waste or the use of materials and reagents, to reduce the analysis time of the samplers (Brockmeyer et al., 2015; Martínez et al., 2017a; Taylor et al., 2019).

20.7 Conclusions and remarks

It is evident that the future of passive sampling lies in the search for alternative materials that are efficient and economical for the capture of pollutants. However, the evolution of the technique goes beyond that, as what is currently being sought is to obtain universal devices (for compounds of different polarity and nature), with low operating costs, reusable, and where, at the level of chemical analysis in the laboratory, the use of solvents or pretreatment steps is minimized. Regarding the universality of the technique, the use of hydrophilic–hydrophobic polymers or mixed receptor phases are options that have so far made it possible to capture compounds of different polarity in the polymer phase.

The resistance and reuse of the polymeric capture material is also something that is sought to reduce costs; for example, Al-Alam et al. (2020), used SiC foams that showed good results for different compounds, but at the same time, they showed that the chemical inertness and mechanical resistance of the material allow its reuse, even with acid treatments. This is advantageous when compared to XAD-2 resins which, despite their good adsorption capacity, tend to deteriorate during the process (Al-Alam et al., 2020).

Valenzuela et al. (2019), for their part, designed a new sampling device for pesticides in surface water and groundwater based on the liquid phase microextraction technique with hollow fiber with direct detection by GC-MS. Basically, the sampler consisted of the use of a small porous polymeric hollow fiber of polyethylene in the shape of a "U," inside which an acceptor solvent with affinity for the test compounds, but immiscible to water, was housed; the sampler presented good results for moderately or highly hydrophobic analytes. This device, by not requiring sample pretreatment (since the solvent is injected directly into the GC), substantially reduces analytical costs, additionally, the isolation and transport of the sample are less problematic and it is possible to reuse and fill the porous fiber for eventual applications.

Finally, something that should also be taken into account is the impact of the manufacture of the adsorbent material, for example, adsorbents produced after the conversion of agricultural waste material have been used for the removal of pollutants (Inyang and Dickenson, 2015), which would lead to thinking that they could be potential sorbents for passive samplers, however, although this approach uses waste material, it requires a pyrolysis stage whose energy expenditure is high; in addition, during this stage polycyclic aromatic hydrocarbons (PAHs) are produced, which also, if the sorbent is contaminated (requiring extensive cleaning), would constitute a source of environmental contamination.

References

Abbasi, Y., Mannaerts, C.M., 2018. Evaluating organochlorine pesticide residues in the aquatic environment of the lake Naivasha river basin using passive sampling techniques. Environ. Monit. Assess. 190, 1–12. https://doi.org/10.1007/s10661-018-6713-4.

Adu-Kumi, S., Kawano, M., Shiki, Y., Yeboah, P.O., Carboo, D., et al., 2010. Organochlorine pesticides (OCPs), dioxin-like polychlorinated biphenyls (dl-PCBs), polychlorinated dibenzo-p-dioxins and polychlorinated dibenzo furans (PCDD/Fs) in edible fish from Lake Volta, Lake Bosumtwi and Weija Lake in Ghana. Chemosphere 81, 675–684. https://doi.org/10.1016/j.chemosphere.2010.08.018.

Aerts, R., Joly, L., Szternfeld, P., Tsilikas, K., de Cremer, K., et al., 2018. Silicone Wristband passive samplers yield highly individualized pesticide residue exposure profiles. Environ. Sci. Technol. 52, 298–307.

Al-Alam, J., Lévy, M., Ba, H., Pham-Huu, C., Millet, M., 2020. Passive air samplers based on ceramic adsorbent for monitoring of organochlorine pesticides, polycyclic aromatic hydrocarbons and polychlorinated biphenyls in outdoor air. Environ. Technol. & Innov. 20, 1–12. https://doi.org/10.1016/j.eti.2020.101094.

Amad, A., Siddique, J.A., Laskar, M.A., Kumar, R., Mohd-Setapar, S.H., et al., 2015. New generation Amberlite XAD resin for the removal of metal ions: a review. J. Environ. Sci. 31, 104–123. https://doi.org/10.1016/j.jes.2014.12.008.

Apell, J.N., Gschwend, P.M., 2017. The atmosphere as a source/sink of polychlorinated biphenyls to/from the lower Duwamish Waterway superfund site. Environ. Pollut. 227, 263–270. https://doi.org/10.1016/j.envpol.2017.04.070.

Berho, C., Claude, B., Coisy, E., Togola, A., Bayoudh, S., et al., 2017. Laboratory calibration of a POCIS-like sampler based on molecularly imprinted polymers for glyphosate and AMPA sampling in water. Anal. Bioanal. Chem. 409, 2029–2035. https://doi.org/10.1007/s00216-016-0150-4.

Berry-Cabán, C.S., 2011. DDT and silent spring: fifty years after. J. Mil. & Veterans' Health 19, 19–24.

Brockmeyer, B., Kraus, U.R., Theobald, N., 2015. Accelerated solvent extraction (ASE) for purification and extraction of silicone passive samplers used for the monitoring of organic pollutants. Environ. Sci. Pollut. Control Ser. 22, 19887–19895. https://doi.org/10.1007/s11356-015-5192-1.

Can-Güven, E., Gedik, K., Kurt-Karakuş, P.B., 2019. Polyurethane foam disk passive sampler derived air concentrations of persistent organic pollutants in an agricultural region with hot climate. Atmos. Pollut. Res. 10, 1913–1920. https://doi.org/10.1016/j.apr.2019.08.004.

Charlestra, L., Courtemanch, D.L., Amirbahman, A., Patterson, H., 2008. Semipermeable membrane device (SPMD) for monitoring PCDD and PCDF levels from a paper mill effluent in the Androscoggin river, Maine, USA. Chemosphere 72, 1171–1180. https://doi.org/10.1016/j.chemosphere.2008.03.057.

Charriau, A., Lissalde, S., Poulier, G., Mazzella, N., Buzier, R., et al., 2016. Overview of the Chemcatcher® for the passive sampling of various pollutants in aquatic environments part A: principles, calibration, preparation and analysis of the sampler. Talanta 148, 556–571. https://doi.org/10.1016/j.talanta.2015.06.064.

Chepchirchir, B.S., Paschke, A., Schüürmann, G., 2017. Passive sampling for spatial and temporal monitoring of organic pollutants in surface water of a rural-urban river in Kenya. Sci. Total Environ. 601–602, 453–460. https://doi.org/10.1016/j.scitotenv.2017.05.143.

Choi, S.D., Baek, S.Y., Chang, Y.S., Wania, F., Ikonomou, M.G., et al., 2008. Passive air sampling of polychlorinated biphenyls and organochlorine pesticides at the Korean Arctic and Antarctic Research stations: implications for long-range transport and local pollution. Environ. Sci. Technol. 42. https://doi.org/10.1021/es801004p.

Chopra, A.K., Sharma, M.K., Chamoli, S., 2011. Bioaccumulation of organochlorine pesticides in aquatic system–an overview. Environ. Monit. Assess. 173 1–4, 905–916. https://doi.org/10.1007/s10661-010-1433-4.

Cornelissen, G., Broman, D., Naes, K., 2010. Freely dissolved PCDDF/F concentrations in the Frierfjord, Norway: comparing equilibrium passive sampling with "active" water sampling. J. Soils Sediments 10, 162–171. https://doi.org/10.1007/s11368-009-0152-3.

Cortés, J., Cobo, M., González, C.M., Gómez, C.D., Abalos, M., et al., 2016. Environmental variation of PCDD/Fs and dl-PCBs in two tropical Andean Colombian cities using passive samplers. Sci. Total Environ. 568, 614–623. https://doi.org/10.1016/j.scitotenv.2016.02.094.

De la Torre, A., Sanz, P., Navarro, I., Martínez, M.A., 2020. Investigating the presence of emerging and legacy POPs in European domestic air. Sci. Total Environ. 746, 1–10. https://doi.org/10.1016/j.scitotenv.2020.141348.

Dhakal, K., Gadupudi, G.S., Lehmler, H.J., et al., 2018. Sources and toxicities of phenolic polychlorinated biphenyls (OH-PCBs). Sci. Pollut. Res. 25, 16277–16290. https://doi.org/10.1007/s11356-017-9694-x.

Donald, C.E., Anderson, K.A., 2017. Assessing soil-air partitioning of PAHs and PCBs with a new fugacity passive sampler. Sci. Total Environ. 596–597, 293–302. https://doi.org/10.1016/j.scitotenv.2017.03.095.

Dong, J., Fan, H., Sui, D., Li, L., Sun, T., 2014. Sampling 4-chlorophenol in water by DGT technique with molecularly imprinted polymer as binding agent and nylon membrane as diffusive layer. Anal. Chim. Acta 822, 69–77. https://doi.org/10.1016/j.aca.2014.03.015.

Dumanoglu, Y., Gaga, E.O., Gungormus, R.C., Sofuoglu, S.C., Obadasi, M., 2018. Spatial and seasonal variations, sources, air-soil exchange, and carcinogenic risk assessment for PAHs and PCBs in air and soil of Kutahya, Turkey, the province of thermal power plants. Sci. Total Environ. 580, 920–935. https://doi.org/10.1016/j.scitotenv.2016.12.040.

Eddy, N.O., 2020. Theoretical chemistry study on the toxicity of some polychlorobiphenyl (PCB) compounds using molecular descriptors. Sci. Afr. 10, 1–12. https://doi.org/10.1016/j.sciaf.2020.e00587.

El-Shenawy, N.S., Nabil, Z.I., Abdel-Nabi, I.M., Greenwood, R., 2010. Comparing the passive and active sampling devices with biomonitoring of pollutants in Langstone and Portsmouth Harbour, UK. J. Environ. Sci. & Technol. 3, 1–17. https://doi.org/10.3923/jest.2010.1.17.

Esteve-Turrillas, F.A., Yusà, V., Pastor, A., de la Guardia, M., 2008. New perspectives in the use of semipermeable membrane devices as passive samplers. Talanta 74, 443–457. https://doi.org/10.1016/j.talanta.2007.06.019.

Esteve-Turrillas, F.A., Pastor, A., de la Guardia, M., 2012. Passive sampling of atmospheric organic contaminants. In: Pawliszyn, J. (Ed.), Comprehensive Sampling and Sample Preparation, Analytical Techniques for Scientists. Elsevier., pp. 201–222. https://doi.org/10.1016/B978-0-12-381373-2.00011-9

Estoppey, N., Omlin, J., Schopfer, A., Esseiva, P., Vermeirssen, L.M., et al., 2015. Low density polyethylene (LDPE) passive samplers for the investigation of polychlorinated biphenyl (PCB) point sources in rivers. Chemosphere 118, 268–276. https://doi.org/10.1016/j.chemosphere.2014.09.032.

Estoppey, N., Schopfer, A., Fong, C., Delémont, O., De Alencastro, L.F., et al., 2016. An in-situ assessment of low-density polyethylene and silicone rubber passive samplers using methods with and without performance reference compounds in the context of investigation of polychlorinated biphenyl sources in rivers. Sci. Total Environ. 572, 794–803. https://doi.org/10.1016/j.scitotenv.2016.07.092.

Fernandes, A.R., Falandysz, J., 2021. Polybrominated dibenzo-p-dioxins and furans (PBDD/Fs): contamination in food, humans and dietary exposure. Sci. Total Environ. 61, 1–18. https://doi.org/10.1016/j.scitotenv.2020.143191.

Fiedler, H., 2007. National PCDD/PCDF release inventories under the Stockholm convention on persistent organic pollutants. Chemosphere 67, S96–S108. https://doi.org/10.1016/j.chemosphere.2006.05.093.

Forbes, P., 2020. Atmospheric chemistry analysis: a review. Anal. Chem. 92, 455–472. https://doi.org/10.1021/acs.analchem.9b04623.

Francisco, A.P., Nardocci, A.C., Tominaga, M.Y., Rodrigues, C., de Assunção, 2017. Spatial and seasonal trends of polychlorinated dioxins, furans and dioxin-like polychlorinated biphenyls in air using passive and active samplers and inhalation risk assessment. Atmos. Pollut. Res. 8, 979–987. https://doi.org/10.1016/j.apr.2017.03.007.

Friedman, C.L., Lohmann, R., 2014. Comparing sediment equilibrium partitioning and passive sampling techniques to estimate benthic biota PCDD/F concentrations in Newark Bay, New Jersey (U.S.A.). Environ. Pollut. 186, 172–179. https://doi.org/10.1016/j.envpol.2013.12.002.

Friedman, C.L., Cantwell, M.G., Lohmann, R., 2012. Passive sampling provides evidence for Newark Bay as a source of polychlorinated dibenzo-p-dioxins and furans to the New York/New Jersey, USA. Atmos. Environ. Toxicol. Chem. 31, 253–261. https://doi.org/10.1002/etc.742.

Garcés, V., Lerma, T., Palencia, M., 2018. Building a probe-type passive sampler to study the contents of heavy metals in the aqueous phase in the soil. J. Sci. Technol. Appl. 4, 4–16. https://doi.org/10.34294/j.jsta.18.4.26.

Godlewska, K., Stepnowski, P., Paszkiewicz, M., 2021. Pollutant analysis using passive samplers: principles, sorbents, calibration and applications. A review. Environ. Chem. Lett. 19, 465–520. https://doi.org/10.1007/s10311-020-01079-6.

Gong, X., Li, K., Wu, C., Wang, L., Sun, H., 2018. Passive sampling for monitoring polar organic pollutants in water by three typical samplers. Trends Environ. Anal. Chem. 17, 23–33. https://doi.org/10.1016/j.teac.2018.01.002.

Greenwood, R., Mills, G.A., Vrana, B., Allan, I., Aguilar-Martínez, G., et al., 2007. Monitoring of priority pollutants in water using chemcatcher passive sampling devices. In: Greenwood, R., et al. (Eds.), Comprehensive Analytical Chemistry, 48. Elsevier, pp. 199–229. https://doi.org/10.1016/S0166-526X(06)48009-0.

Grodtke, M., Paschke, A., Harzdorf, J., Krauss, M., Schüürmann, G., 2020. Calibration and field application of the Atlantic HLB disk containing Chemcatcher® passive sampler – quantitative monitoring of herbicides, other pesticides and transformation products in German streams. J. Hazard Mater. 124538. https://doi.org/10.1016/j.jhazmat.2020.124538.

Guéguen, F., Stille, P., Millet, M., 2013. Optimisation and application of accelerated solvent extraction and flash chromatography for quantification of PCBs in tree barks and XAD-2 passive samplers using GC–ECD with dual columns. Talanta 140–146. https://doi.org/10.1016/j.talanta.2013.02.056.

Guibal, R., Buzier, R., Lissalde, S., Guibaud, G., 2019. Adaptation of diffusive gradients in thin films technique to sample organic pollutants in the environment: an overview of o-DGT passive samplers. Sci. Total Environ. 693, 1–12. https://doi.org/10.1016/j.scitotenv.2019.07.343.

Hayward, S., Gouin, T., Wania, F., 2010. Comparison of four active and passive sampling techniques for pesticides in air. Environ. Sci. Technol. 44, 3410–3416. https://doi.org/10.1021/es902512h.

Hofman-Caris, C.H., Bäuerlein, P.S., Siegers, W.G., Ziaie, J., Tolkamp, H.H., et al., 2015. Water Supply 15, 1207–1219. https://doi.org/10.2166/ws.2015.084.

Huang, B., Lei, C., Wei, C., Zeng, G., 2014. Chlorinated volatile organic compounds (Cl-VOCs) in environment -sources, potential human health impacts, and current remediation technologies. Environ. Int. 71, 118–138. https://doi.org/10.1016/j.envint.2014.06.013.

Huckins, J., Tubergen, M., Manuweera, G., 1990. Semipermeable membrane devices containing model lipid: a new approach to monitoring the bioavailability of lipophilic contaminants and estimating their bioconcentration potential. Chemosphere 20, 533–552. https://doi.org/10.1016/0045-6535(90)90110-F.

Inyang, M., Dickenson, E., 2015. The potential role of biochar in the removal of organic and microbial contaminants from potable and reuse water: a review. Chemosphere 134, 232–240. https://doi.org/10.1016/j.chemosphere.2015.03.072.

Inyinbor, A.A., Adebesin, B.O., Oluyori, A.P., Adelani-Akande, T.A., Dada, A.O., Oreofe, T.A., 2018. Water pollution: effects, prevention, and climatic impact. In: Glavan, M. (Ed.), Water Challenges of an Urbanizing World. IntechOpen., pp. 33–53. https://doi.org/10.5772/intechopen.72018

Jayaraj, R., Megha, P., Sreedey, P., 2016. Organochlorine pesticides, their toxic effects on living organisms and their fate in the environment. Interdiscipl. Toxicol. 9, 90–100. https://doi.org/10.1515/intox-2016-0012.

Jeong, Y., Schäffer, A., Smith, K., 2017. Equilibrium partitioning of organic compounds to OASIS HLB® as a function of compound concentration, pH, temperature and salinity. Chemosphere 174, 297–305. https://doi.org/10.1016/j.chemosphere.2017.01.116.

Kanan, S., Samara, F., 2018. Dioxins and furans: a review from chemical and environmental perspectives. Trends Environ. Anal. Chem. 17, 1–13. https://doi.org/10.1016/j.teac.2017.12.001.

Kaserzon, S.L., Hawker, D.W., Kennedy, K., et al., 2014. Characterisation and comparison of the uptake of ionizable and polar pesticides, pharmaceuticals and personal care products by POCIS and Chemcatchers. Environ. Sci. Process. & Impacts 16, 2517–2526. https://doi.org/10.1039/c4em00392f.

Khairy, M.A., Lohmann, R., 2013. Feasibility of using low density polyethylene sheets to detect atmospheric organochlorine pesticides in Alexandria, Egypt. Environ. Environ. Pollut. 181, 151–158. https://doi.org/10.1016/j.envpol.2013.06.031.

Kjellstrom, T., Lodh, M., McMichael, T., Ranmuthugala, G., Shrestha, R., Kingsland, S., 2006. Air and water pollution: burden and strategies for control. In: Jamison, D.T., Breman, J.G., Measham, A.R., Alleyne, G., Claeson, M., Evans, D.B., Jha, P., Mills, A., Musgrove, P. (Eds.), Disease Control Priorities in Developing Countries, 2nd ed. Washington (DC): The International Bank for Reconstruction and Development/The World Bank, Washington (DC), pp. 817–832.

Klánová, J., Harner, T., 2013. The challenge of producing reliable results under highly variable conditions and the role of passive air samplers in the Global Monitoring Plan. Trends Anal. Chem. 46, 139–149. https://doi.org/10.1016/j.trac.2012.07.021.

Kleanthi, G., Katerina, L., Evaggelia, P., Andreas, L., 2008. Mechanisms of actions and health effects of organochlorine substances: a review. Health Sci. J. 2, 89–99.

Koelmel, J.P., Lin, E.Z., Guo, P., Zhou, J., He, J., et al., 2021. Exploring the external exposome using wearable passive samplers-The China BAPE study. Environ. Pollut. 270, 1–13. https://doi.org/10.1016/j.envpol.2020.116228.

Krasnobaev, A., ten Dam, G., Boerrigter-Eenling, R., Peng, F., van Leeuwen, S.P.J., et al., 2020. Legacy and emerging persistent organic pollutants in Antarctic Benthic Invertebrates near Rothera point, Western Antarctic Peninsula. Environ. Sci. Technol. 54, 2763–2771. https://doi.org/10.1021/acs.est.9b06622.

Lao, W., Hong, Y., Tsukada, D., Maruya, K.A., Gan, J., 2016. A new film-based passive sampler for moderately hydrophobic organic compounds. Environ. Sci. Technol. 50, 13470–13476. https://doi.org/10.1021/acs.est.6b04750.

Lévy, M., Ba, H., Pallares, C., Pham-Huu, C., Millet, M., 2020. Comparison and calibration of diverse passive samplers used for the air sampling of pesticides during a regional sampling monitoring campaign. Atmos. Pollut. Res. 11 https://doi.org/10.1016/j.apr.2020.03.014.

Li, H., Ye, S., Ye, J., Fan, J., Gao, M., Guo, H., 2017. Baseline survey of sediments and marine organisms in Liaohe Estuary: heavy metals, polychlorinated biphenyls and organochlorine pesticides. Mar. Pollut. Bull. 114, 555–563. https://doi.org/10.1016/j.marpolbul.2016.09.002.

Lisouza, F.A., Owuor, P.O., Lalah, J.O., 2020. Sources, distribution, and risk assessment of organochlorine pesticides in Nairobi City, Kenya. J. Environ. Sci. 96, 178–185. https://doi.org/10.1016/j.jes.2020.04.046.

Lohmann, R., Corrigan, B.P., Howsam, M., Jones, K.C., Ockenden, W.A., 2001. Further developments in the use of semipermeable membrane devices (SPMDs) as passive air samplers for persistent organic pollutants: field application in a spatial survey of PCDD/Fs and PAHs. Environ. Sci. Technol. 35 12, 2576–2582. https://doi.org/10.1021/es0001862.

Lompart, M., Celeiro, M., García-Jares, C., Dagnac, T., 2019. Environmental applications of solid-phase microextraction. Trends Anal. Chem. 112, 1–12. https://doi.org/10.1016/j.trac.2018.12.020.

Mari, M., Schuhmacher, M., Feliubadaló, J., Domingo, J.L., 2008. Air concentrations of PCDD/Fs, PCBs and PCNs using active and passive air samplers. Chemosphere 70 (9), 1637–1643. https://doi.org/10.1016/j.chemosphere.2007.07.076.

Martinez, J., Troches, E., Páez, M., 2017a. Evaluation of poly(ethylene-co-vinylacetate) as polymeric adsorbent of polychlorinated biphenyls in water. J. Sci. Technol. Appl. 2, 94–107. https://doi.org/10.34294/j.jsta.17.2.18.

Martínez, J.M., Páez, M.I., Palencia, M.S., Peña, M.R., 2017b. Dispositivos de muestreo de plaguicidas en aguas superficiales basados en adsorbentes de fibra de vidrio-poli(etileno-co-vinilacetato). Revista Latinoamericana de Metalurgia y Materiales 38, 1–12. http://www.rlmm.org/ojs/index.php/rlmm/article/view/838.

Martínez, A., Awad, A.M., Herkert, N.J., Hornbucle, K.C., 2019. Determination of PCB fluxes from Indiana Harbor and Ship Canal using dual-deployed air and water passive samplers. Environ. Pollut. 244, 469–476. https://doi.org/10.1016/j.envpol.2018.10.048.

Meire, R.O., Khairy, M., Targino, A.C., Galvão, P.M., Torres, J.P., et al., 2016. Use of passive samplers to detect organochlorine pesticides in air and water at wetland mountain region sites (S-SE Brazil). Chemosphere 144, 2175–2182. https://doi.org/10.1016/j.chemosphere.2015.10.133.

Mills, G., Fones, G., Booij, K., 2011. Passive sampling technologies. In: Quevauviller, P., Roose, P., Verreet, G. (Eds.), Chemical Marine Monitoring: Policy Framework and Analytical Trends. John Wiley & Sons, Ltd., pp. 397–432. https://doi.org/10.1002/9781119990826.ch13

Mills, G.A., Gravell, A., Vrana, B., Harman, C., Budzinski, H., et al., 2014. Measurement of environmental pollutants using passive sampling devices – an updated commentary on the current state of the art. Environ. Sci. Process. & Impacts 16, 369–373.

Namieśnik, J., Zabiegała, B., Kot-Wasik, A., et al., 2005. Passive sampling and/or extraction techniques in environmental analysis: a review. Anal. Bioanal. Chem. 381, 279–301. https://doi.org/10.1007/s00216-004-2830-8.

Nuro, A., 2018. Introductory chapter: organochlorine. In: Nuro, A. (Ed.), Organochlorine. IntechOpen, pp. 1–6. https://doi.org/10.5772/intechopen.81271.

Okeme, J.P., Nguyen, L.V., Lorenzo, M., Dhal, S., Pico, Y., et al., 2018. Polydimethylsiloxane (silicone rubber) brooch as a personal passive air sampler for semi-volatile organic compounds. Chemosphere 208, 1002–1007. https://doi.org/10.1016/j.chemosphere.2018.05.196.

Oziolor, E.M., Apell, J.N., Winfield, Z.C., Back, J.A., Usenko, S., et al., 2018. Polychlorinated biphenyl (PCB) contamination in Galveston Bay, Texas: comparing concentrations and profiles in sediments, passive samplers, and fish. Environ. Pollut. 236, 609–618. https://doi.org/10.1016/j.envpol.2018.01.086.

O'Toole, S., Metcalfe, C., Craine, I., Gross, M., 2006. Release of persistent organic contaminants from carcasses of lake ontario chinook salmon (*Oncorhynchus tshawytscha*). Environ. Pollut. 140, 102–113. https://doi.org/10.1016/j.envpol.2005.06.019.

Palencia, M., Lerma, T., Córdoba, A., 2016. Polyurethanes with boron retention properties for the development of agricultural fertilization smart systems. J. Sci. Technol. Appl. 1, 39–52. https://doi.org/10.34294/j.jsta.16.1.4.

Palencia, M., 2017. Surface free energy of solids by contact angle measurements. J. Sci. Technol. Appl. 2, 84–93. https://doi.org/10.34294/j.jsta.17.2.17.

Pintado-Herrera, M.G., Lara-Martín, P.A., González-Mazo, E., Allan, I.J., 2016. Determination of silicone rubber and low-density polyethylene diffusion and polymer/water partition coefficients for emerging contaminants. Environ. Toxicol. Chem. 35, 2162–2172. https://doi.org/10.1002/etc.3390.

Potapowicz, J., Lambropoulou, D., Nannou, C., Kozioł, K., Polkowska, Z., 2020. Occurrences, sources, and transport of organochlorine pesticides in the aquatic environment of Antarctica. Sci. Total Environ. 735, 1–12. https://doi.org/10.1016/j.scitotenv.2020.139475.

Rathna, R., Varjani, S., Nakkeeran, E., 2018. Recent developments and prospects of dioxins and furans remediation. J. Environ. Manag. 223, 797–806. https://doi.org/10.1016/j.jenvman.2018.06.095.

Reddy, A.V., Moniruzzaman, M., Aminabhavi, T.M., 2019. Polychlorinated biphenyls (PCBs) in the environment: recent updates on sampling, pretreatment, cleanup technologies and their analysis. Chem. Eng. J. 358, 1186–1207. https://doi.org/10.1016/j.cej.2018.09.205.

Ren, Z., Zhang, B., Lu, P., Li, C., Gao, L., Zheng, M., 2011. Characteristics of air pollution by polychlorinated dibenzo-p-dioxins and dibenzofurans in the typical industrial areas of Tangshan City, China. J. Environ. Sci. 23, 228–235. https://doi.org/10.1016/S1001-0742(10)60425-1.

Reyes-Garcés, N., Giofriddo, E., Gómez-Ríos, G.A., Bojko, B., Sigh, V., et al., 2018. Advances in solid phase microextraction and perspective on future directions. Anal. Chem. 90 https://doi.org/10.1021/acs.analchem.7b04502.

Roach, A.C., Muller, R., Komarova, T., Symons, R., Stevenson, G.J., et al., 2009. Using SPMDs to monitor water column concentrations of PCDDs, PCDFs and dioxin-like PCBs in Port Jackson (Sydney Harbour), Australia. Chemosphere 75, 1243–1251. https://doi.org/10.1016/j.chemosphere.2009.01.071.

Rusiana, T.P., Smedes, F., Klanova, J., Booji, K., Holoubek, I., 2007. Polymer selection for passive sampling: a comparison of critical properties. Chemosphere 68, 1344–1351. https://doi.org/10.1016/j.chemosphere.2007.01.025.

Rusina, T.P., Smedes, F., Klanova, J., 2010a. Diffusion coefficients of polychlorinated biphenyls and polycyclic aromatic hydrocarbons in polydimethylsiloxane and low-density polyethylene polymers. J. Appl. Polym. Sci. 16, 1803–1810. https://doi.org/10.1002/app.31704.

Rusina, T.P., Smedes, F., Koblizkova, M., Klanova, J., 2010b. Calibration of silicone rubber passive samplers: experimental and modeled relations between sampling rate and compound properties. Environ. Sci. Technol. 44, 362–367. https://doi.org/10.1021/es900938r.

Sanders, J.P., Andrade, N.A., Ghosh, U., 2018. Evaluation of passive sampling polymers and nonequilibrium adjustment methods in a multiyear surveillance of sediment porewater PCBs. Environ. Toxicol. Chem. 37, 2487–2495. https://doi.org/10.1002/etc.4223.

Schaeffer, L., 2008. The role of functional groups in drug–receptor interactions. In: Wermuth, C.G., Aldous, D., Raboisson, P., Rognan, D. (Eds.), The Practice of Medicinal Chemistry, fourth ed. Academic Press, pp. 359–378. https://doi.org/10.1016/B978-0-12-417205-0.00014-6.

Schröder, S., San-Román, M.F., Ortiz, I., 2021. Dioxins and furans toxicity during the photocatalytic remediation of emerging pollutants. Triclosan as case study. Sci. Total Environ. 770, 1–10. https://doi-org.bd.univalle.edu.co/10.1016/j.scitotenv.2020.144853.

Schummer, C., Tuduri, L., Briand, O., Appezeller, B.M., Millet, M., 2012. Application of XAD-2 resin-based passive samplers and SPME–GC–MS/MS analysis for the monitoring of spatial and temporal variations of atmospheric pesticides in Luxembourg. Environ. Pollut. 170, 88–94. https://doi.org/10.1016/j.envpol.2012.05.025.

Shen, L., Wania, F., Lei, Y.D., Teixeira, C., Muir, D.C., Bidleman, T.F., 2005. Atmospheric distribution and long-range transport behavior of organochlorine pesticides in North America. Environ. Sci. Technol. 39, 409–420. https://doi.org/10.1021/es049489c.

Silva-Barni, M.F., Gonzalez, M., Miglioranza, K.S., 2019a. Comparison of the epiphyte Tillandsia bergeri and the XAD-resin based passive air sampler for monitoring airborne pesticides. Atmos. Pollut. Res. 10, 1507–1513. https://doi.org/10.1016/j.apr.2019.04.008.

Silva-Barni, M.F., Smedes, F., Fillmann, G., Miglioranza, K.S., 2019b. Passive sampling of pesticides and polychlorinated biphenyls along the Quequén Grande River Watershed, Argentina. Environ. Toxicol. Chem. 38, 340–349. https://doi.org/10.1002/etc.4325.

Singh, Z., Kaur, J., Kaur, R., Singh, S., 2016. Toxic effects of organochlorine pesticides: a review. Am. J. BioSci. 4 (3–1), 11–18. https://doi.org/10.11648/j.ajbio.s.2016040301.13.

Snow, R.W., Amratia, P., Kabaria, C.W., Noor, A.M., Marsh, K., 2012. The changing limits and incidence of malaria in Africa: 1939–2009. Adv. Parasitol. 78, 169–262. https://doi.org/10.1016/b978-0-12-394303-3.00010-4.

Sruthi, S.N., Shyleshchandran, M.S., Mathew, S.P., Ramasamy, E.V., et al., 2017. Contamination from organochlorine pesticides (OCPs) in agricultural soils of Kuttanad agroecosystem in India and related potential health risk. Environ. Sci. Pollut. Control Ser. 24, 969–978. https://doi.org/10.1007/s11356-016-7834-3.

St George, T., Vhalos, P., Harner, T., Helm, P., Wilford, B., 2010. A rapidly equilibrating, thin film, passive water sampler for organic contaminants; characterization and field testing. Environ. Pollut. 159, 481–486. https://doi.org/10.1016/j.envpol.2010.10.030.

Stephens, B.S., Kapernick, A.P., Eaglesham, G., Mueller, J.F., 2009. Event monitoring of herbicides with naked and membrane-covered Empore disk integrative passive sampling devices. Mar. Pollut. Bull. 58, 1116–1122. https://doi.org/10.1016/j.marpolbul.2009.04.012.

Stockholm Convention, 2001. Stockholm Convention on Persistent Organic Pollutants. http://www.pops.int.

Sun, C., Soltwedel, T., Bauerfeind, E., Adelman, D.A., Lohmann, R., 2016. Depth profiles of persistent organic pollutants in the North and Tropical Atlantic Ocean. Environ. Sci. Technol. 50, 6172–6179. https://doi.org/10.1021/acs.est.5b05891.

Taiwo, A.M., 2019. A review of environmental and health effects of organochlorine pesticide residues in Africa. Chemosphere 220, 1126–1140. https://doi.org/10.1016/j.chemosphere.2019.01.001.

Taylor, A.C., Fones, G.R., Vrana, B., Mills, G.A., 2019. Applications for passive sampling of hydrophobic organic contaminants in water-a review. Crit. Rev. Anal. Chem. 1–35. https://doi.org/10.1080/10408347.2019.1675043.

Ter Laak, T.L., Busser, F.J., Hermens, J.L., 2008. Poly(dimethylsiloxane) as passive sampler material for hydrophobic chemicals: effect of chemical properties and sampler characteristics on partitioning and equilibration times. Anal. Chem. 80, 3859–3866. https://doi.org/10.1021/ac800258j.

Toinaga, M.Y., Silva, C.R., Melo, J.P., Niwa, N.A., Plascak, D., et al., 2016. PCDD, PCDF, dl-PCB and organochlorine pesticides monitoring in São Paulo City using passive air sampler as part of the Global Monitoring Plan. Sci. Total Environ. 571, 323–331. https://doi.org/10.1016/j.scitotenv.2016.07.173.

Tromp, P.C., Beeltje, H., Okeme, J.O., Vermeulen, R., Pronk, A., et al., 2019. Calibration of polydimethylsiloxane and polyurethane foam passive air samplers for measuring semi volatile organic compounds using a novel exposure chamber design. Chemosphere 227, 435–443. https://doi.org/10.1016/j.chemosphere.2019.04.043.

Tucca, F., Moya, H., Barra, R., 2014. Ethylene vinyl acetate polymer as a tool for passive sampling monitoring of hydrophobic chemicals in the salmon farm industry. Mar. Pollut. Bull. 88, 174–179. https://doi.org/10.1016/j.marpolbul.2014.09.009.

Valenzuela, E.F., Menezes, H.C., Cardeal, Z.L., 2019. New passive sampling device for effective monitoring of pesticides in water. Anal. Chim. Acta 1054, 26–37. https://doi.org/10.1016/j.aca.2018.12.017.

Valenzuela, E.F., Menezes, H.C., Cardeal, Z.L., 2020. Passive and grab sampling methods to assess pesticide residues in water. A review. Environ. Chem. Lett. 18, 1019–1048. https://doi-org.bd.univalle.edu.co/10.1007/s10311-020-00998-8.

Van Drooge, B.L., Grimalt, J.O., Camarero, L., Catalan, J., Stuchlik, E., Torres, C.J., 2004. Atmospheric semivolatile organochlorine compounds in European high-mountain areas (central pyrenees and high tatras). Environ. Sci. Technol. 38, 3525–3532. https://doi.org/10.1021/es030108p.

Van Drooge, B.L., Grimalt, J.O., Booij, K., Camarero, L., Catalan, J., 2005. Passive sampling of atmospheric organochlorine compounds by SPMDs in a remote high mountain area. Atmos. Environ. 39, 5195–5204. https://doi.org/10.1016/j.atmosenv.2005.05.020.

Vorkamp, K., Odsbjerg, L., Langeland, M., Mayer, P., 2016. Utilizing the partitioning properties of silicone for the passive sampling of polychlorinated biphenyls (PCBs) in indoor air. Chemosphere 160, 280–286. https://doi.org/10.1016/j.chemosphere.2016.06.054.

Vrana, B., Allan, I.J., Greenwood, R., Mills, G.A., Dominiak, E., et al., 2005. Passive sampling techniques for monitoring pollutants in water. Trends Anal. Chem. 24, 845–868. https://doi.org/10.1016/j.trac.2005.06.006.

Vrana, B., Mill, G.A., Dominiak, E., Greenwood, R., 2006. Calibration of the Chemcatcher passive sampler for the monitoring of priority organic pollutants in water. Environ. Pollut. 142, 333–343. https://doi.org/10.1016/j.envpol.2005.10.033.

Vrana, B., Komancová, L., Sobotka, J., 2016. Calibration of a passive sampler based on stir bar sorptive extraction for the monitoring of hydrophobic organic pollutants in water. Talanta 152, 90–97. https://doi.org/10.1016/j.talanta.2016.01.040.

Wania, F., Shunthirasingham, C., 2020. Passive air sampling for semi-volatile organic chemicals. Environ. Sci. Process. & Impacts 22 (10), 1925–2002. https://doi.org/10.1039/D0EM00194E.

Wei, M., Yang, X., Watson, P., Yang, F., Liu, H., 2019. A cyclodextrin polymer membrane-based passive sampler for measuring triclocarban, triclosan and methyl triclosan in rivers. Sci. Total Environ. 648, 109–115. https://doi.org/10.1016/j.scitotenv.2018.08.151.

White, K.B., Kalina, J., Scheringer, M., Přibylová, P., Kukučka, P., et al., 2020. Temporal trends of persistent organic pollutants across Africa after a decade of MONET passive air sampling. Environ. Sci. Technol. https://doi.org/10.1021/acs.est.0c03575.

Yamazaki, K., Itoh, S., Araki, A., Miyashita, C., Minatoya, M., et al., 2020. Associations between prenatal exposure to organochlorine pesticides and thyroid hormone levels in mothers and infants: the Hokkaido study on environment and children's health. Environ. Res. 189, 1–10. https://doi.org/10.1016/j.envres.2020.109840.

Yoonki, M., Jongwon, H., Meehye, L., 2014. Determination of toxic congeners of 17 PCDDs/PCDFs and 12 dl-PCBs using polyurethane foam passive air samplers in ten cities around Seoul. Sci. Total Environ. 491–492, 17–27. https://doi.org/10.1016/j.scitotenv.2014.04.039.

Zeng, L., Li, N., Shao, D., et al., 2014. Concentrations, sources, and risk assessment of polychlorinated biphenyls in vegetables near a waste-Incinerator site, South China. Arch. Environ. Contam. Toxicol. 67, 78–86. https://doi.org/10.1007/s00244-014-0045-3.

Zhang, X., Wong, C., Lei, Y.D., Wania, F., 2012. Influence of sampler configuration on the uptake kinetics of a passive air sampler. Environ. Sci. Technol. 46 (1), 397–403. https://doi.org/10.1021/es203292x.

Zhao, D., Zhang, P., Ge, L., Zheng, G.J., Wang, X., et al., 2018. The legacy of organochlorinated pesticides (OCPs), polycyclic aromatic hydrocarbons (PAHs) and polychlorinated biphenyls (PCBs) in Chinese coastal seawater monitored by semi-permeable membrane devices (SPMDs). Mar. Pollut. Bull. 137, 222–230. https://doi.org/10.1016/j.marpolbul.2018.10.004.

Zubair, M., Adrees, A., 2019. Dioxins and furans: emerging contaminants of air. In: Del Real Olvera, J. (Ed.), Air Pollution - Monitoring, Quantification and Removal of Gases and Particles. IntechOpen, pp. 111–125. https://doi.org/10.5772/intechopen.80680.

Chapter 21

Removal of emergent pollutants of waters

21.1 Emerging pollutants: an overview

Water is one of the most important natural resources for humans, it is an essential resource for life, and is involved in industrial activities, in-home activities, health services, public spaces, commercial establishments, etc. Despite this, it is one of the natural matrices with the highest presence of polluting substances, mostly originating from anthropogenic activities associated with the intensification of industrial processes, population growth, consumer habits, and developing technologies (Jindal and Jain, 2018; Dey et al., 2019; Rathi et al., 2020). Sources of contamination of water effluents are numerous and their impact depends on many factors, from the nature of the pollutant to waste management techniques at the personal, population, government, and industrial levels. Classical examples include: (1) the excessive use of pesticides and fertilizers in the agricultural industry, which can undergo percolation processes and contaminate water bodies, generating alterations in flora and fauna; (2) the mining industry and refineries which release inorganic pollutants, gases, and hydrocarbons; (3) domestic wastewaters and non-specialized wastewater treatment systems, associated with the presence of pathogens and bioactive compounds; and (4) the use of products from the oil industry and their inappropriate disposal, which is related to the release of persistent substances into the environment (Saxena and Bharagava, 2017; Rajput et al., 2020).

With a particular emphasis on heavy metals, pesticides, and hydrocarbons, since the 1960s, the contamination of the hydrosphere has led to progress in the development and implementation of contaminant quantification and remediation methods (Poynton and Robinson, 2018). In this way, it has been possible to establish, for a wide variety of pollutants, their life cycle, effects on the public and environmental health, regulations for the manufacture, use, environmental persistence, bioaccumulation, risks, and criteria for their final disposal (Bandala and Berli, 2018; Poynton and Robinson, 2018). From a historical perspective, this environmental, political, and social consciousness has resulted in decisions such as the prohibition of the manufacture of

Eco-friendly Functional Polymers
https://doi.org/10.1016/B978-0-12-821842-6.00023-3
Copyright © 2021 Elsevier Inc. All rights reserved.

polluting substances such as DDT and polychlorinated biphenyls, but also, in the mid-1990s, in the recognition of emerging pollutants by the US Environmental Protection Agency (USEPA) (Martínez et al., 2017; Poynton and Robinson, 2018). The term emerging pollutant refers to substances, that are not monitored or regulated, incorporated into the environment without anticipating their potential adverse effects on ecological health, and includes drugs, nanomaterials, and all substances of synthetic or natural origin whose concentration has increased to the point of generating harmful effects in ecosystems (Barber, 2014; Sauvé and Desrosiers, 2014; Delgado-Moreno et al., 2021). These substances were initially dismissed as eventual pollutants, and for this reason, emerging pollutants are characterized by their limited regulation in manufacture, use, final disposal, and monitoring (Tang et al., 2020). The above allowed emerging pollutants to be released into the environment with little or no precautions, without complete knowledge of their persistence, accumulation, ability to enter the trophic chain, undergo biotransformation reactions that led to the formation of toxic metabolites, generate diseases, modifications in the relationships of ecosystems, and alteration of the physicochemical and functional properties of water bodies, and therefore, they constitute a latent threat to health, ecosystems, and sustainable development (Bharadwaj, 2018; Poynton and Robinson, 2018; Da Silva Vilar et al., 2021). This general definition allows a wide variety of substances to be grouped under the title of emerging pollutants, including nanomaterials, pharmaceuticals, surfactants, personal care products, toxins, and inorganic ions and oxides, that are different to heavy metals, but with increasing current use in electronics and new technologies for the capture and storage of energy, communications, and those used in the manufacture of so-called smart materials (Poynton and Robinson, 2018).

From the above, it is foreseeable that due to scientific, technological, and industrial developments, the amount of emerging pollutants will increase progressively; consequently, this situation has led to the need to establish regulations, treatment protocols, and, primarily, sustainable and efficient remediation processes (Dey et al., 2019; Livingston et al., 2020). On the other hand, polymers, due to their versatility and structural variability, composition, costs, and other interesting features, have emerged as one of the most promising alternatives for the creation of remediation systems for emerging pollutants. Different approaches can be used to achieve this goal, for instance, separation using polymeric membrane (e.g., ultrafiltration or UF), polymeric adsorbents, molecularly imprinted polymers (MIPs), and even hybrid methods such as liquid-phase polymer-based retention (LPR), soluble polymer coupled to double emulsion (W/O/W) in conjunction with membrane filtration, etc. (Palencia, 2015; Rivas et al., 2016; Martínez et al., 2017; Palencia et al., 2017; Cantarella et al., 2019). A summary of classical and innovative techniques based on polymers, for the removal of pollutants, is given in Table 21.1.

TABLE 21.1 Summary of classical and innovative techniques, based on polymers, for the removal of pollutants.

Technique	Phase active	Example of polymer and pollutant	References
Flocculation	Polyelectrolyte	• Alginate, anionic dialdehyde cellulose • Pollutant: heavy metal ions	Salehizadeh et al. (2018)
	Polyelectrolyte	• Cationic pullulan • Pollutants: pesticides	Macczak et al. (2020)
	Complexing polymer	• Sodium alginate-g-N-vinyl-2-pyrrolidone • Pollutant: Heavy metal ions	Macczak et al. (2020)
Adsorption	Ionic exchange resins	• Resin D301 + epichlorohydrin, dimethyl- and ethylenediamine • Pollutant: Cr(VI)	Han et al. (2020)
	Chelating resins	• Chelating resin functionalized with iminodiacetic acid groups • Pollutants: Pb(II) and Cd(II)	Addala and Belattar (2017)
	MIPs	• Resins based on methacrylic acid and EtG-dimethacrylate • Pollutant: antidepressants	Gornil et al. (2020)
Membrane	Polymer membrane	• Silver-polyacrylonitrile fibers • Pollutant: Nanoparticles from air	Canalli et al. (2019)
	Polymer membrane	• Poly(styrene sulfonate) • Pollutant: nanoparticles from water	Palencia et al. (2014)
LPR	Water-soluble polymer	• Poly(allyl(triethyl)ammonium chloride) • Pollutant: dyes (Acid Blue 129)	Palencia et al. (2017)
LPR + liquid membrane	W/O/W + polymer	• Poly(styrene-co-maleic anhydride) • Pollutant: heavy metal ions	Palencia and Rivas (2011)

EtG, ethylene glycol; *LPR*, liquid-phase polymer-based retention; *MIPs*, molecularly imprinted polymers; *UF*, ultrafiltration; *W/O/W*, double emulsions.

21.2 Remediation of emerging contaminants in water

Emerging pollutants are defined by the US Geological Survey (USGS) as microorganisms and natural or synthetic chemical substances which potentially could generate adverse effects on public and ecosystem health (Poynton and Robinson, 2018). This general definition implies that the substances grouped under the term emerging pollutants do not present a specific physicochemical structure and properties, and a novelty is the inclusion of microbiological entities (Rathi et al., 2020). For example, chemical pollutants include anticorrosive agents such as benzotriazoles, endocrine disruptors such as estradiol, personal care products like siloxanes and TiO_2 nanoparticles, dioxane-like stabilizers, pharmaceuticals such as phenobarbital, plasticizers like bisphenol, fertilizers, and herbicides, to name a few (Salimi et al., 2017; Ramírez-Malule et al., 2020; Delgado-Moreno et al., 2021). Regarding microbiological compounds, these are related to the presence in the natural environment of toxins, viruses, pathogenic bacteria, and extracellular substances with adverse biological activity (Poynton and Robinson, 2018).

One of the main risks associated with emerging pollutants is the interconnection of the routes through which they can reach humans (see Fig. 21.1). For example, in the primary production sectors, the production of food is carried out, which is susceptible to contamination by residuality of veterinary supplies, contamination of animal feed, contamination of irrigation water, etc. In addition, the effluents that supply rural areas, in many cases, are the same as those that supply cities. Industries as a pollution-generating entity takes a back seat, to the extent that they have regulations and control entities that can take measures to reduce pollution. However, the waste generated by the excessive consumption of technology in cities, over-the-counter drugs, under medical treatment or in discharges from hospitals, are constant sources of pollution, whose impact increases with the size of the population and its geographical location with respect to neighboring cities.

On the other hand, emerging pollutants are found in water bodies in reduced concentrations, about from 1.0 ng/L to 1.0 μg/L, which has made their detection in natural environments impossible until improvements in the detection limits of the analytics techniques, which continue to have advances

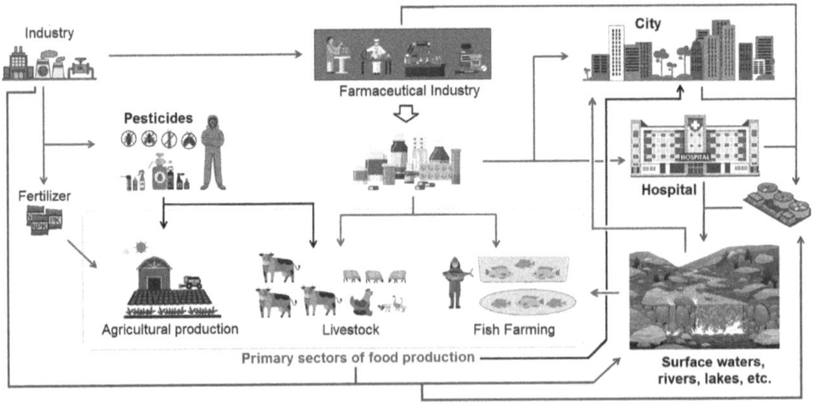

FIGURE 21.1 Illustration of the multiple routes of contamination by emerging pollutants, from industry, cities, food production areas, hospitals, etc.

in the procedures and equipment needed to detect and differentiate these substances (Peralta et al., 2020; Rathi et al., 2020). In addition, emerging pollutants do not necessarily present a continuous and constant release into the natural environment; however, the fact that they are not removed from water bodies through effective treatment generates their permanence, potential bioaccumulation, and contamination of other natural systems, as is the case with soils (Ramírez-Malule et al., 2020; Delgado-Moreno et al., 2021). This generates risks associated with the environmental distribution of emerging pollutants and their adverse interactions with the biotic and abiotic components present in water bodies, constituting a public health problem of considerable dimensions (Tang et al., 2020).

In order to remedy contamination by emerging pollutants, various environmental remediation techniques have been used, mainly based on membrane separation methods, surface adsorption, and ion exchange phenomena (Rodriguez-Narvaez et al., 2017; Zelić et al., 2018; Amrose et al., 2020). In this context, polymers constitute the essential component of most of the emerging pollutant remediation systems, because these macromolecules can be synthesized and functionalized using simple protocols to obtaining better selectivity against a target pollutant, and provide a material with different characteristics and adaptability to much systems (Benitez et al., 2017). The prominent features are mechanical and thermal properties, surface charge, functional groups, dimensions, and morphology, which allow the obtaining of ion exchange resins, MIPs, selective membranes, passive resin-type samplers, interpenetrating polymeric networks, complexing macromolecules, and even composites with active phases for the growth of bioremediation agents, etc. (Córdoba and Palencia, 2017; Garcés et al., 2017, 2018; Palencia et al., 2017; Lerma et al., 2018a,b; Cantarella et al., 2019). The most relevant technologies in the field include the membrane separation methods by ultrafiltration (UF), microfiltration, reverse osmosis, nanofiltration, separation processes by liquid emulsion membranes, LPR, ions exchange resins, polymeric hybrid systems, treatment with nanostructured polymers and/or bioreactors, and with natural or functionalized biomass (Palencia et al., 2015, 2019; Rodríguez-Mozaz et al., 2015; Zhou et al., 2016; Palencia, 2017; Garcés et al., 2018; Li et al., 2018; Luo et al., 2018; Mora et al., 2018). These techniques are in constant development and optimization, along with the generation of new technologies and adequate political regulations to monitor emerging pollutants in water resources, and wastewater and standardized protocols to evaluate their potential risks to organisms (Rathi et al., 2020).

21.3 Removal of pharmaceuticals in water

Pharmaceutical products are substances synthesized to present biological activity in humans and animals, making them compounds widely used in daily life for a diverse variety of beneficial effects, such as antibiotics, anticancer medications, hormonal regulators, antipyretics, anticonvulsants, etc. (Poynton and Robinson, 2018). Nevertheless, since up to 95% of the drug dose or its metabolites can be excreted or discharged into wastewater, such that the presence of pharmaceutical compounds in the environment has increased to levels considered to produce potential adverse effects (Poynton and Robinson, 2018). Therefore, it is necessary to establish an adequate treatment and regulatory standards against the release and discharge of pharmaceutical substances and their derivatives (Bharadwaj, 2018; Rajput et al., 2020).

Several technologies have been applied, with examples including the use of carbon nanotubes for the removal of amoxicillin, ciprofloxacin, ibuprofen/triclosan tetracycline, norfloxacin, etc. (Ji et al., 2010; Mohammadi et al., 2014; Rodriguez-Narvaez et al., 2017), clays like montmorillonite, bentonite, kaolinite, and hectorite have been evaluated for the removal of ciprofloxacin, tetracycline, sulfadimetoxina, ampicillin, ciprofloxacin, etc. (Li et al., 2010, 2011; Parolo et al., 2012; Rodriguez-Narvaez et al., 2017); and pressure-driven membranes (acetaminophen, metoprolol, caffeine, antipyrine, atrazine, naproxen, diclofenac, etc.) (Rodriguez-Narvaez et al., 2017). In Table 21.2 are shown several technologies for the removal of emerging pollutants using polymers, polymeric biomass, or transformation of polymeric biomass directly.

21.4 Other emerging pollutants: nanomaterials, and metallic and metalloid ions

Nanoengineered materials are structures characterized as having at least one of their dimensions in the scale between 1 and 100 nm. With these dimensions, materials show unique physicochemical properties (for more details, see Chapters 7 and 26). Nanomaterials have important applications in academic and industrial development, including biomedicine, agriculture, catalysis, personal hygiene products, water remediation, disease diagnosis, etc. At present, there are more than 1800 commercial products that use nanomaterials as an active component (Sánchez et al., 2018, 2019; Chen et al., 2019). As a consequence, it is expected that the final disposal of nanostructures will focus soon on landfills and natural systems, such as soils (8.0%–28.0%) and water (0.4%–7.0%) (Chen et al., 2019; Mortimer and Holden, 2019). When these nanomaterials come into contact with aqueous media, they undergo agglomeration, complexation, oxidation–reduction, degradation, biotransformation, dissolution, or aging reactions, and they also have the ability to interact with biotic components of natural systems (Mortimer and Holden, 2019). Consequently, nanomaterials have been recognized as potential emerging pollutants due to their ability to interact closely with cellular components, generating the production of reactive oxygen species, lipid peroxidation, protein denaturation, protein corona production, disruption of cell membranes, and in the

TABLE 21.2 Summary of classical and innovative techniques, based on polymers, for the removal of pollutants.

Technique	Detail	Active phase	Emerging pollutant	References
Adsorption	AC	Lignin-activated	Tetracycline	Huang et al. (2014)
		Coconut shell	Tetracycline	Torres et al. (2012)
		Sugar beet pulp	Tetracycline	Torres et al. (2012)
		Olive-waste cake	Ibuprofen, ketoprofen, naproxen, ciclofenac	Baccar et al. (2012)
	PB	Brazilian pepper wood	Culfamethoxazole	Jeong et al. (2009)
		Acid rice husk	Tetracycline	Liu et al. (2012)
Membrane	UF	Poly(vinylpirrolidone)	Tetracycline	Chen et al. (2011)
		Polysulfone	Bisphenol A	Heo et al. (2012)
		Poly(vinylidene fluoride)	Naproxen, ibuprofen, ketoprofen, carbamazepine	McCloskey et al. (2010)
	RO	Aromatic polyamide	Carbamazepine, ibuprofen, naproxen, fenoprofen, ketoprofen	Valladares et al. (2011)
	MBR	Poly(ethersulfone)	Acetaminophen, naproxen, ketoprofen, roxithromycin, sulfamethoxazole, etc.	Tambosi et al. (2010)

AC, activated carbon; *MBR*, membrane bioreactor; *PB*, pyrolytic biochar; *RO*, reverse osmosis; *UF*, ultrafiltration.

case of nanotube-type nanomaterials, asbestos like-toxicity (Palencia et al., 2013, 2015; Iavicoli et al., 2017; Poynton and Robinson, 2018; Soltani and Pouypouy, 2019).

It is relevant to mention that nanostructures not only represent an emerging contaminant threat, but also that they can be used for environmental remediation processes (Bardos et al., 2018; Ingle et al., 2014). This potential application is based on nanomaterials' physicochemical properties and high surface area, and for this, nanoremediation is the most promising technology for environmental remediation employment of zerovalent iron nanoparticles and their macroscopic nanocomposites (Gehrke et al., 2015; Fiorati et al., 2020).

On the other hand, contamination by heavy metals in water systems has been an area of interest widely analyzed in scientific terms, leading to the generation of more efficient detection systems and remediation methods (Garcés et al., 2018). However, some metallic or metalloid ions that are not found in the classification of heavy metals have emerged as water pollutants, as they are used to manufacture medical products, fertilizers, additives, and new technologies, and so their release into natural ecosystems has increased. Some examples are manganese, boron, molybdenum, and antimony ions, which interact with water bodies and bioaccumulate in the trophic chain, generating various disorders such as Parkinson disease, growth retardation, anemia, hypothyroidism, bone and joint deformities, etc. (Amarasiriwardena and Wu, 2011; Pinsino et al., 2012; Palencia et al., 2018; Palencia et al., 2018, 2018).

21.5 Conclusions and remarks

Emerging pollutants are a reality. Over the several decades since this problem was first identified, and studies related to this topic have increased gradually. Initially, emerging pollutants were defined as pharmaceutical products, and their metabolic by-products. However, it is incredible to think that for more than a century, around the world, these products have been increasing continuously. The example of hospitals can be used as a reference to illustrate that the problem has been present for longer than previously thought. These institutions are places where all kinds of drugs are used intensively and, although waste management is very strict, the problem does not lie in the dumping of the used materials, but in the nonmetabolized drugs and by-products of these drugs that are excreted by patients. Clearly, the awareness of this situation has led to the development of science and technology in this field. However, as the scientific community has begun to internalize this problem, the boom of nanotechnology took place, and so, in parallel, another type of emerging pollutant appeared. Likewise, advances in biotechnology and genetic engineering, and also in new technologies used in communications, energy storage systems, and computers, have exacerbated the problem at an alarming rate. Therefore, it is necessary to increase knowledge around this issue, including key aspects such as quantification, understanding of the different dynamics associated with each case, the evaluation of technology to mitigate its impact, and the development of more efficient technologies to remedy the affected environments. In addition, this topic is an area that must necessarily be included in courses related to environmental sciences, analytical chemistry, etc.

References

Addala, A., Belattar, N., 2017. Adsorption of Cd and Pb metal ions onto chelating resin and their application in removal of lead from battery factory wastewaters. Indian J. Chem. Technol. 24, 601–607.

Amarasiriwardena, D., Wu, F., 2011. Antimony: emerging toxic contaminant in the environment. Microchem. J. 97, 1–3. https://doi.org/10.1016/j.microc.2010.07.009.

Amrose, S.E., Cherukumilli, K., Wright, N.C., 2020. Chemical contamination of drinking water in resource-constrained settings: global prevalence and piloted mitigation strategies. Annu. Rev. Environ. Resour. 45, 195–226. https://doi.org/10.1146/annurev-environ-012220-105152.

Baccar, R., Sarr, M., Bouzid, J., Feki, M., Blnquez, P., 2012. Removal of pharmaceutical compounds by activated carbon prepared from agricultural by-product. Chem. Eng. J. 211–212, 310–317. https://doi.org/10.1016/j.cej.2012.09.099.

Bandala, E.R., Berli, M., 2018. Nanomaterials: new agrotechnology tools to improve soil quality? In: López-Valdez, F., Fernández-Luqueño, F. (Eds.), Agricultural Nanobiotechnology, pp. 127–140. https://doi.org/10.1007/978-3-319-96719-6_7.

Barber, L.B., 2014. Emerging contaminants. In: Ahuja, S. (Ed.), Comprehensive Water Quality and Purification, pp. 245–266. https://doi.org/10.1016/b978-0-12-382182-9.00015-3.

Bardos, P., Merely, C., Kvapil, P., Koschitzky, H.P., 2018. Status of nanoremediation and its potential for future deployment: risk-benefit and benchmarking appraisals. Remediat. J. Environ. Cleanup Costs Technol. & Tech. 28 (28), 43–56. https://doi.org/10.1002/rem.21559.

Benitez, E., Lerma, T.A., Córdoba, A., 2017. Making of porous ionic surfaces by sequential polymerization: polyurethanes + grafting of polyelectrolytes. J. Sci. Technol. Appl. 2, 44–53. https://doi.org/10.34294/j.jsta.17.2.13.

Bharadwaj, A., 2018. Bioremediation of xenobiotics. In: Parmar, V., Malhotra, P., Mathur, D. (Eds.), Green Chemistry in Environmental Sustainability and Chemical Education, pp. 1–13. https://doi.org/10.1007/978-981-10-8390-7_1.

Canalli, A.C., Nagarajan, S., de Araújo, B., Guerra, V.G., Lopes, M., et al., 2019. Efficient nanoparticles removal and bactericidal action of electrospun nanofibers membranes for air filtration. Mater. Sci. Eng. C 102, 718–729. https://doi.org/10.1016/j.msec.2019.04.094.

Cantarella, M., Carroccio, S.C., Dattilo, S., Avolio, R., Castaldo, R., Puglisi, C., Privitera, V., 2019. Molecularly imprinted polymer for selective adsorption of diclofenac from contaminated water. Chem. Eng. J. 367, 180–188. https://doi.org/10.1016/j.cej.2019.02.146.

Chen, F., Xiao, Z., Yue, L., Wang, J., Feng, Y., Zhu, X., Wang, Z., Xing, B., 2019. Algae response to engineered nanoparticles. Environ. Sci. Nano 6, 1026–1042. https://doi.org/10.1039/C8EN01368C.

Chen, H., Luo, H., Lan, Y., Dong, T., Hu, B., Wang, Y., 2011. Removal of tetracycline from aqueous solutions using polyvinylpyrrolidone (PVP-K30) modified nanoscale zero valent iron. J. Hazard Mater. 192, 44–53. https://doi.org/10.1016/j.jhazmat.2011.04.089.

Córdoba, A., Palencia, M., 2017. Development of nanostructured crosslinkers with antimicrobial properties for free-radical polymerization. J. Sci. Technol. Appl. 2, 54–64. https://doi.org/10.34294/j.jsta.17.2.14.

Da Silva Vilar, D., Torres, N.H., Bharagava, R.N., Bilal, M., Iqbal, H.M.N., Salazar-Banda, G.R., Barrios, K.I., Ferreira, L.F.R., 2021. Emerging contaminants in environment: occurrence, toxicity, and management strategies with emphasis on microbial remediation and advanced oxidation processes. In: Singh, V.P., Singh, P., Kumar, A., Kumar Mishra, V. (Eds.), Microbe Mediated Remediation of Environmental Contaminants, pp. 1–14. https://doi.org/10.1016/b978-0-12-821199-1.00001-8.

Delgado-Moreno, L., Bazhari, S., Gasco, G., Méndez, A., El Azzouzi, M., Romero, E., 2021. New insights into the efficient removal of emerging contaminants by biochars and hydrochars derived from olive oil wastes. Sci. Total Environ. 752, 141838. https://doi.org/10.1016/j.scitotenv.2020.141838.

Dey, S., Bano, F., Malik, A., 2019. Pharmaceuticals and personal care product (PPCP) contamination - a global discharge inventory. In: Vithanage, M., Prasad, M.N.V., Kapley, A. (Eds.), Pharmaceuticals and Personal Care Products: Waste Management and Treatment Technology, pp. 1–26. https://doi.org/10.1016/b978-0-12-816189-0.00001-9.

Fiorati, A., Grassi, G., Graziano, A., Liberatori, G., Pastori, N., Melone, L., Bonciani, L., Pontorno, L., Punta, C., Corsi, I., 2020. Eco-design of nanostructured cellulose sponges for sea-water decontamination from heavy metal ions. J. Clean. Prod. 246, 119009. https://doi.org/10.1016/j.jclepro.2019.119009.

Garcés, V., Lerma, T.A., Palencia, M., Combatt, E., 2018. Building a probe-type passive sampler to study the contents of heavy metals in the aqueous phase in the soil. J. Sci. Technol. Appl. 4, 4–16. https://doi.org/10.34294/j.jsta.18.4.26.

Garcés, V., Palencia, M., Combatt, E., 2017. Development of bacterial inoculums based on biodegradable hydrogels for agricultural applications. J. Sci. Technol. Appl. 2, 13–23. https://doi.org/10.34294/j.jsta.17.2.10.

Gehrke, I., Geiser, A., Somborn-Schulz, A., 2015. Innovations in nanotechnology for a water treatment. Nanotechnol. Sci. Appl. 6, 1–17. https://doi.org/10.2147/NSA.S43773.

Gornil, T., Shinde, S., Lamovsek, L., Koblar, M., Heath, E., et al., 2020. Molecularly imprinted polymers for the removal of antidepressants from contaminated wastewater. Polymers 13, 120. https://doi.org/10.3390/polym13010120.

Han, S., Zang, Y., Gao, Y., Yue, Q., Zhang, P., et al., 2020. Co-monomer polymer anion exchange resin for removing Cr(VI) contaminants: adsorption kinetics, mechanism and performance. Sci. Total Environ. 709, 136002. https://doi.org/10.1016/j.scitotenv.2019.136002.

Heo, J., Flora, J.R., Her, N., Park, Y.G., Cho, J., Son, A., 2012. Removal of bisphenol A and 17-estradiol in single walled carbon nanotubes-ultrafiltration (SWNTs-UF) membrane systems. Separ. Purif. Technol. 90, 39–52. https://doi.org/10.1016/j.seppur.2012.02.007.

Huang, L., Wang, M., Shi, C., Huang, J., Zhang, B., 2014. Adsorption of tetracycline and ciprofloxacin on activated carbon prepared from lignin with H_3PO_4 activation. Desalination & Water Treat. 52, 2678–2687. https://doi.org/10.1080/19443994.2013.833873.

Iavicoli, I., Leso, V., Beezhold, D.H., Shvedova, A.A., 2017. Nanotechnology in agriculture: opportunities, toxicological implications, and occupational risks. Toxicol. Appl. Pharmacol. 329, 96–111. https://doi.org/10.1016/j.taap.2017.05.025.

Ingle, A., Seabra, A.B., Duran, N., Rai, M., 2014. Nanoremediation: a new and emerging technology for the removal of toxic contaminant from environment. In: Das, S. (Ed.), Microbial biodegradation and bioremediation. Elsevier, pp. 233–250. https://doi.org/10.1016/B978-0-12-800021-2.00009-1.

Jeong, C.Y., Wang, J.J., Dodla, S.K., Eberhardt, T.L., Groom, L., 2009. Effect of biochar amendment on tylosin adsorption-desorption and transport in two different soils. J. Environ. Qual. 41, 1185–1192. https://doi.org/10.2134/jeq2011.0166.

Ji, L., Chen, W., Bi, J., Zheng, S., Xu, Z., et al., 2010. Adsorption of tetracycline on single-walled and multi-walled carbon nanotubes as affected by aqueous solution chemistry. Environ. Toxicol. Chem. 29, 2713–2719. https://doi.org/10.1002/etc.350.

Jindal, D., Jain, I., 2018. Role of companies in environmental management through corporate social responsibility and sustainability policy. In: Parmar, V.S., Malhotra, P., Mathur, D. (Eds.), Green Chemistry in Environmental Sustainability and Chemical Education, pp. 23–42. https://doi.org/10.1007/978-981-10-8390-7_3.

Lerma, T.A., Combatt, E., Palencia, M., 2018a. Soil-mimicking hybrid composites based on clay, polymers and nitrogen-fixing bacteria for the development of remediation systems of degraded soil. J. Sci. Technol. Appl. 4, 17–27. https://doi.org/10.34294/j.jsta.18.4.27.

Lerma, T.A., Martínez, G., Palencia, M., 2018b. Generation of thiolated porous surfaces by interpenetrating polymeric networks: study of their surface properties. J. Sci. Technol. Appl. 3, 56–65. https://doi.org/10.34294/j.jsta.17.3.24.

Li, Z., Chang, P.H., Jean, J.S., Jiang, W.T., Wang, C.J., 2010. Interaction between tetracycline and smectite in aqueous solution. J. Colloid Interface Sci. 341, 311–319. https://doi.org/10.1016/j.jcis.2009.09.054.

Li, Z., Hong, H., Liao, L., Ackley, C.J., Schulz, L.A., et al., 2011. A mechanistic study of ciprofloxacin removal by kaolinite. Colloids Surf. B 88, 339–344. https://doi.org/10.1016/j.colsurfb.2011.07.011.

Li, T., Zhang, L., Ai, W., Dong, W., Yu, Q., 2018. A modified MBR system with post advanced purification for domestic water supply system in 180-day CELSS: construction, pollutant removal and water allocation. J. Environ. Manag. 222, 37–43. https://doi.org/10.1016/j.jenvman.2018.05.023.

Liu, P., Liu, W.J., Jiang, H., Chen, J.J., Li, W.W., et al., 2012. Modification of bio-char derived from fast pyrolysis of biomass and its application in removal of tetracycline from aqueous solution. Bioresour. Technol. 121, 235–240. https://doi.org/10.1016/j.biortech.2012.06.085.

Livingston, A., Trout, B.L., Horvath, I.T., Johnson, M.D., Vaccaro, L., Coronas, J., Babbitt, C.W., Zhang, X., Pradeep, T., Drioli, E., Hayler, J.D., Tam, K.C., Kappe, O., Fane, A.G., Szekely, G., 2020. Challenges and directions for green chemical engineering – role of nanoscale materials. In: Livingston, A., Szekely, G. (Eds.), Sustainable Nanoscale Engineering: from Materials Design to Chemical Processing, pp. 1–18. https://doi.org/10.1016/B978-0-12-814681-1.00001-1.

Luo, M., Lin, H., Li, B., Dong, Y., He, Y., Wang, L., 2018. A novel modification of lignin on corncob-based biochar to enhance removal of cadmium from water. Bioresour. Technol. 259, 312–318. https://doi.org/10.1016/j.biortech.2018.03.075.

Macczak, P., Kaczmarek, H., Ziegler-Borowska, M., 2020. Recent achievements in polymer bio-based flocculants for water treatment. Materials 13, 3951. https://doi.org/10.3390/ma13183951.

Martínez, J.M., Troches, E., Páez, M., 2017. Evaluation of poly(ethylene-co-vinylacetate) as polymeric adsorbent of polychlorinated biphenyls in water. J. Sci. Technol. Appl. 2, 94–107. https://doi.org/10.34294/j.jsta.17.2.18.

McCloskey, B.D., Park, H.B., Ju, H., Rowe, B.W., Miller, D.J., et al., 2010. Influence of polydopamine deposition conditions on pure water flux and foulant adhesion resistance of reverse osmosis, ultrafiltration, and microfiltration membranes. Polymer 51, 3472–3485. https://doi.org/10.1016/j.polymer.2010.05.008.

Mohammadi, A., Kazemipour, M., Ranjbar, H., Walker, R.B., Ansari, M., 2014. Amoxicillin removal from aqueous media using multi-walled carbon nanotubes. Fuller. Nanotub. Carbon Nanostruct. 23, 165–169. https://doi.org/10.1080/1536383X.2013.866944.

Mora, M., Palencia, M., Combatt, E., 2018. Effect of dissolved organic matter and humic substances on transport of ions and low molecular weight molecules by liquid membranes. J. Sci. Technol. Appl. 4, 40–48. https://doi.org/10.34294/j.jsta.18.4.29.

Mortimer, M., Holden, P.A., 2019. Fate of engineered nanomaterials in natural environments and impacts on ecosystems. In: Marmiroli, N., White, J.C., Song, J. (Eds.), Exposure to Engineered Nanomaterials in the Environment, pp. 61–103. https://doi.org/10.1016/b978-0-12-814835-8.00003-0.

Palencia, M., 2015. Liquid-phase polymer-based retention: theory, modeling, and application for the removal of pollutant inorganic ions. J. Chem. 965624. https://doi.org/10.1155/2015/965624.

Palencia, S.L., Buelvas, M.A., Palencia, M.S., 2015. Interaction mechanisms of inorganic nanoparticles and biomolecular systems of microorganisms. Curr. Chem. Biol. 9 https://doi.org/10.2174/2212796809666151022201811.

Palencia, S.L., Combatt, E.M., Palencia, M.S., 2013. Toxicidad de nanopartículas inorgánicas sobre los microorganismos del suelo con importancia agrícola. una revisión. Temas Agrarios 18, 106–122. https://doi.org/10.21897/rta.v18i2.720.

Palencia, M., Córdoba, A., Combatt, E., 2018. Synthesis and characterization of biofouling-resistant nanocomposites based on glucamine polymers and silver/acrylic acid nanoparticles. Polym. Bull. 75, 4555–4569. https://doi.org/10.1007/s00289-018-2286-1.

Palencia, M., Córdoba, A., Melendrez, M.F., 2019. Nanocomposites based on cationic polyelectrolytes and silver nanoparticles: synthesis, characterization, molybdate retention and antimicrobial activity. Arab. J. Chem. 12, 825–834. https://doi.org/10.1016/j.arabjc.2017.11.020.

Palencia, M., Córdoba, A., Rivas, B.L., 2014. Concentration–polarization effect of poly(sodium styrene sulfonate) on size distribution of colloidal silver nanoparticles during diafiltration experiments. Colloid Polym. Sci. 292, 619–626. https://doi.org/10.1007/s00396-013-3096-5.

Palencia, M., Lerma, T.A., Berrio, M.E., 2017. Thin-film composite by in situ polymerization of 4-chloromethyl styrene functionalized with N-methyl-D-glucamine in pore-type microreactors. J. Sci. Technol. Appl. 3, 66–76. https://doi.org/10.34294/j.jsta.17.3.25.

Palencia, M., Martínez, J.M., Arrieta, Á., 2017. Removal of acid blue 129 dye by polymer enhanced ultrafiltration (PEUF). J. Sci. Technol. Appl. 2, 65–74. https://doi.org/10.34294/j.jsta.17.2.15.

Parolo, M.E., Avena, M.J., Pettinari, G.R., Baschini, M.T., 2012. Influence of Ca^{2+} on tetracycline adsorption on montmorillonite. J. Colloid Interface Sci. 368, 420–426. https://doi.org/10.1016/j.jcis.2011.10.079.

Peralta, M.E., Ocampo, S., Funes, E.G., Medina, F.O., Parolo, M.E., Carlos, L., 2020. Nanomaterials with tailored magnetic properties as adsorbents of organic pollutants from wastewaters. Inorganics 8, 24. https://doi.org/10.3390/inorganics8040024.

Palencia, M., Rivas, B.L., 2011. Metal-ion retention properties of water-soluble amphiphilic block copolymer in double emulsion systems (w/o/w) stabilized by non-ionic surfactants. J. Colloid Interface Sci. 363, 682–689. https://doi.org/10.1016/j.jcis.2011.05.089.

Pinsino, A., Matranga, V., Carmela, M., 2012. Manganese: a new emerging contaminant in the environment. In: Srivastava, J. (Ed.), Environmental Contamination, pp. 17–36. https://doi.org/10.5772/31438.

Poynton, H., Robinson, W.E., 2018. Contaminants of emerging concern, with an emphasis on nanomaterials and pharmaceuticals. In: Török, B., Dransfield, T. (Eds.), Green Chemistry, pp. 291–315. https://doi.org/10.1016/B978-0-12-809270-5.00012-1.

Rathi, B.S., Kumar, P.S., Show, P.L., 2020. A review on effective removal of emerging contaminants from aquatic systems: current trends and scope for further research. J. Hazard Mater. 124413. https://doi.org/10.1016/j.jhazmat.2020.124413.

Rajput, P., Thakur, A., Sharma, P.D., 2020. Emerging agro-chemicals contaminants: current status, challenges and technological solutions. In: Prasad, M.N.V. (Ed.), Agrochemicals: Detection, Treatment and Remediation, pp. 117–142. https://doi.org/10.1016/B978-0-08-103017-2.00005-2.

Ramírez-Malule, H., Quiñones-Murillo, D.H., Manotas-Duque, D., 2020. Emerging contaminants as global environmental hazards. A bibliometric analysis. Emerg. Contam. 6, 179–193. https://doi.org/10.1016/j.emcon.2020.05.001.

Rivas, B.L., Sánchez, J., Palencia, M., 2016. Organic membranes and polymers for the removal of pollutants. In: Visakh, P.M., Nazarenko, O. (Eds.), Nanostructured Polymer Membranes, pp. 203–235. https://doi.org/10.1002/9781118831779.ch6.

Rodriguez-Mozaz, S., Ricart, M., Köck-Schulmeyer, M., Guasch, H., Bonnineau, C., Proia, L., de Alda, M.L., Sabate, S., Barceló, D., 2015. Pharmaceuticals and pesticides in reclaimed water: efficiency assessment of a microfiltration–reverse osmosis (MF–RO) pilot plant. J. Hazard Mater. 282, 165–173. https://doi.org/10.1016/j.jhazmat.2014.09.015.

Rodriguez-Narvaez, O.M., Peralta-Hernandez, J.M., Goonetilleke, A., Bandala, E.R., 2017. Treatment technologies for emerging contaminants in water: a review. Chem. Eng. J. 323, 361–380. https://doi.org/10.1016/j.cej.2017.04.106.

Salehizadeh, H., Yan, N., Farnood, R., 2018. Recent advances in polysaccharide bio-based flocculants. Biotechnol. Adv. 36, 92–119. https://doi.org/10.1016/j.biotechadv.2017.10.002.

Salimi, M., Esrafili, A., Gholami, M., Jonidi Jafari, A., Rezaei Kalantary, R., Farzadkia, M., Kermani, M., Sobhi, H.R., 2017. Contaminants of emerging concern: a review of new approach in AOP technologies. Environ. Monit. Assess. 189 https://doi.org/10.1007/s10661-017-6097-x.

Sánchez, Y., García-Quintero, A., Palencia, M., 2018. Determination and distribution of size of inorganic particles by spectral deconvolution of surface plasmon resonance. J. Sci. Technol. Appl. 5, 45–54. https://doi.org/10.34294/j.jsta.18.5.34.

Sánchez, Y., Mora, M., Palencia, M., 2019. A first approach to the development of new drugs with leishmanicidal activity via control of surface properties of silver nanoparticles. J. Sci. Technol. Appl. 6, 118–133. https://doi.org/10.34294/j.jsta.19.6.46.

Saxena, G., Bharagava, R.N., 2017. Organic and inorganic pollutants in industrial wastes, their ecotoxicological effects, health hazards and bioremediation approaches. In: Bharagava, R.N. (Ed.), Environmental Pollutants and Their Bioremediation Approaches, pp. 23–56. https://doi.org/10.1201/9781315173351-3.

Sauvé, S., Desrosiers, M., 2014. A review of what is an emerging contaminant. Chem. Cent. J. 8, 15. https://doi.org/10.1186/1752-153x-8-15.

Soltani, A.M., Pouypouy, H., 2019. Standardization and regulations of nanotechnology and recent government policies across the world on nanomaterials. In: Ghorbanpour, M., Wani, S.H. (Eds.), Advances in Phytonanotechnology, pp. 419–446. https://doi.org/10.1016/b978-0-12-815322-2.00020-1.

Tambosi, J.L., de Sena, R.F., Favier, M., Gebhardt, W., José, H.J., et al., 2010. Removal of pharmaceutical compounds in membrane bioreactors (MBR) applying submerged membranes. Desalination 261, 148–156. https://doi.org/10.1016/j.desal.2010.05.014.

Tang, Y., Zhong, Y., Li, H., Huang, Y., Guo, X., Yang, F., Wu, Y., 2020. Contaminants of emerging concern in aquatic environment: occurrence, monitoring, fate, and risk assessment. Water Environ. Res. 92, 1811–1817. https://doi.org/10.1002/wer.1438.

Torres, J., Gérente, C., Andrès, Y., 2012. Sustainable activated carbons from agricultural residues dedicated to antibiotic removal by adsorption. Chin. J. Chem. Eng. 20, 524–529. https://doi.org/10.1016/S1004-9541(11)60214-0.

Valladares, R., Yangali-Quintanilla, V., Li, Z., Amy, G., 2011. Rejection of micropollutants by clean and fouled forward osmosis membrane. Water Res. 45, 6737–6744. https://doi.org/10.1016/j.watres.2011.10.037.

Zelić, E., Vuković, Ž., Halkijević, I., 2018. Application of nanotechnology in wastewater treatment. Gradevinar 4. https://doi.org/10.14256/JCE.2165.2017.

Zhou, J., Chen, H., Huang, W., Arocena, J.M., Ge, S., 2016. Sorption of atrazine, 17α-estradiol, and phenanthrene on wheat straw and peanut shell biochars. Water Air Soil Pollut. 227, 1–13. https://doi.org/10.1007/s11270-015-2699-5.

Chapter 22

Polymer-based restoration of functional properties of degraded soils

22.1 Soil as a geomimetic model of new multifunctional materials

Soil is a system that, by its origin, formation, and evolution, represents the majority of the terrestrial ecosystem and is of limited extent. It is the natural environment for the growth of plants and in which life develops, and which, once degraded has a difficult and long recovery. It is estimated that around 25% of the soils are degraded globally, that is, they present with alterations in their properties that cause a reduction in their functions. In this sense, it is possible to speak of three different types of degradation, physical, chemical, and biological, which are found depending on the type of alteration produced to the properties of the soil, for example, the loss of structure, and the depletion of nutrients and biological functions (Hatfield et al., 2017). Thus, soil degradation can be a consequence of processes of water or wind erosion, salinization, waterlogging, fire, deterioration of the structure, pollution, etc., which are intensified by anthropogenic factors as a result of intense agricultural and industrial activity. In addition, soil is characterized by being an open system, therefore its degradation is related to and intensifies with climate change, as it is a natural carbon reservoir and a support for biotic fixation systems of energy in chemical form (Scanes, 2018).

At present, given the growing need to increase crop yields for food production and as a traditional practice for the recovery of degraded soils, the use of chemical fertilizers has been promoted, however, their exclusive and excessive use has made them a source of soil deterioration, causing the gradual loss of soil biota, replacing its biological functions, and consequently increasing its dependence on external inputs to maintain its productivity (Rahman and Zhang, 2018). In this sense, in recent decades the application of organic amendments based on composts of vegetable or animal origin, together with tillage, have been the strategies mainly used to reduce soil degradation and maintain its productivity. However, the incorporation of

Eco-friendly Functional Polymers
https://doi.org/10.1016/B978-0-12-821842-6.00013-0
Copyright © 2021 Elsevier Inc. All rights reserved.

exogenous organic matter to soils to achieve the recovery and restoration of its properties is only effective in agricultural soils which retain productive activity, and is not effective for soils in advanced stages of degradation, or those that are highly eroded or in a natural desertification state (Larney and Angers, 2012; Lerma et al., 2018a; Tully and McAskill, 2020). Therefore progress in the development of new alternatives that allow the mitigation of deterioration, rehabilitation, and improvement of the physical, chemical, and biological properties of soils has increased as a primary requirement in the face of the global problem of soil degradation, and the growing expansion of the agricultural frontier.

Thanks to the advances made in the study of supramolecular systems and the understanding of the different interactions and structures of natural systems, they have allowed the necessary foundations for the design of complex bio-inspired supramolecular systems into the functions and structures of nature. Thus, taking soil as a source of inspiration, and considering not only its structure but also its dynamics and functionality resulting from the different relationships that exist in it, geomimesis has emerged as an alternative to advance toward positive geotransformation of degraded soils (Lerma et al., 2018b). (The concept of geomimesis, or geomimicry, was first proposed by Lerma and Palencia, who are researchers from the Research Group in Sciences with Technological Applications of the Universidad del Valle, in Colombia, in 2018). In this way, eco-friendly polymeric systems, particularly supramolecular hybrid materials based on clay–polymer systems, have undergone considerable developments (Fu et al., 2019; Wypych et al., 2018).

The objective of this chapter is to exemplify some investigations focused on the development of geomimetic materials that are able to mimic the structure and functions of soil, and which can be used as an alternative for mitigating the deterioration and loss of soil, while promoting their rehabilitation and improving their physical, chemical, and biological properties. In the first example, Garcés, Palencia, and Combatt describe the synthesis of hydrogels from carbohydrates, polyols, and organic acids of eco-friendly origin, which are used in the construction of bacterial inocula able to preserve beneficial microorganisms in their dormant state, and permit their metabolic activation by the addition of water (Garcés et al., 2017). In the second example, Lerma, Palencia, and Combatt described the construction of hybrid supramolecular systems, based on eco-friendly polymers nanostructured with clays, by covalent and electrostatic interactions, which were structurally and functionally similar to clay–humic acid systems of the soil, and therefore they contributed to the improvement of water retention and cationic exchange properties (Lerma et al., 2018a). Finally, a third example is analyzed from the results described by Lerma, Garcés, and Palencia, in which versions of the two methods described above are integrated (Lerma et al., 2020).

22.2 Hydrogels as artificial systems of microbial growth

Given the increasing need to improve crop yields for the production of food and other agricultural products, different strategies have been developed, including physical, chemical, and biological, in order to create conditions favorable for the growth of the plants in adverse or inappropriate agricultural conditions, for example, the construction of irrigation and drainage systems, the use of fertilizers, and the application of bacterial inocula (Lal, 2015). In particular, among the beneficial effects associated with the use of bacterial inocula are nitrogen fixation, biological control of diseases transmitted by microorganisms, and improvement of mineral absorption. However, in the process of the development of bacterial inocula, the application and survival of the microorganisms is a critical stage in which adequate colonization of the rhizosphere must be achieved. This is difficult to achieve due to the competitive interaction among autochthonous microorganisms and those provided in the inoculum. As a consequence, the environmental conditions must not only be suitable for plant growth but also for the growth of exogenous microorganisms (Thomas and Sekhar, 2016). However, the effectiveness of bacterial inocula can be very limited in soils that are acid, saline, poor in nutrients, with poor organic matter contents, or with a deficit in their water conditions. In addition, the scarcity of soils and their continuous degradation, as well as the expansion of crops to areas not optimal for agriculture due to the need to increase food production, in many cases, make it difficult to use microorganisms that promote plant growth as an agricultural input (Garcés et al., 2017).

On the other hand, hydrogels are absorbent polymeric materials with high water retention capacity due to their highly hydrophilic three-dimensional network. Hydrogels can be obtained by physical or chemical cross-linking of polymer chains, and it is well-known that they absorb large amounts of water in a relatively short time. (Details of hydrogels are given in Chapter 9). Likewise, thanks to the versatility of the synthesis and design of polymeric systems, they can be used for the synthesis of eco-friendly hydrogels from sustainable sources. In general terms, the presence of functional groups able to be easily hydrolyzed by the action of microorganisms gives biodegradability to these materials, minimizing the discharge and accumulation of solid wastes (Ahmed, 2015). Therefore, as an alternative to the use of traditional agricultural amendment strategies and the limitations presented in the use of bacterial inocula, it has been proposed that multifunctional artificial systems be developed for the growth of plant growth-promoting microorganisms, which are based on eco-friendly hydrogels obtained from polyols, carbohydrates, and organic acids of natural origin. For instance, starch can be extracted from different vegetable sources and cross-linked in the presence of agar and glucose by direct heating at low temperature ($<100^{\circ}C$). Microorganisms can be added by inoculation of dried material or by swelling with a suspension of microorganisms. These biomaterials provide optimal conditions for microbial colonization, facilitating their adaptation and growth. In addition, they also act as a water reservoir in soils, whose phytoavailable water potential is modulated by microbial metabolism (Garcés et al., 2017).

22.2.1 Methodological strategy

As an example of a methodological approach, a model like the hydrogel synthesis developed by Garcés et al. is used. This procedure can be divided into three stages: (1) synthesis of the polymeric systems, (2) isolation of the plant growth promoter microorganisms, and (3) preparation of bacterial inocula. In this sense, initially, the polymeric systems developed were obtained from materials obtained from natural and renewable sources. Thus, hydrogels based on citric acid and sorbitol, both products obtained from the sugar industry, were obtained from esterification reactions at high temperatures between their carboxylic acid and hydroxyl groups, respectively. Additionally, a second polymeric system was synthesized from starch extracted from *Solanum tuberosum*, glucose, and carrageenan. Finally, both synthesized hydrogels were chemically and functionally characterized (Garcés et al., 2017). An interesting point to highlight at this stage is the simplicity of the manufacturing procedure.

In the second stage, isolation of the plant growth-promoting microorganisms was performed, in this case, they used the commercial biofertilizer MONIBAC, which is based on Gram-negative bacteria of *Azotobacter chroococcum*. Thus, from a sample of the commercial product, the microorganism was isolated, replicated, and identified by Gram stain. Finally, in the third stage the preparation of the bacterial inocula was carried out. This started with the incorporation of the bacteria, *Azotobacter chroococcum*, to the synthesized polymeric matrices. Thus, depending on the swelling capacity of each hydrogel, the corresponding amounts of bacterial suspension were added to each material. Subsequently, for the inactivation of microbial metabolism, their metabolic activity was suppressed by inducing a state of latency by gradually reducing the water potential of the medium. Finally, they verified the efficacy of the prepared inocula through rehydration of the medium to promote the reactivation of the metabolic activity of the microorganisms (Garcés et al., 2017).

22.2.2 Relevant results and conclusions

In relation to the structural characterization of hydrogels based on citric acid and sorbitol by means of infrared spectroscopy using the attenuated total reflectance technique IR-ATR, and carbon-13 nuclear magnetic resonance (^{13}C-NMR), the characteristic signals of the ester groups were identified, evidencing hydrogel synthesis via esterification reactions. The estimated percentage yield for the production of hydrogels from citric acid and sorbitol was 85%. In this reaction, the high yields obtained were achieved due to the synthesis conditions used, which favor the equilibrium shift toward the formation of products. On the other hand, in relation to the structural characterization of the material based on starch, glucose, and carrageenan, it should be noted that because this hydrogel is not formed by the chemical reaction between its components, but by a mixture of the different constituents, the characteristic signals of their precursors were found in the characterizations by IR-ATR and ^{13}C-NMR. In relation to its yield, they calculated it with respect to the mass of *Solanum tuberosum* used, with 0.19 g of hydrogel/g of potato (Garcés et al., 2017).

On the other hand, thermal characterization was carried out in order to explore the behavior of the polymers at temperatures at which the material could be subjected during its application, for example, forest fires or controlled burns. The reported thermograms of the hydrogels, in general, identified two regions of mass loss. The first region is characteristic for both systems and corresponds to the elimination of water molecules that interact noncovalently with the polymer matrix. The second region, in the case of starch-based hydrogels, is associated with the decomposition of the cross-linked structure of the hydrogel due to the condensation of the hydroxyl groups of the ether linkages and the dehydration of neighboring hydroxyl in the rings causing their breaking off. Similarly, for the hydrogels based on citric acid–sorbitol, the second region of mass loss corresponds to the breaking of the C–O bonds of the ester groups of the matrix, however, these hydrogels showed greater thermal stability as a result of the high degree of chemical cross-linking (Garcés et al., 2017).

In relation to the reported results of the isolation of plant growth-promoting microorganisms, they indicate that the isolation and growth of *Azotobacter chroococcum* in the two studied media were satisfactory. Likewise, studies carried out on the isolated microorganisms show two important characteristics of *Azotobacter chroococcum*: (1) Gram-negative microorganism, due to the pink coloration obtained from the cells at the end of the experiment, and (2) pleomorphic microorganism, that is, different forms of growth were identified (Garcés et al., 2017). On the other hand, once the *Azotobacter chroococcum* suspensions were obtained and incorporated into the hydrogels, they proceeded to induce the latency state of the microorganisms through environmental stress, by eliminating the water from the medium, and forcing the microorganisms to stop their metabolism. Subsequently, the studies showed that after hydration and incubation of the inoculated and dehydrated hydrogels, the reactivation of the growth of the microorganisms was possible. This was verified by the biodegradation of the hydrogels over a period of 15 days after their reactivation (Garcés et al., 2017).

22.3 Nanostructured hydrogels based on hybrid polymer–clay composites

As previously indicated, soil degradation is a generic concept occurring in different ways, and which is due to the simultaneous actions of different factors. It is widely accepted that the loss or removal of topsoil is associated with an instantaneous decrease in soil organic matter, which is a key component associated with very important soil properties, such as pH buffering capacity, cation exchange capacity, soil structure, porosity, and metabolic activity of microorganisms (Murphy, 2015; Wiesmeier et al., 2019). In particular, humified organic matter is the main type of organic matter present in soil contributing to the aforementioned properties, and it is closely related to clay in soil. However, since humified organic matter is the result of subjecting soil to extensive and different biogeochemical processes to produce different types of substances identified according to solubility properties such as humin, fulvic acids, and humic acids (García et al., 2018; Mora et al., 2018; Restrepo et al., 2018), in general, it is accepted that they are a nonrenewable resource that cannot be restored by adding exogenous organic matter (Gerke, 2018).

On the other hand, the combination of clay–polymer compounds has been used in the synthesis and modification of polymers for many applications with the aim of modifying the properties of polymers, for example, strength, stiffness, thermal and oxidative stability, barrier properties, and flame-retardant capacity (Wypych et al., 2018). Using this approach, Lerma et al. (2018a, 2018b), in order to develop remediation strategies based on the positive geotransformation of degraded soils, and taking advantage of the properties and versatility of hybrid polymeric systems, proposed the design and synthesis of clay–polymer composites, analogous to the clay–humic matter structures of soils, which can be used as polymeric soil conditioners for the restoration of soils naturally and anthropogenically degraded by the loss of humified organic matter (Lerma et al., 2018a, 2018b).

22.3.1 Synthesis strategy

The synthesis strategy used for the development of hybrid materials consisted of four stages, (1) synthesis of the cationic monomer N-vinylbenzyl-N-triethylammonium chloride, (2) modification of clays by ionic intercalation and covalent anchoring, (3) synthesis of hybrid composites, and (4) functional characterization of materials. The first stage consisted of the reaction between 4-chloromethylstyrene and triethylamine using acetone as solvent. Thus, through a nucleophilic substitution reaction the cationic monomer N-vinylbenzyl-N-triethylammonium chloride is formed, which is used in the modification of the clays. Next, in the clay modification stage, with the objective of incorporating vinyl groups on its surface, the researchers used two types of interactions to build the hybrid systems, (1) an electrostatic type, which consists of N-vinylbenzyl-N-triethylammonium chloride adsorption to the surface of clays, and (2) a covalent type, which consists of the reaction of the silanol groups on the surface of the clays with trichlorovinylsilane (Lerma et al., 2018a, 2018b).

In the third stage of the development process, the researchers synthesized the hybrid composites by free radical polymerization reactions of acrylic acid in the presence of the electrostatically and covalently modified clays. Finally, as the fourth and last stage, they carried out the functional characterization of the clay–polymer composites through determination of the water retention capacity using the tea bag method; the determination of the cation exchange capacity by the method of 1.0 N ammonium acetate defined by the Agustín Codazzi Geographical Institute (IGAC); and determination of the ion retention capacity through batch-type experiments (K^+, Ca^{2+}, Mg^{2+}, Mn^{2+}, Fe^{2+}, Cu^{2+}, and Zn^{2+}) (Lerma et al., 2018a, 2018b).

22.3.2 Relevant results and conclusions

The synthesis of the cationic monomer used in the modification of clays, that is, functionalization of 4-chloromethyl styrene with triethylamine, showed a yield of 80.5%. Furthermore, the results of elemental analysis and proton NMR, ^{1}H-NMR, of the N-vinylbenzyl-N-triethylammonium chloride verified its chemical and structural composition, respectively. In addition, the chemical reactivity test carried out was positive, and therefore it was verified that the product obtained can easily be polymerized in water by radical polymerization (Lerma et al., 2018a, 2018b).

On the other hand, in relation to the electrostatically and covalently modified clays, the elemental characterization showed for both functionalization techniques an increase in the percentages of nitrogen and carbon in the clay matrices, differing mainly in the nitrogen contents as a consequence of the presence of this element in the synthesized cationic monomer. Furthermore, characterization by IR-ATR showed the characteristic vibration bands of the organic groups anchored on the surface of the clays. In addition, in the results of the X-ray diffraction study, they reported the identification of small variations in the diffraction peaks, attributed to an increase in the interlaminar spaces of the clays as a result of the insertion of organic molecules (Lerma et al., 2018a, 2018b).

Continuing with the compositional and structural analysis of the hybrid systems, particularly the characterization of the synthesized clay–polymer composites, the results of the elemental analysis showed, in all cases, a decrease in the relative amounts of C, N, and H as the percentage of incorporation of modified clay increased; these results were congruent with the increase in the inorganic fraction in the material matrix. Additionally, characterization by IR-ATR allowed observation of the typical vibrations of the different constituent phases of the material, that is, bands associated with the structure of the aluminosilicates of the clays and the vibrations of the different organic groups present in the polymeric phase of composites. Therefore these results suggested that the clay–polymer composites are made up of inorganic and organic phases in a manner analogous to the humin clay–humic/fulvic acid system present in the soil particles (Lerma et al., 2018a, 2018b).

Finally, regarding the functional characterization of the hybrid systems, it was reported that, in relation to the studies of the water retention capacity, in general, all the hybrid materials registered high retention values, at greater than 180%. Furthermore, this parameter depended on the content of acrylic acid and clay in the composites. The results of the tests of the cationic exchange capacity, for the composites based on clays modified by electrostatic interactions showed higher values compared to the materials based on covalently modified clays, which was due to elimination of the sites exchange on the surface of clays as a result of their reaction with trichlorovinylsilane. Finally, evaluation of the ion retention capacity demonstrated a greater retention capacity of macro- and micronutrients in comparison with clays, thanks to the different mechanisms they show, for example, ion exchange reaction, electrostatic and coordination interactions, analogous to those exhibited by soil (Lerma et al., 2018a, 2018b).

22.4 Multi- and bio-functional hybrid polymer hydrogel geomimetics of soil particles

Soils are an essential resource for world agricultural production, and are influenced by anthropogenic and natural factors producing a progressive reduction in their quality. In recent decades, the application of organic matter of natural origin has been the most used strategy to combat soil degradation and maintain its productivity. However, the foregoing is only effective in productive soils and becomes inappropriate for soils with an advanced state of degradation (Lal, 2015). Currently, the use of synthetic polymeric materials as soil conditioners has increased due to the ability to improve the physical properties of the soil regardless of the level of degradation, however, many of these conditioners, being constituted by nonbiodegradable polymers, represent a risk factor for environmental contamination over time, as a consequence of its inappropriate use (Milani et al., 2017). Likewise, significant interest has been placed in the use of biotechnological materials, which are used to increase nitrogen fixation and mineral absorption in soil. However, their implementation is limited as they requires careful control of growth conditions to achieve adequate colonization of the rhizosphere (Wang et al., 2014).

As a consequence of the above, Lerma et al. (2020) integrated the conclusions of the works previously carried out and, in response to the limitations of the systems described, developed new hybrid multifunctional and biofunctional polymeric hydrogels, based on clay–polymer composites and biodegradable polyesters of sorbitol (Lerma et al., 2020). These systems were characterized by containing bentonite, as a functional inorganic phase, and poly(acrylic acid) chains covalently anchored, polymer rich in carboxylic acid groups and highly hydrophilic. Additionally, these systems have a polyester layer, built on the polyacrylic acid chains by means of the in situ esterification of low-molecular-weight, biodegradable and biocompatible molecules, which serve as a medium that favors microbial colonization of soils, that is from a technological approach, these materials allow the manufacture of bacterial inocula. Fig. 22.1 shows a representation of the synthesis process of the hybrid polymer hydrogels proposed (Lerma et al., 2020).

22.4.1 Synthesis strategy

The synthesis strategy used for making the hybrid multifunctional and bio-functional polymeric systems consisted of four stages: (1) superficial modification of the bentonite, (2) synthesis of the composite hydrogels, (3) functional characterization of the materials, and (4) preparation of the bacterial inocula. The first stage described consisted in modifying the bentonite using the co-valent anchoring technique with trichlorovinylsilane for the incorporation of easily polymerizable vinyl groups. This was followed, in the second stage, by two sequential synthesis reactions, polymerization via free radicals and polyesterification at high temperatures, which generated poly(acrylic acid) chains on the modified bentonite and synthesized a polyester matrix, based on citric acid and sorbitol, anchored to it, respectively, in order to obtain bentonite—poly(acrylic acid)—polyester composites (Lerma et al., 2020).

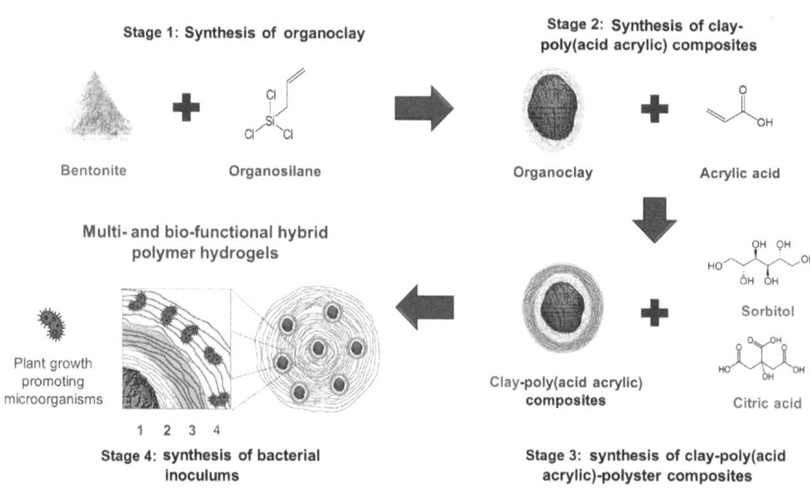

FIGURE 22.1 Representation of the synthesis process of the multi- and bio-functional hybrid polymer hydrogels.

In the third stage, they evaluated the functional properties of the synthesized composite hydrogels through characterization of the water absorption capacity, using the tea bag method, and the cation exchange capacity, using the method of ammonium acetate 1.0 N. In the final step, they proceeded to prepare the bacterial inocula, through the addition of bacterial dispersions of *Azotobacter chroococcum* to the hydrogels obtained, and the consequent induction of the latency state of the microorganisms, by means of the suppression of metabolic activity induced by heat. Finally, they evaluated the bacterial viability of the prepared inoculums through hydration and incubation (Lerma et al., 2020).

22.4.2 Relevant results and conclusions

The results obtained from the characterization by elemental analysis of the organo-clay based on the covalent modification of bentonite with trichlorovinylsilane, evidenced an increase in the contents of the elements of carbon and hydrogen in the organo-clay compared to bentonite, which shows that the insertion of vinyl groups of the trichlorovinylsilane on the bentonite surface was successful. In addition, the results of the characterization by IR-ATR using FEDS analysis, evidenced the characteristic bands of bentonite, as well as the typical vibrations of the vinyl groups in the organo-clay samples analyzed, which confirms the insertion of organic groups on the inorganic particles of aluminosilicate and the formation of organoclay (Lerma et al., 2020).

As was previously mentioned, the production of the hybrid polymer soil conditioners based on clay—polymer composites consisted of two synthesis steps. In the first step, bentonite—poly(acrylic acid) composites rich in carboxylic acid groups were obtained, which can subsequently intervene in the stage of formation of the biodegradable polyester matrix. The details of the characterization of these materials have been previously published (Lerma et al., 2020). On the other hand, the second stage of synthesis of the soil hybrid polymeric hydrogels consisted in the construction of a second polymeric layer of polyester based on citric acid and sorbitol, which is formed on the poly(acrylic acid) chains through reactions of esterification at high temperatures. The results of the structural characterization by FT-IR using FEDS analysis, evidenced the characteristic vibration bands of each of the components of the hybrid system, clay and layers of poly(acrylic acid) and polyester. Likewise, the solid-state ^{13}C-NMR analysis of the hybrid polymeric hydrogels, in general, showed high similarity between the spectra obtained in which the signals of the different carbons present in the structure of the material were identified. However, an additional signal was identified and associated with the formation of carbon—carbon double bonds as a product of collateral reactions during synthesis, that is, alcohol dehydration processes favored by the thermal conditions of the synthesis process of hybrid composites (Lerma et al., 2020).

In relation to the results of the functional characterization, specifically of water absorption capacity, in general, in function of the organic fraction, materials showed a higher water absorption in comparison with bentonite. This behavior is due to the large number of hydrophilic groups present in the organic polymeric fraction that have the ability to interact with the water molecules and retain them in their matrix. On the other hand, the results of the evaluation of the cationic exchange capacity showed an increase in comparison with the average values presented by bentonite, without finding a dependence with respect to its content. The relatively low variation in the cation exchange capacity that the composite polymeric hydrogels presented was due to the type of technique used to obtain the organoclay that involves the reaction between the trichlorovinylsilane and the active exchange sites on the surface of the clay. The foregoing allowed demonstration of the ability of synthesized composites to functionally mimic soil particles, participating not only in cation exchange processes, but also intervening in mediating water content thanks to their high absorption capacity, two essential properties in determining soil quality (Lerma et al., 2020).

For preparation of the bacterial inocula, the high hydrophilicity of the materials was taken advantage of, and the required amounts of the bacterial suspension of *Azotobacter chroococcum* were added for the material to reach its maximum water absorption capacity. Once the *Azotobacter chroococcum* was incorporated, the water was removed from the material by evaporation; with the aim of inducing the bacteria to a latent state, which is reached by the bacteria as a survival mechanism to the environmental stress to which it is subjected. It is highlighted that in the inocula, the population concentration of bacteria per gram of material is of the order of 10^9-10^{10} colony-forming units (CFU); while for natural soils the reports indicate concentration values not greater than 10^4-10^6 CFU/g of soil (Jnawali et al., 2015). This shows an achievement in increasing the population density of *Azotobacter chroococcum* in the materials. Finally, to verify the bacterial viability of the inocula obtained, they were hydrated and subjected to a 24-hour incubation process. In general, the results of bacterial growth in solid agar culture medium showed the ability of all the materials to incorporate and protect *Azotobacter chroococcum* in its polymeric matrix in a dormant state for subsequent metabolic reactivation. The foregoing shows the multifunctionality of the developed bentonite–poly(acrylic acid)–polyester composites, not only potentially contributing to the improvement of the physical and chemical properties of water retention and cationic exchange of the soil, but at the same time allowing the incorporation of plant growth-promoting microorganisms such as *Azotobacter chroococcum* and contributing to its biological restoration (Lerma et al., 2020).

22.5 Conclusions

As a result of the constant degradation suffered by soils, the development of new systems and materials allows the restoration of their physical, chemical, and biological properties. In addition, despite in recent decades the application of organic amendments, synthetic polymeric materials, and biotechnological materials have been widely used with the aim of reducing soil degradation and maintaining its productivity, these strategies in certain settings have proven to be ineffective. Therefore with the advances in the compression of the structures and mechanisms of natural systems, specifically soil, it has been possible to build new eco-friendly hybrid polymeric materials, inspired by the structure and functions of soils, which act as multifunctional materials, potentially contributing to the improvement of soil properties from physical, chemical, and biological points of view.

References

Ahmed, E.M., 2015. Hydrogel: preparation, characterization, and applications: a review. J. Adv. Res. 6 (2), 105–121. https://doi.org/10.1016/j.jare.2013.07.006.

Fu, S., Sun, Z., Huang, P., Li, Y., Hu, N., 2019. Some basic aspects of polymer nanocomposites: a critical review. Nano Mater. Sci. 1 (1), 2–30. https://doi.org/10.1016/j.nanoms.2019.02.006.

Garcés, V., Palencia, M., Combatt, E., 2017. Development of bacterial inoculums based on biodegradable hydrogels for agricultural applications. J. Sci. Technol. Appl. 2, 13–23. https://doi.org/10.34294/j.jsta.17.2.11.

Garcia, A., Combatt, E., Palencia, M., 2018. Structural study of humin and its interaction with humic acids by Fourier-transform mid-infrared spectroscopy. J. Sci. Technol. Appl. 4, 28–39. https://doi.org/10.34294/j.jsta.18.4.28.

Gerke, J., 2018. Concepts and misconceptions of humic substances as the stable part of soil organic matter: a review. Agronomy 8 (5), 76. https://doi.org/10.3390/agronomy8050076.

Hatfield, J.L., Sauer, T.J., Cruse, R.M., 2017. Chapter one - soil: the forgotten piece of the water, food, energy nexus. In: Sparks, D.L. (Ed.), Advances in Agronomy, vol. 143. Academic Press, pp. 1–46. https://doi.org/10.1016/bs.agron.2017.02.001.

Jnawali, A.D., Ojha, R.B., Marahatta, S., 2015. Role of *Azotobacter* in soil fertility and sustainability—a review. Adv. Plants & Agric. Res. 2 (6), 250–253. https://doi.org/10.15406/apar.2015.02.00069.

Lal, R., 2015. Restoring soil quality to mitigate soil degradation. Sustainability 7 (5), 5875–5895. https://doi.org/10.3390/su7055875.

Larney, F.J., Angers, D.A., 2012. The role of organic amendments in soil reclamation: a review. Can. J. Soil Sci. 92 (1), 19–38. https://doi.org/10.4141/cjss2010-064.

Lerma, T.A., Garces, V., Palencia, M., 2020. Novel multi- and bio-functional hybrid polymer hydrogels based on bentonite-poly(acrylic acid) composites and sorbitol polyesters: structural and functional characterization. Eur. Polym. J. 128, 109627. https://doi.org/10.1016/j.eurpolymj.2020.109627.

Lerma, T.A., Palencia, M., Combatt, E.M., 2018a. Soil polymer conditioner based on montmorillonite-poly(acrylic acid) composites. J. Appl. Polym. Sci. 135, 46211. https://doi.org/10.1002/app.46211.

Lerma, T., Combatt, E., Palencia, M., 2018b. Soil-mimicking hybrid composites based on clay, polymers and nitrogen-fixing bacteria for the development of remediation systems of degraded soil. J. Sci. Technol. Appl. 4, 17–27. https://doi.org/10.34294/j.jsta.18.4.27.

Milani, P., França, D., Balieiro, A.G., Faez, R., 2017. Polymers and its applications in agriculture. Polímeros 27 (3), 256–266. https://doi.org/10.1590/0104-1428.09316.

Mora, M., Palencia, M., Combatt, E., 2018. Effect of dissolved organic matter and humic substances on transport of ions and low molecular weight molecules by liquid membranes. J. Sci. Technol. Appl. 4, 40–48. https://doi.org/10.34294/j.jsta.18.4.29.

Murphy, B.W., 2015. Impact of soil organic matter on soil properties—a review with emphasis on Australian soils. Soil Res. 53, 605–635. https://doi.org/10.1071/SR14246.

Rahman, K.M.A., Zhang, D., 2018. Effects of fertilizer broadcasting on the excessive use of inorganic fertilizers and environmental sustainability. Sustainability 10 (3), 759. https://doi.org/10.3390/su10030759.

Restrepo, D., Palencia, M., Palencia, V., 2018. Study by attenuated total reflectance spectroscopy of structural changes of humified organic matter by chemical perturbations via alkaline dissolution. J. Sci. Technol. Appl. 4, 49–59. https://doi.org/10.34294/j.jsta.18.4.30.

Scanes, C.G., 2018. Chapter 18 - impact of agricultural animals on the environment. In: Scanes, C.G., Toukhsati, S.R. (Eds.), Animals and Human Society. Academic Press, pp. 427–449. https://doi.org/10.1016/B978-0-12-805247-1.00025-3.

Thomas, P., Sekhar, A., 2016. Effects due to rhizospheric soil application of an antagonistic bacterial endophyte on native bacterial community and its survival in soil: a case study with *Pseudomonas aeruginosa* from banana. Front. Microbiol. 7, 493. https://doi.org/10.3389/fmicb.2016.00493.

Tully, K.L., McAskill, C., 2020. Promoting soil health in organically managed systems: a review. Organ. Agric. 10, 339–358. https://doi.org/10.1007/s13165-019-00275-1.

Wang, C., Liu, W., Li, Q., Ma, D., Lu, H., Feng, W., Xie, Y., Zhu, Y., Guo, T., 2014. Effects of different irrigation and nitrogen regimes on root growth and its correlation with above-ground plant parts in high-yielding wheat under field conditions. Field Crop. Res. 165, 138–149. https://doi.org/10.1016/j.fcr.2014.04.011.

Wiesmeier, M., Urbanski, L., Hobley, E., Lang, B., von Lützow, M., Marin-Spiotta, E., Wesemael, B., Rabot, E., Ließ, M., Garcia-Franco, N., Wollschläger, U., Vogel, H.J., Kögel-Knabner, I., 2019. Soil organic carbon storage as a key function of soils - a review of drivers and indicators at various scales. Geoderma 333, 149–162. https://doi.org/10.1016/j.geoderma.2018.07.026.

Wypych, F., Bergaya, F., Schoonheydt, R.A., 2018. 10 - from polymers to clay polymer nanocomposites. In: Schoonheydt, R., Johnston, C.T., Bergaya, F. (Eds.), Developments in Clay Science, vol. 9. Elsevier, pp. 331–359. https://doi.org/10.1016/B978-0-08-102432-4.00010-X.

Chapter 23

Immunopharmacological control of blood and hemoderivatives

23.1 Bacteremia during blood storage: an overview

Blood transfusion is a medical process of great importance for patients who require organ transplants, and in cancer treatments, anemia, childbirth, or wound treatments with bleeding, etc. In this type of procedure, each product derived from blood, as well as the blood itself, has a precise indication for its use and specific epidemiological measures in order to prevent its contamination and ensure its immunopharmacological control (Murphy, 2017). Thus, for example, erythrocytes (red blood cells), are used to replace the substantial loss of blood caused by trauma, surgery, organ transplantation, childbirth, leukemia therapies, and thalassemia. The epidemiological measures being used are disinfection of the surface at the time of blood collection, the implementation of hygiene protocols throughout the process, and refrigeration at temperatures between 2 and 4°C for 24 h; however, those red blood cells that have remained outside the correct hypothermic transport and storage conditions for more than 30 min should not be used due to the risk of bacterial contamination, loss of cellular function, and even hemolysis (Klausen et al., 2014). On the other hand, blood plasma is used in cases of massive hemorrhage and should be stored at temperatures of −25°C, avoiding samples being kept out the refrigerator for more than 30 min, mainly, in order to minimize the risk of contamination. However, despite these protocols, several cases have been reported in which strains of *Pseudomonas scepacia* and *Pseudomonas aeruginosa* have proliferated even under refrigeration conditions (Brecher and Shauna, 2005). On the other hand, thrombocytes or platelets are used mainly for the treatment of hematological neoplasms and promotion of blood clotting to stop bleeding. These hemocomponents should be maintained at temperatures not lower than 20°C in order to maintain their hemostatic capacity. However, they are exposed to a potential risk of bacterial contamination when

Eco-friendly Functional Polymers
https://doi.org/10.1016/B978-0-12-821842-6.00004-X
Copyright © 2021 Elsevier Inc. All rights reserved.

stored at a temperature that favors the growth of various bacteria, such as the genera *Staphylococcus*, *Salmonella*, *Enterobacter*, *Serratia*, *Klebsiella*, among etc. (Aubron et al., 2018).

Adverse transfusion reactions can be caused by inadequate operating procedures related to the collection, processing, storage, and distribution of blood. They are classified as acute noninfectious, late noninfectious, and transfusion-transmitted infections. Some reactions are long-term clinical events that affect the recipient's health in a chronic way, causing a global risk associated with transfusion processes. Recently, a study that collected data reported by 25 countries in the database of the International Haemovigilance Network Database determined that the rate of adverse reactions from transfusion of blood products is 660 per 100,000 people, where almost 3% of these cases were classified as acute (Palencia et al., 2016; Frazier et al., 2017). Acute noninfectious reactions include hemolysis, fever, hypervolemia, and pulmonary edema, and they can manifest in the short term; meanwhile noninfectious late reactions can occur days or even years after transfusion, and include hemolytic reactions, alloimmunization to erythrocytes, leukocytes, or platelets (Sahu and Hemlata, 2014; Vasudev et al., 2016).

In particular, transfusion-transmitted infections are caused by bacterial agents present in the blood. The risk of bacterial contamination in blood products is estimated to be three orders of magnitude greater than the risk of infection from viruses such as HIV (human immunodeficiency virus), and the risk of bacterial sepsis is more than two orders of magnitude greater than the risk of viral infections (Liumbruno et al., 2009). Septic reactions caused by the presence of bacteria involve immediate collapse, which is followed by shock and hyperpyrexia, frequent hemorrhagic phenomena, vomiting, diarrhea, hemoglobinuria, kidney failure, and disseminated intravascular coagulation (Delaney et al., 2016). Transfusion-transmitted infections are classified as endogenous and exogenous. In the former, during phlebotomy, bacterial contamination is generated from the entry into the storage bags of bacteria from skin flora of the donor's arm, and from preexisting bacteremia (Panch et al., 2019). On the other hand, in exogenous infections, bacterial contamination is generated by wrong procedures during the processing and conservation of blood and its components (plasma, platelets, red blood cells). They can also be caused by the use of contaminated or inadequately sterilized collection equipment (Kundrapu et al., 2020). Among the main microorganisms causing the bacterial contamination of blood and blood products are Gram-positive bacteria such as *Bacillus cereus*, *Streptococcus* spp., *Staphylococcus aureus*, *Propioni bacteriumacnés*, and Gram-negative bacteria such as *Pseudomonas* spp., *Citrobacter freundi*, *Escherichia coli*, *Ersinia enterocolitica*, *Bartonella*, *Brucella* spp., and *Salmonella* spp. (Damgaard et al., 2015; Ramierz-Arcos and Goldman, 2017).

In this sense, epidemiological measures have been taken to guarantee the safety and efficacy of transfusions, among which the proper processing and storage of blood and its components stand out. Prevention measures have also been taken by diverting the first 10–30 mL of blood from the initial collection. Thus, by diverting and eliminating the first few milliliters of transfused blood, bacterial contamination can be reduced, but not eradicated (Garraud et al., 2018). Furthermore, the number of bacteria is also of importance, and is correlated with the storage time and temperature. Therefore, the storage time of blood products is limited, with platelet concentrates and red blood cells needing to be discarded after 7 and 35 days, respectively (García-Roa et al., 2017). Likewise, the storage of blood products at temperatures close to the environmental temperature, for example, erythrocytes, promotes bacterial proliferation in the concentrates of blood products (Putter and Seghatchian, 2017). Although to a large extent these protocols are effective in optimal operating conditions, in special environments such as areas with difficult access, and with poor biomedical and hospital infrastructure, the effectiveness of these controls may be severely limited (Barro et al., 2018).

In line with the above, in addition to control protocols, photosensitizing agents have been used for antimicrobial control. Examples of these are the Intercept, Mirasol, and Theraflex systems, which, when they are activated by ultraviolet radiation, attack bacterial DNA, killing bacteria; however, activation of these photoinactivators produces by-products that can enter the recipient's bloodstream in blood transfusion. In addition, due to nonselectivity of these agents they can also attack the patient's cellular systems, representing a toxicological risk of carcinogenicity, mutagenicity, and to produce chromosomal aberrations (Drew et al., 2017). Studies carried out indicate that the by-products generated by these agents can be eliminated by adsorption filters, however, as a result of the adsorption of the by-products after several transfer steps, a loss of blood cells is observed, and therefore, total removal of photodegradation by-products cannot be guaranteed (Galli and Bruschi, 2020). Additionally, due to high cost, the application of this technology is currently limited to countries in Europe, the United States, and other developed countries (Salunkhe et al., 2015).

Therefore the need to develop systems that allow the immunohematological control of blood and blood products, efficiently and with the lowest possible risk, is clear. In this way, in order to minimize the risks associated with adverse transfusion reactions, particularly infections caused by bacterial agents present in the blood during transfusion processes, in recent decades, research has been focused on the search for new multiobjective antimicrobial agents based on the development of materials with antimicrobial properties, and in methods of elimination of microorganisms by nonpharmacological strategies. Thus, polymers have emerged as one of the main and most promising types of materials used in medical applications due to their relatively low cost, ease of synthesis and modification, and the ability to generate a wide variety of materials with excellent physical, chemical, and surface properties, and biocompatibility (Asghari et al., 2017; Song et al., 2018).

The objective of this chapter is to describe three different examples of research focused on the development of eco-friendly functional polymer systems with potential applications in the immunopharmacological control of blood and its components. Thus, in the first example, Palencia et al. (2017a), described the synthesis of eco-friendly hydrogels based on cassava starch and used for the controlled release of nanostructured hybrid antibiotics with antibacterial properties. In the second example, the work developed by Palencia et al. described a strategy for modifying poly(vinyl chloride) for the insertion of cationic groups, in order to provide it with antibacterial properties, and by this procedure, to advance the development of techniques for surface-contact pathogen inactivation (Palencia et al., 2019a). Finally, a third example is based on the development of new antibacterial and nonhemolytic materials based on eco-friendly cationic polyurethanes with quaternary ammonium units (Palencia et al., 2020).

23.2 Starch-based hydrogels for the controlled release of nanostructured hybrid antibiotics

Currently, blood transfusion processes are characterized by using different methods that combine serological tests, genomic amplification, and the adequate identification and selection of donors through surveys, or by study of the clinical history of patients. These approaches have made it possible to reduce the risk of acquiring communicable diseases, and type donor–acceptor, during these procedures, for example, human immunodeficiency virus, hepatitis B virus, hepatitis C virus, and syphilis. However, currently, in both high- and low-income countries, bacterial infections are the main risk of acquiring infectious disease during blood-transfusion processes (Heroes et al., 2020). Bacterial infections caused by exogenous agents are mainly related to contamination by transient or commensal bacterial flora present on the donor's skin or by contamination during the collection, processing, and storage process. However, although the elimination of the first few milliliters of blood drawn at the beginning of the apheresis process is used worldwide as a strategy to prevent the introduction of skin bacteria into the units of blood products at the time of venipuncture, there remains a high rate of contamination associated with the movement and entry of bacteria through small pieces of skin produced during the needle insertion process, with irregular skin areas resulting from repeated donations being more likely to produce contamination (Liumbruno et al., 2009).

On the other hand, hydrogels are polymeric materials rich in polar functional groups that give them the ability to absorb large volumes of water. These materials can be constituted through the physical cross-linking of linear polymers or by chemical cross-linking networks, which give them different characteristics and specific properties (Otálora et al., 2019). Under physiological conditions, hydrogels are also capable of retaining biological fluids and possessing a soft rubbery consistency similar to that of tissues, which has made them materials used in various medical and pharmaceutical applications, for example, tissue engineering, medical implants, diagnostic devices, cell immobilization, bone stabilization, biomolecule separation, barrier materials, and drug-delivery systems (Ullah et al., 2015). In this context, with the purpose of mitigating the entry of skin bacteria during apheresis processes, eco-friendly hydrogels based on cassava starch and Carbopol, or poly(acrylic acid), have been developed. These gels can be used as controlled-release systems of nanostructured hybrid drugs with antibacterial properties, and therefore they are potential topical agents for disinfecting skin areas subjected to venipuncture processes (Palencia et al., 2017a).

23.2.1 Synthesis strategy of hydrogels based on starch–poly(acrylic acid)

Starch–poly(acrylic acid) hydrogels are be obtained by the following stages: (1) synthesis of the hydrogel, (2) synthesis of nanostructured agents, and (3) functional characterization of hydrogels (susceptibility to be hydrolyzed, essays of release, and antimicrobial activity tests). For the above, the synthesis of the hydrogel was based on thermal esterification reactions at 110°C, between the hydroxyl groups of the cassava starch chains with the carboxylic acid groups of poly(acrylic acid). The percentage of incorporation of poly-acrylic acid used was ~16%. Subsequently, the antibacterial nanostructured agents (AgNPs-Cysa) were synthesized from the chemical reduction of Ag^+ ions with $NaBH_4$, in the presence of cysteamine as a stabilizing agent at low temperature. Nanoparticle systems were characterized by elemental analysis, dynamic light scattering, and ultraviolet-visible spectroscopy (Palencia et al., 2017a).

23.2.2 Relevant results and conclusions

The structural characterization of the hydrogels was based on the analysis of their elemental composition of carbon and hydrogen, and by spectroscopic analysis through infrared spectroscopy. The results of the carbon and hydrogen contents of the hydrogels reflected an increase in these elements compared to those of starch and were much lower than those presented by poly(acrylic acid). This correlation in the carbon and hydrogen contents is due to the proportions of each precursor used during their synthesis process. Furthermore, infrared spectra show the formation of the ester group by identifying two bands associated with the C=O and C–O–C bonds, at 1700 and 1000 cm^{-1}, respectively. The results show that the end structure of hydrogel is constituted by the starch and poly(acrylic acid) cross-linked via ester-type bonds. In addition, hydrogel was highly hydrophilic, absorbing ~20 times its mass in water (Palencia et al., 2017a).

On the other hand, characterization results of AgNPs-Cysa antimicrobial systems by elemental analysis showed the presence of carbon and sulfur in the nanoparticles. In addition, UV-vis spectra show a characteristic signal of them, that is, surface plasmon resonance band at ~ 388 nm, which is typical of nanoparticles based on silver (Palencia et al., 2017b). The authors reported that the size of nanoparticles was stabilized with cysteamine at 102 ± 75 nm (Palencia et al., 2017a,b).

The functional characterization of the hydrogels began with the evaluation of stability from acidic to alkaline conditions. Hydrolysis percentages correlated with the increase in pH, being low at acidic pH. On the other hand, evaluation of the release of AgNPs-Cysa nanoparticles showed a faster release time, by 5.1 times, compared to cysteamine (Palencia et al., 2017a,b). These results were explained by the difference in the diffusion of analyte through the polymeric matrix, which can be influenced by the electrostatic, physical, and chemical interactions between matrix–analyte and analyte–analyte. In addition, the degree of cross-linking, the swelling capacity, and the chemical potential of the medium are also properties that can affect the release of nanoparticles in hydrogels (Palencia et al., 2014). In this way, a slower initial release of cysteamine in contrast to AgNPs-Cysa was associated with a greater capacity for interaction through hydrogen bond formation with –OH and –COOH groups of the polymer matrix. Finally, the results of the antimicrobial activity test against *Escherichia coli* and *Staphylococcus aureus* reported that all the hydrogels with AgNPs-Cysa show antimicrobial activity against the evaluated strains, being more effective for *E. coli* compared to *S. aureus* (Palencia et al., 2017a,b).

23.3 Functional materials with antibacterial properties for surface-contact pathogen inactivation

Poly(vinyl chloride) (PVC) is a thermoplastic polymer produced from the polymerization of vinyl chloride that is widely used worldwide in various applications, for example, electrical cable coatings, rigid profiles, packaging, and many other consumer products. For the year 2018, a consumption rate of 44.3 mega million tons (MMT) was reported, and it is estimated to reach 59.7 MMT by 2025 (Alsabri and Al-Ghamdi, 2020). PVC has excellent physical and chemical properties, such as thermal stability, mechanical resistance, chemical stability, etc., which have allowed it to be widely used worldwide as a manufacturing material for blood storage bags, as a replacement for the glass bottles which were used in the mid-1950s and that presented great sterility problems. Thus, at present, during the blood transfusion process, blood is stored in sterile and disposable PVC bags, which contain anticoagulant and preservative solutions (e.g., citrate, phosphate, dextrose, and often adenine) (Chiellini et al., 2013; Prowse et al., 2014).

As previously stated, the storage and biosafety status of blood, and its blood products, are of vital importance in many medical processes. Therefore it is important to remember that transfusion-associated bacteremia can also be caused by the inappropriate use of catheters in essential procedures during treatments in complex medical conditions, such as hemodialysis, chemotherapy, administration of intravenous fluids, and parenteral nutrition, etc. Many of the plastic devices used in the above procedures are made from or use plasticized PVC (Zheng et al., 2018). In response to this, a new approach for the inactivation of pathogens through surface contact and electrostatic interactions has been proposed, thus, inherent antibacterial activity is given to materials in contact with the blood or hemocomponents (Palencia et al., 2019a). In Fig. 23.1 the mechanism for surface-contact pathogen inactivation is illustrated.

23.3.1 Methodological approach for making of the PVC-based antibacterial surface

For the development of antibacterial surfaces based on poly(vinyl chloride) with quaternary ammonium groups, the following stages were performed: (1) modification of poly(vinyl chloride) and incorporation of quaternary-ammonium groups, (2) construction of films, (3) evaluation of thermal and surface properties, (4) evaluation of antimicrobial and hemolytic activity, and (5) evaluation of the adsorption of blood components. For the above, the modification of the poly(vinyl chloride) was carried out through sequential reactions. The first consisted of the incorporation of nucleophilic groups of the amino type on the polymer chains through reactions of poly(vinyl chloride)

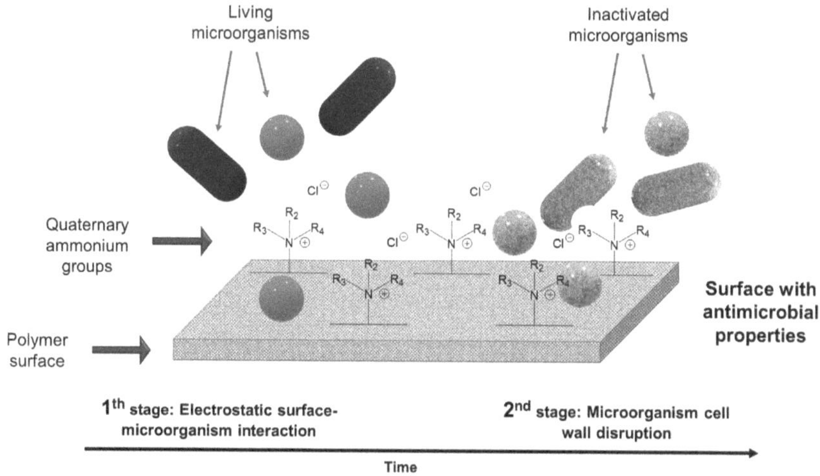

FIGURE 23.1 Illustration of the mechanism for surface-contact pathogen inactivation.

with ethylenediamine in tetrahydrofuran (THF) under reflux. Reaction conditions were evaluated to determine those conditions under which the ethylenediamine reacts through only one of its amino groups. Thus, PVC enriched with secondary amino groups was reacted with 1-bromoethane in THF for the generation of quaternary-ammonium groups by ethylation of the free amino groups. Subsequently, 1×1 cm films were made from quaternized PVC by the dissolution-evaporation technique (Palencia et al., 2019a).

Subsequently, the thermal properties of the materials were evaluated through thermogravimetric analysis and differential scanning calorimetry. Additionally, determination of the surface energies for the compression of the surface processes was carried out by means of contact angle measurements using a stock of three test liquids (Palencia, 2017). Functional characterization assays were mainly the evaluation of antibacterial activity against strains of *E. coli* and *S. aureus*. In addition, the hemolytic properties of cationic PVC were evaluated using the protocol established by ASTM F756-17 (ASTM, 2013). Finally, the effect of the insertion of quaternary ammonium groups on the adsorption of blood plasma and blood cell suspensions was evaluated (Palencia et al., 2019a).

23.3.2 Relevant results and conclusions

Insertion of amino groups on the PVC surface, without cross-linking of polymer chains, was possible using a 2:1 M ratio (ethylenediamine:PVC). Thus, in order to corroborate the changes in the polymer structure and the effect of the inserted cationic groups on the adsorption of blood components, analysis of the functional groups on the polymer, throughout different modification stages, was monitored by infrared spectroscopy and analysis by FEDS deconvolution (Palencia, 2018). For the first modification, infrared analysis was focused on the disappearance of the C–Cl bonds and the appearance of C–N and N–H bonds, while, for the second modification, infrared analysis was focused on the changes related to the disappearance of N–H bonds associated with initially anchored primary amine. In addition, specific vibrations with crystalline or amorphous polymer phases were identified, which allowed estimation of the decrease in crystallinity as a result of the modifications caused on the PVC matrix. The decrease in the degree of crystallinity was 32% in the first stage, and 76.7% in the second stage (i.e., quaternization stage). On the other hand, proton nuclear magnetic resonance allowed the structural verification of the cationic poly(vinyl chloride). Thus, the spectra obtained showed the signals of the protons of the ethylenediamine carbon chain, at a shift between 2.2 and 2.4 ppm, in addition, signals associated to $-CH_2-$ and $-CH_3$ protons were observed, at 1.2 and 3.1 ppm, respectively; these signals are associated with the ethyl groups incorporated during the generation of quaternary ammonium groups. In conclusion, the obtaining of cationic PVC by means of sequential nucleophilic substitution reactions was demonstrated. Additionally, by elemental analysis it was estimated that the percentage of substitution of the Cl atom in the PVC was only 14.9%, whereas the percentage of quaternization of the amino groups was only 40.5% with respect to its precursor; therefore only 6.7% of repeat units from PVC were modified (Palencia et al., 2019a).

On the other hand, thermogravimetric characterization of the cationic PVC showed three exothermic degradation phases. The first degradation phase was associated with the degradation of quaternary ammonium groups and ethylenediamine side chains around 150–240°C. The second and third degradation phases were observed around 300–380°C and 420–460°C, respectively, and were associated with the degradation of the hydrocarbon phase of the polymer. In conclusion, it was evidenced that the thermal stability of cationic PVC is lower than the thermal stability of PVC, however, this change does not affect its application, which is expected to be at room temperature (common temperatures of the storage and sterilization processes are between −25 and 30°C, respectively). The values of the contact angle of the cationic PVC showed a decrease in the hydrophobicity of cationic PVC compared with PVC, that is, the angle changed from 92.1 ± 5.6 to 66.8 ± 4.7, suggesting that, as expected, the surface of the cationic PVC is more hydrophilic than PVC. Estimation of the surface energies showed an increase in the polar contributions, a fact that is consistent with the formation of quaternary ammonium groups (Palencia et al., 2019a).

Finally, functional characterization of the materials was based on three specific properties: (1) antibacterial properties, (2) hemolytic activity, and (3) adsorption of blood components. Antibacterial tests showed a substantial increase in the antimicrobial activity of cationic PVC compared to PVC (inhibition percentages were 32% and 38% against *S. aureus* and *E. coli*, respectively). It was concluded that antimicrobial activity is mediated by the electrostatic interaction of the surface cationic groups with the bacterial cell walls, which promote the cell lysis and consequently their inactivation. In relation to the hemolytic activity of the materials, the hemolysis percentage was 7.2% for the cationic PVC, which, according to the values established by the ASTM F-756, means that this material is classified as a hemolytic material. However, this classification is in constant discussion and the acceptance ranges vary depending on the specific application. In this way, hemolytic activity of cationic PVC is also associated with electrostatic interactions between the cationic quaternary ammonium groups of polymer with the negatively charged membrane of the erythrocytes. Finally, the results of the infrared spectroscopic analysis of the cationic PVC films in contact with the blood products are consistent with the analysis of surface energies, where it is concluded that the insertion of the quaternary ammonium groups changes the surface properties of the polymer, increasing its hydrophilicity, and promoting the adsorption of blood components (Palencia et al., 2019a).

23.4 Antibacterial and nonhemolytic materials based on eco-friendly cationic polyurethanes

Over the few last decades, polymeric materials, both natural and synthetic, have been widely recognized in the field of biomedicine thanks to their wide availability as a resource, their relatively low cost, their ease of synthesis and modification, and their ability to generate a wide variety of new functional materials (Akindoyo et al., 2016). In particular, polyurethanes are a family of polymers discovered in the 1930s by Professor Dr. Otto Bayer, generally formed from the reaction between hydroxyl groups (−OH) with isocyanates (−NCO) to produce urethane groups (R−CON−R). Polyurethanes also can contain other groups such as ethers, esters, urea, and aromatic units depending on the nature of the precursors and the synthesis conditions. Due to the wide variety of sources from which polyurethanes can be synthesized, and their specific applications, they can be grouped into rigid, flexible, thermoplastic, water-based, binders, coatings, adhesives, sealants, and elastomers (Joseph et al., 2018; Otálora et al., 2019).

In particular, polyurethanes, unlike other polymers, have mechanical and physical properties comparable to those of tissues, which, together with low platelet adhesion and low in vitro protein adsorption, allows them to be used in the biomedical industry (Clauser et al., 2016). However, although polyurethanes can have excellent mechanical and biocompatibility properties, they can also act as surfaces that promote microbial infections; mainly those that are used in invasive procedures or in direct contact with biomedical or biological components, for example, catheters, hospital bedding, wound dressings, short-term implants, etc. In this way, contamination by microorganisms, specifically bacteria, is one of the main problems associated with various medical procedures due to the drug resistance that these have developed and the formation of biofilms on medical surfaces (Francolini and Piozzi, 2016).

On the other hand, different investigations have reported the effectiveness of polymeric materials with cationic groups, ammonium, and phosphonium, as potential antimicrobial agents (Xue et al., 2015; Palencia et al., 2019b). Based on the ease of synthesis and modification offered by polyurethanes, by means of an adequate selection of precursors or modifying agent, it is possible to modulate their properties to obtain structural and functional materials (Druvari et al., 2018; Palencia et al., 2020), therefore functional polyurethanes can be synthesized to incorporate in their structure cationic groups in order to produce antibacterial materials. Therefore these materials emerge as a promising alternative for the development of biomedical surfaces used in contact with blood (Palencia et al., 2020).

23.4.1 Methodological approach

Cationic polyurethanes with quaternary ammonium groups can be obtained by the following methods: (1) synthesis of the compound N-carboxymethyl-N,N,N-triethylammonium chloride, (2) synthesis of cationic polyurethanes, and (3) functional characterization of materials (thermal, antibacterial, and hemolytic characterization tests). For this, the synthesis of N-carboxymethyl-N,N,N-triethylammonium chloride was based on two reaction steps. First was the reaction among ethyl bromoacetate and triethylamine to obtain esters with one quaternary ammonium group, and second was acid hydrolysis with HCl to produce one carboxylic acid. Thus, a molecule with a low molecular weight, one cationic group, and one carboxylic acid is obtained. For the synthesis of cationic polyurethanes, by a tricomponent reaction in the absence of solvent, 4,4′-methylene-bis-phenylisocyanate, mannitol, and N-carboxymethyl-N,N,N-triethylammonium chloride were polymerized (Palencia et al., 2020). On the other hand, functional characterization of the materials was carried out by evaluation of their thermal, antibacterial, and hemolytic properties. Thermal characterization was carried out by thermogravimetric analysis and differential scanning calorimetry. Antibacterial properties were evaluated through the percentage of inhibition of bacterial growth using *E. coli* and *S. aureus* as microorganism models. Finally, evaluation of the hemolytic properties of cationic polyurethanes was carried out according to ASTM F756-17 (Standard Practice for Assessment of Hemolytic Properties of Materials) using the cyanmethemoglobin method (ASTM, 2013; Palencia et al., 2020).

23.4.2 Relevant results and conclusions

For the development of cationic polyurethanes, in the first instance, the synthesis of a bifunctional molecule with quaternary ammonium groups and carboxylic acids was required, which is essential because this actively acts in urethanization reactions and the inactivation of pathogens. The results showed that the synthesis of N-carboxymethyl-N,N,N-triethylammonium chloride can be easily done with a percentage of total synthesis yield of about 60%. In addition, the melting point of the compound was $188.1 \pm 0.6°C$. By infrared spectroscopy and nuclear magnetic resonance the respective structure was verified. In this way, displacement of the carbonyl group band, $\sim 20\ cm^{-1}$, was observed after obtaining the compound, and the respective signal was located at $\sim 1714\ cm^{-1}$. The above was explained as a consequence of the delocalization of π electrons from the carbonyl group. Likewise, infrared study at $1080\ cm^{-1}$ identified the stretching vibration signal of the CN bond, which was related to the quaternary ammonium group formed. In addition, by studies of proton nuclear magnetic resonance, the signals at 1.23 and 3.51 ppm were identified, which are associated with the protons of the $-CH_3$ and $-CH_2-$, respectively; these signals correspond to the protons of the ethyl groups incorporated (Palencia et al., 2020).

Cationic polyurethanes obtained through three-component reactions were structurally characterized by infrared spectroscopy and solid-state carbon-13 nuclear magnetic resonance. Thus, by infrared analyses, in the region of vibrations of the carbonyl group, two signals associated with the C=O bond of the urethane and uretidione groups were identified. In addition, a third signal was identified at values lower than 1700 cm^{-1} associated with the stretching vibrations of the C=O bond of the amide group. On the other hand, in the spectra obtained by carbon-13 nuclear magnetic resonance of the cationic polyurethanes, typical signals of the different carbons from polyurethane samples were identified without the incorporation of N-carboxymethyl-N,N,N-triethylammonium chloride; likewise, the signals between 10–20 and 50–65 ppm were attributed to the $-CH_2-$ and $-CH_3$ carbons, respectively, and associated with the ethyl groups of the cationic compound incorporated in the polymeric matrix. Regarding the description of the physical properties of polyurethanes, powders insoluble in water were obtained in the solid state; with particle sizes between 5 and 500 μm (Palencia et al., 2020). All cationic materials obtained showed a single degradation stage at temperatures between 160 and 340°C. However, it was found that the initial degradation temperature of cationic polyurethanes was a function of the mole fraction of N-carboxymethyl-N,N,N-triethylammonium chloride in the polymeric matrix, therefore, as a consequence of the thermal lability of the ammonium groups, the compounds with the highest degree of insertion of this type of groups were those that presented lower initial degradation temperatures. In addition, by differential scanning calorimetry, three different endothermic processes were identified, which were associated with the decomposition of the polymeric matrix (Palencia et al., 2020).

On the other hand, N-carboxymethyl-N,N,N-triethylammonium chloride exhibited a percentage of inhibition of around 92% for both types of bacterial strains. However, cationic polyurethanes also showed an increase in their activity compared to their noncationic counterparts, being greater than 25%. Finally, the results of the hemolytic activity tests showed a dependence in the hemolytic activity as a function of the content of N-carboxymethyl-N,N,N-triethylammonium chloride in the polymeric matrix. Thus, the percentages of hemolysis reported for the materials with different levels of insertion of cationic groups used were between 2% and 18%. It was concluded that hemolytic activity increases as a consequence of electrostatic interactions between cationic groups and the erythrocyte membrane, generating their lysis and the consequent release of hemoglobin (Palencia et al., 2020).

References

Akindoyo, J.O., Beg, M.D.H., Ghazali, S., Islam, M.R., Jeyaratnam, N., Yuvaraj, A.R., 2016. Polyurethane types, synthesis and applications - a review. RSC Adv. 6 (115), 114453–114482. https://doi.org/10.1039/C6RA14525F.

Alsabri, A., Al-Ghamdi, S.G., 2020. Carbon footprint and embodied energy of PVC, PE, and PP piping: perspective on environmental performance. Energy Rep. 6 (8), 364–370. https://doi.org/10.1016/j.egyr.2020.11.173.

Asghari, F., Samiei, M., Adibkia, K., Akbarzadeh, A., Davaran, S., 2017. Biodegradable and biocompatible polymers for tissue engineering application: a review. Artif. Cells, Nanomed. & Biotechnol. 45 (2), 185–192. https://doi.org/10.3109/21691401.2016.1146731.

ASTM F756-13, 2013. Standard Practice for Assessment of Hemolytic Properties of Materials. ASTM International, West Conshohocken, PA. https://doi.org/10.1520/F0756-13.

Aubron, C., Flint, A., Ozier, Y., McQuilten, Z., 2018. Platelet storage duration and its clinical and transfusion outcomes: a systematic review. Crit. Care 22, 185. https://doi.org/10.1186/s13054-018-2114-x.

Barro, L., Drew, V.J., Poda, G.G., Tagny, C.T., El-Ekiaby, M., Owusu-Ofori, S., Burnouf, T., 2018. Blood transfusion in sub-Saharan Africa: understanding the missing gap and responding to present and future challenges. Vox Sang. 113 (8), 726–736. https://doi.org/10.1111/vox.12705.

Brecher, M.E., Shauna, N.H., 2005. Bacterial contamination of blood components. Clin. Microbiol. Rev. 18 (1), 195–204. https://doi.org/10.1128/CMR.18.1.195-204.2005.

Chiellini, F., Ferri, M., Morelli, A., Dipaola, L., Latini, G., 2013. Perspectives on alternatives to phthalate plasticized poly (vinyl chloride) in medical devices applications. Prog. Polym. Sci. 38 (7), 1067–1088. https://doi.org/10.1016/j.progpolymsci.2013.03.001.

Clauser, J., Gester, K., Steinseifer, U., Sonntag, S.J., 2016. Regulating blood cell adhesion via surface modification of polyurethanes. In: Cooper, S.L., Guan, J. (Eds.), Advances in Polyurethane Biomaterials. Woodhead Publishing., pp. 287–318. https://doi.org/10.1016/B978-0-08-100614-6.00010-X

Damgaard, C., Magnussen, K., Enevold, C., Nilsson, M., Tolker-Nielsen, T., Holmstrup, P., Nielsen, C.H., 2015. Viable bacteria associated with red blood cells and plasma in freshly drawn blood donations. PLoS One 10 (3), e0120826. https://doi.org/10.1371/journal.pone.0120826.

Delaney, M., Wendel, S., Bercovitz, R.S., Cid, J., Cohn, C., Dunbar, N.M., Apelseth, T.O., Popovsky, M., Stanworth, S.J., Tinmouth, A., Watering, L., Waters, J.H., Yazer, M., Ziman, A., 2016. Transfusion reactions: prevention, diagnosis, and treatment. Lancet 388 (10061), 2825–2836. https://doi.org/10.1016/S0140-6736(15)01313-6.

Drew, V.J., Barro, L., Seghatchian, J., Burnouf, T., 2017. Towards pathogen inactivation of red blood cells and whole blood targeting viral DNA/RNA: design, technologies, and future prospects for developing countries. Blood Transfus. 15 (6), 512–521. https://doi.org/10.2450/2017.0344-16.

Druvari, D., Koromilas, N.D., Bekiari, V., Bokias, G., Kallitsis, J.K., 2018. Polymeric antimicrobial coatings based on quaternary ammonium compounds. Coatings 8 (1), 8. https://doi.org/10.3390/coatings8010008.

Francolini, I., Piozzi, A., 2016. Antimicrobial polyurethanes for intravascular medical devices. In: Cooper, S.L., Guan, J. (Eds.), Advances in Polyurethane Biomaterials. Woodhead Publishing., pp. 349–385. https://doi.org/10.1016/B978-0-08-100614-6.00012-3

Frazier, S.K., Higgins, J., Bugajski, A., Jones, A.R., Brown, M.R., 2017. Adverse reactions to transfusion of blood products and best practices for prevention. Crit. Care Nurs. Clin. 29 (3), 271–290. https://doi.org/10.1016/j.cnc.2017.04.002.

Galli, L., Bruschi, F., 2020. New strategies for the control of infectious and parasitic diseases in blood donors: the impact of pathogen inactivation methods. EuroBiotech J. 4 (2), 53–66. https://doi.org/10.2478/ebtj-2020-0007.

García-Roa, M., Vicente-Ayuso, M., Bobes, A.M., Pedraza, A.C., González-Fernández, A., Martín, M.P., Saéz, I., Seghatchian, J., Gutiérrez, L., 2017. Red blood cell storage time and transfusion: current practice, concerns and future perspectives. Blood Transfus. 15 (3), 222–231. https://doi.org/10.1016/10.2450/2017.0345-16.

Garraud, O., Cognasse, F., Laradi, S., Hamzeh-Cognasse, H., Peyrard, T., Tissot, J.D., Fontana, S., 2018. How to mitigate the risk of inducing transfusion-associated adverse reactions. Transfus. Clin. Biol. 25 (4), 262–268. https://doi.org/10.1016/j.tracli.2018.07.006.

Heroes, A.S., Ndalingosu, N., Kalema, J., Luyindula, A., Kashitu, D., Akele, C., Kabinda, J., Lagrou, K., Vandekerckhove, P., Jacobs, J., Lunguya, O., 2020. Bacterial contamination of blood products for transfusion in the Democratic Republic of the Congo: temperature monitoring, qualitative and semi-quantitative culture. Blood Transfus. 18 (5), 348–358. https://doi.org/10.2450/2020.0108-20.

Joseph, J., Patel, R.M., Wenham, A., Smith, J.R., 2018. Biomedical applications of polyurethane materials and coatings. Trans. IMF 96 (3), 121–129. https://doi.org/10.1080/00202967.2018.1450209.

Klausen, S.S., Hervig, T., Seghatchian, J., Reikvam, H., 2014. Bacterial contamination of blood components: Norwegian strategies in identifying donors with higher risk of inducing septic transfusion reactions in recipients. Transfus. Apher. Sci. 51 (2), 97–102. https://doi.org/10.1016/j.transci.2014.08.007.

Kundrapu, S., Srivastava, S., Good, C.E., Lazarus, H.M., Maitta, R.W., Jacobs, M.R., 2020. Bacterial contamination and septic transfusion reaction rates associated with platelet components before and after introduction of primary culture: experience at a US Academic Medical Center 1991 through 2017. Transfusion 60 (5), 974–985. https://doi.org/10.1111/trf.15780.

Liumbruno, G.M., Catalano, L., Piccinini, V., Pupella, S., Grazzini, G., 2009. Reduction of the risk of bacterial contamination of blood components through diversion of the first part of the donation of blood and blood components. Blood Transfus. 7 (2), 86–93. https://doi.org/10.2450/2008.0026-08.

Murphy, M.F., 2017. The epidemiology of transfusion: where blood goes and why we should care about it. Editorial 57, 2821–2823. https://doi.org/10.1111/trf.14385.

Otálora, A., Palencia, M., Lerma, T., 2019. Synthesis and characterization of polyurea-base hydrogels by multicomponent polycondensation of 1,6-hexametilenediisocyanate, sorbitol and cysteine. J. Sci. Technol. Appl. 7, 5–16. https://doi.org/10.34294/j.jsta.19.7.47.

Palencia, M., 2018. Functional transformation of Fourier-transform mid-infrared spectrum for improving spectral specificity by simple algorithm based on wavelet-like functions. J. Adv. Res. 14, 53–62. https://doi.org/10.1016/j.jare.2018.05.009.

Palencia, M., 2017. Surface free energy of solids by contact angle measurements. J. Sci. Technol. Appl. 2, 84–93. https://doi.org/10.34294/j.jsta.17.2.17.

Palencia, M.S., Berrio, M.E., Palencia, S.L., 2017b. Effect of capping agent and diffusivity of different silver nanoparticles on their antibacterial properties. J. Nanosci. Nanotechnol. 17 (8), 5197–5204. https://doi.org/10.1166/jnn.2017.13850.

Palencia, M., Córdoba, A., Melendrez, M.F., 2019b. Nanocomposites based on cationic polyelectrolytes and silver nanoparticles: synthesis, characterization, molybdate retention and antimicrobial activity. Arab. J. Chem. 12 (6), 825–834. https://doi.org/10.1016/j.arabjc.2017.11.020.

Palencia, M., Córdoba, A., Rivas, B.L., 2014. Concentration–polarization effect of poly (sodium styrene sulfonate) on size distribution of colloidal silver nanoparticles during diafiltration experiments. Colloid Polym. Sci. 292 (3), 619–626. https://doi.org/10.1007/s00396-013-3096-5.

Palencia, M., Lerma, T.A., Afanasjeva, N., 2019a. Antibacterial cationic poly (vinyl chloride) as an approach for in situ pathogen-inactivation by surface contact with biomedical materials. Eur. Polym. J. 115, 212–220. https://doi.org/10.1016/j.eurpolymj.2019.03.041.

Palencia, M., Lerma, T.A., Arrieta, Á.A., 2020. Antibacterial and non-hemolytic cationic polyurethanes with N-carboxymethyl-N, N, N-triethylammonium groups for bacteremia-control in biomedical-using materials. Mater. Today Commun. 22, 100708. https://doi.org/10.1016/j.mtcomm.2019.100708.

Palencia, M., Berrio, M., Melendrez, M., 2016. Nanostructured polymer composites with potential applications into the storage of blood and hemoderivates. J. Sci. Technol. Appl. 1, 4–14. https://doi.org/10.34294/j.jsta.16.1.1.

Palencia, M., Lerma, T., Combatt, E., 2017a. Hydrogels based in cassava starch with antibacterial activity for controlled release of cysteamine-silver nanostructured agents. Curr. Chem. Biol. 11 (1), 28–35. https://doi.org/10.2174/2212796810666161108152319.

Panch, S.R., Montemayor-Garcia, C., Klein, H.G., 2019. Hemolytic transfusion reactions. N. Engl. J. Med. 381 (2), 150–162. https://doi.org/10.1056/NEJMra1802338.

Putter, J.S., Seghatchian, J., 2017. Cumulative erythrocyte damage in blood storage and relevance to massive transfusions: selective insights into serial morphological and biochemical findings. Blood Transfus. 15 (4), 348–356. https://doi.org/10.2450/2017.0312-16.

Prowse, C.V., De Kort, D., Hess, J.R., Van der Meer, P.F., 2014. Commercially available blood storage containers. Vox Sang. 106 (1), 1–13. https://doi.org/10.1111/vox.12084.

Ramírez-Arcos, S., Goldman, M., 2017. Bacterial contamination. In: Murphy, M.F., Roberts, D.J., Yazer, M.H. (Eds.), Practical Transfusion Medicine. John Wiley & Sons, pp. 168–175. https://doi.org/10.1002/9781119129431.ch16.

Sahu, S., Hemlata, V.A., 2014. Adverse events related to blood transfusion. Indian J. Anaesth. 58 (5), 543–551. https://doi.org/10.4103/0019-5049.144650.

Salunkhe, V., Van der Meer, P.F., de Korte, D., Seghatchian, J., Gutiérrez, L., 2015. Development of blood transfusion product pathogen reduction treatments: a review of methods, current applications and demands. Transfus. Apher. Sci. 52 (1), 19–34. https://doi.org/10.1016/j.transci.2014.12.016.

Song, R., Murphy, M., Li, C., Ting, K., Soo, C., Zheng, Z., 2018. Current development of biodegradable polymeric materials for biomedical applications. Drug Des. Develop. & Ther. 12, 3117–3345. https://doi.org/10.2147/DDDT.S165440.

Ullah, F., Othman, M.B.H., Javed, F., Ahmad, Z., Akil, H.M., 2015. Classification, processing and application of hydrogels: a review. Mater. Sci. Eng. C 57, 414–433. https://doi.org/10.1016/j.msec.2015.07.053.

Vasudev, R., Sawhney, V., Dogra, M., Raina, T.R., 2016. Transfusion-related adverse reactions: from institutional hemovigilance effort to National Hemovigilance program. Asian J. Transfus. Sci. 10 (1), 31–36. https://doi.org/10.4103/0973-6247.175391.

Xue, Y., Xiao, H., Zhang, Y., 2015. Antimicrobial polymeric materials with quaternary ammonium and phosphonium salts. Int. J. Mol. Sci. 16 (2), 3626–3655. https://doi.org/10.3390/ijms16023626.

Zheng, Y., He, L., Asiamah, T.K., Otto, M., 2018. Colonization of medical devices by staphylococci. Environ. Microbiol. 20 (9), 3141–3153. https://doi.org/10.1111/1462-2920.14129.

Chapter 24

Surface for spectral sensors in microbiological analysis

24.1 Introduction: polymer-based spectral sensors

Biopolymers are highly versatile materials considered to be the better alternative for replacing petroleum-derived polymers in many respects, with the most important being that they are obtained from renewable sources. As it was discussed in earlier chapters of this book, biopolymers can be defined from a broad perspective such as synthetic biopolymers (i.e., polymers obtained by synthesis of biomolecules or biomolecule-based synthetic polymers) and natural biopolymers (i.e., polymers obtained directly from natural sources, including in this category, chemically modified natural biopolymers) (Garcés and Chate-Galvis, 2020). On the other hand, petroleum-derived polymers can be structural and functional materials. In the latter case, due to their structural characteristics, biopolymers are more similar to the functional petroleum-derived polymers and, therefore, it is expected that they can be more easily replaced by biopolymers for functional applications. In previous chapters, the fundamental aspects of biopolymers were developed from an environmental approach and from a general applied vision, and as a consequence, these aspects are not discussed here (for more details, see Chapters 1–3). In this chapter, a specific application is addressed in order to show the versatility of biopolymers in the different fields of application. Thus, this chapter is focused on a specific analytical application: the construction of sensing surfaces in microbiological analysis.

A spectral sensor is a material able to produce a reproducible response as a result of an analytical target. In general, an analytical objective can be any quantifiable factor or properties of the study system. Polymers can be used to produce an analytical response in multiple cases depending on the structural properties (Cichosz et al., 2018; Latif et al., 2018). For example, conductive polymers can be used to identify change in the electrical conductivity of dissolutions in the food industry or for environmental and agricultural applications (Arrieta-Almario et al., 2018; Khalifa and Abdllah, 2019), but nonconductive polymers can be used to identify changes in dielectric response (Palencia et al., 2010); also, molecularly imprinted polymers have been used

Eco-friendly Functional Polymers
https://doi.org/10.1016/B978-0-12-821842-6.00009-9
Copyright © 2021 Elsevier Inc. All rights reserved.

for the detection of metabolites by surface sensing in conductometric sensors (Latif et al., 2018; Sorer et al., 2018). In the case of spectral response, depending on the analytical technique used, a specific response is expected, and as a consequence, the structural characteristics cannot be generalized. For example, depending on the crystalline domains of polymer material, they can be active to ultraviolet region, or by the addition of nanoparticles, dyes, pigments, or any other chromophore, and the polymer can be active to visible region (Cichosz et al., 2018). Here, it is of interest to obtain the spectral response in the infrared region, specifically, the mid-infrared region (4000 to 500 cm^{-1}).

24.2 Study of the mid-infrared vibrational spectrum of pathogenic bacterial biofilms: context and problematics

Mid-infrared spectroscopy (mid-IR spectroscopy) is an analytical technique which has been used in many science and technology fields, in addition, it is considered to be a fundamental tool for the identification of functional groups of pure and simple systems. The range of wavenumbers used in the mid-IR spectroscopy are from ~500 to ~4000 cm^{-1}, corresponding to the region of the electromagnetic spectrum where vibrations are associated with the main functional groups and typical bonds of organic molecules (Dadd et al., 2000; Huang et al., 2006; Alvarez-Ordoñez et al., 2011; Ojeda and Dittrich, 2012; Prakash et al., 2013; Khatoon et al., 2018). However, currently its application for the study of multicomponent complex systems has been increased as a result of the generalized use of the Fourier transform algorithm at the level of equipment instrumentation, specifically in signal processing, the use of several chemometric tools, and the different available transmissions including by mid-IR spectroscopy, mid-IR spectroscopy attenuated total reflectance (ATR), and mid-IR spectroscopy by diffuse reflectance infrared Fourier transform (or DRIFT). However, with the advent of AT and DRIFT, traditional incompatibility problems associated with the moisture or water content of samples are outdated (Li et al., 2018; Restrepo et al., 2018; Singhalage et al., 2018; Otálora and Palencia, 2019; Palencia et al., 2019a,b; Anaya-Tatis et al., 2019).

When mid-IR spectra of bacteria are analyzed, the analysis of spectra usually is directed at the differentiation of bacteria, identification and monitoring of bacteria on substrates with low spectral complexity, and monitoring of significant surface changes via spectral changes (Dadd et al., 2000; Huang et al., 2006; Khatoon et al., 2018; Li et al., 2018; Singhalage et al., 2018). In particular, for microbiological analysis, the main limitations of mid-IR spectroscopy are: high spectral similitude between spectra of different bacteria, high signal overlap and widening of adjacent signals, incompatibility for the acquisition of spectra into samples with content of intrinsic water, background signals of substrate producing "false" signals not associated with the sample, signals with a very weak intensity resulting in a low signal-to-noise ratio (SNR), as well as excessive spectral noise (Alvarez-Ordoñez et al., 2011; Ojeda and Dittrich, 2012; Prakash et al., 2013). For resolution enhancement of mid-IR spectra, the most common technique is the Fourier deconvolution or Fourier self-deconvolution algorithm (Friesen and Michaelian, 1991; Vazhnova and Lukyanov, 2013; Palencia, 2018). However, although from a conceptual approach the Fourier self-deconvolution is simple, its application is limited by the complexity of computation, the appearance of negative intensities from calculations, the high sensitivity to noise, and the appearing of "false" signals resulting in mathematical arguments without physical meaning (Friesen and Michaelian, 1991). Recently, a simple method for the deconvolution of signals and the increase of spectral resolution have been developed, based on the transformation of the spectrum with the use of a derivative algorithm enhanced by functional transformations that was named functionally enhanced derivative spectroscopy (FEDS) (Palencia and Martínez, 2017; Palencia, 2018). Details of FEDS are shown in Section 24.3.

The objective of this chapter is to evaluate the ability of FEDS to increase the spectral resolution and differentiation of mid-IR spectra of bacterial biofilms which are understood to be microbiological models of complex systems. In this way, it is proposed that implementation of the FEDS algorithm is a simple and easy mid-IR spectrum analysis tool for the microbiological analysis and study of bacterial systems.

24.3 Methodological strategy

The methodological strategy is based on the use of porous polymeric bilayer systems for surface sensing. These bilayer systems were made of two layers, one of polypropylene acting as a support, and on top of it, a thin layer of cellulose ($\sim$500 μm). The thin layer of cellulose was modified by urethanization with methylene-diphenyl-isocyanate (MDI) to generate a characteristic signal in the mid-IR spectrum, which is not observed in the mid-IR spectra of bacteria. This signal corresponds to unreacted isocyanate groups of the MDI. Thus, two types of isocyanate groups can be identified: (1) surface isocyanates available to react with biomolecules on the bacterial surface, humidity from the environment, or any other species with carboxylic acid, hydroxyl, or amine groups; and (2) isocyanates inside cellulose pores, which remain on the surface of the fiber, but are not accessible to react with groups on the bacterial surface. In this way, these internal groups generate a characteristic signal that can be standardized to obtain information from the biofilm deposited on the surface (e.g., thickness of biofilm). As bacterial models were used four bacterial strains: *E. coli* ATCC 25922 and *S. aureus* ATCC 25923, which are typical models of Gram-negative and Gram-positive bacteria, respectively, and two strains of *H. pylori* corresponding to *H. pylori* ATCC J99 and *H. pylori* ATCC 43504. Bacterial strains were grown by inoculating using standard procedures.

In particular, *H. pylori* is a human pathogen with medical interest for its association with gastric cancer (Palencia et al., 2016a; Stave et al., 2017).

Bacterial biofilms were deposited on modified-cellulose membranes using a spectral marker (CISM, Mindtech s.a.s., Colombia). The thickness was studied by depth of penetration (dp) from the changes observed on the intensity signal of the spectral marker (Palencia et al., 2016b). After obtaining the spectra, in quadruplicate, it is necessary to process them through the following sequence of steps: (1) spectral smoothing to eliminate noise, (2) autoscaling of the data, and (3) transformation of the data by means of FEDS (Palencia-Luna et al., 2020a,b, 2021). A schematic illustration of methodological approach is shown in Fig. 24.1A.

24.4 Results and discussion: analysis of bacterial biofilms by mid-IR spectroscopy

Characteristic mid-IR spectra of bacteria and the cellulose-modified support (CMS) are shown in Fig. 24.1B. To ease the interpretation, several analysis regions were defined ($4000-2500\ cm^{-1}$, $2500-2100\ cm^{-1}$, and $2100-1300\ cm^{-1}$). In the range $4000-2500\ cm^{-1}$, signals related to O—H, N—H, and C—H bonds are expected; however, these signals are highly overlapped and the performing of specific assignations is very difficult. In general, it was evidenced that new or important information is not obtained from analysis of this region. The main reason for this is the strong overlap of signals (Palencia-Luna et al., 2020b). In the range $2500-2100\ cm^{-1}$, the signal from the spectral marker was easily identified (α); this signal allows obtaining of information about biofilm through the changes in its intensity. Thus, the analysis of changes in intensity of α for biofilm spectra, compared with the corresponding signal in the CMS, evidenced that vibrations from spectra of bacteria are related to the bacterial surface and not the CMS. In general, the smaller the signal α is, the lower is the contribution of CMS on the signal of biofilm, and the better is the quality of information obtained from the analysis of the spectrum (Palencia-Luna et al., 2020a,b).

From the depth of penetration values for biofilms, it was concluded that the IR beam struck the biofilm and reached the CMS, suggesting that bacterial mid-IR spectra could be strongly contaminated with signals from CMS; however, in contrast, the decrease in signal α demonstrates that the CMS signals do not significantly affect the spectra. Four implicit assumptions can be used to understand this result: (1) the density and chemical composition of the biofilm do not vary substantially within a single sample or between different samples; (2) the penetration depth for the biofilm, at the wavelength of the single peak, is greater than the thickness of the biofilm; (3) the penetration depth is the same at different points in the sample; and (4) the biofilm is chemically stable during performing of the testing (Kane et al., 2009). From the above, it is concluded that the values of Δy cannot be interpreted as a linear

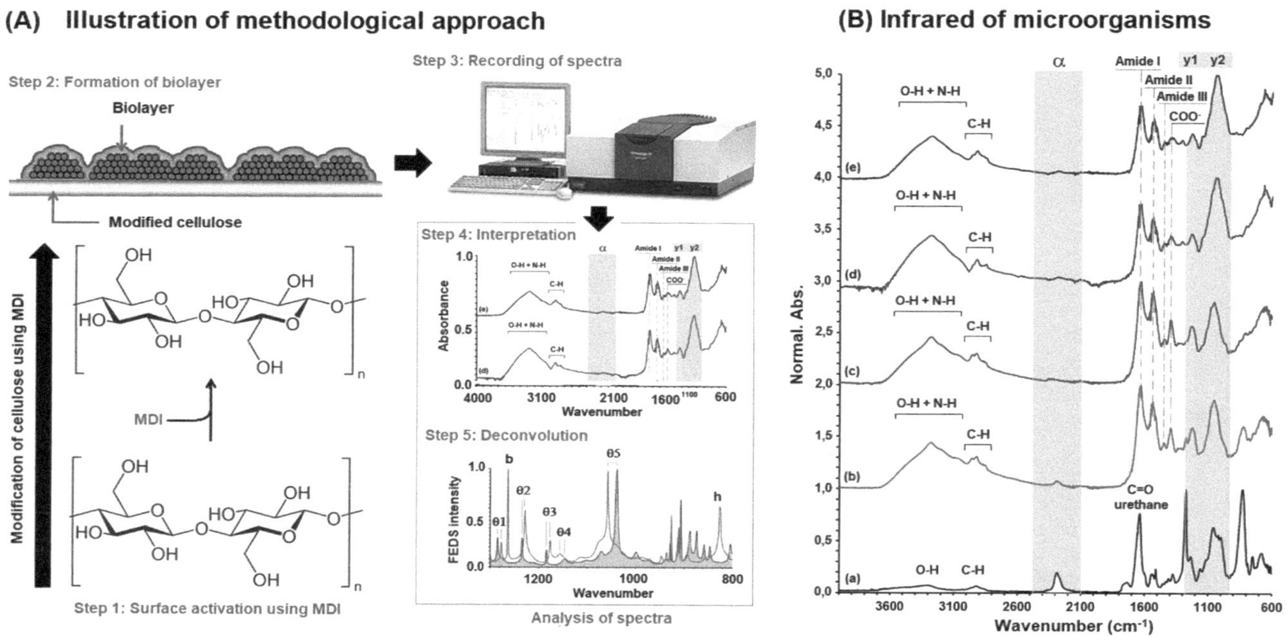

FIGURE 24.1 (A) Illustration of the methodological approach for the use of cellulose thin film as the sensing surface, and (B) mid-IR spectra of target bacteria and sensing surface: (a) modified cellulose, (b) *E. coli*, (c) *S. aureus*, (d) *H. pylori* 43504, and (e) *H. pylori* J99.

length scale, that is to say, it is incorrect to indicate that the thickness of the biofilm of *H. pylori* 43504 is 1.76 times the thickness of the *S. aureus* biofilm. However, note that the Δy is the result of the conjunction of several parameters affecting the values of dp since, for biofilm, dp is not constant nor can it be assumed to be homogeneous. Therefore, from values of Δy it is inferred that biofilm of *H. pylori* strains has greater cohesion at the molecular level and, consequently, greater resistance to be penetrated by the IR beam (Palencia-Luna et al., 2020b). The authors suggested that the values of Δy can be interpreted as evidence that the *H. pylori* biofilms are more persistent as they would be more intimately linked at the molecular level in comparison with the *E. coli* and *S. aureus* biofilms, and therefore, it is suggested that they have a greater capacity to adaptation to and protection of the environment. However, previous inferences should be coherent with the analysis of the interaction from vibrational spectra being necessary for the correct identification of displacement of signals. From this, it was concluded that the analysis of biofilm is possible without interferences or "waste" signals from CMS.

In a third region, 1300–2100 cm^{-1}, characteristic signals related to vibrations of peptide units are usually described as amide I, II, and III. Thus, C=O stretching in amides occurs at 1647 cm^{-1} (amide I band), N–H bending in amides at 1548 cm^{-1} (amide II band), and C–N stretching in amides at 1453 cm^{-1} (amide III band). In addition, symmetric stretching for deprotonated carboxylate groups at 1405 cm^{-1} is clearly identified in the spectra. In a fourth region, 1000–1300 cm^{-1}, at least two signals can be easily identified at ~1230 and ~1060 cm^{-1}, which are associated with stretching of P=O in phosphates and stretching of P=O of phosphodiester and phosphorylated proteins (in Fig. 24.1B these signals are denoted as y1 and y2, respectively) (Palencia-Luna et al., 2020a,b, 2021).

So far, it has been evidenced that mid-IR spectra of different bacteria show high-grade similitude in hindering their differentiation and, as a consequence, limiting the application of mid-IR spectroscopy for the comparative study of bacterial biofilms. However, from spectral correlation analysis of normalized spectra it can be seen that the spectral differentiation should be possible in different parts of the spectra (an illustration of the correlation analysis is shown in Fig. 24.2A for *E. coli* and *H. pylori* 43504). From the correlation analysis it was observed that bacteria from the same species have a higher correlation, though, in principle, large values of Pearson correlation coefficients (R2) were obtained in all cases (i.e., values of R2 larger than 0.80). These large values of R2 were explained by the compensation of contributions by symmetry effects (Wiedermann and Hagmann, 2015). The disadvantage of the use of R2 is that it is a descriptor that is very sensitive to the values of absorbances, but also, in cases where there is a low degree of spectral similarity, by correlation analysis it is not possible to obtain information on the factors producing the differences and, as previously indicated, it is sensitive to the overlapping of signals (Mainali et al., 2017; Coenen and Weitz, 2018; Armstrong, 2019).

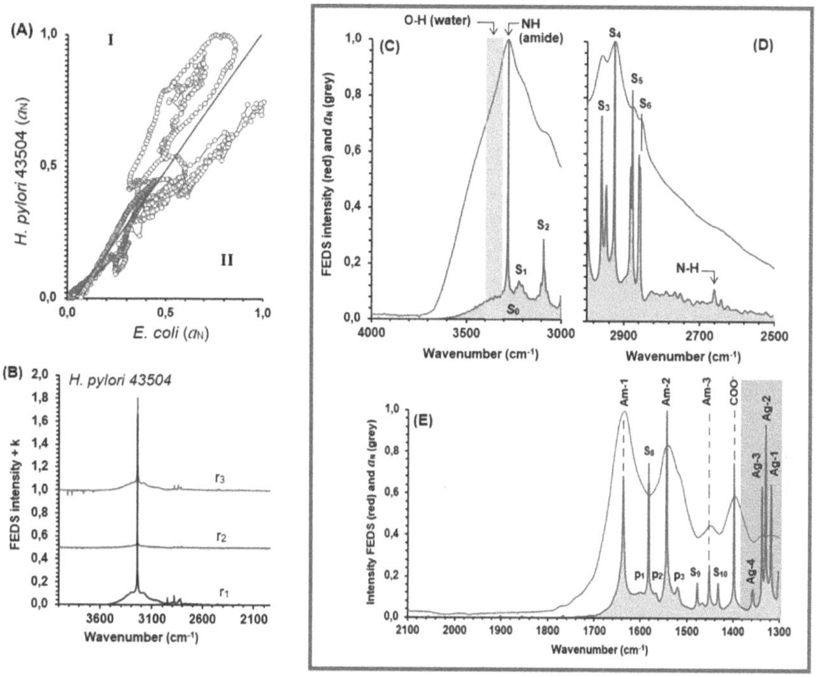

FIGURE 24.2 (A) Correlation analysis by Pearson's coefficient for *E. coli* and *H. pylori* 43504. (B) Illustration of the reproducibility for FEDS spectra of *E. coli*. FEDS spectra of *E. coli* obtained by spectral fractionation: (C) 4000–3000 cm^{-1}, (D) 3000–2500 cm^{-1}, and (E) 2100–1300 cm^{-1}.

On the other hand, a critical point for the application of deconvolution by FEDS into microbiology is the spectral reproducibility of mid-IR spectra and FEDS spectra. High reproducibility was observed in all cases (see an example in Fig. 24.2B). An effect of the normalization factor on FEDS intensity was evidenced since these spectra were transformed using all of the experimental spectral range. This suggests that the better alternative for obtaining an adequate FEDS intensity is implementing the technique of fractionation and reconstruction of the spectra. Basically, this consists in performing the partition of the spectrum into regions or analysis windows, after carrying out transformation by the FEDS algorithm, and finally, performing reconstitution of the spectrum in the case where it is necessary due to the analysis being performed using the analysis window (see Fig. 24.2C–E). The effect of the previous procedure is that if there is a very intense signal into the infrared spectrum, this will not affect the intensity of other signals by a reduction of their intensities during the FEDS transformation. A complete analysis of the FEDS spectra has been published previously (Palencia-Luna et al., 2020b).

24.5 Conclusions and remarks

Beyond the uses related to the biodegradability and biocompatibility of biopolymers, these biomaterials show other important features which can be used for innovative applications. As was evidenced, biopolymers have high potential for use as functional polymers in specialized applications such as those discussed in this chapter. Specifically, cellulose thin films can be modified, by insertion of isocyanate groups, for making sensing surfaces, which can be used for the spectral analysis of biofilms. These cellulose–isocyanate sensing surfaces have been shown to be useful, in conjunction with FEDS, for the applications described herein.

References

Alvarez-Ordoñez, A., Mouwen, D., Lopez, M., Prieto, M., 2011. Fourier transform infrared spectroscopy as a tool to characterize molecular composition and stress response in foodborne pathogenic bacteria. J. Microbiol. Methods 84, 369–378. https://doi.org/10.1016/j.mimet.2011.01.009.

Anaya-Tatis, L.R., Libreros, K.H., Palencia, V.J., Atencio, V.J., Palencia, M., 2019. Mid-infrared spectral characterization of fish scales: "Bocachico" (Prochilodus magdalenae) by functionally-enhanced derivative spectroscopy (FEDS) - a methodological approach. J. Sci. Technol. Appl. 6, 28–39. https://doi.org/10.34294/j.jsta.19.6.39.

Armstrong, R.A., 2019. Should Pearson's correlation coefficient be avoided? Ophthalmic Physiol. Opt. 39, 316–327. https://doi.org/10.1111/opo.12636.

Arrieta-Almario, A.A., Palencia-Luna, M.S., Arrieta-Torres, P.L., 2018. Determination of adulter in milk through the use of a portable voltammetric electronic tongue. Revista Mexicana de Ingeniería Química 17, 877–884.

Cichosz, S., Masek, A., Zaborski, M., 2018. Polymer-based sensors: a review. Polym. Test. 67, 342–348. https://doi.org/10.1016/j.polymertesting.2018.03.024.

Coenen, A.R., Weitz, J.S., 2018. Limitations of correlation-based inference in complex virus-microbe communities. mSystems 3. https://doi.org/10.1128/mSystems.00084-18 e00084-18.

Dadd, M., Sharp, D., Pettman, A., Knowles, C., 2000. Real-time monitoring of nitrile biotransformations by mid-infrared spectroscopy. J. Microbiol. Methods 41, 69–75. https://doi.org/10.1016/S0167-7012(00)00138-X.

Friesen, W., Michaelian, K., 1991. Deconvolution and curve-fitting in the analysis of complex spectra: the CH stretching region in infrared spectra of coal. Appl. Spectrosc. 45, 50–56. https://doi.org/10.1366/0003702914337812.

Garcés, V., Chate-galvis, N., 2020. Biosynthesis and biotransformations of biopolymers: starch and poly(hydroxyalkanoates). J. Sci. Technol. Appl. 9, 4–9. https://doi.org/10.34294/j.jsta.20.9.60.

Huang, W., Hopper, D., Goodacre, R., Beckmann, M., Singer, A., Draper, J., 2006. Rapid characterization of microbial biodegradation pathways by FT-IR spectroscopy. J. Microbiol. Methods 67, 273–280. https://doi.org/10.1016/j.mimet.2006.04.009.

Kane, S., Ashby, P., Prutt, L., 2009. ATR-FTIR as a thickness measurement technique for hydrated polymer-on-polymer coatings. J. Biomed. Mater. Res. B Appl. Biomater. 91B, 613–620. https://doi.org/10.1002/jbm.b.31436.

Khalifa, M.E., Abdllah, A.B., 2019. Molecular imprinted polymer based sensor for recognition and determination of profenofos organophosphorous insecticide. Biosens. & Bioelectron. X. 2, 100027. https://doi.org/10.1016/j.biosx.2019.100027.

Khatoon, Z., McTiernan, C., Suuronen, E., Mah, T.F., Alarcon, E., 2018. Bacterial biofilm formation on implantable devices and approaches to its treatment and prevention. Heliyon 4, e01067. https://doi.org/10.1016/j.heliyon.2018.e01067.

Latif, U., Ping, L., Dickert, F.L., 2018. Conductometric sensor for PAH detection with molecularly imprinted polymer as recognition layer. Sensors 18, 713. https://doi.org/10.3390/s18030767.

Li, Z., Chen, S., Xu, C., Ju, L., Li, F., 2018. Rapid subtyping of pathogenic and nonpathogenic Vibrio parahaemolyticus by Fourier transform infrared spectroscopy with chemometric analysis. J. Microbiol. Methods 155, 70–77. https://doi.org/10.1016/j.mimet.2018.11.003.

Mainali, K.P., Bewick, S., Thielen, P., Mehoke, T., Breitwieser, F.P., et al., 2017. Statistical analysis of co-occurrence patterns in microbial presence-absence datasets. PLoS One 12. https://doi.org/10.1371/journal.pone.0187132 e0187132.

Ojeda, J., Dittrich, M., 2012. Fourier transform infrared spectroscopy for molecular analysis of microbial cells. Methods Mol. Biol. 881, 187–211. https://doi.org/10.1007/978-1-61779-827-6_8.

Otálora, A., Palencia, M., 2019. Application of functionally-enhanced derivative spectroscopy (FEDS) to the problem of the overlap of spectral signals in binary mixtures: triethylamine-acetone. J. Sci. Technol. Appl. 6, 96–107. https://doi.org/10.34294/j.jsta.19.6.44.

Palencia, M., 2018. Functional transformation of Fourier-transform mid-infrared spectrum for improving spectral specificity by simple algorithm based on wavelet-like functions. J. Adv. Res. 14, 53–62. https://doi.org/10.1016/j.jare.2018.05.009.

Palencia, S., Lagos, G., García, A., 2016a. Fungi-bacterium interactions: *Helycobacter pyroli - Candida albicans*. J. Sci. Technol. Appl. 1, 53–69. https://doi.org/10.34294/j.jsta.16.1.5.

Palencia, M., Lerma, T., Palencia, V., 2016b. Description of fouling, surface changes and heterogeneity of membranes by color-based digital image analysis. J. Membr. Sci. 510, 229–237. https://doi.org/10.1016/j.memsci.2016.02.057.

Palencia, M., Lerma, T., Afanasjeva, N., 2019a. Antibacterial cationic poly(vinyl chloride) as an approach for in situ pathogen-inactivation by surface contact with biomedical materials. Eur. Polym. J. 115, 212–220. https://doi.org/10.1016/j.eurpolymj.2019.03.041.

Palencia, M., Mora, M., Lerma, T., 2019b. Study of time-dependent diffusion coefficient membrane systems and its application to extractions from sugarcane bagasse. J. Sci. Technol. Appl. 6, 40–52. https://doi.org/10.34294/j.jsta.19.6.40.

Palencia, M., Rivas, B.L., Pereira, E., Arrieta, A., 2010. Polyelectrolyte adsorption study on polyethersulfone membrane during polymer-enhanced ultrafiltration by electrochemical impedance spectroscopy. Polym. Bull. 65, 145–156. https://doi.org/10.1007/s00289-009-0236-7.

Palencia-Luna, S.L., García, A., Palencia, M., 2020a. Study of mid-infrared vibrational spectrum of pathogenic bacterium biolayers by Functionally-Enhanced Derivative Spectroscopy (FEDS). J. Sci. Technol. Appl. 9, 34–44. https://doi.org/10.34294/j.jsta.20.9.64.

Palencia-Luna, S.L., García, A., Palencia, M., 2020b. Vibrational spectrum characterization of outer surface of *Helicobacter pylori* biofilms by Functional-Enhanced Derivative Spectroscopy. J. Chil. Chem. Soc. 65, 5015–5022. https://doi.org/10.4067/S0717-97072020000405015.

Palencia-Luna, S.L., García, A., Palencia, M., 2021. Mid-infrared vibrational spectrum characterization of the outer surface of *Candida albicans* by functionally-enhanced derivative spectroscopy. J. Appl. Spectrosc. 88, 161–167.

Palencia, V.J., Martínez, J.M., 2017. Spectral data transformation analysis by simple mathematical functions: Fourier transform infrared spectrum. J. Sci. Technol. Appl. 3, 44–55. https://doi.org/10.34294/j.jsta.17.3.23.

Prakash, J., Kar, S., Lin, C., Chen, C., Chang, Y., Jean, J., Kulp, T., 2013. Identification and discrimination of bacteria using Fourier transform infrared spectroscopy. Spectrosc. Acta Part A: Mol. & Biomol. Spectrosc. 116, 478–484. https://doi.org/10.1016/j.saa.2013.07.062.

Restrepo, D.F., García, A., Combatt, E.M., 2018. Infrared spectroscopy in the analysis of cation exchange capacity of soils - a review. J. Sci. Technol. Appl. 5, 55–65. https://doi.org/10.34294/j.jsta.18.5.35.

Singhalage, D., Seneviratne, G., Madawala, S., Manawasinghe, I., 2018. Characterization of structural properties of fungal-bacterial biofilms by fourier transform infrared spectroscopy. Ceylon J. Sci. 47, 77–83. https://doi.org/10.4038/cjs.v47i1.7490.

Sorer, C.S., Coldrick, Z., Tate, D.J., Marsden, J., Grieve, B., 2018. Towards phosphate detection in hydroponics using molecularly imprinted polymer sensors. Sensors 18, 531. https://doi.org/10.3390/s18020531.

Stave, K., Berrio, M.E., Palencia, S., 2017. Methods of diagnosis of *Helicobacter pylori* in children: current status and prospects. J. Sci. Technol. Appl. 12, 24–43. https://doi.org/10.34294/j.jsta.17.2.12.

Vazhnova, T., Lukyanov, D., 2013. Fourier self-deconvolution of the IR spectra as a tool for investigation of distinct functional groups in porous materials: brønsted acid sites in zeolites. Anal. Chem. 85, 11291–11296. https://doi.org/10.1021/ac4020337.

Wiedermann, W., Hagmann, M., 2015. Asymmetric properties of the Pearson correlation coefficient: correlation as the negative association between linear regression residuals. Commun. Stat. Theor. Methods 45, 6263–6283. https://doi.org/10.1080/03610926.2014.960582.

Chapter 25

Separation of secondary metabolites and bioactive substances from agricultural residues

25.1 Introduction: from agricultural wastes to bioactive raw materials

25.1.1 Agricultural waste problem

The globalization of the planet, the growth of its population, the consequent need to produce large quantities of food and, the prioritization of the need over environmental health, are factors that have contributed to the development of a modern linear production agriculture system. Currently, agriculture generates tons of waste that are generally destined for the generation of energy in industrial boilers or they are sent to landfills. It has been reported that up to a third of agricultural production is discarded, including agro-industrial waste and products that do not meet quality standards (Torres-León et al., 2018; Zuin and Ramin, 2018). Within the European Union, approximately 700 million tons of agricultural waste are generated annually (between the total losses of food produced by households, the food industry, and through the distribution chain) (Fritsch et al., 2017). In addition, the United Nations forecasts that waste generation will increase by 70% in the next 40 years (Chiocchio et al., 2021). However, these residues are an important source of bioactive compounds, that is, compounds with biological activity, such as antimicrobial, antioxidant, antiinflammatory, cardioprotective, and anticancer activity, among others (Chiocchio et al., 2021; Das et al., 2021). Therefore, instead of being discarded, interest has now been aroused in reintroducing the useful part of the residue in the production chain with new purposes, especially in the pharmaceutical, cosmetic, and food industries (Castro-Muñoz et al., 2016; Jimenez-Lopez et al., 2020).

The methods that allow the extraction and separation of these secondary metabolites or active substances, in general, must be green methods.

Eco-friendly Functional Polymers
https://doi.org/10.1016/B978-0-12-821842-6.00017-8
Copyright © 2021 Elsevier Inc. All rights reserved.

Conventional extraction and separation methods are laborious, require large volumes of organic solvents, and apply high temperatures that, in addition to causing high energy consumption, can cause denaturation or degradation of bioactive compounds (Ren et al., 2013). In order to solve these limitations in the separation of metabolites and bioactive substances, a few years ago the research and development of new environmentally friendly separation methods was promoted. Among these is membrane separation technology, which operates in aqueous or hydroalcoholic media, at room temperature, without the need for heating and, therefore, with lower energy consumption (Panzella et al., 2020).

In membrane separation technology, the separation process is mainly physical and both permeate and retentate fractions can be recovered. Therefore, it is a promising approach for concentrating and separating bioactive compounds (Argenta and De Paula Scheer, 2020; Castro-Muñoz and Yañez-Fernandez, 2015; Zhu and Bai, 2017). In the next subsection, the definition of membrane, its classification, main characteristics, and methods of operation are introduced. It continues with a brief definition and generalities of the secondary metabolites and bioactive substances. Then, there are descriptions given of two examples of current processes that involve the use of filtration membranes for the separation and concentration of bioactive compounds from agricultural residues. This contextualizes the problem with that waste, the methodological strategy used for the separation process, and the results obtained. The chapter ends with some conclusions about the efficiency of membrane technology in separation processes.

25.1.2 Polymeric membrane technology

A membrane is defined by the IUPAC as "a structure with lateral dimensions much greater than its thickness, through which a mass transfer can occur under a variety of driving forces" (Koros et al., 1996). Consequently, a membrane is a thin planar structure or interface that separates two phases and allows mass transfer between the phases (Asad et al., 2020). Commonly, membranes have different classifications according to their characteristics, for example, according to their origin they can be classified into biological membranes and synthetic or artificial membranes (Palencia, 2015). The latter in turn are classified into two main geometries: flat sheet membranes and cylindrical or hollow fiber membranes (El-Zanati et al., 2020). Furthermore, membranes can be classified according to their structure as asymmetric or symmetric membranes. Asymmetric membranes consist of two layers, the upper one being a very thin dense layer, commonly called the skin layer or active layer, and it is this layer that determines the permeation properties. Meanwhile the lower layer is a porous underlayer, commonly called a support layer, and is a passive layer that provides mechanical resistance to the membrane (Abdullah et al., 2018). The most commonly used membranes for bioactive substance

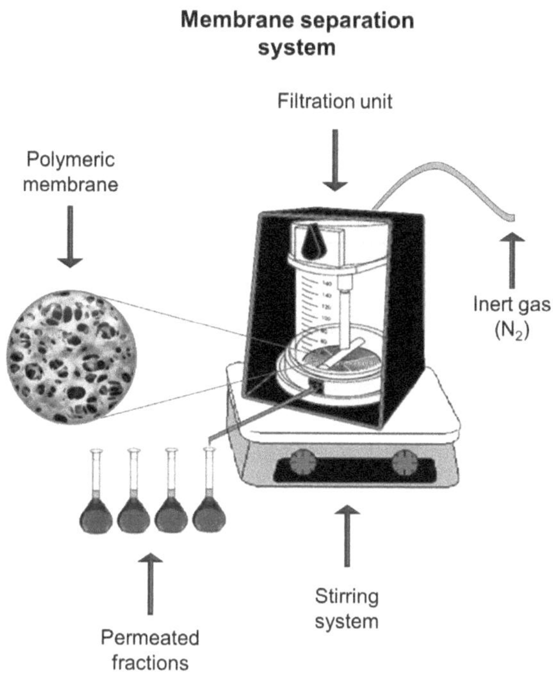

FIGURE 25.1 Representation of a membrane separation system.

separation processes are micro-, ultra-, and nanofiltration porous membranes, which are structurally and functionally analogous to a conventional filter. This type of membrane has a rigid and hollow structure (that is, interconnected and randomly distributed pores) (Yu et al., 2018; Koop et al., 2021). The porous membranes mentioned above involve pressure-actuated separation methods. To carry out the separation processes using this porous membrane technology, equipment such as the one shown in Fig. 25.1 must be used. It is equipped with a pressure source, generally an inert gas such as nitrogen, a filtration unit (cell filter and membrane unit), a magnetic stirrer, and vials to collect the permeate.

Once the filtration proceeds, the feed stream (the initial solution) is separated into two effluent streams: the permeate, which is the portion of the fluid that has passed through the membrane, and the retained or concentrated part, which is the portion that contains the constituents that have been rejected or retained by the membrane (Palencia, 2015; Abdullah et al., 2018). Therefore, during the membrane separation process, three fractions can be identified: feed, permeate, and retained.

In pressure-actuated membrane separation methods, two types of filtration are distinguished according to the direction of flow: conventional filtration or dead end, and tangential filtration or cross-flow filtration, the first consists of a flow perpendicular to the surface of the membrane where the only outlet for upstream fluid is through the membrane (Hasan et al., 2013; Wibisono et al.,

2014). Meanwhile, cross-flow filtration or cross-flow filtration consists of the passage of flow through a membrane module in which the fluid on the upstream side of the membrane moves parallel to the membrane surface, and the fluid on the downstream side of the membrane moves away from the membrane in the direction perpendicular to the membrane surface (Cassano et al., 2018; Tylkowski and Tsibranska, 2017; Zhu and Bai, 2017).

25.1.3 Secondary metabolites and bioactive substances

In plants, secondary metabolites are commonly synthesized to facilitate the interaction with the biotic environment by increasing the ability of the producing organism, for example, providing odor, color, and taste to attract symbiotic insects, and for the establishment of a mechanism for defense against surrounding organisms (for example, deterring herbivores) (Câmara et al., 2021; Jurić et al., 2020; Kumar et al., 2021).

These metabolites can be classified into three main groups: alkaloids, terpenoids, and phenolic compounds (Peters et al., 2019). Alkaloids are mainly made up of nitrogen and are often highly toxic. Some examples include morphine that comes from *Papaver somniferum* (or roost) and caffeine. Terpenoids are the largest group of secondary metabolites, and are made of units of isoprene, which is a very volatile colorless liquid produced in chloroplasts and released by leaves. Some examples of terpenoids are phytosterols such as campesterol and stigmasterol, which are widely used in the pharmaceutical industry due to their ability to inhibit the adsorption of cholesterol in the intestine (Tolve et al., 2020). Finally, phenolic compounds are compounds with at least one aromatic hydrocarbon ring with one or more hydroxyl groups (−OH) in their structure. Although phenolic compounds have similar chemical structures, there are some differences that allow them to be subdivided into two classes: flavonoids and nonflavonoids (Tsimogiannis and Oreopoulou, 2019).

Regarding the therapeutic actions in different studies, it has been shown that phenolic compounds such as gallic, p-coumaric, chlorogenic, and anacardic acids have antioxidant properties, that is, they protect against oxidative stress. In addition, they have antiinflammatory and anticarcinogenic properties, among other biological properties. The phenolic compound widely used as a standard in antioxidant tests such as Folin Ciocalteu is gallic acid, due to its ability to capture free radicals (Baptista et al., 2018; Mubofu and Mgaya, 2018; Gao et al., 2019).

25.2 Applications

25.2.1 Use of artichoke waste

The artichoke (*Cynara scolymus* L., family Asteraceae) is a perennial herbaceous plant native to the Mediterranean area, but now widely cultivated throughout the world (Jiménez-Moreno et al., 2019). The edible parts of the plant are large immature inflorescences, called capitula or heads. The internal and external bracts are discarded, corresponding to 60%–85% of the harvested product. However, the phenolic composition of this residue is similar to the edible parts of the plant, and so, countries such as France, Spain, and Italy, where 80% of the total world production is consumed, have been interested in taking advantage of these by-products that currently exceed one million tons annually (Lavecchia et al., 2019). Therefore, it is important to look for alternatives to recover phenolic compounds from artichoke waste. Recently, Rabelo et al. used polymeric membranes in order to study the feasibility of separating these bioactive substances, taking into account that phenolic acids from artichoke have a molecular weight of between 300 and 600 g/mol. In addition, they evaluated the use of eco-friendly solvents such as water and ethanol in the extraction and concentration processes to enable the future use of these compounds in products for the food, pharmaceutical, and cosmetic industries.

Next, the methodological strategies that the authors carried out to achieve the separation of phenolic compounds from hydroalcoholic extracts of artichoke are described. Also, the results and conclusions obtained are reported (Rabelo et al., 2016).

25.2.2 Methodological strategy for the use of artichoke residues

The research carried out by Rabelo et al. for the separation of phenolic compounds present in artichoke residues is detailed below, using supported polymeric membranes (Rabelo et al., 2016). To start the extraction process, the artichoke residues (external and internal bracts) were crushed and subjected to an ultrasonic extraction process with a power of 0, 240, 480, and 720 W, using different ethanolic solutions as solvent (0%, 25%, 50%, and 75%) for 60 min.

Once the extracts rich in phenolic compounds were obtained, the concentration method was carried out using a tangential filtration system (INVICT, MENTEST). Three supported membranes of different porosity were used (NF270, DK, and DL). All consisted of supported nanofiltration membranes, where the support is composed of polysulfone and an active layer of polyamide. The geometry of these membranes was a flat sheet, with an effective permeation area of 7.7×10^{-3} m^2 and a maximum liquid recirculation flow rate of 60 L/h. The system pressure was kept at 20 bar and a temperature of 25°C. The initial feed volume was 900 mL and the concentration process was carried out up to a volume reduction factor of 2.5.

Before starting the permeation processes, the membranes were washed, hydrated, and conditioned with the solvent. Each process was repeated in duplicate with new membranes so as not to alter the subsequent results. The membranes were characterized by the following techniques before and after each experiment in order to be able to contrast the changes that occurred on their surface. The characterization was started by atomic force microscopy

(Park Scientific Instruments), where the surface roughness was analyzed using the Gwyddion 2.39 software (Czech Metrology Institute). Fourier transform infrared spectroscopy coupled with attenuated total reflectance (ATR) was then used to investigate the functional groups and molecular structures on membrane surfaces. Finally, the change in surface hydrophilicity was monitored by measuring the contact angle by the sessile drop method using a tensiometer (Teclis, Tracker).

On the other hand, the total phenolic content of each retentate and permeate was quantified in triplicate using a gallic acid calibration curve. The antioxidant capacity was determined by DPPH (2,2-diphenyl-1-picryl-hydrazil) and FRAP (ferric reducing antioxidant power) expressing both results as mg of Trolox/g of dry matter. The chlorogenic acid content was determined by Dionex chromatography (LC Standard Ultimate 3000) using a Poroshel 120 EC-C18 reversed-phase column and a diode matrix detector (Dionex, DAD-3000). Quantification was carried out at a wavelength of 324 nm. Finally, the size and size distribution of the feed and the concentrated sample were measured by laser diffraction (Mastersizer 2000, Malvern Instruments), while the size distribution was determined using the dynamic light-scattering technique (Zetasizer Nano-ZS, Malvern Instruments). The volumetric flow of the permeate (J_v, L/m^2h) was also monitored using the following equation:

$$J_v = \left(\frac{m_p}{\rho \cdot t \cdot A_p} \right) \tag{25.1}$$

where m_p is the collected mass of the permeate (g), ρ is the density (g/L), t is the time (h), and A_p is the area of permeability of the membrane (m^2).

On the other hand, determination of the decrease in flow (D_f) with respect to the time was calculated using:

$$D_f = \left(\frac{J_i - J_s}{J_i} \right) \cdot 100 \tag{25.2}$$

where J_i and J_s are the initial flow and average permeate flow.

The retention coefficient (R) was calculated using the equation:

$$\% R = \left(1 - \frac{C_p}{C_A} \right) \cdot 100 \tag{25.3}$$

where C_p and C_A are the solute concentration in the permeate and in the food, respectively.

Finally, the volumetric concentration factor (VCF) was given by:

$$VCF = \left(\frac{V_i}{V_f} \right) \tag{25.4}$$

where V_i and V_f are the starting and ending volumes, respectively.

25.2.3 Results of the separation of phenolic compounds present in artichoke residues

The results of the physicochemical properties of the extracts obtained show that the most suitable combination of the process to obtain the highest chlorogenic acid was ultrasound extraction at a frequency of 20 kHz and power of 240 W. Ethanol was used as a solvent at 50%. The results related to the extraction process (content of total phenols and antioxidant capacity) showed that a greater amount of phenolic compounds was obtained when the 75% ethanolic solution was used, because ethanol reduces the dielectric constant of the medium, facilitating the extraction of these compounds. However, the retention processes showed that it is more convenient to use 50% solutions, since a higher proportion of alcohol leads to a larger effective pore diameter, and therefore a lower obstacle to the transport of phenolic acids. However, after several permeations and at the end of the concentration process, both the 50% ethanolic extract and the 75% extract showed a 30% increase in the chlorogenic acid content with respect to the initial concentration in the food. Regarding the flow, the lowest permeate flow was observed and, consequently, the longest residence time of the bioactive compounds within the filtration cell, for the 75% ethanolic solutions, due to the fact that the membranes are hydrophilic and they form hydrogen bonds with ethanol, hindering the permeation of water.

In the three polymeric membranes evaluated, the roughness was initially determined by AFM. It was found that the one with the greatest roughness was the one with flux reduced at the beginning of its permeation, as, according to Vrijenhoek, in the initial stages of filtration, the particles accumulated in the "valleys" of the rough membranes, resulting in "valley clogging" (Vrijenhoek et al., 2001). The susceptibility to fouling in the membranes follows the order NF270 > DL > DK. This behavior can be explained because the NF270 membranes, being made up of piperazine-based semiaromatic polyamides, have a higher content of carboxylic acid groups, and, therefore, greater hydrophilicity and ability to form hydrogen bonds with various compounds in solution, facilitating encrustation of polysaccharides on the surface of the membrane. This was verified with the signals corresponding to C—O—C and C—O—P bonds observed in the infrared spectra of the membranes, which also explains the decrease in the permeate flux. Therefore, the previous simplification of the extract must be sought, if possible with a microfiltration membrane.

On the other hand, DK membrane showed less susceptibility to fouling, with a more stable flow and short concentration time. From permeations through it, retention of more than 95% of chlorogenic acid was obtained for extracts with 50% ethanol. From the results obtained, the authors concluded that it is possible through the use of polymeric nanofiltration membranes, NF270, DK, and DL, to concentrate hydroalcoholic extracts of artichoke residues in a fast, simple, and eco-friendly way.

25.3 Use of waste from wine industries

The wine industry produces a large amount of organic and inorganic waste in a short period of time, creating serious environmental problems due to its composition (Maicas and Mateo, 2020; Rani et al., 2020). The most significant residues are represented by grape pomace (a solid material separated from the must in the pressing stage), grape seeds, and lees (residues formed at the bottom of the wine containers after fermentation, during storage, or after wine filtration) (Arboleda Meija et al., 2019). Global grape production in 2017 was around 73.3 million tons. It has been estimated that 0.13 t of pomace, 0.06 t of lees, 0.03 t of bunches, and 1.65 m^3 of wastewater are generated for each ton of processed grapes (Oliveira and Duarte, 2016). These residues contain a high concentration of bioactive molecules that can be used to obtain extracts or semifinished products of interest for food, nutraceutical, and pharmaceutical applications, in particular, wine lees (Arboleda Meija et al., 2019).

Conventional treatments of winery waste are becoming increasingly expensive, requiring significant amounts of effort, resources, and energy for safe discharge into the environment. For example, the usual purification process to obtain phenolic compounds from these residues is through the use of polymeric absorbent resins designed to eliminate undesirable compounds, such as fats, proteins, and sugars, which interact with phenolic compounds, reducing their antioxidant activity (Kammerer and Gajdos, 2005). Phenolic compounds are eluted from the columns using hydroalcoholic solutions and then dried by evaporation techniques to obtain a specific product. This multistep process is limited by several drawbacks, including the high use of solvents, the low flexibility of the methodology in relation to the composition of the final product, and the large volume of brine used for the regeneration of resins.

Due to the potential use of these residues, the wine industries are strongly motivated to find new technological approaches capable of reducing the impact of agro-industrial residues on the environment and provide additional sources of income from their industrial exploitation (Kalli et al., 2018). In this context, Arboleda et al. applied the concept of biorefinery for the recovery of phenolic compounds from red wine lees; the details of the research, methodological strategies, and results will be described below.

25.3.1 Methodological strategy for the use of wine waste

The red wine lees obtained after fermentation of grape must were stored at −20°C and then dried at 50°C for 24 h before use. Polyphenol extraction was carried out using a domestic microwave oven (Modelo, JT 356 WH, Whirlpool) using 1 g of red wine lees with 10 mL of solvent (75% ethanol, 1% hydrochloric acid in water) and subjected to different microwave powers (range: 90–350 W) and exposure times (range: 0.5–3 min). A total of 12 replicate trials were carried out. Then, the extracts obtained were simplified using a microfiltration membrane made of polyvinylidene fluoride (PVDF) with a pore size of 0.15 μm. The clarified extracts were used as feed in the filtration system, applying the no-leak filtration mode. The system consisted of a magnetic stirrer, a pressure regulator, a manometer, a pressure source (N2), a stirred cell (300 mL) at high pressure (up to 68.9 bar) (Sterlitech HP 4750, Kent, WA), and a membrane area. In this section, three other membranes were used also, an ultrafiltration membrane (UF, called Etna 01 PP) with a molecular weight cut-off (MWCO) of 1000 Da and two nanofiltrations (NF, called NFT50 and Desal DK) with MWCO of 150–300 Da. All had an effective membrane area of 13.85 cm^2. For the MF, UF, and NF systems, pressures of 2, 10, and 24 bar were used. The flow was determined using Eq. (25.1).

The fouling index (*FI*) of the investigated membranes, expressed as a percentage drop in water permeability, was estimated by measuring the water permeability before and after the treatment of the wine lees extracts, according to the following equation:

$$FI = \left(1 - \frac{L_{w1}}{L_{w0}}\right) \cdot 100 \quad (25.5)$$

where L_{w0} and L_{w1} are the permeability to pure water before and after treatment with the extracts.

All experiments were performed using 250 mL of feeding solution until a permeate volume of 70 mL was collected to reach a volume concentration factor (*VCF*) of 3.5. The performance of the selected membranes was evaluated in terms of productivity (permeate flux and retention capacity of phenolic compounds, and fouling index). The following determinations and quantifications were made on the food, retentate, and permeate samples. The pH (HANNA 209, Merk), turbidity (Aqualytic compact, AL250T-IR), total polyphenols (measured by the direct method at 280 nm and the Folin-Ciocalteu method at 750 nm were determined using a Shimadzu UV-Vis spectrophotometer UV-160A), proanthocyanidins [spectrophotometric assay with the reagent 4-dimethylaminocinnamaldehyde (DMAC)] and quantification of sugars (D-glucose and D-fructose by enzymatic assay with hexokinases and phosphoglucose isomerase, the NADPH formed throughout the reactions was measured at 340 nm, Megazyme). Once the separation process was completed, the permeate was analyzed (determined and quantified) by high-performance liquid chromatography (HPLC) using reference standards. Finally, after performing all the quantifications, the rejection of selected membranes toward specific compounds was determined according to Eq. (25.3).

25.3.2 Results of the separation of phenolic compounds present in red wine lees

The quantification of total phenols in the initial extract that was used as the feeding phase in the filtration process, yielded a concentration of 1939 mg/L by the direct method, and a concentration of 933 mg/L by the method indirect of Folin-Ciocalteu. This is due to the fact that after wine fermentation, the lees have a high protein content as a result of yeast residues, which generates interferences and overestimations in both results. In the direct method, these interferences are due to the ability of other wine constituents (nucleotides, proteins, peptides, and amino acids) to adsorb this wavelength. Meanwhile using the Folin-Ciocalteu method, the main drawback is due to the ability of some wine components to donate electrons, also causing an overestimation.

By quantifying total sugars, it was found that the sample initially contained a total 610 mg/L of glucose and fructose. Regarding the pH of the sample, an acidic environment of approximately 2 was determined, due to the use of hydrochloric acid to facilitate the obtaining of phenolic acids such as syringic acid, gallic acid, etc. The presence of these phenolic compounds and three others were verified by analyzing the sample by HPLC. A flavanol derivative catechin, gallic acid, catechin, syringic acid, and a derivative of gallocatechin were identified. After subjecting this sample to the permeation process through the microfiltration membrane, the permeate flux and fouling of the membrane were monitored. The initial permeate flow, which was initially 12 L/m^2h, decreased to 2.5 L/m^2h over a period of 350 min, where the greatest change was observed in the first 50 min where its flow decreased by approximately 50%, which was a consequence of the concentration of the retained sample and, therefore, the fouling of the membrane.

Then, this extract, clarified using the microfiltration membrane, was permeated by three other nanofiltration membranes, showing a correlation between the MWCO of the membranes, the permeate flux, and their hydrophilicity. That is, there was a higher flux for membranes with a higher MWCO, as is the case with Etna. Meanwhile the small differences in the flow of the NFT and Desal DK membranes can be inferred by their different affinities toward water, with the highest flow being that which presents a lower contact angle with water (21.2 ± 0.9), that is, the NFT. In contrast, Desal DK reported higher contact angles (28.1 ± 4.8), as a reflection of its lower hydrophilicity. However, both membranes have the ability to form hydrogen bonds between the hydroalcoholic solution and the amide groups of the polymeric membranes.

Subsequently, the fouling index of the membranes was calculated from the measurement of their water permeability before and after the permeation of each extract. According to these measurements, the FSMO microfiltration membrane was the one that suffered the most fouling, as it yielded an FI of 50% as a consequence of the first clarification step of the sample, where its turbidity was reduced by 77%. Then, the membrane that was the most dirty was the NFT50, with an FI of 32.1%. This was followed by the ETNA 01 PP membrane with a fouling index of 18.4%, as a consequence of the particular composition of this membrane, which is manufactured by grafting a thin layer of hydrophilic cellulosic polymer on a polyvinylidene fluoride material and finally, the one that was least dirtied was the nanofiltration membrane that did not have the Desal DK cellulosic polymer layer with an FI of 15.5%. The contrast between the results of the physicochemical tests obtained for the initial sample, the retained one, and the permeate, accompanied by the results of the high-efficiency liquid chromatography, ensure that the phenolic compounds were separated in the permeate fraction, mainly in the clarified extract, thanks to a size exclusion process. This is consistent with the low molecular weights of phenolic compounds. For example, catechin and gallic acids have molecular weights of 290 and 170 Da, which are much smaller compared to the pore sizes of the microfiltration membranes used to obtain the first extract, allowing their easy diffusion and the separation of larger molecules. Therefore, the rejection of selected membranes toward specific compounds was strictly correlated with their nominal MWCO. Taking into account that the membrane with the lowest MWCO was the NFT50, with a value of 150 Da, the vast majority of compounds, including phenolic, sugar, and others, have a higher molecular weight, so greater retention was achieved of various compounds with a higher fouling index.

25.4 Conclusions

Membrane separation technology is a promising alternative for the simplification of complex mixtures of secondary metabolites and bioactive substances that are present in natural extracts. This is due to the fact that these compounds have a molecular weight in the order of 100–300 Da, which can be separated by a size exclusion mechanism using porous polymeric membranes, whether using micro-, ultra-, and/or nanofiltration. The main disadvantage of membrane separation processes is fouling of their surface, which is a consequence of various factors, such as the deposition of larger species that block the pores, affinity between the solutes and the surface of the membrane, concentration polarization, high membrane roughness, lack of agitation, etc. Therefore, these aspects must be controlled to avoid the decrease in permeate flux and, therefore, result in an increase in the efficiency of the membrane. Membrane separation technology is an emerging but promising technique that should continue to be researched, in order to potentiate its applications in different scientific areas.

References

Abdullah, N., Rahman, M., Dzarfan, M., Jaafar, J., Ismail, A., 2018. Membranes and membrane processes: fundamentals. In: Basile, A., Mozia, S., Molinari, R. (Eds.), Current Trends and Future Developments on (Bio-) Membranes. Elsevier, pp. 45–70. https://doi.org/10.1016/B978-0-12-813549-5.00002-5.

Argenta, A., De Paula Scheer, A., 2020. Membrane separation processes applied to whey: a review. Food Rev. Int. 36, 499–528. https://doi.org/10.1080/87559129.2019.1649694.

Arboleda Meija, J., Parpinello, G., Versari, A., Conidi, C., Cassano, A., 2019. Microwave-assisted extraction and membrane-based separation of biophenols from red wine lees. Food Bioprod. Process. 117, 74–83. https://doi.org/10.1016/j.fbp.2019.06.020.

Asad, A., Sameoto, D., Sadrzadeh, M., 2020. Overview of membrane technology. In: Sadrzadeh, M., Mohammadi, T. (Eds.), Micro and Nano Technologies, Nanocomposite Membranes for Water and Gas Separation. Elsevier, pp. 1–28. https://doi.org/10.1016/B978-0-12-816710-6.00001-8.

Baptista, A., Vilela, R., Bressan, J., Gouveia, M., 2018. Antioxidant and antimicrobial activities of crude extracts and fractions of cashew (*Anacardium occidentale* L.), Cajui (*Anacardium microcarpum*), and Pequi (*Caryocar brasiliense* C.): a systematic review. Oxid. Med. & Cell. Longev. 1–13. https://doi.org/10.1155/2018/3753562.

Câmara, J., Albuquerque, B., Aguiar, J., Corrêa, R., Gonçalves, J., Granato, D., Pereira, J., Barros, L., Ferreira, I., 2021. Food bioactive compounds and emerging techniques for their extraction: polyphenols as a case study. Foods 10, 37. https://doi.org/10.3390/foods10010037.

Cassano, A., Conidi, C., Ruby-Figueroa, R., Castro-Muñoz, R., 2018. Nanofiltration and tight ultrafiltration membranes for the recovery of polyphenols from agro-food by-products. Int. J. Mol. Sci. 19, 351. https://doi.org/10.3390/ijms19020351.

Castro-Muñoz, R., Yañez-Fernandez, J., 2015. Valorization of nixtamalization wastewaters (Nejayote) by integrated membrane process. Food Bioprod. Process. 95, 7–18. https://doi.org/10.1016/j.fbp.2015.03.006.

Castro-Muñoz, R., Yáñez-Fernández, J., Fíla, V., 2016. Phenolic compounds recovered from agro-food by-products using membrane technologies: an overview. Food Chem. 213, 753–762. https://doi.org/10.1016/j.foodchem.2016.07.030.

Chiocchio, I., Mandrone, M., Tomasi, P., Marincich, L., Poli, F., 2021. Plant secondary metabolites: an opportunity for circular economy. Molecules 26, 495. https://doi.org/10.3390/molecules26020495.

Das, S., Singh, V., Upadhyay, N., Singh, B., Prasad, J., Tiwari, S., Dubey, N., 2021. Secondary metabolites of higher plants as green preservatives of herbal raw materials and their active principles during postharvest processing. In: Mandal, S., Chakraborty, R., Sen, S. (Eds.), Evidence Based Validation of Traditional Medicines. Springer, pp. 261–277. https://doi.org/10.1007/978-981-15-8127-4_13.

El-Zanati, E., Farg, E., Taha, E., El-Gendi, A., Abdallah, H., 2020. Preparation and characterization of different geometrical shapes of multi-bore hollow fiber membranes and application in vacuum membrane distillation. J. Anal. Sci. & Technol. 11. https://doi.org/10.1186/s40543-020-00244-4.

Fritsch, C., Staebler, A., Happel, A., Cubero, M., Aguiló-Aguayo, I., et al., 2017. Processing, valorization and application of bio-waste derived compounds from potato, tomato, olive and cereals: a review. Sustainability 9, 1492. https://doi.org/10.3390/su9081492.

Gao, M., Xu, Q., He, Q., Sun, Q., Zeng, W., 2019. A theoretical and experimental study: the influence of different standards on the determination of total phenol content in the Folin–Ciocalteu assay. J. Food Meas. & Charact. 13, 1349–1356. https://doi.org/10.1007/s11694-019-00050-6.

Hasan, A., Peluso, C., Hull, T., Fieschko, J., Chatterjee, S., 2013. A surface-renewal model of cross-flow microfiltration. Braz. J. Chem. Eng. 30, 167–186. https://doi.org/10.1590/S0104-66322013000100019.

Jimenez-Lopez, J., Fraga-Corral, M., Carpena, M., García-Oliveira, P., Echave, J., Pereira, G., Lourenço-Lopes, C., Prieto, M., Simal-Gandara, J., 2020. Agriculture waste valorisation as a source of antioxidant phenolic compounds within a circular and sustainable bioeconomy. Food & Funct. 11, 4853–4877. https://doi.org/10.1039/D0FO00937G.

Jiménez-Moreno, N., Cimminelli, M., Volpe, F., Ansó, R., Esparza, I., Mármol, I., Rodriguez-Yoldi, M., Ancín-Azpilicueta, C., 2019. Phenolic composition of artichoke waste and its antioxidant capacity on differentiated Caco-2 cells. Nutrients 11, 1723. https://doi.org/10.3390/nu11081723.

Jurić, S., Sopko, K., Król-Kilińska, Ż., Žutić, I., Fabek, S., Đermić, E., Topolovec-Pintarić, S., Vinceković, M., 2020. The enhancement of plant secondary metabolites content in *Lactuca sativa* L. by encapsulated bioactive agents. Sci. Rep. 10, 3737. https://doi.org/10.1038/s41598-020-60690-3.

Kalli, E., Lappa, I., Bouchagier, P., Tarantilis, P., Skotti, E., 2018. Novel application and industrial exploitation of winery by-products. Bioresour. & Bioprocess. 5, 46. https://doi.org/10.1186/s40643-018-0232-6.

Kammerer, D., Gajdos, J., 2005. Recovery of anthocyanins from grape pomace extracts (*Vitis vinifera* L. cv. *Cabernet Mitos*) using a polymeric adsorber resin. Eur. Food Res. Technol. 220, 431–437. https://doi.org/10.1007/s00217-004-1078-z.

Koros, W., Ma, Y., Shimidzu, T., 1996. Terminology for membranes and membrane processes. Pure Appl. Chem. 7, 1479–1489. https://doi.org/10.1351/pac199668071479.

Koop, B., Knapp, M., Di Luccio, M., Pinto, V., Tormen, L., Ayala, G., Rodrigues, A., 2021. Bioactive compounds from Jambolan (*Syzygium cumini* (L.)) extract concentrated by ultra- and nanofiltration: a potential natural antioxidant for food. Plant Foods Hum. Nutr. 42 https://doi.org/10.1007/s11130-021-00878-8.

Kumar, P., Dey, S., Dwivedi, P., 2021. Plant- and microbes-mediated secondary metabolites: remunerative venture for discovery and development. In: Yadav, A., Singh, J., Singh, C., Yadav, N. (Eds.), Current Trends in Microbial Biotechnology for Sustainable Agriculture, Environmental and Microbial Biotechnology. Springer. https://doi.org/10.1007/978-981-15-6949-4_15.

Lavecchia, R., Maffei, G., Paccassoni, F., Piga, L., Zuorro, A., 2019. Artichoke waste as a source of phenolic antioxidants and bioenergy. Waste & Biomass Valoriz. 10, 2975–2984. https://doi.org/10.1007/s12649-018-0305-y.

Maicas, S., Mateo, J., 2020. Sustainability of wine production. Sustainability 12, 559. https://doi.org/10.3390/su12020559.

Mubofu, E., Mgaya, J., 2018. Chemical valorization of cashew nut shell waste. Top. Curr. Chem. 376, 1–15. https://doi.org/10.1007/s41061-017-0177-9.

Oliveira, M., Duarte, E., 2016. Integrated approach to winery waste: waste generation and data consolidation. Front. Environ. Sci. Eng. 10, 168–176. https://doi.org/10.1007/s11783-014-0693-6.

Palencia, M., 2015. Liquid-phase polymer-based retention: theory, modeling, and application for the removal of pollutant inorganic ions. J. Chem. 2015, 1–9. https://doi.org/10.1155/2015/965624.

Panzella, L., Moccia, F., Nasti, R., Marzorati, S., Verotta, L., Napolitano, A., 2020. Bioactive phenolic compounds from agri-food wastes: an update on green and sustainable extraction methodologies. Front. Nutr. 7, 60. https://doi.org/10.3389/fnut.2020.00060.

Peters, K., Treutler, H., Döll, S., Kindt, A., Hankemeier, T., Neumann, S., 2019. Chemical diversity and classification of secondary metabolites in nine bryophyte species. Metabolites 9, 222. https://doi.org/10.3390/metabo9100222.

Rabelo, R., Machado, M., Martínez, J., Hubinger, M., 2016. Ultrasound assisted extraction and nanofiltration of phenolic compounds from artichoke solid wastes. J. Food Eng. 178, 170–180. https://doi.org/10.1016/j.jfoodeng.2016.01.018.

Rani, J., Rautela, A., Kumar, S., 2020. Chapter 4 - biovalorization of winery industry waste to produce value-added products. In: Krishnaraj, N., Sani, R. (Eds.), Biovalorisation of Wastes to Renewable Chemicals and Biofuels. Elsevier, pp. 63–85. https://doi.org/10.1016/B978-0-12-817951-2.00004-3.

Ren, Q., Xing, H., Bao, Z., Su, B., Yang, Q., Yang, Y., Zhang, Z., 2013. Recent advances in separation of bioactive natural products. Chin. J. Chem. Eng. 21, 937–952. https://doi.org/10.1016/S1004-9541(13)60560-1.

Tolve, R., Cela, N., Condelli, N., Di Cairano, M., Caruso, M., Galgano, F., 2020. Microencapsulation as a tool for the formulation of functional foods: the phytosterols' case study. Foods 9, 470. https://doi.org/10.3390/foods9040470.

Torres-León, C., Ramírez-Guzman, N., Londoño-Hernandez, L., Martinez-Medina, G., Díaz-Herrera, R., Navarro-Macias, V., Alvarez-Pérez, O., Picazo, B., Villarreal-Vázquez, M., Ascacio-Valdes, J., Aguilar Cristóbal, N., 2018. Food waste and byproducts: an opportunity to minimize malnutrition and hunger in developing countries. Front. Sustain. Food Syst. 2, 1–52. https://doi.org/10.3389/fsufs.2018.00052.

Tsimogiannis, D., Oreopoulou, V., 2019. Chapter 16 - classification of phenolic compounds in plants. In: Watson, R. (Ed.), Polyphenols in Plants. Academic Press, pp. 263–284. https://doi.org/10.1016/B978-0-12-813768-0.00026-8.

Tylkowski, B., Tsibranska, I., 2017. Polymer application for separation/filtration of biological active compounds. Phys. Sci. Rev. 2, 1–12. https://doi.org/10.1515/psr-2017-0022.

Vrijenhoek, E., Hong, S., Elimelech, M., 2001. Influence of membrane surface properties on initial rate of colloidal fouling of reverse osmosis and nanofiltration membranes. J. Membr. Sci. 188, 115–128. https://doi.org/10.1016/S0376-7388(01)00376-3.

Wibisono, Y., Cornelissen, E., Kemperman, A., van der Meer, W., Nijmeijer, K., 2014. Two-phase flow in membrane processes: a technology with a future. J. Membr. Sci. 453, 566–602. https://doi.org/10.1016/j.memsci.2013.10.072.

Yu, W., Liu, T., Crawshaw, J., Liu, T., Graham, N., 2018. Ultrafiltration and nanofiltration membrane fouling by natural organic matter: mechanisms and mitigation by pre-ozonation and pH. Water Res. 139, 353–362. https://doi.org/10.1016/j.watres.2018.04.025.

Zhu, X., Bai, R., 2017. Separation of biologically active compounds by membrane operations. Curr. Pharmaceut. Des. 23, 218–230. https://doi.org/10.2174/1381612822666161027153823.

Zuin, V., Ramin, L., 2018. Green and sustainable separation of natural products from agro-industrial waste: challenges, potentialities, and perspectives on emerging approaches. Top. Curr. Chem. 376, 3. https://doi.org/10.1007/s41061-017-0182-z.

Chapter 26

Hydrogels for separation and delivery of antibacterial inorganic nanoparticles

26.1 Inorganic nanoparticles

Inorganic nanoparticles (INPs) are particulate chemical entities with different natures, sizes, and shapes, that are usually dispersed or destabilized colloids, but also, can be solids or even aerosols. The main condition to define a material as a nanoparticle, or nanomaterial, is that its size, in at least one dimension, must be in the "nanoscale" (i.e., 1−100 nm). Thus, when matter is in the nanoscale new properties appears which can be used for novel applications. The nanoscale is not a physical limit, but rather, a convention that is, in fact, widely accepted, however, in many studies, the size of INPs can exceed 100 nm (Palencia et al., 2015; Jiang et al., 2017; Kaur and Kaur, 2020; Meena et al., 2020).

Nanomaterials can be organic, inorganic, or hybrid nanomaterials, in addition, they are obtained with different sizes and shapes, including fibers, bars, cubes, wires, quantum dots, and amorphous shapes. Sometimes, researchers subjectively assign names to these shapes, therefore, in the language of nanotechnology it is possible to find "nanoonions," "nanoflores," "nanostars," etc. (Ealia and Saravanakumar, 2017; Guilguer-Casagrande and de Lima, 2019; Khan et al., 2019). Usually, the properties of nanoparticles are defined by their small size and large surface area in comparison with the bulk material, however, it is important to indicate that size, volume, and surface area are physical descriptions of material entities confined to occupying a determined space and not the origin of one or more specific properties, which should be explained in term of its nature. Therefore, the origin of quantum effects in the interface related to thermodynamic, optical, magnetic, dielectric, etc., properties when matter is "grouped" in a nanometric size level is the subjacent cause of the nanoscale properties. All the properties of interest of nanoparticles can be described in terms of the surface-area-to-volume ratio, and consequently, of their shape and size, since these two parameters define this ratio. These characteristics can be explained due to strong dependence of

Eco-friendly Functional Polymers
https://doi.org/10.1016/B978-0-12-821842-6.00025-7
Copyright © 2021 Elsevier Inc. All rights reserved.

the nanoscale-level properties on the dimensions and orientation of the particles that make up the material. In addition, the small size is related to unbalance in the cohesion forces at the interface and to dependence of the melting point on the cohesive energy, which is the energy required to divide the material into its individual constituent elements. Other characteristics, such as the optical properties, are also described by the surface-area-to-volume ratio, since these properties are determined by the electronic structure, which depends on the surface atoms which are significantly affected by the shape. Similar descriptions have been made for other properties such as dielectric and magnetic properties. Consequently, as expected, the nanomaterial properties will show an average behavior defined by the size and shape distribution of the particles that make it up. This impacts the quality of the material compared to a specific property, since the characteristics defining the efficiency of that property are affected by the width of the distribution, which greatly depends on the synthesis process (Ghost, 2015; Palencia et al., 2015; Ealia and Saravanakumar, 2017; Jiang et al., 2017; Guilguer-Casagrande and de Lima, 2019; Khan et al., 2019; Kaur and Kaur, 2020; Meena et al., 2020).

Two general pathways are possible for the production of INPs. The first is to start with small particles and add them in such a way that their size increases until a larger particle is obtained, through this strategy physical methods can be used (for example, by means of temperature changes that lead to nucleation and growth processes), chemicals (for example, through oxidation—reduction reactions), or a combination of these. The action of microorganisms for the synthesis of nanoparticles has been included within chemical methods because biotransformation processes occur through changes in the nature of matter. The second route for obtaining INPs is through the degradation or division of a large material into smaller parts. It can also be carried out by chemical (e.g., acid dissolution, hydrolysis, and thermal degradation) or physical (e.g., electric arc discharge, laser ablation, and direct mechanization processes) methods. Many examples are available in a large number of publications with a focus on the basic science and applications in engineering, in multiple fields (Palencia et al., 2015; Palencia et al., 2016a,b; Córdoba et al., 2017; Palencia et al., 2017a,b; Ealia and Saravanakumar, 2017; Palencia et al., 2018; Sanchez et al., 2018; Palencia et al., 2019; López et al., 2019; Villalobos et al., 2019a; Villalobos et al., 2019b). In addition, the combination of INPs with polymers is usually referred to as nanocomposites where the nanoparticle phase is dispersed in a polymeric phase (Fu et al., 2019; Omanović-Mikličanin et al., 2020), and core—shell nanomaterials where the nanoparticles are covered by a polymer (Chatterjee et al., 2014; Chiozzi and Rossi, 2020). Some examples and applications of nanoparticles and nanomaterials are summarized in Table 26.1.

At this point it is important to build an image of an INP in order to improve the understanding of them and their properties. An INP can be imagined as an

TABLE 26.1 Examples of nanoparticles and nanomaterials, and their synthesis routes and applications.

Inorganic core	Nanomaterial	Synthesis routes	Application	References
Ag	PVC–AgNPs	Chemical reduction	Storage of blood and hemoderivatives	Palencia et al. (2016a,b)
	AgNPs–AA	Chemical reduction	Antimicrobial agent	Córdoba et al. (2017)
	AgNPs	Chemical reduction	Studies in basic science	Sanchez et al. (2018)
	PVBNG–AgNPs	Chemical reduction	Membrane with antibiofouling properties	Palencia et al. (2018)
	PVBTMA–AgNPs	Chemical reduction	Molybdate retention and antimicrobial activity	Palencia et al. (2019)
Cu	CuNPs	Chemical reduction	Catalyst for "click" reaction	Sarkar et al. (2008)
	CuNPs	Chemical reduction	Plasmonic solar cell	Parvee et al. (2016)
	CuNPs	Chemical reduction	Studies in basic science	Suares-Cerda et al. (2017)
ZnO	ZnO–nanorod	Atomic layer deposition	Studies in basic science	López et al. (2019)
	ZnO–nanobars	Hydrothermal synthesis with NaOH	Studies in basic science	Villalobos et al. (2019b)
TiO_2	EPM–ZnONPs	Hydrothermal synthesis with NaOH	Nanoreinforcement of structural materials	Villalobos et al. (2019a)
	EPM–TiO_2NPs	Hydrothermal synthesis	Nanoreinforcement of structural materials	Villalobos et al. (2019a)
	TiO_2–nanotubes	Hydrothermal synthesis	Studies in basic science	Villalobos et al. (2019b)

AA, acrylic acid; *AgNPs*, silver nanoparticles; *CuNPs*, copper nanoparticles; *ERM*, epoxy resin matrix; *PVBNG*, poly(N-vinylbenzyl-N-methyl-D-glucamine); *PVBTMA*, poly(N-vinylbenzyl-N-triethylammonium chloride); *PVC*, poly(vinyl chloride); *TiO_2NPs*, titanium dioxide nanoparticles; *ZnONPs*, zinc oxide nanoparticles.

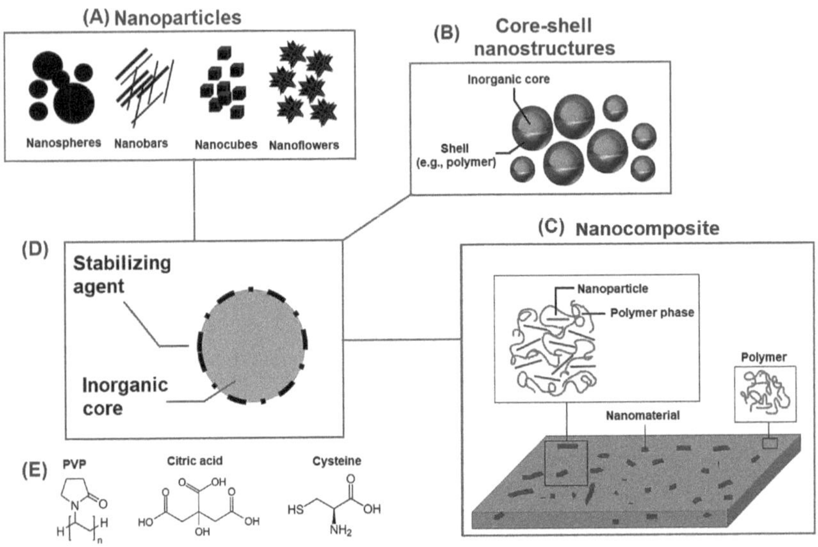

FIGURE 26.1 Different shapes of nanoparticles (A), and nanostructures: core–shell nanostructures (B) and polymer nanocomposites (C). Illustration of the main component of inorganic nanoparticles in colloidal state (D) and example of typical stabilizing or capping agents (E).

inorganic phase stabilized at the interfacial level (see Fig. 26.1). This stabilization is normally carried out through the use of organic molecules such as polymers and low-molecular-weight molecules, but it can also be the result of surface oxidation phenomena that give a surface charge to the particles, preventing their nucleation and further growth beyond the limits of the nanoscale (Palencia et al., 2015; Ealia and Saravanakumar, 2017; Guilguer-Casagrande and de Lima, 2019). Several examples of nanoparticles and nanomaterials, and their respective stabilizing agents, are summarized in Table 26.2.

26.1.1 Antibacterial inorganic nanoparticles: silver nanoparticles (AgNPs)

Antibacterial properties of INPs are associated with the ability to halt or inhibit bacterial growth (i.e., bactericide activity and bacteriostatic activity, respectively).

The main problem related to antimicrobial substances is the production of resistance, and in this context, INPs, like silver nanoparticles (AgNPs), are characterized as excellent antimicrobial agents for a wide spectrum of bacteria, including *S. aureus*, *E. coli*, *H. pylori*, *S. aeruginosa*, and *B. subtilis*, among others. Other examples of INPs with antibacterial activity are oxides of silver, titanium, iron, copper, nickel, zinc, calcium, aluminum, magnesium, and metals like gold and copper (Palencia et al., 2015; Kaur and Kaur, 2020).

TABLE 26.2 Examples of nanoparticles and nanomaterials, and their respective stabilizing agents.

Inorganic core	Types of material	Stabilizing agent	References
Ag	Nanocomposite	N-methyl-D-glucamine	Palencia et al. (2016a,b)
	Nanocomposite	Cysteine	Palencia et al. (2016a,b)
	Nanocomposite	Cysteamine	Palencia et al. (2016a,b)
	Nanoparticles	Acrylic acid	Córdoba et al. (2017)
	Nanoparticle	Sodium citrate	Sanchez et al. (2018)
	Nanocomposite	Acrylic acid	Palencia et al. (2018)
	Nanocomposite	Acrylic acid	Palencia et al. (2019)
Cu	Nanoparticles	Poly(N-vinyl-2-pyrrolidone)	Sarkar et al. (2008)
	Nanoparticles	Hydrazine	Parvee et al. (2016)
	Nanoparticles	Cyclodextrins	Suares-Cerda et al. (2017)
TiO_2	Nanocomposite	Epoxy resin matrix	Villalobos et al. (2019a)
	Nanoparticles	No identified	Villalobos et al. (2019b)
ZnO	Nanocomposite	Epoxy resin matrix	Villalobos et al. (2019a)
	Nanoparticles	No identified	Villalobos et al. (2019b)

AgNPs, Silver nanoparticles.

Factors describing the antibacterial activity of INPs are supported by the interaction between bacterial biomolecules and nanoparticles, characteristics of the bacteria, characteristics of the medium, and properties of the nanoparticles. Factors related to bacterial biomolecules and nanoparticle interaction include the ability to produce the denaturalization of proteins, induce changes in metabolic cycles by blocking of active sites, or the chemical transformation of chemical substances directly related to their metabolic functions; in addition, the disruption of membrane or intracellular structures, immobilization of nanoparticles on the cell surface, blocking of membrane channels, and others mechanisms are possible (Palencia et al., 2015, 2017a; Hoseinzadeh et al., 2017; Wang et al., 2017; Kaur and Kaur, 2020).

The main characteristics associated with bacteria are connected to the cell wall, bacterial growth rate, and biofilm formation. It has been suggested that a lower concentration of peptidoglycan in the cell wall could be associated with the susceptibility of bacteria to undergo disruption of the cell envelope; in addition, slow-growing bacteria have been identified as more susceptible to antibiotics and INPs. On the other hand, biofilms are an efficient mechanism for adaptation and survival in adverse conditions. During biofilm formation a set of biomolecules forming an extracellular matrix is secreted by the bacteria. Thus, extracellular DNA, exo-polysaccharides, proteins, and glycopeptides change the environmental conditions during biofilm formation, including surfaces and fluid medium. In this way, these components increase the viscosity, affecting the diffusion of INPs, or react with the nanoparticles by immobilizing them and limiting their accessibility to the bacterial cell (Palencia et al., 2015; Espinosa et al., 2020; Kaur and Kaur, 2020). This is strongly related to the effect of the medium. In general, the accessibility of INPs to bacterial cells is regulated by the chemical and physical barriers along the propagation pathway. Therefore, a medium with high viscosity will limit diffusion, as will a medium that strongly interacts with the particles (Palencia et al., 2015).

Factors associated with INPs include the chemical composition (e.g., AgNPs are expected to show higher antibacterial activity compared with hydroxyapatite nanoparticles due to the biocompatibility of the latter), the size and shape, which are two aspects described in the previous section, and concentration (Ferraz et al., 2004; Palencia et al., 2015).

26.2 Hydrogels for delivery of antibacterial inorganic nanoparticles

Biopolymer-based hydrogels can be made from multiple polymer systems. In this section, they will be used as a model to show their applications in the nanotechnology field (for details about hydrogels see Chapter 9).

Hydrogels can be used for the delivery of bioactive substances including antibiotics, nanoparticles, metabolites, hybrid drugs, etc. The most common mechanisms are passive diffusion, diffusion-controlled, swelling-controlled, and degradation-controlled release. The antibacterial activity of AgNPs was shown inSection 26.1.1. However, because INPs can be easily functionalized, they can act as nanocarriers for antibacterial drugs in order to achieve a synergistic effect. In this way, nanostructured antibacterial cocktails can be easily manufactured by immobilizing a drug on the surface of INPs. In this context, cysteamine is an amino-thiol used for the treatment of cystic fibrosis and cystinosis, and recently, its use as an antimicrobial and antibiofilm agent has been described. Thus, AgNPs can act as carriers on the surface cysteamine molecules, and can avoid biofilm formation to ease the bactericide activity for both components (Charrier et al., 2014; Devereux et al., 2015;

Palencia et al., 2017b). Other research has been directed at the combined use of INPs and antibiotics, such as streptomycin, gentamicin, erythromycin, chloramphenicol, kanamycin, neomycin, ciprofloxacin, norfloxacin, and ampicillin (Burygin et al., 2009; Prema and Thangapandiyan, 2013; Fayaz et al., 2010; Palencia et al., 2017b).

26.2.1 Methodological strategy and results

Cassava starch and poly(acrylic acid) (PAA) were used for the synthesis of hydrogel. Silver nitrate was used as a source of Ag^+ ions, which were reduced using sodium borohydride to obtain AgNPs in colloidal state which were stabilized separately using cysteamine and citric acid (AgNPs-cysteamine and AgNPs-citrate, respectively). AgNPs were characterized by elemental analysis, dynamic light scattering (DLS), and ultraviolet-visible spectroscopy in order to identify the band associated with surface plasmon resonance (SPR) which is characteristic of AgNPs (Palencia et al., 2017b).

On the other hand, hydrogel was synthesized by cross-linking of polymers by esterification of hydroxyl groups from starch and carboxylic acid groups from PAA (mass ratio 5:1 starch:PAA). Hydrogel was dried, milled, and sieved to obtain a particulate material. Typical structural characterization was performed by elemental analysis and infrared spectroscopy by attenuated total reflectance (ATR-FTIR).Functional characterization then was directed toward the determination of water absorption capacity by the gravimetric method. Hydrolysis of starch-PAA hydrogels was studied by batch experiments as a function of pH (pH: 5.0, 6.0, 7.0, 8.0, and 9.0) for 48 h. The hydrolysis percentage was gravimetrically determined. AgNPs were introduced into the hydrogel by flow of the dispersion as the material was swollen by immersion in the colloidal dispersion. Delivery experiments were carried out by swelling of hydrogel at maximal water absorption capacity using dispersions of AgNPs-citrate and AgNPs-cysteamine. Releases of AgNP-cysteamine and AgNP-citrate were monitored by titration with formaldehyde (i.e., released amine) and atomic absorption spectroscopy (i.e., released silver) (Palencia et al., 2017b).

As bacterial models, *E. coli* (ATCC 25922) and *S. aureus* (strain obtained from clinical samples) were used, these were grown using tryptic soy broth as the liquid phase. Later, inocula were standardized using McFarland scale (0.5 CFU/mL) with respect to the barium sulfate standards. Starch-PAA hydrogels loaded with AgNPs-cysteamine and AgNPs-citrate were placed in contact with previously sterilized culture medium (7 mL) containing 10 μL of a standardized bacterial dispersion. The samples were incubated at 37°C for 24 h. Additional experiments were analogous to previously described using PAA, cysteamine, citric acid, AgNPs-cysteamine, AgNPs-citrate, starch, hydrogel without AgNPs, and ampicillin. All experiments were performed in triplicate (Palencia et al., 2017b).

Details of the characterization results are described in Palencia et al. (2017b). The water absorption capacity of the starch-PAA hydrogel decreased compared to the PAA (i.e., ~ 4450% for the PAA vs. ~ 1979 for the hydrogel). Therefore, although the hydrophilicity decreased due to cross-linking and the subsequent formation of ester groups, the material continued to show hydrogel properties, a fact that was explained by the presence of −COOH from PAA and −OH groups from starch in the hydrogel structure. On the other hand, in these experiments, the presence of cysteamine on the surface of the AgNPs was easily determined by elemental analysis and the identification and quantification of the sulfur content. It was demonstrated that hydrolyzable hydrogels can be easily synthesized from PAA and cassava starch, which is a biopolymer obtained from tropical crops. In addition, potential drugs based on AgNPs-cysteamine with antibacterial properties were synthesized by chemical reduction of silver ions and cysteamine. Thus, antibacterial properties against *E. coli* and *S. aureus* were evidenced at the studied concentrations. In addition, it was proposed that antimicrobial activity could be modulated by control of the concentration during the swelling process (Palencia et al., 2017b).

26.3 Development of hydrogels for massive synthesis of AgNPs

The manufacture of colloidal particles in large quantities differs significantly from experiments carried out in the laboratory. These differences are not new, it is common that an experiment that works satisfactorily in the laboratory must be evaluated and redesigned when it is desired to scale it up. These differences lie mainly in the fact that the experiments at the laboratory level are carried out under controlled conditions, and the same control level in many cases cannot be guaranteed at an industrial level. The reasons are various and vary for each case, but normally these include difficulty in guaranteeing the homogeneity of the system, a decrease in the efficiency of the heat transfer and diffusion processes, a decrease in the quality of the reagents, and the need to decrease the costs in order to make the process economically viable.

In the synthesis of AgNPs in large-scale processes, it is normal to obtain a high polydispersity of sizes and shapes, and, as previously explained, as these distributions widen, the quality of the nanometric properties of interest decreases, and consequently, high coefficients of variation and a decrease in reproducibility are observed. To address this problem, post-synthesis separation methods have been designed, which, instead of being addressed under the conditions of synthesis, are focused on fractionation of nanoparticles in the function of their size. For the above, the creation of noncross-linked hydrodynamic hydrogels has been proposed and evaluated (Palencia et al., 2014a,b). These hydrogels can be obtained from polyelectrolyte solutions, and therefore, biopolymers such as alginate, chitosan, cellulose, and quaternized chitosan, as well as synthetic derivatives of biomolecules, with neutral, ionic, or ionizable groups, can be used for this purpose (Rivas et al., 2011; Palencia, 2015). To carry out these hydrogels, ultrafiltration systems are used, and the phenomenon of concentration polarization is simultaneously used for the same purpose. Concentration polarization is a concentration gradient that arises in pressure-driven membrane separation processes by the action of flow and driving force. When this phenomenon occurs in polymer solutions, the proximity of the chains causes the solution to behave like a gel, whose density increases as it approaches the surface of the membrane. Thus, the greater the polarization of the polymer concentration, the greater is the proximity of the chains and the lower is the space available for the passage of particles. Concentration polarization can be modified depending on operating variables such as pressure, stirring speed, and concentration. For these variables the effect is as follows: (1) if pressure is increased then the concentration polarization increases, (2) if stirring speed is decreased then the concentration polarization also is increased, and (3) when the experiment is performed at high polymer concentrations, the concentration polarization is increased (Palencia et al., 2016a,b). In conclusion, when the flow of nanoparticles passes through this hydrodynamic layer resulting in concentration polarization, larger particles will be retained in the upper part of the gel, while the smaller ones will be able to pass through the membrane, passing to the permeate stream. Subsequently, by modifying the operation variables, it is possible to collect a second larger fraction with a larger particle size, and so on (Palencia et al., 2014a,b).

26.4 Conclusions and remarks

Hydrogels are promising materials with great potential as multifunctional materials. They can be obtained from polymers derived from petroleum, from biopolymers, and from a combination of these. Their main characteristic is the ability to absorb water in a proportion higher than their equivalent mass. Although their use in biomedicine, agriculture, food, and pharmaceutical industries is extensive, they can be used for applications directed toward the release of antibacterial INPs, and also, they can be used in nanoparticle manufacturing processes through the creation of hydrodynamic gels. In this context, biopolymer-based hydrogels are eco-friendly materials from a functional point of view, but also, they are biodegradable materials, and as a consequence, it is possible to infer that their residuality is extremely low and in many cases it is expected to be negligible. Finally, it is important to emphasize that the combined use of biopolymers and nanoparticles allows the construction of multifunctional materials that are nanotechnologically designed and eco-friendly.

References

Burygin, G.L., Khlebtsov, B.N., Shantrokha, A.N., Dykman, L.A., Bogatyrev, V.A., Khlebstov, N.G., 2009. On the enhanced antibacterial activity of antibiotics mixed with gold nanoparticles. Nanoscale Res. Lett. 4 https://doi.org/10.1007/s11671-009-9316-8, 794-01.

Charrier, C., Rodger, C., Robertson, J., Kowalczuk, A., Shand, N., Frase-Pitt, D., Mercer, D., ONeil, D., 2014. Cysteamine (Lynovex®), a novel mucoactive antimicrobial & antibiofilm agent for the treatment of cystic fibrosis. Orphanet J. Rare Dis. 9 https://doi.org/10.1186/s13023-014-0189-2, 189-00.

Chatterjee, K., Sarkar, S., Jagajjanani, K., Paria, S., 2014. Core/shell nanoparticles in biomedical applications. Adv. Colloid Interface Sci. 209, 8–39. https://doi.org/10.1016/j.cis.2013.12.008.

Chiozi, V., Rossi, F., 2020. Inorganic–organic core/shell nanoparticles: progress and applications. Nanoscale Adv. 2, 5050–5105. https://doi.org/10.1039/D0NA00411A.

Cordoba, A., Palencia, M., Córdoba, A., Palencia, M., 2017. Development of nanostructured crosslinkers with antimicrobial properties for free-radical polymerization. J. Sci. Technol. Appl. 2, 54–64. https://doi.org/10.34294/j.jsta.17.2.14.

Devereux, G., Fraser-Pitt, D., Robertson, J., Devlin, E., Mercer, D., O'Neil, D., 2015. Cysteamine as a future intervention in cystic fibrosis against current and emerging pathogens: a patient-based ex vivo study confirming its antimicrobial and mucoactive potential in sputum. Ebiomedicine 2, 1507–1512. https://doi.org/10.1016/j.ebiom.2015.08.018.

Ealia, S.A.M., Saravanakumar, M.P., 2017. A review on the classification, characterisation, synthesis of nanoparticles and their application. IOP Conf. Ser. Mater. Sci. Eng. 263, 032019. https://doi.org/10.1088/1757-899X/263/3/032019.

Espinosa, A., Palencia, S., García, A., Palencia, M., 2020. Polymicrobial biofilms: fundamentals, diseases, and treatments - a review. J. Sci. Technol. Appl. 8, 4–54. https://doi.org/10.34294/j.jsta.20.8.54.

Fayaz, A.M., Balaji, K., Girilal, M., Tadav, R., Thangavelu, P., Venketesan, R., 2010. Biogenic syn-thesis of silver nanoparticles and their synergistic effect with antibiotics: a study against gram-positive and gram-negative bacteria. Nanomed. Nanotechnol. Biol. Med. 6, 103–109. https://doi.org/10.1016/j.nano.2009.04.006.

Ferraz, M.P., Monteiro, F.J., Manuel, C.M., 2004. Hydroxyapatite nanoparticles: a review of preparation methodologies. J. Appl. Biomater. Funct. Mater. 2, 74–80. https://doi.org/10.1177/228080000400200202.

Fu, S., Sun, Z., Li, Y., Hu, N., 2019. Some basic aspects of polymer nanocomposites: a critical review. Nano Mater. Sci. 1, 2–30. https://doi.org/10.1016/j.nanoms.2019.02.006.

Ghosh, C.K., 2015. Quantum effect on properties of nanomaterials. In: Sengupta, A., Sarkar, C. (Eds.), Introduction to Nanoengineering Materials, vol. 2015. Springer, pp. 73–111. https://doi.org/10.1007/978-3-662-47314-6_5.

Guilguer-Casagrande, M., de Lima, R., 2019. Synthesis of silver nanoparticles mediated by fungi: a review. Front. Bioeng. & Biotechnol. 7, 287. https://doi.org/10.3389/fbioe.2019.00287.

Hoseinzadeh, E., Makhdoumi, P., Taha, P., Hossini, H., Stelling, J., Kamal, M.A., Ashraf, G.M., 2017. A review on nano-antimicrobials: metal nanoparticles, methods and mechanisms. Curr. Drug Metabol. 18, 120–128. https://doi.org/10.2174/1389200217666161201111146.

Jiang, C.C., Cao, Y.K., Xiao, G.Y., Zhu, R.F., Lu, Y.P., 2017. A review on the application of inorganic nanoparticles in chemical surface coatings on metallic substrates. RSC Adv. 7, 7531–7539. https://doi.org/10.1039/C6RA25841G.

Kaur, R., Kaur, G., 2020. Review article on inorganic nanoparticles as antibacterial agents. Int. J. Res. Appl. Sci. Eng. Technol. 8, 944–948. https://doi.org/10.22214/ijraset.2020.29964.

Khan, I., Saeed, K., Khan, I., 2019. Nanoparticles: properties, applications and toxicities. Arab. J. Chem. 12, 908–931. https://doi.org/10.1016/j.arabjc.2017.05.011.

López, J., Castro, V., Solis, F., Gutierrez, C., Meléndrez, M., Martinez, E., Peréz, E., Fundora, A., 2019. Thermal treatment of ZnO seed layer into atomic layer deposition to improve crystal quality of ZnO nanorod arrays. J. Sci. Technol. Appl. 6, 108–117. https://doi.org/10.34294/j.jsta.19.6.45.

Meena, J., Gupta, A., Ahuja, R., Singh, M., Bhaskar, S., Panda, A., 2020. Inorganic nanoparticles for natural product delivery: a review. Environ. Chem. Lett. 18, 2107–2118. https://doi.org/10.1007/s10311-020-01061-2.

Omanović-Mikličanin, E., Badnjević, A., Kazlagić, A., Hajlovac, M., 2020. Nanocomposites: a brief review. Health Technol. 10, 51–59. https://doi.org/10.1007/s12553-019-00380-x.

Palencia, M., 2015. Liquid-phase polymer-based retention: theory, modeling, and application for the removal of pollutant inorganic ions. J. Chem. 965624. https://doi.org/10.1155/2015/965624.

Palencia, M., Berrio, M.E., Melendrez, M.F., 2016a. Nanostructured polymer composites with potential applications into the storage of blood and hemoderivates. J. Sci. Technol. Appl. 1, 4–14. https://doi.org/10.34294/j.jsta.16.1.1.

Palencia, M., Berrio, M.E., Palencia, S.L., 2017a. Effect of capping agent and diffusivity of different silver nanoparticles on their antibacterial properties. J. Nanosci. Nanotechnol. 17, 1–8. https://doi.org/10.1166/jnn.2017.13850.

Palencia, M., Córdoba, A., Rivas, B.L., 2014a. Concentration–polarization effect of poly(sodium styrene sulfonate) on size distribution of colloidal silver nanoparticles during diafiltration experiments. Colloid Polym. Sci. 292, 619–626. https://doi.org/10.1007/s00396-013-3096-5.

Palencia, M., Córdoba, A., Vera, M., 2016b. Membrane technology and chemistry. In: Visakh, P.M., Nazarenko, O. (Eds.), Nanostructured Polymer Membranes, vol. 1, pp. 27–54.

Palencia, M., Córdoba, A., Melendrez, M.F., 2019. Nanocomposites based on cationic polyelectrolytes and silver nanoparticles: synthesis, characterization, molybdate retention and antimicrobial activity. Arab. J. Chem. 12, 825–834. https://doi.org/10.1016/j.arabjc.2017.11.020.

Palencia, M., Córdoba, A., Combatt, E., 2018. Synthesis and characterization of biofouling-resistant nanocomposites based on glucamine polymers and silver/acrylic acid nanoparticles. Polym. Bull. 75, 4555–4569. https://doi.org/10.1007/s00289-018-2286-1.

Palencia, M., Lerma, T.A., Combatt, E.M., 2017b. Hydrogels based in cassava starch with anti-bacterial activity for controlled release of cysteamine-silver nanostructured agents. Curr. Chem. Biol. 11, 28–35. https://doi.org/10.2174/2212796810666161108152319.

Palencia, M., Rivas, B.L., Valle, H., 2014b. Size separation of silver nanoparticles by dead-end ultrafiltration: description of fouling mechanism by pore blocking model. J. Membr. Sci. 455, 7–14. https://doi.org/10.1016/j.memsci.2013.12.051.

Palencia, S.L., Buelvas, A.M., Palencia, M.S., 2015. Interaction mechanisms of inorganic nanoparticles and biomolecular systems of microorganisms. Curr. Chem. Biol. 9, 11–23. https://doi.org/10.2174/2212796809666151022201811.

Parvee, F., Sannkki, B., Mandke, M.V., Pathan, H.M., 2016. Copper nanoparticles: synthesis methods and its light harvesting performance. Sol. Energy Mater. Sol. Cell. 144, 371–382. https://doi.org/10.1016/j.solmat.2015.08.033.

Prema, P., Thangapandiyan, S., 2013. In-vitro antibacterial activity of gold nanoparticles capped with polysac-charide stabilizing agents. Int. J. Pharm. Pharmaceut. Sci. 5, 310–311.

Rivas, B.L., Pereira, E.D., Palencia, M., Sanchez, J., 2011. Water-soluble functional polymers in conjunction with membranes to remove pollutant ions from aqueous solutions. Prog. Polym. Sci. 36, 294–322. https://doi.org/10.1016/j.progpolymsci.2010.11.001.

Sanchez, Y., García-Quintero, A., Palencia, M., Sanchez, Y., García, A., Palencia, M., 2018. Determination and distribution of size of inorganic particles by spectral deconvolution of surface plasmon resonance. J. Sci. Technol. Appl. 5, 45–54. https://doi.org/10.34294/j.jsta.18.5.34.

Sarkar, A., Mukherjee, T., Kapoor, S., 2008. PVP-stabilized copper Nanoparticles: a reusable catalyst for "click" reaction between terminal alkynes and azides in nonaqueous solvents. J. Phys. Chem. 112, 3334–3340. https://doi.org/10.1021/jp077603i.

Suares-Cerda, J., Espinoza-Gómez, H., Alonso-Nuñez, G., Rivero, I.A., Gochi-Ponce, Y., Flores-López, L.Z., 2017. A green synthesis of copper nanoparticles using native cyclodextrins as stabilizing agents. J. Saudi Chem. Soc. 21, 341–348. https://doi.org/10.1016/j.jscs.2016.10.005.

Villalobos, R., Medina, C., Canales, C., Melendrez, M., Flores, P., 2019a. Development of nanocomposite materials with thermostable matrix from nanoreinforcement of titanium dioxide (TNPS and TNTS) and zinc oxide (ZNPS and ZNBS) on epoxy resin matrix. J. Sci. Technol. Appl. 6, 65–78. https://doi.org/10.34294/j.jsta.19.6.42.

Villalobos, R., Rojas, D., Diaz, A., Ramirez, J., Melendrez, M., Flores, P., 2019b. Synthesis and characterization of 1D metal oxide nano-structures: titanium dioxide nanotubes (TNTS) and zinc oxide nanobars (ZNBS). J. Sci. Technol. Appl. 6, 79–95. https://doi.org/10.34294/j.jsta.19.6.43.

Wang, L., Hu, C., Shao, L., 2017. The antimicrobial activity of nanoparticles: present situation and prospects for the future. Int. J. Nanomed. 12, 1227–1249. https://doi.org/10.2147/IJN.S121956.

Chapter 27

Polymeric supports for grown of beneficial microorganisms in agriculture

27.1 Sustainable agriculture: food security and conservation of the environment

Sustainability is a quality of ecosystems, describing how biological entities remain productive over time. Biological entities should be understood as simple biological systems (i.e., individuals and small groups forming a small population in the ecosystem), but also, complex biological systems (i.e., communities, the ecosystem itself, and a set of possible interactions between external biotic and abiotic factors with the ecosystems, but also, interactions between two ecosystems). Therefore, sustainability describes the balance of one or more species with the ecosystemic resources of its environment (Jelti et al., 2020; Liu et al., 2020; Fehmi et al., 2021). On the other hand, sustainable development is usually defined as a development that, satisfying current needs, does not affect the ability of future generations to satisfy their own needs (Mourad et al., 2020). However, this concept is vague, and even ambiguous, from a practical approach, as it is unclear about how that goal is implemented and achieved (Lichtfouse et al., 2009).

Here, we introduce a concept of sustainable development based on the generic concept, with the intention of giving clarity, scope, and ease of understanding. Thus, sustainable development is the eco-friendly growth of any entity, biological or not, whose progress does not affect the capacity of existing ecosystem services and dynamics, and consequently, does not affect their availability for use by future generations. A biological entity can be an individual, a population, or a community, since at the ecosystem level the effect of the individual on the environment is usually relatively small, it is not appropriate to restrict the term to an individual entity. Meanwhile, a nonbiological entity can be an industrial process, an industry, a city, or even a policy, legislation, or line of thought (Jelti et al., 2020; López-Serrano et al., 2020; Liu et al., 2020; Mourad et al., 2020; Rocchi et al., 2020; Serebrennikov et al., 2020; Ungureanu et al., 2020; Fehmi et al., 2021; Ferraz et al., 2021).

Eco-friendly Functional Polymers
https://doi.org/10.1016/B978-0-12-821842-6.00021-X
Copyright © 2021 Elsevier Inc. All rights reserved.

Sustainable agriculture is understood by some authors as a set of agriculture management strategies for satisfactory societal development, in terms of food quality and/or environment protection (Serebrennikov et al., 2020; Ungureanu et al., 2020; Ferraz et al., 2021), whereas for others it is the ability of agricultural systems to maintain crop productivity over the long term, or the capacity of agricultural systems to adapt to future changes (López-Serrano et al., 2020; Rocchi et al., 2020).

Sustainable agriculture requirements are based on three approaches or dimensions: (1) the environmental, (2) economical, and (3) social dimensions, including energy and mass aspects, from a global point of view, that is, taking into consideration the acquisition of inputs that enter the process, the generation of products, and the emission of waste (Fehmi et al., 2021; Ferraz et al., 2021). In other words, sustainable agriculture must be, in the short, medium, and long term, economically viable (i.e., production costs must be lower than marketing income), environmentally safe (e.g., avoiding soil physical, chemical, and biological degradation, protecting surface water and groundwater, without generating emissions and waste that cannot be easily assimilated by ecosystems, providing safe food, using renewable energies, and with the ability to have less impact on climate change) and socially just (e.g., appropriate approaches in terms of wealth distribution, accessibility to food, contributing to rural employment and development, adequate offering of rural landscape for urban people, etc.). Although several indicators have been developed to evaluate and characterize agricultural sustainability, there is no consensus regarding the criteria or parameters to measure it (Ungureanu et al., 2020; Fehmi et al., 2021; Ferraz et al., 2021).

On the other hand, the current and future challenges to agriculture are circumscribed in its inherent purpose (i.e., food production), and its relationship with the environment (i.e., environmental impact). These have been described to be, mainly: (1) achieving increased crop production without an expansion of the agricultural frontier, which would imply the transformation of natural systems into agricultural production systems; (2) achieving a decrease and efficient use of fertilizers which are released into the environment. The excessive use of fertilizers is related to processes of salinization of soils and eutrophication of waters; (3) minimizing and implementing efficient use of pesticides. When pesticides are released into the environment in an uncontrolled way they impact on the fauna and flora. In addition, they can be mobilized to areas other than the application site; (4) making an efficient use of water resources; (5) implementing resilient systems to new scenarios triggered by climate change; and (6) developing inclusive and equitable production systems that offer opportunities for growth in rural areas. Other challenges can be identified, however, for more details it is recommended to consult the aforementioned (Combatt et al., 2017a,b; García et al., 2018; Lerma et al., 2018; Mora et al., 2018; Restrepo et al., 2018; Calicioglu et al., 2019; Ramakrishna et al., 2019; Veeck et al., 2020).

27.2 Plant growth-promoting microorganisms (PGPMs)

Soil is a heterogeneous and dynamic system, characterized by having a biotic component and an abiotic component. In addition, its composition includes an organic phase formed by living organic matter (e.g., microorganisms, roots of plants, insects, etc.), fresh organic matter (e.g., leaves and biomass), humified organic matter (e.g., humin, fulvic and humic acids), soluble organic matter (e.g., water-soluble organic compounds of different origins), but also, it includes an inorganic phase (e.g., clays, salts, water). The processes occurring in the soil are very varied, ranging from physical processes, like diffusion, and chemical process, like redox reactions, to geological and bio-geochemical processes. In terms of properties, their description is also complex, because simultaneously soils are pH buffer systems, colloidal systems, ion exchangers, porous systems, etc. In addition, they are the substrate for the growth of plants, and the natural reservoir of nutrients for their growth (Combatt et al., 2016, 2017; McKenzie et al., 2018).

Soil microorganisms play an important role in soil dynamics, in particular with regard to the regulation of soil fertility, the nutrient cycle, symbiotic processes with plants, etc. In this context, it is possible to highlight a very particular type of microorganisms called, genetically, plant growth-promoting microorganisms (PGPMs), a name which refers to a set of diverse microorganisms, which include bacteria and rhizobacteria (Ramakrishna et al., 2019; Denaya et al., 2021).

Multiple studies have described that PGPMs are capable of acting on soil dynamic processes and physiological aspects of plants in order to attenuate the adverse effects of biotic and abiotic stress. In this way, it has been described that PGPMs improve the efficiency of water use, fix atmospheric nitrogen, promote the dissolution of minerals from the solid phase of the soil, regulate the photosynthesis process, and improve the hydraulic conductivity of the roots, among other relevant contributions to the growth of plants. In addition, colonization by PGPMs is related to changes in the metabolism, signaling, and hormonal homeostasis of plants. It is currently well known that PGPMs, as a result of their metabolic processes, produce phytohormones, for example, auxins, which influence and improve plant growth (Garcés et al., 2017; Berrio et al., 2018; Khan et al., 2020; Rodríguez et al., 2020). On the other hand, the adverse effects on the quality of the soil and natural aquifers of the use of chemical fertilizers, pesticides, and agrochemicals lead to a significant impact on the environmental quality, the health of soils, and the ecosystem services of agricultural and surrounding areas. As a substitute for chemical fertilizers, PGPMs emerge as an economically viable and sustainable technology (Mushtaq et al., 2021; Sabarinathan et al., 2021). This is evidenced by the large number of scientific publications associated with this field (Bernard, 2012; Umesha et al., 2018; Debnath et al., 2019; Mitra et al., 2019; Naik et al., 2019; Pandey et al., 2019; Ruiu, 2020; Shilev, 2020).

27.3 Polymer-based bacterial inocula

In the technology of bacterial inocula, the application and survival of the microorganisms is a critical stage in which adequate colonization of the rhizosphere must be achieved. However, this stage is limited by competition with autochthonous microorganisms; therefore, the environmental conditions must not only be appropriate for the growth of plants, but also for the growth of exogenous PGPMs, which is a fact that is not necessarily compatible for the wide range of agricultural conditions. It is well known that the effectiveness of bacterial inocula can be very limited in acidic, saline, nutrient-poor soils or with a deficit in their hydric conditions (Mould et al., 2005; Garcés et al., 2017). In addition, the scarcity of soils and their continuous degradation, as well as the expansion of crops to areas not optimal for agriculture, due to the need to achieve a constant increase in terms of food production, in many cases, make it difficult to use PGPMs as an agricultural supply (Perrotta et al., 2017; Poursat et al., 2019; Nagymate et al., 2020).

There are different methods for the application of microorganisms to the plant (e.g., inoculation of the seed, construction of beds on the ground, foliar spraying, and root immersion) (Mould et al., 2005). One of the most appropriate techniques for the introduction of microorganisms in soils is coating of the seeds with the appropriate bacteria; in this case, the bacteria need to have adequate field conditions to achieve adherence to the seed and constitute a numerically viable population. However, although the immersion technique is appropriate for seeds, it is not very viable for other reproduction techniques such as asexual multiplication by cuttings (Mould et al., 2005; Garcés et al., 2017).

Polymers for agricultural applications include different types of petroleum-derived polymers, biopolymers, and mixtures of these, such as hydrogels, polyelectrolytes, polymer–clay hybrid materials, etc. (see Chapter 12).

In particular, hydrogels are absorbent materials able to absorb high amounts of water in a relatively short time, and they can maintain a swollen state even under the application of weak pressure. These materials are used in areas such as agriculture, water treatment, medicine, etc. (see Chapter 9). In agriculture, for example, the use of hydrogels has been reported for the controlled release of water and moisture conservation under conditions of low soil water potential (Neethu et al., 2018; Amir et al., 2020; Martino et al., 2021). From a chemical point of view, hydrogels are cross-linked or difficult-dissolving macromolecules, with a large number of polar groups and whose structural composition can be controlled for modulation of their physical and chemical properties. Consequently, the synthesis of biodegradable hydrogels is possible, which is characterized by having functional groups in their structure that can be easily hydrolyzed by the action of microorganisms, and in this way, these materials are excellent candidates for the construction of artificial systems of growth and adaptation of PGPMs (Palencia et al., 2016; Sukriti et al., 2017; Zhang et al., 2018). Examples of hydrogels used as support of PGPMs are shown in Table 27.1. To exemplify the use of hydrogels in agriculture, we focus on the development of polymer-based bacterial inocula for the administration of PGPMs.

TABLE 27.1 Examples of hydrogels used as support of PGPMs.

Polymer matrix	Microorganisms	References
Carboxymethyl cellulose + corn starch	Diazotrophic bacteria, *Gluconacetobacter diazotrophicus*, *H. rubrisubalbicans*, *Azospirillum amazonense*, *Burkholderia tropica*	Flores da Silva et al. (2012)
Hydroxy-propylmethyl cellulose	*Azotobacter chroococcum*	Bashan et al. (2016)
Alginate	*Pseudomonas putida*, *Bacillus subtilis*, *B. megaterium*, *Azospirillum lipoferum*, *Beauveria bassiana*	Bashan et al. (2016)
Poly(vinyl alcohol) nanofibers	*Pantoea agglomerans*, *Burkholderia caribensis*	De Gregorio et al. (2017)
Starch + glucose	*Azotobacter chroococcum*	Garcés et al. (2017)
Poly(vinylpyrolidone) + poly(ethylene glycol) + gum arabic + alginate	*Pseudomonas fluorescens*	Biradar and Santhosh (2018)
Chitosan + starch	*Azospirillum brasilense*	Pérez et al. (2018)
Alginate + starch	*Rhodopseudomonas palustris*	Rohman et al. (2020)
Poly(acrylic acid) + cellulose	*Pseudomonas*	Chaín et al. (2020)
Cross-linked chitosan	Microalgae (e.g., *Lobosphaera*)	Vasilieva et al. (2020)
Bentonite + poly(acrylic acid) + sorbitol polyester	*Azotobacter chroococcum*	Lerma et al. (2020)

27.3.1 Development of bacterial inocula from bio-based hydrogels: a methodological approach

The experiments on the development of bacterial inocula based on polymeric hydrogels contain at least three steps: (1) the synthesis and characterization of the hydrogel, (2) preparation of the bacterial inoculum, and (3) evaluation of the inoculum.

The synthesis and characterization of the hydrogel is highly dependent on the raw materials. By way of illustration, the preparation of hydrogels based on polyesters of sorbitol and citric acid, and hydrogels based on starch and PDA (or potato-dextrose-agar) are used as models. In the first case, for the polyesters of sorbitol and citric acid, the hydrogels are prepared in the absence of water, by melting the sorbitol at 95°C, and the subsequent addition of the sorbitol to melted sorbitol. Heating is carried out for 3 h. The end of the process involves washing with distilled water, filtering, and drying at 50°C for 4 h. The percentage yield can be easily determined by gravimetric analysis, and the characterization is directed toward monitoring of the changes associated with functional groups (i.e., carboxylic acid, hydroxyl, and ester). Characterization can be easily carried out by elemental analysis and infrared spectroscopy. However, additional techniques also can be employed, such as nuclear magnetic resonance, differential scanning calorimetry, thermogravimetric analysis, etc. For hydrogels determination of the degree of swelling can be easily determined by gravimetric analysis. On the other hand, for the synthesis of hydrogels based on starch and PDA, a starch solution is prepared and mixed with dextrose and agar. Subsequently, they are heated to 40°C for 5 h. The characterization techniques are the same as those previously indicated for citric acid and sorbitol polyesters (Garcés et al., 2017).

In the second stage, the preparation of the inocula must be carried out, for which the PGPMs must be clearly identified. In this case, we will illustrate the methodology with *Azotobacter chroococcum*. For this microorganism, the standard growth medium is Mueller Hinton Agar. Its growth is carried out for 24 h at 30°C. These bacteria are highly negative, therefore, the Gram stain is a quick, simple, and useful technique for observing and characterizing them. After, microorganisms are cultured in a liquid medium (e.g., soybean trypticase), and the colony-forming units are standardized using the McFarland scale. Later, hydrogels are inoculated into the material; for that, several methods can be used: by immersing of dry material in the liquid medium, by injecting of the culture medium into the swollen hydrogel, or by adding small drops to the material. Each of these procedures has its own peculiarities, advantages, and disadvantages, and the best selection will be linked to the response of the microorganism in the stage of reactivation. Since the medium is a suitable environment for their growth, desiccation is a common stage for the preservation of the inoculum. Various studies must be carried out in order to identify the best growing conditions, including analysis of the effects of pH, ionic strength, incorporation of nutrients, etc., on growing and reactivation (Garcés et al., 2017).

Finally, it must be verified that the microorganisms can be metabolically reactivated, in a simple way and under the conditions of use. In this way, microorganisms, when they leave the state of latency and activate their metabolism, fulfill their function. At this point, the reactivation is usually evaluated in vitro and at the greenhouse level, with plants as biological indicators of the effectiveness of the inoculum (Garcés et al., 2017). Fig. 27.1 shows, as an example, photographs of the evaluation process of the in vitro reactivation of microorganisms for each of the above-mentioned hydrogels.

27.4 Conclusions and remarks

Bacterial inocula are a promising alternative for the substitution of agrochemicals. They are of biological origin, with low cost, and do not generate pollution. However, their effectiveness is limited as, although they show excellent results in controlled laboratory conditions and in greenhouse experiments, at the field level, the results are not as satisfactory. These results with low efficiency have been associated with factors specific to the environment in which the PGPMs must be developed. Normally, laboratory-level experiments do not include competitive effects, nor possible interactions with other microorganisms. Additionally, in the case of liquid inocula, these are not exempt from losses due to washing. Clearly, biopolymer-based systems

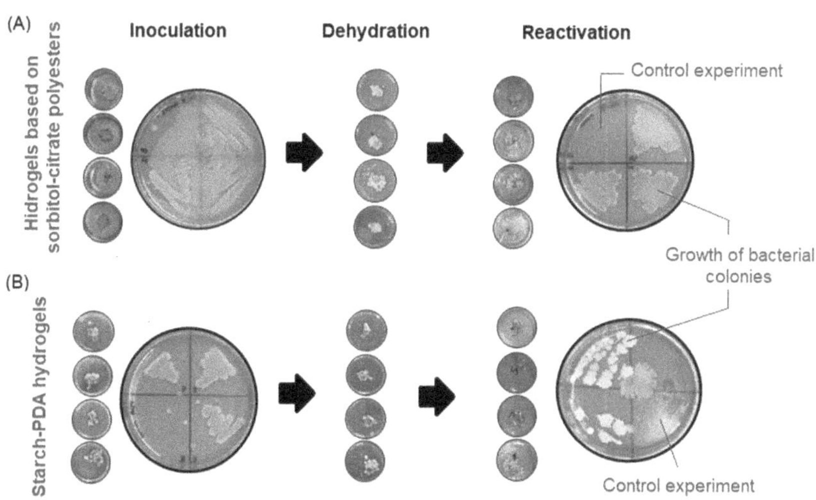

FIGURE 27.1 Photographs of the evaluation process of the in vitro reactivation of *Azotobacter chroococcum* in hydrogels based on citrate-sorbitol polyesters (A) and starch-PDA (B).

offer certain advantages over liquid inocula. Their main advantages are, among others, to be a reservoir of water, a source of metabolic energy, and a protective matrix allowing microorganisms to grow inside them, but also, they can be manufactured from biodegradable materials by means of relatively simple synthetic processes. Thus, although the technology of bacterial inocula is not new, it still has to overcome multiple challenges to achieve the substitution of conventional chemical fertilizers.

References

Amir, M., Nishat, N., Jadoun, S., Zaid, M., 2020. Polysaccharide based superabsorbent hydrogels and their methods of synthesis: a review. Carbohydr. Polym. Technol. & Appl. 1, 100014. https://doi.org/10.1016/j.carpta.2020.100014.

Bashan, Y., de-Bashan, L.E., Prabhu, S.R., 2016. Superior polymeric formulations and emerging innovative products of bacterial inoculants for sustainable agriculture and the environment. In: Singh, H.B., et al. (Eds.), Agriculturally Important Microorganisms Commercialization and Regulatory Requirements in Asia. Springer, pp. 15–46. https://doi.org/10.1007/978-981-10-2576-1_2.

Bernard, G., 2012. Plant growth-promoting bacteria: mechanisms and applications. Sci. Tech. Rep. 963401. https://doi.org/10.6064/2012/963401.

Berrio, M., Palencia, S., Lerma, T., Mora, M., 2018. Bacterial colonization modelling on soil particles: effect of humic acids on the formation of nitrogen-fixing bacteria biofilms. J. Sci. Technol. Appl. 5, 33–44. https://doi.org/10.34294/j.jsta.18.5.33.

Biradar, B.J., Santhosh, G.P., 2018. Role of polymeric additives in formulation, shelflife and bioefficacy of liquid inoculant of *Pseudomonas fluorescens*. Indian J. Pure & Appl. Biosci. 6, 123–133. https://doi.org/10.18782/2320-7051.6822.

Calicioglu, O., Flammini, A., Bracco, S., Bellu, L., Sims, R., 2019. The future challenges of food and agriculture: an integrated analysis of trends and solutions. Sustainability 11, 222. https://doi.org/10.3390/su11010222.

Chaín, J.M., Tubert, E., Graciano, C., Castagno, L.N., Recchi, M., et al., 2020. Growth promotion and protection from drought in *Eucalyptus grandis* seedlings inoculated with beneficial bacteria embedded in a superabsorbent polymer. Sci. Rep. 10, 1–17. https://doi.org/10.1038/s41598-020-75212-4.

Combatt, E., Palacio, D., Mercado, J., 2016. Evaluation of capacity for the extraction of Ca^{2+} and Mg^{2+} by the use of four extracting solutions. J. Sci. Technol. Appl. 1, 70–76. https://doi.org/10.34294/j.jsta.16.1.6.

Combatt, E., Mercado, J., Begambre, A., Kerguelen, L., Borja, K., 2017a. Chloride solutions in the extraction of calcium and magnesium from soils with alkaline characteristics. J. Sci. Technol. Appl. 2, 4–12. https://doi.org/10.34294/j.jsta.17.2.10.

Combatt, E., Polo, V., Racero, L., Cantero, M., 2017b. Process of soil degradation as a result of agricultural activities of different zones in the Monteria municipality, Department of Cordoba-Colombia. J. Sci. Technol. Appl. 3, 5–14. https://doi.org/10.34294/j.jsta.17.3.20.

Debnath, S., Rawat, D., Mukherjee, A.K., Adhikary, S., Kundu, R., 2019. Applications and constraints of plant beneficial microorganisms in agriculture. Biostimulants Plant Sci. 1–25. https://doi.org/10.5772/intechopen.89190.

De Gregorio, P.R., Michavela, G., Ricciardi, L., de Souza, C., Pomares, M.F., et al., 2017. Beneficial rhizobacteria immobilized in nanofibers for potential application as soybean seed bioinoculants. PLoS One 12, e0176930. https://doi.org/10.1371/journal.pone.0176930.

Denaya, S., Yulianti, R., Pambudi, A., Effendi, Y., 2021. Novel microbial consortium formulation as plant growth promoting bacteria (PGPB) agent. IOP Conf. Ser. Earth Environ. Sci. 637, 012030. https://doi.org/10.1088/1755-1315/637/1/012030.

Fehmi, H., Hunt, D., Rogers, C.D., 2021. Exploring urban sustainability understanding and behaviour: a systematic review towards a conceptual framework. Sustainability 13, 1139. https://doi.org/10.3390/su13031139.

Ferraz, D., Falguera, F.P.S., Mariano, E.B., Hartmann, D., 2021. Linking economic complexity, diversification, and industrial policy with sustainable development: a structured literature review. Sustainability 13, 1265. https://doi.org/10.3390/su13031265.

Flores da Silva, M., de Souza, C., Jansen, P., Ribeiro, G., Gouvea, N., et al., 2012. Survival of endophytic bacteria in polymer-based inoculants and efficiency of their application to sugarcane. Plant Soil 356, 231–243. https://doi.org/10.1007/s11104-012-1242-3.

Garcés, V., Palencia, M., Combatt, E., 2017. Development of bacterial inoculums based on biodegradable hydrogels for agricultural applications. J. Sci. Technol. Appl. 2, 13–23. https://doi.org/10.34294/j.jsta.17.2.11.

Garcia, A., Combatt, E., Palencia, M., 2018. Structural study of humin and its interaction with humic acids by Fourier-transform mid-infrared spectroscopy. J. Sci. Technol. Appl. 4, 28–39. https://doi.org/10.34294/j.jsta.18.4.28.

Jelti, F., Allouhi, A., Büker, M.S., Saadani, R., Jamil, A., 2020. Renewable power generation: a supply chain perspective. Sustainability 12, 1271. https://doi.org/10.3390/su13031271.

Khan, N., Bano, A., Curá, J.A., 2020. Role of benefical microorganisms and salicylic acid in improving rainfed agriculture and future food safety. Microorganisms 8, 1018. https://doi.org/10.3390/microorganisms8071018, 2–22.

Lerma, T., Combatt, E., Palencia, M., 2018. Soil-mimicking hybrid composites based on clay, polymers and nitrogen-fixing bacteria for the development of remediation systems of degraded soil. J. Sci. Technol. Appl. 4, 17–27. https://doi.org/10.34294/j.jsta.18.4.27.

Lerma, T., Garces, V., Palencia, M., 2020. Novel multi- and bio-functional hybrid polymer hydrogels based on bentonite-poly(acrylic acid) composites and sorbitol polyesters: structural and functional characterization. Eur. Polym. J. 128, 109627. https://doi.org/10.1016/j.eurpolymj.2020.109627.

López-Serrano, M.J., Velasco-Muños, J., Aznar-Sánchez, J.A., Román-Sánchez, I.M., 2020. Sustainable use of wastewater in agriculture: a bibliometric analysis of worldwide research. Sustainability 12, 8948. https://doi.org/10.3390/su12218948.

Lichtfouse, E., Mireille, N., Debaeke, P., Souchere, V., Alberola, C., Menassieu, J., 2009. Agronomy for sustainable agriculture: a review. Agron. Sustain. Dev. 29, 1–6. https://doi.org/10.1007/978-90-481-2666-8_1.

Liu, F., Lai, K.H., Cai, W., 2020. Responsible production for sustainability: concept analysis and bibliometric review. Sustainability 12, 1275. https://doi.org/10.3390/su13031275.

Martino, A., Khan, Y., Durpekova, S., Sedlarik, V., Elich, O., Cechmankova, J., 2021. Ecofriendly renewable hydrogels based on whey protein and for slow release of fertilizers and soil conditioning. J. Clean. Prod. 285, 124848. https://doi.org/10.1016/j.jclepro.2020.124848.

McKenzie, V.J., Kueneman, J.G., Harris, R.N., 2018. Probiotics as a tool for disease mitigation in wildlife: insights from food production and medicine. Ann. N. Y. Acad. Sci. 1–10. https://doi.org/10.1111/nyas.13617.

Mitra, D., Andjelkovic, A., Panneerselvam, P., Manisha, Senapati, A., et al., 2019. Plant growth promoting microorganisms (PGPMs) helping in sustainable agriculture: current perspective. Int. J. Agric. Sci. Vet. Med. 7, 50–74.

Mora, M., Palencia, M., Combatt, E., 2018. Effect of dissolved organic matter and humic substances on transport of ions and low molecular weight molecules by liquid membranes. J. Sci. Technol. Appl. 4, 40–48. https://doi.org/10.34294/j.jsta.18.4.29.

Mould, F.L., Kliem, K.E., Mauricio, R.M., 2005. In vitro microbial inoculum: a review of its function and properties. Anim. Feed Sci. Technol. 123–124, 31–50. https://doi.org/10.1016/j.anifeedsci.2005.04.028.

Mourad, K., Hasan, S., Avery, H., 2020. The role of citizen science in sustainable agriculture. Sustainability 12, 10375. https://doi.org/10.3390/su122410375.

Mushtaq, N., Faizan, S., Hussain, A., 2021. Role of microorganisms as biofertilizer. In: Microbiota and Biofertilizers. Springer, Cham, pp. 83–98. https://doi.org/10.1007/978-3-030-48771-3_6.

Naik, K., Mishra, S., Srichandan, H., Kumar, P., Kumar, P., 2019. Plant growth promoting microbes: potential link to sustainable agriculture and environment. Biocatal. & Agric. Biotechnol. 21, 101326. https://doi.org/10.1016/j.bcab.2019.101326.

Nagymate, Z., Jurecska, L., Romsics, C., Toth, F., Bodai, V., et al., 2020. Preparation and characterization of site-specific dechlorinating microbial inocula capable of complete dechlorination enriched in anaerobic microcosms amended with clay mineral. World J. Microbiol. Biotechnol. 36 https://doi.org/10.1007/s11274-020-2806-7.

Neethu, T.M., Kumar, P., Kaswala, A., 2018. Prospects and applications of hydrogel technology in agriculture. Int. J. Curr. Microb. & Appl. Sci. 7, 3155–3162. https://doi.org/10.20546/ijcmas.2018.705.369.

Pandey, A., Tripathi, A., Srivastava, P., Kumar, K., Dikshit, A., 2019. Plant growth-promoting microorganisms in sustainable agriculture. In: Kumar, A., Kishore, A., Kumar, K. (Eds.), Role of Plant Growth Promoting Microorganisms in Sustainable Agriculture and Nanotechnology. Elsevier, pp. 1–19. https://doi.org/10.1016/B978-0-12-817004-5.00001-4.

Palencia, M., Lerma, T., Córdoba, A., 2016. Polyurethanes with boron retention properties for the development of agricultural fertilization smart systems. J. Sci. Technol. Appl. 1, 39–52. https://doi.org/10.34294/j.jsta.16.1.4.

Pérez, J.J., Francois, N.J., Maroniche, G.A., Borrajo, M.P., Pereyra, M.A., Creus, C.M., 2018. A novel, green, low-cost chitosan-starch hydrogel as potential delivery system for plant growth-promoting bacteria. Carbohydr. Polym. 1–22. https://doi.org/10.1016/j.carbpol.2018.07.084.

Perrotta, A.R., Kumaraswarmy, R., Bastidas-Oyanedel, J.R., Alm, E.J., Rodríguez, J., 2017. Inoculum composition determines microbial community and function in an anaerobic sequential batch reactor. PLoS One 12, e0171369. https://doi.org/10.1371/journal.pone.0171369.

Poursat, B.A.J., van Spanning, R., de Voogt, P., Parsons, J.R., 2019. Implications of microbial adaptation for the assessment of environmental persistence of chemicals. Crit. Rev. Environ. Sci. Technol. 49, 2220–2255. https://doi.org/10.1080/10643389.2019.1607687.

Ramakrishna, W., Yadav, R., Li, K., 2019. Plant growth promoting bacteria in agricultura: two sides of coin. Appl. Soil Ecol. 138, 10–18. https://doi.org/10.1016/j.apsoil.2019.02.019.

Restrepo, D., Palencia, M., Palencia, V., 2018. Study by attenuated total reflectance spectroscopy of structural changes of humified organic matter by chemical perturbations via alkaline dissolution. J. Sci. Technol. Appl. 4, 49–59. https://doi.org/10.34294/j.jsta.18.4.30.

Rocchi, L., Boggia, A., Paolotti, L., 2020. Sustainable agricultural systems: a bibliometrics analysis of ecological modernization approach. Sustainability 12, 9635. https://doi.org/10.3390/su12229635.

Rodríguez, M., Torres, M., Blanco, L., Bejar, V., Sampedro, I., Llamas, I., 2020. Plant growth-promoting activity and quorum quenching-mediated biocontrol of bacterial phytopathogens by *Pseudomonas segetis* strain P6. Sci. Rep. 10, 1–12. https://doi.org/10.1038/s41598-020-61084-1.

Rohman, S., Kaewtatip, K., Kantachote, D., Tantirungkij, M., 2020. Encapsulation of *Rhodopseudomonas palustris* KTSSR54 using beads from alginate/starch blends. J. Appl. Polym. Sci. 1–9. https://doi.org/10.1002/app.50084.

Ruiu, L., 2020. Plant-growth-promoting bacteria (PGPB) against insects and other agricultural pests. Agronomy 10, 861. https://doi.org/10.3390/agronomy10060861.

Sabarinathan, K.G., Gomathy, M., Kumar, D.A., Kannan, R., Aiyanathan, K.E.A., 2021. Cyanobacteria-mediated bioremediation of problem soils. Microbial rejuvenation of polluted environment. Microorg. Sustain. 25, 141–152. https://doi.org/10.1007/978-981-15-7447-4_5.

Serebrennikov, D., Thorne, F., Kallas, Z., McCarthy, S.N., 2020. Factors influencing adoption of sustainable farming practices in europe: a systemic review of empirical literature. Sustainability 12, 9719. https://doi.org/10.3390/su12229719.

Shilev, S., 2020. Plant-growth-promoting bacteria mitigating soil salinity stress in plants. Appl. Sci. 10, 7326. https://doi.org/10.3390/app10207326.

Sukriti, Kaith, B.S., Jindal, R., 2017. Controlled biofertilizer release kinetics and moisture retention in gum xanthan-based IPN. Iran. Polym. J. (Engl. Ed.) 26, 563–577. https://doi.org/10.1007/s13726-017-0539-8.

Umesha, S., Singh, P.K., Singh, R., 2018. Microbial biotechnology and sustainable agriculture. Biotechnol. Sustain. Agric. 185–205. https://doi.org/10.1016/b978-0-12-812160-3.00006-4.

Ungureanu, N., Vladut, V., Voicu, G., 2020. Water scarcity and wastewater reuse in crop irrigation. Sustainability 12, 9055. https://doi.org/10.3390/su12219055.

Vasilieva, S., Lobakova, E., Grigoriev, T., Selyakh, I., Semenova, L., Chivkunova, O., Gotovtsev, P., Antipova, C., Zagoskin, Y., Scherbakov, P., Lukyanov, A., Lukanina, K., Solovchenko, A., 2020. Bio-inspired materials for nutrient biocapture from wastewater: microalgal cells immobilized on chitosan-based carriers. J. Water Process Eng. 1–11. https://doi.org/10.1016/j.jwpe.2020.101774.

Veeck, G., Veeck, A., Yu, H., 2020. Challenges of agriculture and food systems issues in China and the United States. Geogr. & Sustain. 1, 109–117. https://doi.org/10.1016/j.geosus.2020.05.002.

Zhang, J., Zhao, T., Sun, B., Song, S., Guo, H., Shen, H., Wu, Y., 2018. Effects of biofertilizers and super absorbent polymers on plant growth and soil fertility in the arid mining area of Inner Mongolia, China. J. Mt. Sci. 15, 1920–1935. https://doi.org/10.1007/s11629-017-4801-5.

Chapter 28

Polymeric supports for growth of probiotic microorganisms

28.1 Microbiota and probiotics

Over the last century, for several decades the term "bacteria" was considered synonymous with disease or poor health. Nevertheless, advances in molecular microbiology combined with the development of bioinformatics tools, engineering applied to microbial systems, have made possible a more structured and specific study of microorganisms. Another important factor that has contributed to this development is the use of "high-performance technologies" These include genomics, proteomics, and metabolomics, which allow the identification, characterization, and quantification of the entire repertoire of genes, proteins, and metabolites found in living organisms (Casado-Vela et al., 2011). At present, these facts have permitted re-evaluation of the role played in health by the microbiota, which is defined as the community of microorganisms, including bacteria, viruses, fungi, and protozoa, living in a host (Dumas et al., 2018; Liu et al., 2018). Thus, the number of microorganisms colonizing the human body, including the intestine, skin, and other mucosal environments is an integral part of human microbiome, and this concept is extended to animals and even plants in specific regions of the symbiotic interaction such as the rhizosphere (Zheng et al., 2020). In humans, the gut microbiota is increasingly observed to be a symbiotic partner for the maintenance of health. However, multiple factors are associated with the presence of microorganisms in humans, for example, it is well known that the homeostasis of the intestinal microbiota depends on the characteristics of the host (e.g., age, sex, genetic background), on environmental conditions (e.g., stress, drugs, gastrointestinal surgery, infectious and toxic agents), and socioeconomic conditions which are associated with the quality and type of food, quality and availability of water, access to drugs and medical attention, etc. Experimental data obtained from animals, but also observational studies in obese patients, suggest that the composition of the gut microbiota is a factor that characterizes obese versus lean individuals, diabetic versus nondiabetic patients, and patients with liver diseases such as nonsteatohepatitis (i.e., a type of fatty liver disease). Interestingly, changes in gut microorganisms can be reversed through diet and are

Eco-friendly Functional Polymers
https://doi.org/10.1016/B978-0-12-821842-6.00008-7
Copyright © 2021 Elsevier Inc. All rights reserved.

related with weight loss. Qualitative and quantitative changes in the intake of specific components of food, such as fatty acids, carbohydrates, micronutrients, prebiotics, and probiotics, do not only have consequences on the composition of the gut microbiota, but also can modulate the expression of genes in host tissues such as liver, adipose tissue, intestine, and muscle. In addition, the development of fat mass and metabolic alterations associated with the function of the intestinal barrier and systemic immunity can be promoted or decreased (Delzenne et al., 2011; Everard and Cani, 2013; Dumas et al., 2018; Britton et al., 2020).

After having recognized the importance of the symbiotic interactions between humans and their microbiome, it is clear that the human body is an ecosystem for microorganism growth which is modulated by an important balance between the health of the host and the nature of the microorganisms. In this context, in terms of the intestinal microbiota, probiotics can be defined as microorganisms interacting positively with humans and generating a health benefit. Although in general terms they are biologically safe, it is true that there remain questions regarding their effectiveness. However, it is clear that the market for probiotics is growing, and consequently, the increase in their consumption has increased interest in the efficacy of probiotics, their uses, and in the development of new technologies associated with them (Casado-Vela et al., 2011; Dumas et al., 2018; Liu et al., 2018; Zheng et al., 2020). The different fields where probiotic research has increased notably include medicine (Stavropoulou and Bezirtzoglou, 2020; Carvalho et al., 2021), veterinary medicine (Ahasan et al., 2015), poultry (Macit et al., 2021), agriculture (McKenzie et al., 2018), and aquaculture (Klakegg et al., 2019, 2020; Anas et al., 2021), among others.

28.2 Probiotics used for health and disease treatment

Few treatment strategies for human diseases have received the level of attention and research in the last 20 years that probiotic-based treatments have gained (Liu et al., 2018; Stavropoulou and Bezirtzoglou, 2020). The World Health Organization (WHO) defines probiotics as any living organisms conferring a benefit for the health of the host when administered in adequate doses [i.e., at least 10^6 viable colony-forming units (CFU)/g] (Reid, 2016; Dimidi et al., 2017; Fijan, 2014). However, it is important to clarify that dead bacteria and their components can also exhibit probiotic properties, and consequently, in a strict sense, the WHO definition can be identified as inexact (Plaza-Diaz et al., 2019), therefore, the concept of probiotics includes a wide number of fermentable foods, pills, powders, and liquid drops. Some examples of probiotic microorganisms are *Lactobacillus rhamnosus*, *Lactobacillus reuteri*, *Lactobacillus casei*, *Lactobacillus paracasei*, *Lactobacillus acidophilus*, *Lactobacillus plantarum*, *Lactobacillus bulgaris*, *Lactococcus lactis*, *Saccharomyces boulardii*, *Bacillus coagulans*, *Bacillus clausii*, *Bifidobacterium infantis*, *Bifidobacterium longum*, *Bifidobacterium infantomy*, etc. In addition, many probiotics contain two or more individual species (Liu et al., 2018). The main action modes of probiotics include improvement of the epithelial barrier, greater adhesion to the intestinal mucosa and concomitant inhibition of pathogen adhesion, competitive exclusion of pathogenic microorganisms, production of antimicrobial substances, and modulation of the immune system (Bermudez-Brito et al., 2012). Perhaps one of the most common applications of probiotics is to restore, and improve, the intestinal microbiota in order to correct digestive disorders, including health alterations of nosocomial origin, diarrhea resulting from the action of antibiotics, allergic disorders such as atopic dermatitis and allergic rhinitis in babies. Additionally, probiotics are used as co-adjuncts in the treatment of metabolic disorders such as nonalcoholic fatty liver disease, type II diabetes, obesity, and metabolic syndrome (Plaza-Diaz et al., 2019).

Currently, the microbiota—host balance is accepted as an important factor associated with the state of well-being and health because it is accepted that intestinal microbiota interact with different organs such as the liver and brain (e.g., autism, mood, and brain development) (Ruiz et al., 2016). Probiotics are used to replace beneficial microorganisms that are naturally found in a certain niche, but whose number has been depleted, usually due to some external factor, and consequently triggers the disease. Note that it is a different situation to the proliferation of opportunistic microorganisms that, due to a decrease in the host's immune system, triggers a negative microbiological condition. For example, the use of *Lactobacillus crispatus* in the vagina is used to counteract bacterial vaginosis or an increase in the population of *E. coli* in the bladder. On the other hand, probiotic strains are selected because they have pathogen-counteracting properties; in this way, with the goal of restoring health and ideally allowing indigenous beneficial microbes to return, probiotics can be used. An example is the oral administration of *Lactobacillus rhamnosus* GR-1 and *Lactobacillus reuteri* RC-14, which are not very prevalent species in the vagina, but whose results have made evident that their administration favors the return of autochthonous microorganisms (e.g., anaerobic microorganisms like *Prevotella*, *Gardnerella vaginalis*, and *Atopobium vaginae*) (Puebla-Barragán and Reid, 2019).

28.3 Biopolymers for the encapsulation of probiotic strains

Much of the applied research on biopolymers, in the health field, is directed at the manufacture of medical devices, hospital furniture surfaces, dental restorations, surgical equipment, health care products, and hygienic applications (for example, wastewater treatment systems, textile manufacturing, food packaging, and storage, etc.) (for details, see Chapters 6, 16, 17, and 18). In all the above cases, the focus or line of research is directed at the elimination of microorganisms to guarantee asepsis and microbiological safety (Jain et al., 2014; Hook et al., 2012; Blok et al., 2020). The motivations for these investigations range from the control and prevention of infections in wounds and tissues to the treatment, control, and development of biosafety systems in the face of large pandemics such as influenza, swine flu, severe acute respiratory syndrome (SARS), and the COVID-19 pandemic, among others (Taubenberger and Morens, 2006; Morens and Fauci, 2007; Myers et al., 2007; Dandagi and Byahatti, 2011; Mohanty et al., 2020; Sharma et al., 2020; Xue et al., 2020).

On the contrary, the use of biopolymers for their application in probiotics is directed toward the development of systems acting as vehicles for their administration, growth, and in some specific cases, for the biological generation of metabolites. Behind this interest in research on biomaterials applied to probiotic systems, it is important to highlight three key aspects: (1) the demand for alternatives that minimize serious side effects and increase the efficiency of treatments, (2) a growing interest in synthetic drug reduction, and in this area, a more eco-friendly look at products of natural origin, and (3) the clinical evidence of the positive impact of probiotic therapies, as well as their enormous potential in terms of designs and alternatives (Puebla-Barragan and Reid, 2019). Specifically, among the main drawbacks associated with probiotics is the preparation of different formulations that guarantee the survival of bacteria under the conditions of processing and storage, but also, during their application through different routes of administration, with the oral route being the most common; therefore, probiotic microorganisms must survive the action of hydrolytic enzymes, the extreme acidity of the stomach, and the action of bile salts present in the human intestine (Asgari et al., 2020; Tilak et al., 2020).

The viability of microorganisms is not easy, since it must guarantee optimal growth conditions, allow the passage of microorganisms from metabolically active cells to cellular systems in a state of latency, and allow their metabolic activation by some physiological stimuli. Therefore, biopolymers emerge as a suitable material for the production of biopolymer-based probiotic systems because they offer the possibility to be designed and modified, besides their properties such as biocompatibility, low or null toxicity, and easy bioassimilation (Asgari et al., 2020; Tilak et al., 2020). The most commonly used technology for the making of biopolymer-based probiotic systems is encapsulation, which consists of the use of the biopolymer as a vehicle or carrier material, with biodegradability characteristics for the action of pH or enzymes (Asgari et al., 2020).

Conventional techniques for making powders such as lyophilization and spray drying do not allow the viability of the probiotic cells due to the effects of temperature and dehydration at the cellular level. Therefore, the viable cell count in probiotic formulations is strongly associated with the carrier material, and consequently, the biopolymeric material plays an important role in the composition of the probiotic system (Tilak et al., 2020). Due to the small cell size, this encapsulation technology has advanced toward the manufacture of microparticles and nanoparticles, which represents an increase in the efficiency of transport and release. The main type of biopolymer used for this application is polysaccharides, which provide specific characteristics that allow their classification into: (1) ion-induced gelatinous polysaccharides (e.g., alginate, pectine, and carrageenan), polysaccharides that reinforce the structure (e.g., gellan gum), enteric dissolvable polysaccharides (e.g., cellulose acetate propionate), polysaccharides that interact electrostatically (e.g., chitosan and alginate) and prebiotics, which are biopolymeric fibers that act as stimulators of the microbiota (Zhang et al., 2018; Zupančič et al., 2018; He et al., 2021; Ma et al., 2020; Tilak et al., 2020) (see Table 28.1).

28.4 Immobilization and encapsulation of probiotic strains

Cell immobilization techniques and encapsulation are often based on the biomimicry of nature, since many microorganisms have the ability to adhere through different mechanisms and metabolic processes, allowing them to form biofilms and survive on different types of surfaces and conditions. The terms immobilization and encapsulation are largely equivalent terms, and consequently, they are often used synonymously. However, microencapsulation refers to a process of forming a continuous coating around the biological entity of interest, or what is the same in terms of the final product, is the inclusion of the biological entity of interest in a continuous phase. On the other hand, immobilization has a more superficial character, although it can include the absorption of the biological entity in the polymer matrix (Frakolaki et al., 2020).

Macro-, micro-, and nanoencapsulation provide protection to microbiological cultures when they have a high sensitivity to high levels of oxygen, mechanization processes during manufacturing, storage conditions such as freezing, and as indicated in the previous section, inherent conditions of transit through the human gastrointestinal tract (Mitropoulou et al., 2013; Chavarri et al., 2012). A capsule, regardless of its size, is a compartment that includes a porous or shell-shaped solid structure, containing a liquid, gas, or colloidal phase. In probiotic formulations, the outer part of the capsule has the function of acting as a barrier for the protection of probiotic cells from the external environment, while the metabolites and nutrients act as a reservoir. Encapsulation facilitates the handling of probiotics and their controlled dosage. For applications in the food industry, microencapsulated probiotics are generally added to different dairy and nondairy products. Furthermore, usually the carrier materials used to encapsulate probiotics are biopolymers, mainly proteins or carbohydrates (Asgari et al., 2020; Anal and Singh, 2007).

TABLE 28.1 Examples of biopolymers, modified polymers, and synthetics polymers used for probiotic encapsulation.

Polymer material	Polymer type	Probiotic strains	Description and application	References
PEO nanofibers	Synthetic polymer	*Bacillus* sp. (Cepa 25.2.M)	Periodontal disease treatments	Zupančič et al. (2018)
		L. plantarum	Restoring the balance of the microbiota	Škrlec et al. (2019)
PVA	Synthetic polymer	*L. gasseri CRL1320*	Prevention or treatment of urogenital infections	Silva et al. (2020)
		L. rhamnosus CRL1332	Prevention or treatment of urogenital infections	Silva et al. (2020)
Cellulose nanofibers	Biopolymers	*L. plantarum*	Restoration of gut microbiota	Zhang et al. (2018)
Bacterial cellulose nanofibers	Biopolymers	*L. acidophilus 016*	Multiple applications	Tilak et al. (2020)
Alginate nanofibers	Biopolymers	*L. plantarum*	Restoration of gut microbiota	Zhang et al. (2018)
Cellulose macrogels	Biopolymers	*L. plantarum*	Prevention or treatment by bioactive agents	Luan et al. (2017)
Cellulose acetate membranes	Modified biopolymers	*L. plantarum*	Immobilization of biofilms for the production of metabolites	Hu et al. (2019)
PVA + alginate nanofibers	Mixtures	*L. rhamnosus*	Microbiological stability of fish fillets in the food industry	Ceylan et al. (2018)
Gum arabic and pullulan	Mixtures	*Lactobacillus* sp.	Protection of probiotics	Ma et al. (2020)

POE, poly(ethylene oxide); *PVA*, poly(vinyl alcohol).

The selection of the encapsulation method depends on the average particle size required, the physical and chemical properties of the biopolymer, the application and administration pathway, the release mechanism, and the cost of the procedure; furthermore, the process must not be aggressive in order to ensure the viability of the encapsulated cells (Burgain et al., 2011; Asgari et al., 2020; Rodrigues et al., 2020). Since encapsulation systems based on biopolymers are suitable environments for microbial growth, encapsulation procedures must be carried out in environments with a high degree of microbiological control to avoid contamination of the product with other mechanisms that act as competing entities or infections. Among the main techniques of microencapsulation of probiotics are spray drying, spray freeze drying, spray cooling, emulsion-based techniques, extrusion, agglomeration and fluidized bed coating, freezing, and vacuum drying, among others (see Table 28.2 and Fig. 28.1).

TABLE 28.2 Encapsulation techniques of probiotic strains.

Encapsulation techniques	Advantages	Disadvantages	References
Spray drying	High production rates at relatively low operating costs. The resulting powders are stable and easily applicable. The process can be operated continuously	Most probiotic strains do not survive well in high temperatures and dehydration during the process. Low stability during storage	Chavarri et al. (2012), Martín et al. (2015), Arslan-Tontul and Erbas (2017)
Spraying by lyophilization	Provides a controlled size, larger specific surface area than spray-dried capsules. The capsules can be encased in an additional shell that uses the fluid bead method to provide protection against harsh environmental conditions	High energy use, long processing time, and cost is 30–50 times more expensive than spray drying using maltodextrin as wall matrix	Martín et al. (2015)

Continued

TABLE 28.2 Encapsulation techniques of probiotic strains.—cont'd

Encapsulation techniques	Advantages	Disadvantages	References
Spray cooling	This is considered the cheapest encapsulation technology that has the possibility of manufacturing on an industrial scale. Adequate technique for encapsulating heat-sensitive agents	The microparticles produced include a low capacity for encapsulation and ejection of core material during storage	Pedroso et al. (2012), Rodrigues et al. (2020)
Emulsification	Small capsule size. Improve the viability of microorganism cells	Expensive technology. Instability of the emulsion. Need for vigorous shaking. Random incorporation of cells into capsules. Inability to sterilize vegetable oil	Alemzadeh and Oryan (2020)
Coacervation	Relatively simple inexpensive process. Allows the incorporation of a large number of microorganisms in relation to the encapsulant	This is a batch process. Needs an additional drying process	Alemzadeh and Oryan (2020)
Extrusion	Flavor and volatile material stability. High performance and lower production rates. Ecological technique because now it does not produce large effluents due to the low humidity process	Thin wall, high porosity. The extrusion process involves high temperature (80–150°C) applied for a short period (1 min)	Alemzadeh et al. (2020), Bamidele and Emmambux (2020)

Among the applications of biopolymers the manufacture of drug-delivery systems (see Chapter 18) and their use for probiotic formulations stand out, which is the central element of this chapter (Haghshenas et al., 2015; Rodrigues et al., 2020). Regarding the type of polymers, as shown in

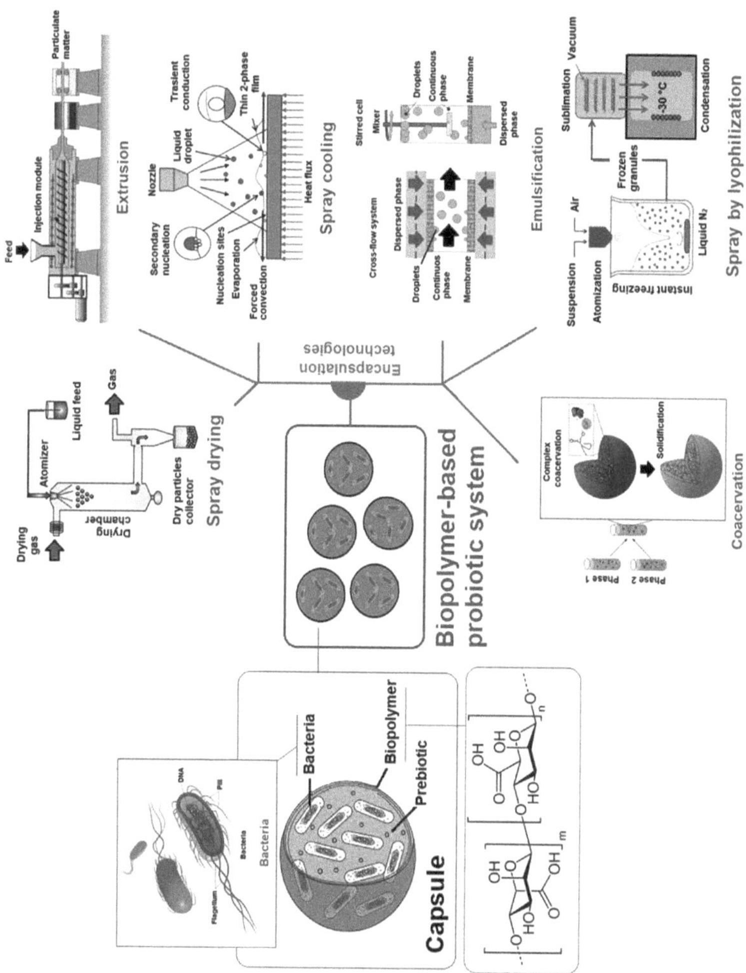

FIGURE 28.1 Illustration of the main encapsulation techniques of probiotics.

Table 28.1, these can be natural, synthetic, natural modified by synthetic procedures, polymer mixtures, composite materials, etc. (Namazi, 2017; Khoshnevisan et al., 2018). In particular, the mixtures of polymers are aimed at obtaining multifunctional capsules in which each polymeric component contributes to one or more aspects of the application. For example, mixtures of alginate and pullulan would allow combining in a single capsule a nontoxic, thermosetting, and biocompatible biopolymer, which is a safe food additive, with a polymer that has a prebiotic functionality (Haghshenas et al., 2015). However, not all have ideal properties for use as an encapsulation matrix by all techniques. For example, extrusion processes involve the transition of the polymer from its solid state to its molten state, for which it must be subjected to high temperatures that can lead to the modification of its functional groups through thermal degradation. In addition, it is a technique limited to thermoplastic polymers, that is, noncross-linked polymers. However, it is also not a viable technique for polymers classified as pseudo-thermosets, like polymers with a high degree of branching. For these highly branched polymers, the movements of the highly coiled branches are restricted even though there is no chemical crossover between the chains, therefore, polymer thermally degrades instead of melting (Prestinaci et al., 2020; Delfi et al., 2020; Getahun et al., 2020).

28.5 Conclusions and remarks

Probiotics are a biotechnological therapy that have emerged as an alternative for different treatments. Currently these treatments have focused on the digestive and urinary tracts. However, new applications that broaden their spectrum, mechanism, and mode of action are expected to emerge in the medium term as the potential they offer is enormous, as well as being an eco-friendlier strategy compared to alternatives based on synthetic drugs.

The selection of both the technique and the encapsulation material must be made ensuring sufficient viability, mechanical stability, protection, and compatibility, in order to guarantee adequate application of the encapsulated cells. In addition, it is clear that, with the rise of biopolymeric materials, there are new advances and redirections in encapsulation technology, applications, and the multifunctionality of these systems.

References

Ahasan, A., Agazzi, A., Invernizzi, G., Bontempo, V., Savoini, G., 2015. The beneficial role of probiotics in monogastric animal nutrition and health. J. Dairy Vet. & Anim. Res. 2, 116–132. https:/doi.org/10.15406/jdvar.2015.02.00041.

Alemzadeh, E., Oryan, A., 2020. Application of encapsulated probiotics in health care. J. Exp. Pathol. 1, 16–21. https://doi.org/10.33696/pathology.1.003.

Alemzadeh, I., Hajiabbas, M., Pakzad, H., Dehkordi, S.S., Vossoughi, A., 2020. Encapsulation of food components and bioactive ingredients and targeted release. Int. J. Eng. 33, 1–11. https://doi.org/10.5829/IJE.2020.33.01A.01.

Anal, A.K., Singh, H., 2007. Recent advances in microencapsulation of probiotics for industrial applications and targeted delivery. Trends Food Sci. Technol. 18, 240–251. https://doi.org/10.1016/j.tifs.2007.01.004.

Anas, A., Sukumaran, V., Nampullipurackal Devarajan, D., Maniyath, S., Chekidhenkuzhiyil, J., Mary, A., Kuttan, S.P., Tharakan, B., 2021. Probiotics inspired from natural ecosystem to inhibit the growth of Vibrio spp. causing white gut syndrome in *Litopenaeus vannamei*. 3 Biotech 11, 66. https://doi.org/10.1007/s13205-020-02618-2.

Arslan-Tontul, S., Erbas, M., 2017. Single and double layered microencapsulation of probiotics by spray drying and spray chilling. LWT - Food Sci. Technol. (Lebensmittel-Wissenschaft -Technol.) 81, 160–169. https://doi.org/10.1016/j.lwt.2017.03.060.

Asgari, S., Pourjavadi, A., Licht, T.R., Boisen, A., Ajalloueian, F., 2020. Polymeric carriers for enhanced delivery of probiotics. Adv. Drug Deliv. Rev. 161–162, 1–21. https://doi.org/10.1016/j.addr.2020.07.014.

Bamidele, O.P., Emmambux, M.N., 2020. Encapsulation of bioactive compounds by "extrusion" technologies: a review. Crit. Rev. Food Sci. Nutr. 1–19. https://doi.org/10.1080/10408398.2020.1793724.

Bermudez-Brito, M., Plaza-Díaz, J., Muñoz-Quezada, S., Gómez-Llorente, C., Gil, A., 2012. Probiotic mechanisms of action. Ann. Nutr. Metabol. 160–174. https://doi:10.1159/000342079.

Blok, A.J., Gurnani, P., Xenopoulos, A., Burroughs, L., Duncan, J., Urbanowicz, R.A., Alexander, M.R., 2020. Polymer microarrays rapidly identify competitive adsorbents of virus-like particles. Biointerphases 15, 061005. https://doi:10.1116/6.0000586.

Britton, G.J., Contijoch, E.J., Spindler, M.P., Aggarwala, V., Dogan, B., Bongers, G., Faith, J.J., et al., 2020. Defined microbiota transplant restores Th17/RORγt$^+$ regulatory T cell balance in mice colonized with inflammatory bowel disease microbiotas. Proc. Natl. Acad. Sci. U. S. A. 1–10. https://doi:10.1073/pnas.1922189117.

Burgain, J., Gaiani, C., Linder, M., Scher, J., 2011. Encapsulation of probiotic living cells: from laboratory scale to industrial applications. J. Food Eng. 104, 467–483. https://doi.org/10.1016/j.jfoodeng.2010.12.031.

Casado-Vela, J., Arancha, C., Gómez del Pulgar, M.T., Lacal, J.C., 2011. Approaches for the study of cancer: towards the integration of genomics, proteomics and metabolomics. Clin. Transl. Oncol. 13, 617. https://doi.org/10.1007/s12094-011-0707-9.

Carvalho, F.M., Teixeira-Santos, R., Filipe, J.M., Mergulhão, Gomes, L.C., 2021. The use of probiotics to fight biofilms in medical devices: a systematic review and meta-analysis. Microorganisms 9, 27. https://doi.org/10.3390/microorganisms9010027.

Ceylan, Z., Meral, R., Karakaş, C.Y., Dertli, E., Yilmaz, M.T., 2018. A novel strategy for probiotic bacteria: ensuring microbial stability of fish fillets using characterized probiotic bacteria-loaded nanofibers. Innovat. Food Sci. Emerg. Technol. 48, 212–218. https://doi.org/10.1016/j.ifset.2018.07.002.

Chavarri, M., Izaskun Marañon, I., Villaran, M.C., 2012. In: Rigobelo, E.C. (Ed.), Encapsulation Technology to Protect Probiotic Bacteria. IntechOpen. https://doi.org/10.5772/50046.

Dandagi, G., Byahatti, S., 2011. An insight into the swine-influenza A (H1N1) virus infection in humans. Lung India 28, 34–38. https://doi.org/10.4103/0970-2113.76299.

Delfi, M., Ghomi, M., Zarrabi, A., Mohammadinejad, R., Baghban, Z., Ashrafizadeh, M., et al., 2020. Functionalization of polymers and nanomaterials for biomedical applications: antimicrobial platforms and drug carriers. Prosthesis 2, 117–139. https://doi.org/10.3390/prosthesis2020012.

Delzenne, N.M., Neyrinck, A.M., Cani, P.D., 2011. Modulation of the gut microbiota by nutrients with prebiotic properties: consequences for host health in the context of obesity and metabolic syndrome. Microb. Cell Factories 10, S10. https://doi.org/10.1186/1475-2859-10-S1-S10.

Dimidi, E., Christodoulides, S., Scott, S.M., Whelan, K., 2017. Mechanisms of action of probiotics and the gastrointestinal microbiota on gut motility and constipation. Adv. Nutr. Int. Rev. J. 8, 484–494. https://doi.org/10.3945/an.116.014407.

Dumas, A., Bernard, L., Poquet, Y., Lugo-Villarino, G., Neyrolles, O., 2018. The role of the lung microbiota and the gut-lung axis in respiratory infectious diseases. Cell Microbiol. 20 https://doi.org/10.1111/cmi.12966 e12966.

Everard, A., Cani, P.D., 2013. Diabetes, obesity and gut microbiota. Best Pract. Res. Clin. Gastroenterol. 27, 73–83. https://doi.org/10.1016/j.bpg.2013.03.007.

Fijan, S., 2014. Microorganisms with claimed probiotic properties: an overview of recent literature. Int. J. Environ. Res. Publ. Health 11, 4745–4767. https://doi.org/10.3390/ijerph110504745.

Frakolaki, G., Giannou, V., Kekos, D., Tzia, C., 2020. A review of the microencapsulation techniques for the incorporation of probiotic bacteria in functional foods. Crit. Rev. Food Sci. Nutr. 1–22. https://doi.org/10.1080/10408398.2020.1761773.

Getahun, H., Smith, I., Trivedi, K., Paulin, S., Balkhy, H.H., 2020. Tackling antimicrobial resistance in the COVID-19 pandemic. Bull. World Health Organ. 98 https://doi.org/10.2471/BLT.20.268573, 442–442A.

Haghshenas, B., Abdullah, N., Nami, Y., Radiah, D., Rosli, R., Khosroushahi, A.Y., 2015. Microencapsulation of probiotic bacteria *Lactobacillus plantarum* 15HN using alginate-psyllium-fenugreek polymeric blends. J. Appl. Microbiol. 118, 1048–1057. https://doi.org/10.1111/jam.12762.

He, C., Sampers, I., Raes, K., 2021. Dietary fiber concentrates recovered from agro-industrial by-products: functional properties and application as physical carriers for probiotics. Food Hydrocoll. 111, 106175. https://doi.org/10.1016/j.foodhyd.2020.106175.

Hook, A.L., Chang, C.Y., Yang, J., Luckett, J., Cockayne, A., Atkinson, S., Mey, Y., et al., 2012. Combinatorial discovery of polymers resistant to bacterial attachment. Nat. Biotechnol. 30, 868–875. https://doi.org/10.1038/nbt.2316.

Hu, M.X., Li, J.N., Guo, Q., Zhu, Y.Q., Niu, H.M., 2019. J. Agric. Food Chem. 67, 3198–3208. https://doi.org/10.1021/acs.jafc.8b05024.

Jain, A., Duvvuri, L.S., Farah, S., Beyth, N., Domb, A.J., Khan, W., 2014. Antimicrobial polymers. Adv. Healthc. Mater. 3, 1969–1985. https://doi.org/10.1002/adhm.201400418.

Khoshnevisan, K., Maleki, H., Samadian, H., Shahsavari, S., Sarrafzadeh, M.H., Larijani, B., Dorkoosh, F.A., Haghpanah, V., Khorramizadeh, M.R., 2018. Cellulose acetate electrospun nanofibers for drug delivery systems: applications and recent advances. Carbohydr. Polym. 198, 131–141. https://doi.org/10.1016/j.carbpol.2018.06.072.

Klakegg, Ø., Myhren, S., Juell, R.A., Aase, M., Salonius, K., Sørum, H., 2019. Improved health and better survival of farmed lumpfish (*Cyclopterus lumpus*) after a probiotic bath with two probiotic strains of *Aliivibrio*. Aquaculture 518, 734810. https://doi:10.1016/j.aquaculture.2019.734810.

Klakegg, Ø., Salonius, k., Nilsen, A., Fulberth, M., Sørum, H., 2020. Enhanced growth and decreased mortality in Atlantic salmon (*Salmo salar*) after probiotic bath. J. Appl. Microbiol. 129, 146–160. https://doi.org/10.1111/jam.14649.

Liu, Y., Tran, D.Q., Rhoads, J.M., 2018. Probiotics in disease prevention and treatment. J. Clin. Pharmacol. 58, S164–S179. https://doi:10.1002/jcph.1121.

Luan, Q., Zhou, W., Zhang, H., Bao, Y., Zheng, M., Shi, J., Huang, F., 2017. Cellulose-based composite macrogels from cellulose fiber and cellulose nanofiber as intestine delivery vehicles for probiotics. J. Agric. Food Chem. 66, 339–345. https://doi:10.1021/acs.jafc.7b04754.

Ma, J., Xu, C., Yu, H., Feng, Z., Yu, W., Gu, L., Hou, J., 2020. Electro-encapsulation of probiotics in gum Arabic-pullulan blend nanofibres using electrospinning technology. Food Hydrocoll. 111, 106381. https://doi:10.1016/j.foodhyd.2020.106381.

Macit, M., Karaoglu, M., Celebi, S., Esenbuga, N., Yoruk, M.A., Kaya, A., 2021. Effects of supplementation of dietary humate, probiotic, and their combination on performance, egg quality, and yolk fatty acid composition of laying hens. Trop. Anim. Health Prod. 53, 63. https://doi.org/10.1007/s11250-020-02546-6.

Martín, M.J., Lara-Villoslada, F., Ruiz, M.A., Morales, M.E., 2015. Microencapsulation of bacteria: a review of different technologies and their impact on the probiotic effects. Innovat. Food Sci. Emerg. Technol. 2013, 15–25. https://doi.org/10.1016/j.ifset.2014.09.010.

Mitropoulou, G., Nedovic, V., Goyal, A., Kourkoutas, Y., 2013. Immobilization technologies in probiotic food production. J. Nutr. & Metabol. 1–15. https://doi.org/10.1155/2013/716861.

Mohanty, S.K., Satapathy, A., Naidu, M.M., Mukhopadhyay, S., Sharma, S., Barton, L.M., Stroberg, E., Duval, E.J., Pradhan, D., Tzankov, A., Parwani, A.V., 2020. Severe acute respiratory syndrome coronavirus-2 (SARS-CoV-2) and coronavirus disease 19 (COVID-19) – anatomic pathology perspective on current knowledge. Diagn. Pathol. 15, 103. https://doi.org/10.1186/s13000-020-01017-8.

Morens, D.M., Fauci, A.S., 2007. The 1918 influenza pandemic: insights for the 21st century. J. Infect. Dis. 195, 1018–1028. https://doi.org/10.1086/511989.

McKenzie, V.J., Kueneman, J.G., Harris, R.N., 2018. Probiotics as a tool for disease mitigation in wildlife: insights from food production and medicine. Ann. N. Y. Acad. Sci. 1429, 18–30. https://doi.org/10.1111/nyas.13617.

Myers, K.P., Olsen, C.W., Gray, G.C., 2007. Cases of swine influenza in humans: a review of the literature. Clin. Infect. Dis. 44, 1084–1088. https://doi.org/10.1086/512813.

Namazi, H., 2017. Polymers in our daily life. Bioimpacts 7, 73–74. https://doi.org/10.15171/bi.2017.09.

Puebla-Barragan, S., Reid, G., 2019. Thirty-year evolution of probiotic therapy. Microbial Cell. 6, 184–196. https://doi.org/10.15698/mic2019.04.673.

Pedroso, D., Thomazini, M., Heinemann, R.J.B., Favaro-Trindade, C.S., 2012. Protection of Bifidobacterium lactis and *Lactobacillus acidophilus* by microencapsulation using spray-chilling. Int. Dairy J. 26, 127–132. https://doi.org/10.1016/j.idairyj.2012.04.008.

Plaza-Diaz, J., Ruiz-Ojeda, F.J., Gil-Campos, M., Gil, A., 2019. Mechanisms of action of probiotics. Adv. Nutr. S49–S66. https://doi.org/10.1093/advances/nmy063.

Prestinaci, F., Pezzotti, P., Pantosti, A., 2020. Antimicrobial resistance: a global multifaceted phenomenon. Pathog. Glob. Health 109, 309–318. https://doi.org/10.1179/2047773215Y.0000000030.

Reid, G., 2016. Probiotics: definition, scope and mechanisms of action. Best Pract. Res. Clin. Gastroenterol. 30, 17–25. https://doi.org/10.1016/j.bpg.2015.12.001.

Rodrigues, F.J., Cedran, M.F., Bicas, J.L., Sato, H.H., 2020. Encapsulated probiotic cells: relevant techniques, natural sources as encapsulating materials and food applications – a narrative review. Food Res. Int. 137, 1–16. https://doi.org/10.1016/j.foodres.2020.109682.

Ruiz, L., Hidalgo, C., Blanco-Míguez, A., Lourenço, A., Sánchez, B., Margolles, A., 2016. Tackling probiotic and gut microbiota functionality through proteomics. J. Proteom. 147, 28–39. https://doi.org/10.1016/j.jprot.2016.03.023.

Silva, J.A., De Gregorio, P.R., Rivero, G., Abraham, G., Nader-Macías, M.E.F., 2020. Immobilization of vaginal *Lactobacillus* in polymeric nanofibers for its incorporation in vaginal probiotic products. Eur. J. Pharmaceut. Sci. 156, 105563. https://doi.org/10.1016/j.ejps.2020.105563.

Sharma, A., Tiwari, S., Deb, M.K., Marty, J.L., 2020. Severe acute respiratory syndrome coronavirus -2 (SARS-CoV-2): a global pandemic and treatments strategies. Int. J. Antimicrob. Agents 56, 106054. https://doi.org/10.1016/j.ijantimicag.2020.106054.

Škrlec, K., Zupančič, Š., Prpar Mihevc, S., Kocbek, P., Kristl, J., Berlec, A., 2019. Development of electrospun nanofibers that enable high loading and long-term viability of probiotics. Eur. J. Pharm. Biopharm. 136, 108–119. https://doi.org/10.1016/j.ejpb.2019.01.013.

Stavropoulou, E., Bezirtzoglou, E., 2020. Probiotics in medicine: a long debate. Front. Immunol. 11, 2192. https://doi.org/10.3389/fimmu.2020.02192.

Taubenberger, J.K., Morens, D.M., 2006. 1918 influenza: the mother of all pandemics. Emerg. Infect. Dis. 12, 15–22. https://doi.org/10.3201/eid1201.050979.

Tilak, J., Sanjeev, B., Marimuthu, S., Uthandi, S., 2020. Nano fibra de celulosa bacteriana (BCNF) como soporte portador para la inmovilización del probiótico, *Lactobacillus acidophilus* 016. Carbohydr. Polym. 250, 116965. https://doi.org/10.1016/j.carbpol.2020.116965.

Xue, X., Ball, J.K., Alexander, C., Alexander, M.R., 2020. all surfaces are not equal in contact transmission of SARS-CoV-2. Matter 3, 1433–1441. https://doi.org/10.1016/j.matt.2020.10.006.

Zhang, H., Yang, C., Zhou, W., Luan, Q., Li, W.L., Deng, Q., Huang, F., 2018. A pH-responsive gel macrosphere based on sodium alginate and cellulose nanofiber for potential intestinal delivery of probiotics. ACS Sustain. Chem. Eng. 6, 13924–13931. https://doi.org/10.1021/acssuschemeng.8b02237.

Zheng, D., Liwinski, T., Elinav, E., 2020. Interaction between microbiota and immunity in health and disease. Cell Res. 30, 492–506. https://doi.org/10.1038/s41422-020-0332-7.

Zupančič, Š., Rijavec, T., Lapanje, A., Petelin, M., Kristl, J., Kocbek, P., 2018. Nanofibers with incorporated autochthonous bacteria as potential probiotics for local treatment of periodontal disease. Biomacromolecules 19, 4299–4306. https://doi.org/10.1021/acs.biomac.8b01181.

Chapter 29

Polymeric drug carriers against *Helicobacter pylori*

29.1 Introduction: microbiological characteristics of *H. pylori* and its importance in health

Helicobacter pylori is a Gram-negative pathogenic bacterium, with spiral-like shape, four to six flagella, characterized by being slow-growing and requiring a microaerophilic atmosphere, and also able to colonize the gastric mucosa. For its survival in the extremely acid environment of the stomach it has developed an adaptation system based on the transformation of its environment through the action of extracellular enzymes called urease (Palencia et al., 2016; Graham and Miftahussurur, 2018). A brief historical background on *H. pylori* was previously published (Stave et al., 2017). In the cited paper it is shown that the first references on this microorganism in scientific publications were in Germany, in 1889 and 1906, that is, more than a century ago. Currently, it is recognized that *H. pylori* has been identified all around the world, with different levels of prevalence depending on multiple factors (Khoder et al., 2019). Nowadays, infection by *H. pylori* occurs mainly in childhood, with a higher prevalence in developing countries (Stave et al., 2017). In addition to the health effects listed above, at the extraintestinal level, particularly in children, infection *H. pylori* has been identified and associated with subnormal growth, malnutrition, and iron-deficiency anemia (Yang, 2016; Stave et al., 2017; Zhang et al., 2020).

This bacterium promotes the development of gastrointestinal diseases, such as gastritis, gastric or duodenal ulcer, mucosa-associated lymphoid tissue (MALT) lymphoma, and gastric cancer. Its infection is characterized as starting with a first stage of infiltration in gastric epithelial cells. Later, there is subsequent activation of defense mechanisms to neutralize the infection, such as cell exfoliation, regenerative variations, and cellular and humoral immune responses. In this stage, a set of bacterial products promotes the secretion of proinflammatory mediators and the secretion of neutrophil chemoattractant factors (e.g., IL-8 or interleukin-8, that is a chemotactic and inflammatory cytokine) (Nardone and Compare, 2015; Palencia et al., 2016; Jarzab et al., 2020). In this process, a small percentage of cells control and neutralize the

Eco-friendly Functional Polymers
https://doi.org/10.1016/B978-0-12-821842-6.00006-3
Copyright © 2021 Elsevier Inc. All rights reserved.

bacterial infection, however, a large percentage of cells remain infected for many years, affecting the tissue and producing changes in the morphology of gastric epithelial cells (Nardone and Compare, 2015; Huang et al., 2016). Due to its ability to interact, for long periods of time, that is, decades, with the gastric mucosa without causing damage to the host, *H. pylori* has been described as part of the transient human gastric microbiome, nevertheless, it has been demonstrated that *H. pylori* is a predisposing factor for gastric cancer, being defined by multiple authors as a biocarcinogen, or a biological entity capable of promoting the development of cancer. Gastric adenocarcinomas related to *H. pylori* are developed through the so-called "Correa cascade" which consists of progressive transformations of the gastric mucosa, starting with chronic gastritis and degenerating into atrophic gastritis, intestinal metaplasia, dysplasia, and finally carcinoma (Jarzab et al., 2020). Consistent with the above, according to recent statistical data, *H. pylori* has been described as the third most common cause of cancer-related death in the world, with more than 720,000 deaths per year (Caguazango, 2020; Sukri et al., 2020).

29.1.1 Drugs against *H. pylori*: an overview of pharmacological treatments

The standard protocol for the treatment of *H. pylori* infection is based on the "triple therapy," which includes omeprazole, which is a proton pump inhibitor, and two antibiotics (i.e., amoxicillin and clarithromycin or metronidazole) (see Fig. 29.1A). Although clarithromycin is the most used antibiotic against *H. pylori*, the bacterial eradication rate has decreased from 88% to 20% due to resistance to this drug (Caguazango, 2020; Kipritci et al., 2020; Marques et al., 2020). Therefore, when the failure of the first treatment occurs, alternative treatments must be implemented, which consist mainly of a change in the antibiotics. A substitute treatment is "quadruple therapy," which is based on the use of at least one proton-pump inhibitor, bismuth salts, metronidazole, and tetracycline. However, this treatment is associated with a high incidence of side effects.

Another alternative is the use of levofloxacin, rabeprazole, and tinidazole or amoxicillin in combination with the triple therapy. In quadruple therapy, furazolidone has been used as a replacement for metronidazole. However, the standard procedure is based on replacing the antibiotics when initial treatments fail. Some alternatives to substituted antibacterial drugs are rifabutin, levofloxacin, and furazolidone (Siddalingam and Chidambaram, 2014; Kim et al., 2015; Hu et al., 2020).

In 2017, the World Health Organization (WHO) published a list of "priority pathogens" including antibiotic-resistant bacteria, and particularly, clarithromycin-resistant strains of *H. pylori* were classified as high-risk microorganisms for human health (Saracino et al., 2020). The main reasons

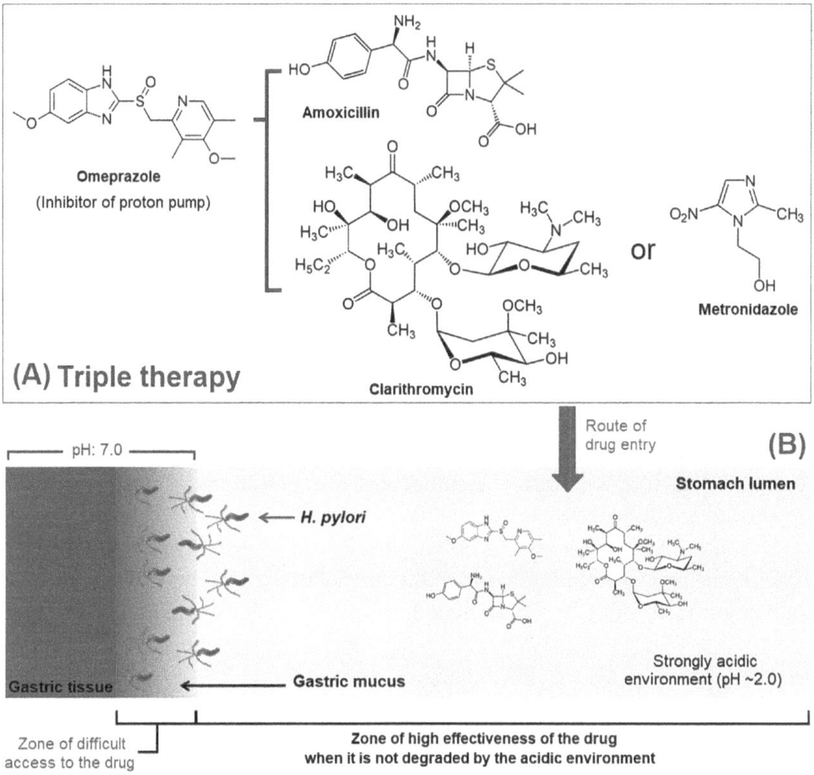

FIGURE 29.1 (A) Chemical structure of drugs used in the triple therapy. (B) Illustration of the route of drug entry: pharmaceutical products enter the zone characterized by a strongly acidic environment. If drugs are not stable to acid pH, then they can be degraded. Stable drugs diffuse to reach the target point, but they can only act efficiently in the external part of the gastric mucosa, being ineffective in eliminating the fraction of the bacterial population colonizing the interior of the gastric mucosa.

identified to explaining the failure of numerous therapeutic approaches against *H. pylori* are related with: (1) the degradation of many antibiotics by the extreme acidity of the stomach, (2) short half-life times of antibacterial drugs in the gastric region, and (3) the fact that the effects of the drugs administered to prevent bacterial cell reproduction are not effective against *H. pylori*, because the bacteria are immersed in the gastric mucosa, which acts as a protective barrier preventing the access of most antibiotics (Gottesmann et al., 2020; Raafat et al., 2021). Furthermore, although the study of *H. pylori* has been almost exclusively in its planktonic form, there is evidence suggesting that *H. pylori* can form biofilms, which are often associated with chronic infectious diseases, and consequently, due to biofilm providing additional protection to the bacteria, the effectivity of antimicrobial drug is expected to be strongly limited (Hathroubi et al., 2018; Espinosa et al., 2020; Jarzab et al., 2020). As a consequence, new therapeutic strategies are required to guarantee the efficiency of anti-*H. pylori* drugs and avoid the advance of bacterial resistance to pharmacological treatments.

29.2 Polymers as carriers of anti-*H. pylori* drugs

Not being toxic is one of the main requirements that a polymer must meet to be used as a drug carrier or in a drug release-controlled system, in addition, once its function is fulfilled, it must be digestible or, in contrast, it must be able to continue through the digestive tract to be excreted. Therefore, biopolymers are undoubtedly the most suitable candidates for this type of application. They can be obtained from different sources, and can be structurally modified in order to improve their use for specific applications (Jain et al., 2009; Arora et al., 2012; Arif et al., 2021). More details of biopolymers for this application are shown in Chapter 18.

In Table 29.1 a summary of biopolymers used in drug transport is shown. In addition, their presentation and main characteristics in terms of the desired application have been briefly included.

29.3 Conclusions and remarks

One of the great problems facing the world's population is the decrease in therapeutic strategies to treat bacterial, fungal, and viral infections. The main cause is the exponential and significant increase in resistance to antimicrobials. In the particular case of *H. pylori*, there is a large number of investigations that show the resistance developed to certain antibiotics, especially clarithromycin. In addition to the other reasons mentioned above, one of the identified causes is treatment failure, which gives the microorganism the opportunity to adapt, to guarantee its survival.

It is in this context that biopolymers emerge as a novel and versatile option to enhance the effective application of drugs, either by targeting the active site of infection, transport through the gastric mucosa, increasing drug residence times, etc. As has been shown, several investigations have as their objective the use of polymers to apply formulations of antibiotics. However, it is only the beginning of the design of new strategies. *H. pylori* has been proven to be very resistant, with a high capacity for adaptation, and although knowledge about it has advanced, there remains great uncertainty about many molecular and metabolic aspects of this bacterium.

TABLE 29.1 Polymers used as carriers for the administration of drugs against *H. pylori*.

Polymer	Drug	Presentation	Characteristics	References
Alginate	Amoxicillin-trihydrate	Floating hollow pearls	Release-controlled system to increase the residence time in the stomach. It eradicates more than 95% of *H. pylori* after 10 h	Raafat et al. (2021)
Chitosan modified with lipid grafts	It is used as glycolipid antimicrobial biosurfactant	Nanoparticles	Greater antimicrobial effect against *H. pylori*, with a minimum inhibitory concentration of 132 μg/mL and a biofilm inhibition capacity of 99%	Arif et al. (2021)
Apple pectin	Amoxicillin	Liposomes	It has been shown to be effective for the encapsulation, stability, and release of amoxicillin	Gottesmann et al. (2020)
Chitosan-poly(ethylene glycol)	Clarithromycin	Nanoparticles	It has been shown to be effective for the eradication of *H. pylori* biofilm. In addition, it inhibits adhesion. It has shown good ability to penetrate the gastric mucosa	Li et al. (2019)
Poly(acrylic acid) + poly(allylamine hydrochloride)	Amoxicillin and metronidazole	Nanocapsules	They have been shown to be useful as vehicles for the sustained release of the drugs into simulated gastric fluid. Furthermore, in situ adhesion and agglutination tests have been shown to be satisfactory	Jain et al. (2009)

Continued

TABLE 29.1 Polymers used as carriers for the administration of drugs against *H. pylori*.—cont'd

Polymer	Drug	Presentation	Characteristics	References
Chitosan-alginate polyelectrolyte complex	Amoxicillin	Nanoparticles	Mucopenetrating delivery system with low mucoadhesion	Arora et al. (2011)
Ethyl cellulose + carbopol 934P	Cucurcumin	Microparticles	Mucoadhesive microparticulate based on curcumin for drug delivery system	Ali et al. (2014)
Ethyl cellulose	Clarithomycin	Nanoparticles	Particles are dispersed well in water and simulated gastric fluid. Minimum inhibitory concentration of 0.09 −0.18 μg/mL against four strains of *H. pylori*	Pan-In et al. (2014)
Rhamnolipid + phospholipids	Amoxicillin	Lipid polymer nanoparticles containing pectin sulfate	The size of nanoparticles was around 200 nm. Particles were negatively charged and showed sustained and complete drug release within 24 h	Cai et al. (2015)

References

Ali, M.S., Pandit, V., Jian, M., Dhar, K.L., 2014. Mucoadhesive microparticulate drug delivery system of curcumin against *Helicobacter pylori* infection: design, development and optimization. J. Adv. Pharm. Technol. Res. (JAPTR) 5, 48–56. https://doi.org/10.4103/2231-4040.126996.

Arif, M., Sharaf, M., Samreen, Khan, S., Chi, Z., Liu, C.G., 2021. Chitosan-based nanoparticles as delivery-carrier for promising antimicrobial glycolipid biosurfactant to improve the eradication rate of *H. pylori* biofilm. J. Biomater. Sci. Polym. Ed. 1–19. https://doi.org/10.1080/09205063.2020.1870323.

Arora, S., Gupta, S., Narang, R., Budhiraja, R., 2011. Amoxicillin loaded chitosan-alginate polyelectrolyte complex nanoparticles as mucopenetrating delivery system for *H. pylori*. Sci. Pharm. 79, 673–694. https://doi.org/10.3797/scipharm.1011-05.

Arora, S., Bisen, G., Budhiraja, R., 2012. Mucoadhesive and muco-penetrating delivery systems for eradication of *Helicobacter pylori*. Asian J. Pharm. 6, 1–14. https://doi.org/10.4103/0973-8398.100127.

Caguazango, J.C., 2020. Ecological models of gastric microbiota dysbiosis: *H. pylori* and gastric carcinogenesis. Med. Microecol. 3, 1–7. https://doi.org/10.1016/j.medmic.2020.100010.

Cai, J., Huang, H., Song, W., Hu, H., Chen, J., et al., 2015. Preparation and evaluation of lipid polymer nanoparticles for eradicating *H. pylori* biofilm and impairing antibacterial resistance in vitro. Int. J. Pharm. 495, 728–737. https://doi.org/10.1016/j.ijpharm.2015.09.055.

Espinosa, A., Palencia, S., García, A., Palencia, M., 2020. Polymicrobial biofilms: fundamentals, diseases, and treatments - a review. J. Sci. Technol. Appl. 8, 4–54. https://doi.org/10.34294/j.jsta.20.8.54.

Gottesmann, M., Goycoolea, F.M., Steinbacher, T., Menogni, T., Hensel, A., 2020. Smart drug delivery against *Helicobacter pylori*: pectin-coated, mucoadhesive liposomes with anti-adhesive activity and antibiotic cargo. Appl. Microbiol. Biotechnol. 104, 5943–5957. https://doi.org/10.1007/s00253-020-10647-3.

Graham, D., Miftahussurur, M., 2018. *Helicobacter pylori* urease for diagnosis of *Helicobacter pylori* infection: a mini review. J. Adv. Res. 13, 51–57. https://doi.org/10.1016/j.jare.2018.01.006.

Hathroubi, S., Stephanie, L., Servetas, S.L., Windham, I., Merrell, D.S., Ottemann, K.M., 2018. *Helicobacter pylori* biofilm formation and its potential role in pathogenesis. Microbiol. Mol. Biol. Rev. 82, 1–12. https://doi.org/10.1128/MMBR .00001-18.

Hu, Y., Zhu, Y., Lu, N.H., 2020. Recent progress in *Helicobacter pylori* treatment. Chin. Med. J. 133, 335–343. https://doi.org/10.1097/CM9.0000000000000618.

Huang, Y., Wang, Q.L., Cheng, D.D., Xu, W.T., Lu, N.H., 2016. Adhesion and invasion of gastric mucosa epithelial cells by *Helicobacter pylori*. Front. Cell. & Inf. Microbiol. 6, 159. https://doi.org/10.3389/fcimb.2016.00159.

Jain, P., Jain, S., Prasad, K.N., Jain, S.K., Vyas, S.P., 2009. Polyelectrolyte coated multilayered liposomes (nanocapsules) for the treatment of *Helicobacter pylori* infection. Mol. Pharm. 6, 1–11. https://doi.org/10.1021/mp8002539.

Jarzab, M., Posselt, G., Meisner-Kober, N., Wessler, S., 2020. *Helicobacter pylori*-derived outer membrane vesicles (OMVs): role in bacterial pathogenesis? Microorganisms 8, 1–19. https://doi.org/10.3390/microorganisms8091328.

Khoder, G., Sualeh, J., Mahmoud, I., Soliman, S.S., Burucoa, C., 2019. Prevalence of *Helicobacter pylori* and its associated factors among healthy asymptomatic residents in the United Arab Emirates. Pathogens 8, 44. https://doi.org/10.3390/pathogens8020044.

Kim, S.Y., Choi, S.J., Ching, J.W., 2015. Antibiotic treatment for *Helicobacter pylori*: is the end coming? World J. Gastrointest. Pharmacol. Therapeut 6, 183–198. https://doi.org/10.4292/wjgpt.v6.i4.183.

Kipritci, Z., Gurol, Y., Celik, G., 2020. Antibiotic resistance results of *Helicobacter pylori* in a University Hospital: comparison of the hybridization test and real-time polymerase chain reaction. Int. J. Microbiol. 2020, 1–5. https://doi.org/10.1155/2020/8853298.

Li, P., Chen, X., Shen, Y., Li, H., Zou, Y., Yuan, G., Hu, P., Hu, H., 2019. Mucus penetration enhanced lipid polymer nanoparticles improve the eradication rate of *Helicobacter pylori* biofilm. J. Contr. Release 300, 52–63. https://doi.org/10.1016/j.jconrel.2019.02.039.

Marques, A.T., Vítor, J.M.B., Santos, A., Oleastro, M., Vale, F.F., 2020. Trends in *Helicobacter pylori* resistance to clarithromycin: from phenotypic to genomic approaches. Microb. Genom. 6 https://doi.org/10.1099/mgen.0.000344.

Nardone, G., Compare, D., 2015. The human gastric microbiota: is it time to rethink the pathogenesis of stomach diseases? Unit. Eur. Gastroenterol. J. 3, 255–260. https://doi.org/10.1177/2050640614566846.

Palencia, S.L., Lagos, G., García, A., 2016. Fungi-bacterium interactions: *Helycobacter pyroli - Candida albicans*. J. Sci. Technol. Appl. 1, 53–69. https://doi.org/10.34294/j.jsta.16.1.5.

Pan-In, P., Banlunara, W., Ethyl, C.N., 2014. Cellulose nanoparticles: clarithomycin encapsulation and eradication of *H. pylori*. Carbohydr. Polym. 109, 22–27. https://doi.org/10.1016/j.carbpol.2014.03.025.

Raafat, A.I., Kamal, H., Sharada, H.M., Elhalin, A.A., Mohamed, R.D., 2021. Radiation development of gastroretentive amoxicillin trihydrate floating-alginate based beads for the treatment of *Helicobacter pylori*. Radiat. Phys. Chem. 179 https://doi.org/10.1016/j.radphyschem.2020.109268.

Saracino, I.M., Fiorini, G., Zullo, A., Pavoni, M., Saccomanno, L., Vaira, D., 2020. Trends in primary antibiotic resistance in *H. pylori* strains isolated in Italy between 2009 and 2019. Antibiotics 9, 1–11. https://doi.org/10.3390/antibiotics9010026.

Siddalingam, R., Chidambaram, K., 2014. *Helicobacter pylori* - Terapia actual y estrategias terapéuticas futuras, Tendencias en la infección por *Helicobacter pylori*. Bruna Maria Roesler. https://doi.org/10.5772/58338. IntechOpen.

Stave, K., Berrio, M.E., Palencia, S.L., 2017. Methods of diagnosis of *Helicobacter pylori* in children: current status and prospects. J. Sci. Technol. Appl. 2, 24–43. https://doi.org/10.34294/j.jsta.17.2.12.

Sukri, A., Hanafiah, A., Zin, N.M., Kosai, N.R., 2020. Epidemiology and role of *Helicobacter pylori* virulence factors in gastric cancer carcinogenesis. J. Pathol. Microb. & Inmunol. 128, 150–161. https://doi.org/10.1111/apm.13034.

Yang, Y., 2016. Updates on the diagnosis of *Helicobacter pylori* infection in children: what are the differences between adults and children? Pediatr. Gastroenterol Hepatol. & Nutr. 19, 96–103. https://doi.org/10.5223/pghn.2016.19.2.96.

Zhang, Q., Wu, W., Zhang, J., Xia, X., 2020. Eradication of *Helicobacter pylori*: the power of nanosized formulations. Nanomedicine 15, 527–542. https://doi.org/10.2217/nnm-2019-0329.

Index

Note: 'Page numbers followed by "f" indicate figures and "t" indicate tables.'

A

Accelerated solvent extraction (ASE), 315
Accident prevention, inherently safer chemistry for, 16–17
Acetobacter xylinum, 98
Adequate control of plagues, 184–185
Adhesion, 248–249
  promoter, 248–249
Adhesive, 248–249
Adipose-derived stem cells, 261–262
Adsorption mechanism, 249–250
Advanced fiber applications, 285–287
Adverse transfusion reactions, 356
AgNPs-Cysa antimicrobial systems, 360
Agricultural waste
  applications, 384–387
  artichoke waste
    methodological strategy, 385–386
    phenolic compounds separation, 387
    uses of, 384–385
  asymmetric membranes, 382–383
  bioactive substances, 384
  conventional extraction, 381–382
  cross-flow filtration, 383–384
  membrane separation technology, 382
  polymeric membrane technology, 382–384
  pressure-actuated membrane separation methods, 383–384
  secondary metabolites, 384
  separation methods, 381–382
  wine industries
    methodological strategy, 388–389
    phenolic compounds separation, 389–391
    waste from, 388–391
Agriculture
  functional polymers in
    adequate control of plagues, 184–185
    adequate seed and crop conservation, 187
    efficient fertilization, 185–186
    suitable soils for agriculture, 182–183
    use of water resources, 183–184
  problems in, 181
Agrobacterium, 94
Agrotex, 286–287
Aliphatic polyesters, 59–60
Alkaline treatment, 114
Antibacterial and nonhemolytic materials, 364–366
Antibacterial inorganic nanoparticles, 400–402
Antimicrobial surfaces/platforms, 266–267
Aquatic ecosystems, 202–204
Aromatic groups, 54–55
Artichoke waste
  methodological strategy, 385–386
  phenolic compounds separation, 387
  uses of, 384–385
Aspergillus niger, 91–92
Asymmetric membranes, 382–383
Atom, 121–122
  economy, 8–9
Attenuated total reflectance (ATR-FTIR), 401
Azotobacter chroococcum, 351, 413f

B

Bacterial inocula, 412–413
Bioactive substances, 384
Bio-based hydrogels, 412–413
Bio-based monomers and polymers, 89–92
Bio-based polymers, 112–113
Biodegradable functional polymers, 51–54
Biodegradable hydrogels, 410–411
Biodegradable polymers
  biodegradable functional polymers, 51–54
  biodegradable polymers and their potential applications, 62–66
  markets, 77–78
  physicochemical properties, 58–60
  potential applications, 62–66
  structural and morphological characteristics, 54–58

Biodegradable polymers (*Continued*)
transformation technologies, 60–62
Bioinspiration-solution-problem, 126–127
Biological biopolymer, 91
Biologic polymer market, 75–77
Biomimetic supramolecular systems, 125–128
Biomimicry
approaches to biomimetic design, 126–127
polymer-based biomimetic supramolecular systems, 127–128
Bioplastics, 23–25
Biopolymer-based drug-delivery systems, 273–277
Biopolymers, 23–25, 27t, 258–266
applications of, 426–428
biosynthesis of, 92–97
biosynthesis of polyhydroxyalkanoates (PHAs), 97
biosynthesis of starches, 94–96
cellulose biosynthesis, 93–94
chitin biosynthesis, 96–97
biotransformations of, 97–98
collagen matrices, 264–265
diagnosis and detection, 265–266
directly by living organisms, 29
extracted from biomass, 26–28
hydroxyethyl cellulose, 265
silver nanoparticles, 265
synthesized from biomass monomers, 28–29
three-dimensional (3D) printing, 263–264
tissue engineering, 260–264, 262t
wound dressing, 264–265
Bio-specific interactions, 124–125
BioSwitch, 98
Biosynthetic process, 94–95
Biotransformations of biopolymers, 97–98
Biotransformation strategy, 98
Biphasic and triphasic samplers, 306–309
Blood transfusion
adverse transfusion reactions, 356
antibacterial and nonhemolytic materials, 364–366
cationic polyurethanes, 365–366
eco-friendly cationic polyurethanes, 364–366
immunohematological control, 357
methodological approach, 364–365
nanostructured hybrid antibiotics, 358–360
N-carboxymethyl-N,N, N-triethylammonium chloride, 366
photosensitizing agents, 357
poly(vinyl chloride) (PVC), 360–363
Pseudomonas aeruginosa, 355–356
Pseudomonas scepacia, 355–356
PVC-based antibacterial surface, 361–362
results, 359–360
septic reactions, 356
starch-based hydrogels, 358–360
starch–poly (acrylic acid), 359
surface-contact pathogen inactivation, 360–363
transfusion-transmitted infections, 356
Bombyx mori, 29

C

Carbohydrates, 91
Cationic polyurethanes, 365–366
Cellular adhesion, 249–250
Cellular systems, 128
Cellulose biosynthesis, 93–94
Ceramic matrix composites, 107–108
Chemical degradation, 129
Chemical industry, 7t
Chemical recycling, 37
plastic classification according to, 38–45
Chemical species, soils, 194–196
Chitin biosynthesis, 96–97
Chloroethene, 40–41
Chondroitin sulfate, 261
Clay-alginate hybrid particles, 171f
Clay-polymer supramolecular systems, 131–132
Click chemistry, 158–162, 160f
Closed-loop recycling, 36
Clothtex, 286–287
Coating–gas surface, 248
Collagen matrices, 27–28, 264–265
Colloidal gel, 141, 402
Concentration polarization, 402–403
Conventional extraction, 381–382
Conventional techniques, 422–423
Copolymer, 235–236
Coupling agents, 248–249
Cross-flow filtration, 383–384
Crystalline properties, 249–250
Crystallographic bonding, 249–250

D

Degradation design, 15–16

Difference scanning calorimetry (DSC), 59–60
Diffusion and entanglement, 249–250
Diglycidyl ether of bisphenol A (DGEBA), 45
Dioxins, 301
Dynamic covalent materials, 42–43
Dynamic light scattering (DLS), 401

E

Eco-friendly cationic polyurethanes, 364–366
Eco-friendly chemical transformations
  in current environment, 155
  green chemistry, click chemistry, and eco-sustainable development, 158–159
  principles of click chemistry, 159–162, 160f
  principles of green chemistry, 158–159
Eco-friendly composites and nanocomposites, 111–112
  chemical and physical treatments for, 113–116
  classification of composites, 107–111
    classification based on type of matrix, 107–109
    shape and size of reinforcement agent, 109–111
  composites and nanocomposites, 105–107
  natural polymer matrices and reinforcements used, 112–113
Eco-friendly composites development, chemical and physical treatments for, 113–116
Eco-friendly hydrogels
  applications of, 149
  classification of, 143–146
    function and nature of external stimulus, 146
    nature of cross-linking, 145
    nature of the monomer, 145–146
    origin, 143–144
    polymer composition, 144
  hydrogels, 141–143
  synthesis of hydrogels, 146–147
Eco-friendly polymer-based fibers, 288–289
Eco-friendly polymer coatings, 252–253
Eco-friendly polymer market
  biodegradable polymer markets, 77–78
  biologic polymer market, 75–77
  recyclable polymer markets, 78
Eco-friendly polymer surfaces, 250–251
Eco-friendly supramolecular systems
  biomimicry
    approaches to biomimetic design, 126–127
    polymer-based biomimetic supramolecular systems, 127–128
  geomimicry
    clay-polymer supramolecular systems, 131–132
    mimetic systems of structure and/or function of soils, 132–134
    soil composition and interactions, 130–131
    soil degradation as current global problem, 129
  polymeric systems, 123–125
  supramolecular systems, 121–123
Eco-sustainable development, 158–159
Ecotex, 286–287
Edible polymers, 225–226
  applications
    edible polymer-like emulsifying food additives, 225–226
    functional edible films and coatings on food packaging, 215–222
    functional food, 222–225
    synthetic meat, 226–227
  in food industry, 209–210
  polysaccharides, 211
  proteins, 212–213
  synthetics and composites, 213–215
E-factor, 6, 7t
Emerging pollutants
  innovative techniques, 329t
  liquid-phase polymer-based retention (LPR), 328
  metallic and metalloid ions, 332–334
  molecularly imprinted polymers (MIPs), 328
  nanoengineered materials, 332–333
  nanomaterials, 332–334
  pharmaceuticals removal in water, 331–332
  remediation in water, 329–331, 330f
  remediation methods, 327–328
  remedy contamination, 331
  simple protocols, 331
  smart materials, 327–328
  ultrafiltration (UF), 331
  US Environmental Protection Agency (USEPA), 327–328
  US Geological Survey (USGS), 329–330
  water, 327

Encapsulation, 423–428
biopolymers, 421–423
techniques, 425t–426t, 427f
Energy efficiency, design for, 12
Environmental applications
functional polymers for environmental monitoring, 194–200
monitoring of air, 196–198
monitoring of chemical species in soils, 194–196
monitoring of water, 198–200
functional polymers for environmental remediation, 200–204
of aquatic ecosystems, 202–204
of terrestrial ecosystems, 201–202
problems in, 193–194
Environmental monitoring, functional polymers for, 194–200
monitoring of air, 196–198
monitoring of chemical species in soils, 194–196
monitoring of water, 198–200
Environmental remediation, functional polymers for, 200–204
of aquatic ecosystems, 202–204
of terrestrial ecosystems, 201–202
Enzymatic biotransformation, 98
Escherichia coli, 360
Esterification, 114
Etherification, 114

F

Fiber-reinforced composite materials, 109
Fibers, surfaces, 167–173
Food conservation, 407–408
Food industry, polymers in, 209–210
Food packaging, coatings on, 215–222
Food security, 407–408
Functional edible films, 215–222
Functional fibers, 287t–288t
Functional food, 222–225
Functional polymers, 238
agriculture
adequate control of plagues, 184–185
adequate seed and crop conservation, 187
availability and efficient use of water resources, 183–184
availability of suitable soils for agriculture, 182–183
efficient fertilization, 185–186
for environmental monitoring, 194–200
monitoring of air, 196–198
monitoring of chemical species in soils, 194–196
monitoring of water, 198–200
for environmental remediation, 200–204
of aquatic ecosystems, 202–204
of terrestrial ecosystems, 201–202
Furans, 301

G

General-purpose crystal polystyrene or polystyrene (GPPS), 42
Geomimetic material, 202, 342
Geomimetic model, 341–342
Geomimetic systems, 133f, 134
Geomimicry
clay-polymer supramolecular systems, 131–132
mimetic systems of structure and/or function of soils, 132–134
soil composition and interactions, 130–131
soil degradation as current global problem, 129
Geotex, 286–287
Global Monitoring Plan (GMP), 298
Gluconacetobacter, 94
Graft polymers, 115
Green chemistry, 3–5, 158–159, 160f
Green pharmacy concept, 272

H

Hazardous chemical synthesis, 9–10
Headspace (HS), 315
Helicobacter pylori
amoxicillin, 434
Correa cascade, 433–434
levofloxacin, 434
microaerophilic atmosphere, 433
microbiological characteristics, 433–436
mucosa-associated lymphoid tissue (MALT) lymphoma, 433–434
polymers, 436
quadruple therapy, 434
rabeprazole, 434
tinidazole, 434
triple therapy, 435f
World Health Organization (WHO), 434–436
High-density polyethylene (HDPE), 38–39

High-performance technologies, 419–420
Homopolymer, 235–236
Human immunodeficiency virus (HIV), 356
Humic substances, 130–131
Humified organic matter and mineral components, 131
Hybrid fibers, 289
Hydrodynamic properties, 173
Hydrogels, 358–359
  antibacterial inorganic nanoparticles, 400–402
  attenuated total reflectance (ATR-FTIR), 401
  colloidal particles, 402
  concentration polarization, 402–403
  dynamic light scattering (DLS), 401
  inorganic nanoparticles (INPs), 395–398
  methodological strategy, 401–402
  nanoflores, 395–396
  nanomaterials, 397t
  nanoonions, 395–396
  nanoparticles, 397t
  nanostars, 395–396
  poly(acrylic acid) (PAA), 401
  silver nanoparticles (AgNPs), 398–400, 402–403
  surface-area-to-volume, 395–396
  surface plasmon resonance (SPR), 401
Hydrogen bonding, 249–250
Hydrogen-catalyzed dechlorination, 41
Hydroxyethyl cellulose, 265

I

Immobilization, 423–428
Inorganic fraction of soil, 130
Inorganic nanoparticles (INPs), 395–398
Interphase formation, 249–250
Ion-conductive polymericelectrolytes, 65
Isocyanates, 114

L

Liquid-phase polymer-based retention (LPR), 328
Llignin-based thermoplastics, 159
Low-density polyethylene (LDPE), 38–39

M

Mass spectrometry (MS), 314–315
Matter organization levels, 121–123
Mechanical interlocking, 249–250
Mechanical/solvent extraction methods, 115
Meditex, 286–287
Membrane separation technology, 382
Metallic/metalloid ions, 332–334
Metal matrix composites, 108–109
Miconia acuminifera Triana, 28
Microbiological/biological adhesion, 249–250
Microbiological compounds, 329–330
Microbiota, 419–420
Microorganisms viability, 422
Microwave-assisted extraction (MAE), 315
Mid-IR spectroscopy, bacterial biofilms by, 374–377
Molecularly imprinted polymers (MIPs), 328
Monophasic samplers, 304–306
Montmorillonite particles, 169–170
Mucosa-associated lymphoid tissue (MALT) lymphoma, 433–434
Multifunctional structural polymers, 238–242

N

Nanocomposites, 106–107, 110
Nanoengineered materials, 332–333
Nanoflores, 395–396
Nanomaterials, 332–334, 397t
Nanoonions, 395–396
Nanoparticles, 397t
Nanostars, 395–396
Nanostructured hybrid antibiotics, 358–360
Natural fibers, 171–172, 290f
Natural polymer-based fibers, 288
Natural polymer matrices and reinforcements, 112–113
N-carboxymethyl-N,N,N-triethylammonium chloride, 366
N-vinylbenzyl-N-triethylammonium chloride, 347

O

Organochlorine compounds (OCs)
  accelerated solvent extraction (ASE), 315
  chemical analysis, 314–316
  chemicals by passive samplers accumulation, 302–303
  dioxins, 301
  factors, 309–314
  functional polymers using passive samplers, 303–309, 311t–313t
    biphasic and triphasic samplers, 306–309

Organochlorine compounds (OCs) (*Continued*)
monophasic samplers, 304–306
furans, 301
general context, 299–301
Global Monitoring Plan (GMP), 298
headspace (HS), 315
mass spectrometry (MS), 314–315
microwave-assisted extraction (MAE), 315
organochlorine pesticides (OCPs), 300
polychlorinated biphenyls, 301
polyurethane foam (PUF), 298
pressurized solvent extraction (PSE), 315
Organochlorine pesticides (OCPs), 300
Organosilanes, 132

P

Particle-reinforced composites, 109–110
Pathogenic bacterial biofilms, mid-infrared vibrational spectrum of, 372–373
Peroxidesoxidizing agents, 114–115
Pharmaceutical applications
biopolymer-based drug-delivery systems, 273–277
biopolymers in, 272
delivery pathways, 273–277
emerging pollutants, 271–272
green pharmacy concept, 272
stimuli-responsive biopolymers, 277–279
Phenolic compounds separation, 387
Photocatalytic processes, 173
Photosensitizing agents, 357
Planet Earth, 3–4
Plant growth-promoting microorganisms (PGPMs), 409
Plasma treatment, 115–116
PMMA, 44–45
Pollutant-releasing vehicles, 238–239
Pollution prevention, real-time analysis for, 16–17
Poly(acrylic acid) (PAA), 401
Poly(ethylene) (PE), 64, 235–236
Poly(ethylene terephthalate) (PET), 238–239
Poly(ethylene-vinyl-acetate) (EVA), 238–239, 242
Poly(styrene) (PS), 238–241
Poly(vinyl chloride) (PVC), 236–238, 241, 360–363
Poly(vinylidene fluoride) (PVDF), 257–258
Polyamides, 44–45
Polychlorinated biphenyls, 301
Polyhydroxyalkanoates (PHAs), 97
Polymer-based bacterial inocula, 410–413, 411t
Polymer-based fibers, 288
Polymer-based particles, 169–173
Polymer-based restoration, degraded soils
Azotobacter chroococcum, 351
hybrid polymer–clay composites, 346–348
hydrogels, 343–345
methodological strategy, 344
results, 344–345
multi- and bio-functional hybrid polymer hydrogel geomimetics, 348–351
relevant results, 350–351
synthesis strategy, 349–350
nanostructured hydrogels, 346–348
relevant results, 347–348
synthesis strategy, 346–347
new multifunctional materials, geomimetic model of, 341–342
Polymer-based spectral sensors, 371–372
Polymer-clay multilayer systems, 122
Polymer–coating interface, 248
Polymer flexibility, 55–57
Polymer green chemistry
atom economy, 8–9
catalysis, 14–15
design
degradation, 15–16
energy efficiency, 12
designing safer chemicals, 10–11
green chemistry, 3–5
less hazardous chemical synthesis, 9–10
principles of polymer synthetic green chemistry, 5–17
real-time analysis for pollution prevention, 16–17
reduce derivatives, 13–14
use of renewable feedstocks, 12–13
use safer solvents and auxiliaries, 11–12
Polymeric biomaterials
antimicrobial surfaces/platforms, 266–267
biopolymer materials, 258–266
collagen matrices, 264–265
diagnosis and detection, 265–266
hydroxyethyl cellulose, 265
silver nanoparticles, 265
three-dimensional (3D) printing, 263–264
tissue engineering, 260–264, 262t
wound dressing, 264–265
medicine and biomedicine, 257–258

poly(vinylidene fluoride) (PVDF), 257–258
synthetic polymers, 258
Polymeric hydrogels, 24–25
Polymeric materials, 10–11, 358–359
Polymeric matrix composites, 107
Polymeric membrane technology, 382–384
Polymeric systems, 123–125
Polymer market, 73–75
ecofriendly, 75–78
economic impact, 78–80
perspectives, 80–81
Polymer morphology, 55–57
Polymer recycling, basic principles of, 35–37
Polymer surface, 248
Polymer synthetic green chemistry, principles of, 5–17
Polymolecular assemblies, 123–124
Polypropylene, 38–39
Polysaccharide-protein complexes, 214
Polysaccharides, 211
Polyurethane foam (PUF), 298
Polyvinyl chloride (PVC), 40–41
Pressure-actuated membrane separation methods, 383–384
Pressurized solvent extraction (PSE), 315
Primary recycling, 36
Probiotic microorganisms
biopolymers, applications of, 426–428
conventional techniques, 422–423
encapsulation, 423–428
biopolymers, 421–423
techniques, 425t–426t, 427f
health and disease treatment, 420–421
high-performance technologies, 419–420
immobilization, 423–428
microbiota, 419–420
microorganisms viability, 422
qualitative and quantitative changes, 419–420
severe acute respiratory syndrome (SARS), 421–422
Problem-bioinspiration-solution, 126
Production of wastes, 6
Productive innovation strategies, 36
Protein biopolymer, 92–93
Protein biopolymers, 91
Proteins, 212–213
Protex, 286–287
Pseudomonas aeruginosa, 266, 355–356
Pseudomonas scepacia, 355–356
PVC-based antibacterial surface, 361–362

Q

Quadruple therapy, 434

R

Rabeprazole, 434
Real-time analysis for pollution prevention, 16–17
Recyclable polymer
chemistry
basic principles of polymer recycling, 35–37
plastic classification according to chemical recycling, 38–45
markets, 78
Reduce derivatives, 13–14
Remedy contamination, 331
Renewable feedstocks
use of, 12–13

S

Safer chemicals, designing, 10–11
Safer solvents and auxiliaries, 11–12
Satisfactory coating, 248–249
Scaffold-based method, 227f
Secondary metabolites, 384
Seed/crop conservation, 187
Separation methods, 381–382
Septic reactions, 356
Severe acute respiratory syndrome (SARS), 421–422
Sheet-reinforced composite, 110
Silane, 115
Silver nanoparticles (AgNPs), 265, 398–400, 402–403
Simple protocols, 331
Smart materials, 327–328
Society of the Plastic Industry (SPI), 37
Soil composition/interactions, 130–131
Soil degradation, 129
Soil-inspired geomimetic system, 133f
Soil microorganisms, 409
Solid-state fermentation, 98
Spectra fiber, 286
Spectral sensors surface, microbiological analysis
methodological strategy, 373–374

Spectral sensors surface, microbiological analysis (*Continued*)
- mid-IR spectroscopy, bacterial biofilms by, 374–377
- pathogenic bacterial biofilms, mid-infrared vibrational spectrum of, 372–373
- polymer-based spectral sensors, 371–372

Staphylococcus aureus, 360
Starch-based hydrogels, 358–360
Starches
- biosynthesis of, 94–96

Starch–poly(acrylic acid), 359
Step-growth polymerizations, 8–9
Stimuli-responsive biopolymers, 277–279
Structural polymer, 235–236
Suitable soils for agriculture, 182–183
Supramolecular system, 121–123
Surface, 247–248
- condition and shape, 249–250
- free energy, 249–250
- wetting, 249–250

Surface-area-to-volume, 395–396
Surface-contact pathogen inactivation, 360–363
Surface plasmon resonance (SPR), 401
Sustainable agriculture
- *Azotobacter chroococcum*, 413f
- bacterial inocula, 412–413
- bio-based hydrogels, 412–413
- biodegradable hydrogels, 410–411
- food conservation, 407–408
- food security, 407–408
- plant growth-promoting microorganisms (PGPMs), 409
- polymer-based bacterial inocula, 410–413, 411t
- soil microorganisms, 409

Syndiotactic polymer, 57–58
Synthetic biodegradable polymers, 52–53
Synthetic biopolymers, 91–92
Synthetic meat, 226–227
Synthetic polymers, 35, 258, 288
Synthetics and composites, 213–215

T

Terrestrial ecosystems, 201–202
Tertiary/chemical recycling, 36–37
Tetrahydrofuran (THF), 361–362
Textile applications
- advanced fiber applications, 285–287
- agrotex, 286–287
- clothtex, 286–287
- eco-friendly polymer-based fibers, 288–289
- Ecotex, 286–287
- functional fibers, 287t–288t
- geotex, 286–287
- high-performance clothing, 286
- Meditex, 286–287
- natural and synthetic fibers, 286t
- natural fibers, 290f
- Protex, 286–287
- spectra fiber, 286

Thermogravimeter analysis (TGA), 59–60
Three-dimensional (3D) printing, 263–264
Tinidazole, 434
Tissue engineering, 260–264, 262t
Transfusion-transmitted infections, 356
Triple therapy, 435f

U

Ubiquitous materials, 5–6
Ultrafiltration (UF), 331
US Environmental Protection Agency (USEPA), 327–328
US Geological Survey (USGS), 329–330

V

van Oss–Chaudhury–Good (vOCG) model, 248
Vegetable fibers, 171–172

W

Water resources, availability and efficient use of, 183–184
Wine industries
- methodological strategy, 388–389
- phenolic compounds separation, 389–391
- waste from, 388–391

World Health Organization (WHO), 434–436
Wound dressing, 264–265

X

Xerogel, 141

FSC
www.fsc.org
MIX
Papier | Fördert gute Waldnutzung
FSC® C083411

Kolibri 360 GmbH
Ferdinand-Jühlke-Straße 7
99095 Erfurt, Deutschland
produktsicherheit@kolibri360.de